Sustainable Development Goals Series

The **Sustainable Development Goals Series** is Springer Nature's inaugural cross-imprint book series that addresses and supports the United Nations' seventeen Sustainable Development Goals. The series fosters comprehensive research focused on these global targets and endeavours to address some of society's greatest grand challenges. The SDGs are inherently multidisciplinary, and they bring people working across different fields together and working towards a common goal. In this spirit, the Sustainable Development Goals series is the first at Springer Nature to publish books under both the Springer and Palgrave Macmillan imprints, bringing the strengths of our imprints together.

The Sustainable Development Goals Series is organized into eighteen subseries: one subseries based around each of the seventeen respective Sustainable Development Goals, and an eighteenth subseries, "Connecting the Goals," which serves as a home for volumes addressing multiple goals or studying the SDGs as a whole. Each subseries is guided by an expert Subseries Advisor with years or decades of experience studying and addressing core components of their respective Goal.

The SDG Series has a remit as broad as the SDGs themselves, and contributions are welcome from scientists, academics, policymakers, and researchers working in fields related to any of the seventeen goals. If you are interested in contributing a monograph or curated volume to the series, please contact the Publishers: Zachary Romano [Springer; zachary.romano@springer.com] and Rachael Ballard [Palgrave Macmillan; rachael.ballard@palgrave.com].

Sinan Küfeoğlu

Emerging Technologies

Value Creation for Sustainable
Development

 Springer

Sinan Küfeoğlu
Department of Engineering
University of Cambridge
Cambridge, UK

This book is an open access publication.

ISSN 2523-3084   ISSN 2523-3092   (electronic)
Sustainable Development Goals Series
ISBN 978-3-031-07129-4   ISBN 978-3-031-07127-0   (eBook)
https://doi.org/10.1007/978-3-031-07127-0

This Springer imprint is published by the registered company Springer Nature Switzerland AG
The registered company address is: Gewerbestrasse 11, 6330 Cham, Switzerland

Foreword by Li Wan

Technological innovation has been instrumental in the evolution of human societies. For example, the advent of motor vehicles and container ships has enabled the movement of physical goods and people at ever-increasing speed and ever-decreasing cost. It has been transforming not only the spatial structure of cities but also the political and economic geography across regions, countries and continents. It has been argued that the rate of technological changes has accelerated in the twenty-first century, epitomised by the prevalence of artificial intelligence (AI), which results in existential concerns over the survival of humanism against dataism. Narratives on the role and impact of digital technology seem much polarised, where technology (perhaps more specifically, AI) has been either celebrated as the saviour for sustainability or condemned as a looming tyranny. Perhaps we need more than narratives, but evidence-based investigations and deliberations to find ways forward.

Dr Sinan Küfeoğlu's book represents a formidable endeavour to investigate the nexus of value-based innovation, entrepreneurial ecosystems and sustainable development. It starts by introducing a theoretical framework for understanding the value creation of/through innovation, supporting mechanisms and impact. An overview of 34 emerging technologies of varying levels of maturity is then presented, including several widely perceived megatrends. He would impress the readers by then presenting an admirable series of case studies, covering 650 innovative companies selected from 51 countries worldwide and examining their specific technological expertise and value-based business model in relation to the United Nations Sustainable Development Goals.

I worked with Dr Küfeoğlu in the 'Digital Cities for Change' programme at the Cambridge Centre for Smart Infrastructure and Construction, University of Cambridge. His extensive research experience on innovation and entrepreneurial ecosystems has contributed greatly to the project. His expertise on business models and finance for technological innovation largely enhanced our understanding of digitalisations in cities as complex socio-technical transition processes.

We would need technological innovation to address the unprecedented challenges of our time such as global warming and widening disparities incurred by the COVID-19 pandemic and geo-political conflicts. To this end, bridging the silos between technology innovators, financiers, regulators and consumers has never been more urgent. This book of pragmatism is an excellent start.

Assistant Professor, Department of Land Economy   Li Wan
University of Cambridge
Cambridge, UK

Foreword by Yu Wang

I met Sinan Küfeoğlu when I was at Cambridge University as a visiting scholar in 2018. At the time, we were working together at the Energy Policy Research Group (EPRG). Sinan impressed me with his diligence, quick thinking and decisive action. After two years of the epidemic, I was delighted and encouraged to receive Sinan's masterpiece during the Chinese Spring Festival.

As a legally binding international treaty on climate change, the Paris Agreement's goal is to limit global warming to well below 2, preferably to 1.5 °C, compared to pre-industrial levels. Although countries submit nationally determined contributions and plans for climate action, the knowledge of available technologies to support the carbon neutrality achievement is still badly needed. Sinan's book perfectly combines the macro sustainable development goals with micro concrete technologies very well and smoothly.

First of all, the author clearly explains the definition of technological innovation and analyses the mechanism of technology innovation, from idea to application, the financing and funding and the specific commercialization model needed in the process of technology commercialization.

At the micro-level, 34 emerging technologies are sorted out and evaluated comprehensively. Listed in this book, some of them are directly related to low-carbon energy transformation and sustainable development, while others are identified with great significance to the global carbon neutrality and sustainable development from the system point of view.

Based on the above emerging technologies assessment, the author combines each available technology with specific Sustainable Development Goals and attaches success cases to illustrate how the promotion and application of these technologies contribute to the realization of the Sustainable Development Goals at the macro level.

I think this is a valuable guideline and reference to policymakers, managers and researchers. Which helps policymakers draw blueprints and specific roadmap, helps enterprise decision-makers determine the future direction of technology development and action planning and helps researchers solve problems from a more systematic perspective.

Institute of Energy, Environment and Economy Yu Wang
Tsinghua University
Beijing, China

Foreword by John Seed

Constrained government budgets resulting from the impacts of Covid-19 threaten the transition to the green and sustainable infrastructure needed to achieve the Sustainable Development Goals (SDGs). Moreover, as identified during the recent COP26, we are at a critical juncture in the global decarbonisation transition – this decade must mark the turning point during which the transition takes hold and accelerates. For multilateral development banks (MDBs) like the European Bank for Reconstruction and Development (EBRD), it is, therefore, all the more important to ensure that sustainability is at the heart of every investment we make. EBRD identified in 2020 the role that technological innovation played in improving sustainable development and included this as a core pillar in the Banks Strategic Capital Framework. The identification of emerging innovative technologies that genuinely improve sustainability is, therefore, an essential first step in improving green and sustainable investment outcomes.

The SDGs now represent a global common 'language' to express sustainability impacts and contributions. There is increasing demand from stakeholders, including global investors, to identify how investments supporting transition align with and contribute to SDGs and the broad development impact spectrum. MDBs have been working to coordinate a common approach on SDG achievements from investments, and their board investment approval processes now increasing require SDG contribution justifications. Understanding the economic, environmental and social value that innovative technologies can bring to investments in the SDG contribution language helps greatly in achieving this.

Cities are not only at the centre of the Covid-19 crisis but also play a critical role in the transition to low-carbon economies. Cities are where an ever-growing majority of people in the world live at close quarters and are where most traffic, industry and commerce are located. Hence, most of the greenhouse gas emissions in the world come from cities – making them at the same time the biggest contributor to the climate crisis and the biggest opportunity to solve it.

Now Covid-19 has cast into a still sharper focus the need for urban infrastructure everywhere to be underpinned by smart technology that intelligently supports cities' needs while keeping communities safe. This is the reason that the G20 group of leading nations' Infrastructure Working Group responded to the onset of the Covid-19 by prioritising their 'InfraTech' workstream in 2020 that looks at incorporating smart technology into infrastructure. This

follows evidence that more digitally enhanced, smarter, more innovative cities have been able to mitigate Covid-19 impacts quicker and more efficiently
than those that aren't.

Emerging technologies also have big green benefits, as integrating innovative solutions into urban infrastructure can result in 10–15% of greenhouse
gas emissions savings. This is the reason for the EBRD's recent decision to
integrate smartly into its flagship Green Cities urban sustainability programme. Since 2016, this EBRD €2 billion flagship programme has brought
together 52 cities that want to upgrade their infrastructure for the twenty-first
century – the aim is to have 100 cities on board by 2024. The EBRD helps
each one with both municipal investments and technical support to develop
tailor-made programmes for green infrastructure projects.

The EBRD's experience has been that cities start to think about building
innovative technology solutions into urban infrastructure in one of two ways.
Either they are led towards smart integration by public-sector champions with
experience or visions of city and community benefit, or, more commonly,
they gain direct experience of how such solutions make life more efficient
simply via private-sector 'leapfrogging' through the introduction of new digital applications.

Hence, for EBRD, emerging technology integration will not only include
recommendations on how investments will benefit from such innovations but
also include an evaluation of the maturity of the city municipal government
to adopt these technologies. This will ensure that any technology applications
that are specified for new infrastructure projects will be of true benefit for the
city and their communities, and value for money will be achieved. The knowledge and data contained within this 'Emerging Technologies: Value Creation
for Sustainable Development' will greatly assist EBRD not only in identifying viable established innovations that deliver sustainable outcomes but also
in justifying their inclusion in each investment through the identification and
quantification of SDG contributions.

Head of Project Preparation and Implementation,   John Seed
Sustainable Infrastructure Group
European Bank for Reconstruction and Development (EBRD)
London, UK

Foreword by Aaron Praktiknjo

As societies, we have the responsibility to allow our future generations a chance for a good livelihood on our planet. While important, this responsibility for sustainability encompasses not only ecological but also extends to economic and social aspects. The 193 member states of the United Nations (UN) have committed themselves to this responsibility by agreeing on 17 Sustainable Development Goals (SDGs) until 2030.

Germany has formulated a Sustainable Development Strategy in line with the 17 SDGs. Innovative sustainable technologies play a crucial role in Germany's strategy to reach these goals. The future technologies package, with a funding volume of EUR 50 billion, aims at fostering these sustainable innovations in five domains: (1) mobility, (2) energy and climate, (3) digitalisation, (4) education and research and (5) healthcare.

In this book, Sinan Küfeoğlu analyses the potential of 34 emerging technologies for the 17 SDGs. However, this book does not limit itself to sustainable technologies but also includes an in-depth review of business models of 650 innovative companies. The lessons learnt in this book will be a guiding source for entrepreneurs, policymakers and regulators, and all those who want to learn about modern business development with emerging technologies to help contribute to a sustainable future.

School of Business and Economics   Aaron Praktiknjo
RWTH Aachen University
Aachen, Germany

Foreword by Alex O'Cinneide

I had the great pleasure of working with Sinan for a while in Cambridge, where his different approach and background to our shared research agenda in the area of energy transition allowed me to think about the field in a very new way. When combined with aspects of social science and policy, his hard science and engineering knowledge led to many great conversations on topics where his depth of expertise (and sense of humour) allowed those chats to develop in many interesting ways. Our many conversations have improved my background as an investor in renewables and clean technologies. He is regarded by both his peers and his students as a radical thinker on the most important of topics. It is clear that we now need to embark upon a radical departure from the present socio-technical paths throughout the world's energy structures if we are to achieve sustainability and low-carbon goals. The quest for a bold energy system transformation has underscored the significant realisation gap between sustainability goals and current untenable paths. This transition creates a systemic task that our societies must meet, and Sinan's career is focused on helping that task, and this book is an important tool within that work.

Renewable energy policies have been created and implemented worldwide for much of the last 30 years, and the role of emerging technologies has been key to the development of those policies. Those policies and the work the UN has undertaken with their Sustainable Development Goals have been key in helping guide those policies and, therefore, the actual on the ground development. Although those goals are often treated as dry legal documents, they are critical support for how actors should use and develop technologies for our fight against climate change and the transformation of the energy system. The investigation of the innovation journey highlights the challenges facing society due to climate change. This book – addressing innovation as a concept, a detailed review of various emerging technologies and then a compression with the goals themselves – should provide a valuable tool for researchers, policymakers and the critical industry in pushing forward our transition to a low-carbon society.

Alex O'Cinneide
CEO
Gore Street Capital
London, UK

Foreword by Seungwan Kim

It has been my pleasure to know Sinan Küfeoğlu since 2018 as a colleague and friend. We met as post-doctoral researchers at Judge Business School, University of Cambridge. I was thrilled when Sinan told me about his academic interests and achievements because they were almost the same as mine unbelievably. Our daily discussions in the office mainly were about climate change, sustainability development, the future of zero-carbon energy systems, energy transactions and all kinds of related emerging technologies. After not that long academic visit, we continued our research inspired by our discussions back in 2018 on our own paths.

In the November of the last year, South Korea claimed a goal of 2050 Carbon Neutrality and enhanced Nationally Determined Contributions (NDCs) with a strong will of mitigating climate change. Since then, diverse opinions have been expressed from the political world, academia, media and industry professionals. The most important one of the several issues is finding a new way of sustainable development that can simultaneously reduce carbon emission and economic development.

Many policymakers and researchers in South Korea think that emerging technologies introduced in this book are keys to a new future of the Korean economy with a high-tech manufacturing base. In addition, creating new jobs for the young and transitioning to a start-up economy from family-owned conglomerates, called Chaebols, are also important matters for the sustainable development of the South Korean economy ecosystem.

This book will be a guiding source for those who want to learn a comprehensive foundation of innovative economics and their business practices, to overview up-to-date emerging technologies and their relationship with 17 United Nations Sustainable Development Goals and to search a variety of real cases of innovative companies for each SDG.

I am delighted to endorse *Emerging Technologies: Value Creation for Sustainable Development* as an inspiring book quenching readers' thirst with a big picture of a sustainable future.

Assistant Professor, Department of   Seungwan Kim
Electrical Engineering
Chungnam National University
Daejeon, South Korea

Foreword by Soysal Değirmenci

Artificial Intelligence (AI) is a field aimed to devise systems with 'human-like' intelligence. These systems can range from playing board games such as Go or predicting whether there is a cat in a given image or not. Machine learning (ML), a sub-field of AI, uses data to build such systems that can learn and improve in tasks of interest.

Thanks to digital transformation, a tremendous amount of data has been available for academia and industry to research and improve systems' capabilities. This, along with advances in computing and AI research, enabled companies and researchers to build more capable AI models than ever before. AI algorithms can rapidly build and improve systems that can perform tasks that would otherwise require human intelligence. Now, we are at an important junction where some of such systems can perform at a human level or even better than a human level in some cases.

AI is already a part of our everyday lives. The apps we use on our smartphones, ads we see on the internet, TV shows we are recommended on streaming platforms and credit scores we receive all use a form of AI. Amazon leverages AI and ML in many domains to delight its customers. These include but are not limited to search, recommendations, logistics, stopping fraud and abuse and conversational voice assistants.

This book presents a comprehensive overview of how emerging technologies, including AI, can generate value for sustainable development. I highly recommend this book to readers interested in understanding the emerging technology landscape and how they relate to the UN sustainable development goals.

Machine Learning Scientist   Soysal Değirmenci
Amazon
San Diego, CA, USA

Contents

About the Author

Sinan Küfeoğlu is working as the International Outstanding Research Fellow at the Scientific and Technological Research Council of Turkey on his project 'Digitalisation in Energy Sector: Digital Solutions and New Business Models'. He is a senior research fellow at the Oxford Institute for Energy Studies. He also works as a research associate at the Cambridge Centre for Smart Infrastructure and Construction, Department of Engineering, University of Cambridge. Furthermore, he is leading the collaboration with the European Bank for Reconstruction and Development (EBRD) in preparing recommendations for boosting the digital resilience of critical infrastructure. This work is a part of the EBRD's Digital Pathways agenda. He was an adviser at the United Nations Institute for Training and Research (UNITAR), CIFAL Istanbul, for integrating United Nations Sustainable Development Goals into university education and provided consultancy to the World Bank in the field of application of Machine Learning in electric power system.

Dr Küfeoğlu completed his DSc and MSc degrees in Electrical Engineering at Aalto University, Finland in 2015 and 2011, respectively. He got his BSc degree in Electrical and Electronics Engineering Department from Middle East Technical University, Ankara, Turkey, in 2009. His research interests include energy futures, sustainable development, energy economics and technology policy.

Innovation, Value Creation and Impact Assessment

1

Abstract

Emerging technologies can be defined as a set of technologies whose development and application areas are still expanding rapidly, and their technical and value potential is still largely unrealised. Naturally, this leads to a vivid innovation environment for these technologies. In this book, tech-savvy people can easily read and understand the working principles of 34 different emerging technologies. And then, they can see in what areas these technologies are used and how they can create value. Moreover, the book starts with an "Innovation Journey" chapter. This chapter focuses on innovation and how ideas are converted into value and business. By value, we mean monetary, environmental and social value. In addition, for entrepreneurs and start-ups, we also show the funding and financing mechanisms for innovative ideas.

Keywords

Value creation · Impact assessment · Business model · Innovation

The author would like to acknowledge the help and contributions of Emre Çağatay Gümüş, Abdullah Talip Akgün, Ali Kağan Önver, Nisa Erdem, Mert Yavaşca and Mehmet Akif Ekrekli in completing this chapter.

Climate change and sustainability are alarming subjects in the world. On the other hand, mitigating the adverse effects of climate change, protecting the economies and supporting green business growth are other concerns as well. "Emerging Technologies: Value Creation for Sustainable Development" compiles 34 emerging technologies and investigates their use cases in 17 United Nations Sustainable Development Goals fields. After reviewing thousands of companies worldwide, we explore the business models of 650 noteworthy and innovative companies from 51 countries. These models demonstrate how technologies are converted to value to support sustainable development. The technologies are the ones with market diffusion. Thus, we deliberately omitted the research and development stage ones that are not commercially available. The book focuses on the use of emerging technologies to support sustainable development. Authorities and policymakers will investigate how technology is utilised to foster any of the 17 goals by seeing the novel and innovative business models all around the world.

One purpose is to attract the attention of the technological world. Emerging technologies can be defined as a set of technologies whose development and application areas are still expanding rapidly, and their technical and value potential is still largely unrealised. Naturally, this leads to a vivid innovation environment for these technologies. In this book, tech-savvy people can easily

S. Küfeoğlu, *Emerging Technologies*, Sustainable Development Goals Series,
https://doi.org/10.1007/978-3-031-07127-0_1

read and understand the working principles of 34 different emerging technologies. And then, they can see in what areas these technologies are used and how they can create value. Moreover, the book starts with an "Innovation Journey" chapter. This chapter focuses on innovation and how ideas are converted into value and business. By value, we mean monetary, environmental and social value. In addition, for entrepreneurs and start-ups, we also show the funding and financing mechanisms for innovative ideas. On the other hand, the companies and investors will be able to see what sort of skills and competencies are used in various fields. They can follow bright and successful use cases and see how they can expand their businesses in other sectors. The 650 companies include international corporations as well as numerous bright and promising start-ups. This will give corporations an opportunity for acquisition and investment opportunities.

This book aims to present that emerging technologies can create economic, environmental and social value to achieve United Nations Sustainable Development Goals. The book is organised as follows: Chap. 1 investigates the innovation journey by focusing on the innovation process, supporting mechanisms for innovation, funding and financing mechanisms, business model theory and value creation and, finally, impact assessment. Chapter 2 presents 34 emerging technologies, including 3D Printing, 5G, Advanced Materials, Artificial Intelligence, Autonomous Vehicles, Big Data, Biometrics, Bioplastics, Biotech and Biomanufacturing, Blockchain, Carbon Capture and Storage, Cellular Agriculture, Cloud Computing, Crowdfunding, Cybersecurity, Datahubs, Digital Twins, Distributed Computing, Drones, Edge Computing, Energy Storage, Flexible Electronics and Wearables, Healthcare Analytics, Hydrogen, Internet of Behaviours, Internet of Things, Natural Language Processing, Quantum Computing, Recycling, Robotic Process Automation, Robotics, Soilless Farming, Spatial Computing and Wireless Power Transfer. The following chapters briefly provide information about 17 United Nations (UN) Sustainable Development Goals (SDGs) and present the business models of 650 use cases from 51 countries

worldwide. The 17 SDGs with their respected UN SDG numbers are: (1) No Poverty; (2) Zero Hunger; (3) Good Health and Well-being; (4) Quality Education; (5) Gender Equality; (6) Clean Water and Sanitation; (7) Affordable and Clean Energy; (8) Decent Work and Economic Growth; (9) Industry, Innovation and Infrastructure; (10) Reduced Inequality; (11) Sustainable Cities and Communities; (12) Responsible Consumption and Production; (13) Climate Action; (14) Life Below Water; (15) Life On Land; (16) Peace, Justice and Strong Institutions; and (17) Partnerships for the Goals. Finally, we conclude the book with a conclusion and a brief discussion.

1.1   Innovation

Innovation means creative destruction or a process of industrial evolution (Schumpeter 1942). Innovation can be achieved by new products or processes, manufacturing methods, markets and supply chains and new organisational structures and more (Joseph Schumpeter 1934). From the built environment perspective, social innovation has come forth as a tool to address challenges faced by the environment and societies and foster sustainable development (Horgan and Dimitrijević 2018). Moreover, social innovation is necessary to respond to "an unmet social need" (Murray et al. 2010). Procurement of innovation can be done by purchasing the process of innovation or its outcomes. On the other hand, public procurement is a term used to represent the purchase of goods, services and works by government units and state-owned institutions (Crisan 2020). The public acquisition enables public institutions to perform their functions. The procurement of innovation can be used to achieve social outcomes (Crisan 2020). As a result of those benefits, public institutions are in favour of innovation and public procurement. To take advantage of those benefits, fostering innovation is a reasonable strategy for public management.

Demand-side policies to trigger innovation, including public procurement, can be applied through the whole life cycle of innovation.

Innovation life cycle follows the pattern of identifying problems, generating ideas, developing proposals, implementing projects, evaluating projects and finally diffusing lessons (OECD and Statistical Office of the European Communities 2005). Public procurement can be used as an innovation fostering tool in two ways: public procurement of innovation (PPI) and pre-commercial procurement (PCP). PPI refers to buying an innovation that is new to the market or is not commercialised, whereas PCP means buying an innovation that does not exist in the market and is looking for help in its R&D (Rolfstam 2012). PPI is realised when products are ready to enter the market or already in the market in small sizes (European Commission 2020). The innovation process of goods or services does not end. Publicly procured goods or services may have a chance to gain scale in time, such as production in high volumes, acquisition of new customers and procurement overseas. Corporate growth can be determined by the market's available resources and consumption rates (Bettencourt et al. 2007). When the solution posed by the project has other buyers both from public and private institutions, production and consumption will increase by nature. The next question that should be examined is why growth in PPI projects is beneficial for governments? The answer is that the scaling of PPI projects leads to a more stable and stronger market (European Commission 2020). Public procurement of innovation will endorse innovation through (Lember et al. 2011):

(i) Creating new markets for products and services
(ii) Creating demand and providing use cases
(iii) Providing a testbed for innovative ideas
(iv) Encouraging innovation by leading the market

In addition to the increase in production and innovation, companies with public sector support for innovative solutions can become more successful in exportation. It means that growth from globalisation in those small- and medium-sized enterprises (SMEs) is more likely to happen (Love and Roper 2015). Scaling and growth are crucial in terms of sustainable innovation. A successful local start-up with an innovative solution should extend its products and/or services overseas for growth.

A broader definition of PPI refers to a policy instrument that supports innovation and includes small- and medium-sized enterprises (SMEs) in the public procurement process (Radoslav Delina 2021). Public procurement of innovation is a powerful tool for regional authorities since it encourages innovation in small firms, which is easier than PPI at large firms and is helpful for economic growth in those firms (Crisan 2020). Based on earlier research, several conditions to achieve an effective public procurement are identified as (1) expertise in PPI laws and procedures, (2) strategic management of PPI, (3) market interaction, (4) risk management, (5) coordination and communication and (6) capacity building (Radoslav Delina 2021).

On the other hand, as catalysers of the innovation ecosystem, cities can serve as testbeds to trial innovative and novel ideas, services and solutions. As we will see in section Innovation Districts, various innovative solutions need city infrastructure to test their ideas, services or products. Therefore, cities often step forth as an ideal place to provide this support and see the innovation's performance in the whole phases of the innovation journey.

When it comes to the efficient use of resources, we must underline technology diffusion. By diffusion, we mean one technological solution to a problem that might answer fully or partially other problems as well. This tells us that one particular technological application may diffuse to other areas where it was not intended at the beginning of the design phase. We reviewed numerous approaches and wished to highlight the outcome-based approach which is adopted by the London Office for Technology & Innovation (LOTI). Figure 1.1 shows the LOTI outcome-based approach ideology.

The LOTI approach starts with looking for real-life outcomes that the local authorities wish to achieve. Then they go back to discovering the problems preventing these outcomes. After defining and developing phases, the process is

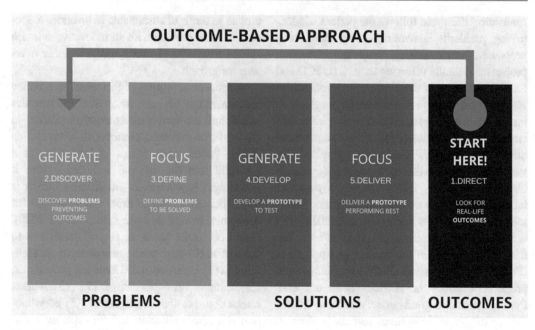

Fig. 1.1 The outcome-based approach of the London Office for Technology & Innovation. (LOTI 2021)

achieved by delivering a prototype that performs the best. We reviewed widely adopted mechanisms globally and briefed some of the prominent ones in the following section.

1.1.1 Mechanisms for Supporting Innovation

1.1.1.1 Innovation Districts

These districts are characterised as places where cutting-edge anchor organisations and companies cluster and interact with start-ups, business incubators and accelerators. They are also functionally compact, usable for transport and technically wired and offer lodging, office and retail for mixed use (Brookings 2014). Figure 1.2 illustrates the components of an innovation district.

1.1.1.2 Regulatory Sandboxes

Regulatory sandboxes allow a targeted testing atmosphere under a particular testing strategy for new products, facilities or business models, which typically entail a degree of regulatory leniency along with some protections (PARENTI 2020). Table 1.1 summarises the potential benefits of regulatory sandboxes.

1.1.1.3 Living Labs

With the collaboration and co-creation of customers, collaborators and other parties, living labs bring innovation from the R&D teams of businesses to real-life environments. Utiliser-driven, enabler-driven, provider-driven and user-driven living labs are four different types of networks characterised by open innovation. The typology is based on interviews with participants in Finland, Sweden, Spain and South Africa from 26 living laboratories. Companies will benefit from learning the features of each type of living laboratory; this information can help them recognise which actor drives creativity, predict future consequences and determine what kind of function they can play in a "living laboratory". Living labs are networks that can help us build technologies that are superior to consumer requirements that can be easily upgraded to the global market. Table 1.2 represents the details of these living labs (Leminen et al. 2012).

In some countries or in different contexts, the concept of living labs and innovation districts might be used in similar fashions. For example, in Finland, Forum Virium Helsinki is a typical living lab and innovation ecosystem usually held for the innovation districts. One of the prominent characteristic differences between living labs and

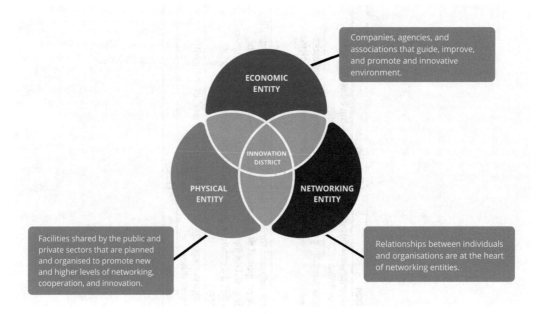

Fig. 1.2   Components of an innovation district. (Brookings 2014)

Table 1.1   Potential benefits of regulatory sandboxes (PARENTI 2020)

Regulator	Innovator	Consumers
Knowledge and experimentation guide long-term decision-making	Streamline the authorisation process to minimise time to market	Encourage the launch of new, probably safer goods
Signal dedication to imagination and education	Reduce legal confusion, such as the banning of emerging technology and business models	Increasing access to financial goods and services
Foster contact and interaction with investors in the business	Gather feedback on regulatory criteria and hazards	
Improve laws that preclude beneficial invention by many	Increase capital access	

innovation districts is that the former is more product-focused, while the latter is more focused on market creation. Innovation districts step forth in transforming products that come out of the living lab into marketable, feasible and useful assets for the residents of the cities (Cosgrave et al. 2013).

1.2   How to Finance and Fund Innovation

The approval of an innovative idea in the eyes of the public is somewhat reflected in the market's reaction to the idea. For this reason, a test almost all entrepreneurs face when formulating an innovation is the task of finding essential funding. Depending on the business model, an innovation may require and take advantage of different funding options, each with its advantages and disadvantages. An innovation with a focus on improving a problem the public faces, despite a lower profitability prediction, might be more eligible in obtaining grants provided by governments, intergovernmental organisations, universities, non-governmental organisations and entities or a hybrid of these mentioned bodies. Yet, a new business with a clearer high-growth business plan that would potentially provide a high return to investors on their investment could attract different kinds of funds, such as those made by venture capitalists or angel investors. Alternatively, a small company formed with smaller short-term goals or ones that require smaller amounts of initial investment might benefit by taking out a small bank loan or asking

Table 1.2 Characteristics of different types of living labs (Leminen et al. 2012)

	Utiliser-driven	Enabler-driven	Provider-driven	User-driven
Purpose	Strategic R&D activity with preset objectives	Strategy development through action	Operations development through increased knowledge	Problem-solving by collaborative accomplishments
Organisation	Network forms around a utiliser, who organises action for rapid knowledge results	Network forms around a region (regional development) or a funded project (e.g. public funding)	Network forms around a provider organisation(s)	Network initiated by users lacks formal coordination mechanisms
Action	Utiliser guides information collected from the users and promotes knowledge creation that supports the achievement of pre-set goals	Information is collected and used together, and knowledge is co-created in the network	Information is collected for immediate or postponed use; new knowledge is based on the information that the provider gets from the others	Information is not collected formally and builds upon users' interests; knowledge is utilised in the network to help the user community
Outcomes	New knowledge for product and business development	Guided strategy changes into a preferred direction	New knowledge supporting operations development	Solutions to users' everyday-life problems
Lifespan	Short	Short/medium/long	Short/medium/long	Long

family and friends for funds. All the mentioned funding options each have their advantages and disadvantages, and we will try to mention some of them independently.

1.2.1 Loans as a Means of Financing Innovation

One of the main tools of financing used throughout many sectors of the business world is loans. A loan is described as "an amount of money that is borrowed… and has to be paid back" (Cambridge Dictionary 2021). Most scenarios under these loans take place either by obtaining a loan from a personal relative or getting credit from a bank. With smaller amounts required to finance innovation, this method would be highly effective, as it does not have any implication on the business itself, and there is a definite method of paying back. Furthermore, different forms of financing have been present in the start-up ecosystem. The National Endowment for Science, Technology and the Arts (Nesta) is an organisation that promotes innovation. Out of several different financing options Nesta provides, project-specific loans and convertible loans to equity options are two we find important to mention. Although details vary with each project and each agreement, a "repayable grant" option used with project-specific loans states that a loan is only required to be paid back if the innovation is successful. Some agreements also include a zero-interest loan structure (Nesta 2018).

1.2.2 Grants as an Incentive to Fuel Innovation

A widely used form of financing that is generally beneficial to new innovative ideas is the obtaining of grants. The State of Queensland states that "…grants have specific application requirements and usually relate to… certain stages of the business cycle…" (Queensland Government 2021). This is the case with many models of grant funding systems. For example, the European Commission's Innovation Fund states that proj-

ects that will be awarded will be selected on the following criteria: "effectiveness of greenhouse gas emissions avoidance, degree of innovation, project maturity, scalability, and cost-efficiency" (European Commission 2021). Similarly, the European Investment Bank InnovFin Energy Demonstration Projects (EDP), which is a product produced by the European Investment Bank that supports innovations, points out several different eligibility criteria (European Investment Bank 2021). InnovFin EDP's funding eligibility requirements, among others, include the innovativeness of the project, readiness for demonstration at scale and prospects of bankability.

1.2.2.1 Example Case Study for Eligibility and Procedures for Grant Approval

In November 2018, the Government of Canada launched the Sustainable Development Goals Funding Program to further the world's goal of achieving the 17 Sustainable Development Goals (SDGs) by 2030, advancement of research and increasing partnerships (Government of Canada 2020). All members of the United Nations had committed to the same goals and had mobilised similar financial incentives for the advancement of the 17 SDGs. This specific Canadian funding program was open for application to nonprofit organisations, networks or committees, research organisations and institutes, for-profit organisations and many more. If you could describe how your project proposal would advance 2 of the 17 SDGs, require less than $100,000 and last no longer than 12 months, you would be eligible to apply for this funding program.

While the contents of the project are of utmost importance, similar to the eligibility requirements, the application assessments are clear. Firstly, the fund requires your project to explicitly identify what problem your project will solve and procure reasoning for your methodology. Secondly, the project application should provide clear and specific timelines and descriptions for every aspect of activities. Later on, the application then requires expected and desired outcomes of the project that link to the SDGs. Furthermore, and more importantly, the application requires you to include result measurement indicators. Lastly, a detailed report indicating required funds and how they correspond to different costs is required.

Grants from such financiers have many similar properties to this case. While it is extremely important to straighten out bureaucratic and procedural details, many funds intend to incentivise innovations without deterring bright and entrepreneurial minds.

1.2.3 Role of Venture Capitals and Business Angels in Innovation Financing

Startup Genome's Global Startup Ecosystem Report from 2019 indicates that 11 in 12 new businesses fail (Startup Genome 2019). These new businesses with high risks might not be able to take out huge amounts of loans from banks without any collateral. In such cases, taking out loans would not be a feasible and sensible option to finance your innovation. Venture capital meets the needs that not many institutions can, "…as traditional financing such as bank loans became more complex to attract, the development of alternative investments, like seed and start-up capital investments, crowdfunding, venture capital, and business angels, became a bold topic" (Dibrova 2015). Although, for VCs to fill in this void of high-risk investments, VCs expect a high enough return to "…attract private equity funds, attractive returns for its participants, and sufficient upside potential to entrepreneurs to attract high-quality ideas that will generate high returns" (Zider 1998). Robinson's (1987) study indicates that personal motivation, organisation skills and executive experience were the most critical criteria of VC's choice of investment in an investee company/person. Moreover, in a growth industry, substantial growth objectives by a complete management team with enough expertise had a very important role in the decision. Albeit these criteria, it is accepted that VC investments tend to be risky investments (Schilit 1993). Thus, informational asymmetries, a high amount of required capital and very important business risk in ven-

ture capital led to the requirement of a higher overall return by venture capital companies (Manigart et al. 2002). From the perspective of an innovative business idea leader, the pressure caused by the expectation of a lucrative exit option may be a disadvantage. Along with giving up some equity and voting power to the venture capitalist, independence in setting a course for your company may be curtailed. Furthermore, reports indicate that the agency problem associated with venture capital firms means that each venture capital professional may only spend less than 2 hours per week on a specified investment (Zider 1998). On the other hand, business angels provide extensive mentoring services as they invest their own money into the business; they see a potential of profit. "Business angels are individuals who offer risk capital to unlisted firms…" (Politis 2008). A comparison with venture capital is that business angels give more importance to the entrepreneur and "investor fit" issue (Mason and Stark 2004).

1.2.4 Venture Capitals

Venture capital (VC) is an important source of capital for small businesses that has steadily grown into a major part of funders' portfolios (Javed et al. 2019). Fresh and creative businesses that cannot obtain conventional financing (such as bank loans) can find VC investments a valuable option (Bellucci et al. 2021). VCs can invest in innovation and firms at several stages of the innovation journey. VC investment certainly helps in scaling and growth of the innovative ideas as they come into the picture in these late phases. Lerner and Nanda (2020) list three concerning matters regarding the role of VCs in the financing of innovation:

1. The very narrow band of technological innovations that fit the requirements of institutional venture capital investors
2. The relatively small number of venture capital investors who hold and shape the direction of a substantial fraction of capital that is deployed into financing radical technological change

3. The relaxation in recent years of the intense emphasis on corporate governance by venture capital firms

The European VC funds, with an average of €56 million, are much smaller than the US VC funds, which are €156 million on average (VentureEU 2020). Furthermore, venture capitalists invested about €6.5 billion in the EU compared to €39.4 billion in the USA in 2016 (VentureEU 2020). We may conclude that Europe needs bigger VC funds to support R&D and innovation.

1.2.5 Crowdfunding as a Modern Alternative to Financing Innovation

Crowdfunding had emerged and developed within the Internet community, mostly within the creative industries such as the arts and media; thus, it was mostly unnoticed by the outside world (Hemer 2011). Accordingly, data shows that the interest revolving around the concept of crowdfunding has emerged largely post-2010 and peaked around 2015, as can be seen in Fig. 1.3 (Google Trends 2021). Consequently, today, crowdfunding has become a prevalent mode of financing and has a very wide usage across several sectors. Today, crowdfunding is a method for funding new ventures that allows founders to request funding from many people in exchange for future products or equity (Mollick 2014). There are two major forms of crowdfunding: entrepreneurs soliciting individuals either to pre-order the product or an advancement of a fixed amount of money in exchange for equity (Belleflamme et al. 2014). While some innovators use crowdfunding platforms such as Kickstarter and GoFundMe to utilise early adopter financing for their needs, some tend to rely on this method as a continuous financing tool similar to a subscription model. In this regard, the dynamic nature of this form of financing creates new uses constantly.

1.2.6 Public-Private Partnerships

Public-private partnerships (PPPs) are a collaboration between a governmental agency and a private sector corporation that can be used to fund, develop and run projects such as public transit networks, parks and conference centres. Financing a project in a public-private collaboration will help it get done faster or even get it started in the first place. Tax or other operating income compromises, liability insurance and partial ownership interests to nominally public services and land are all standard features of public-private partnerships (Brock 2021). The smart city is a holistic concept for dealing with urbanisation problems in modern cities. Public-private partnerships (PPPs) are a blueprint for the public and private sectors to cooperate on designing and implementing smart city infrastructure projects (Liu et al. 2020).

Apart from the mechanisms we list here, there might be other useful approaches in the procurement and support of innovation. Further tools and mechanisms for the funding and financing of innovation will be explained in the following sections.

1.3 The Business Model Theory

1.3.1 What Is a Business Model?

A business model has no universally accepted definition. The origin of the business model definition goes back to Drucker's (1994) definition of this model with four fundamental questions:

- Who is the customer?
- What do customers value?
- What are our revenue streams?
- What economic reasoning explains how we can provide value to customers at a reasonable price?

Figure 1.4 summarises the business model core components.

These components are developed to solve these core concerns for every company.

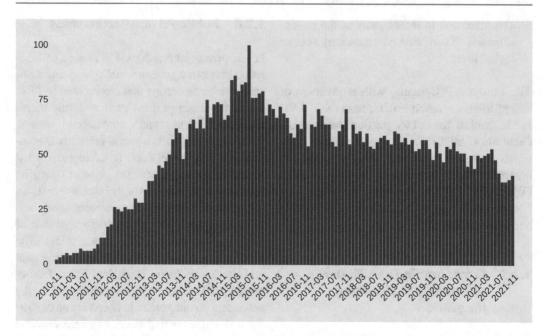

Fig. 1.3 Global search interest over time – "crowdfunding". (Google Trends 2021)

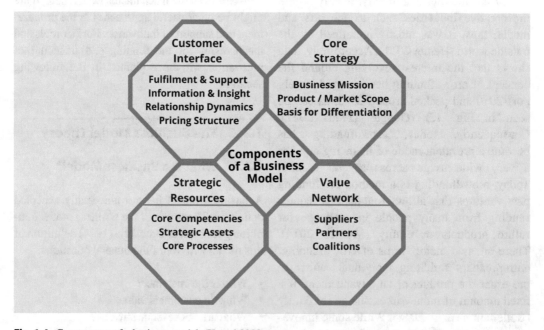

Fig. 1.4 Components of a business model. (Hamel 2000)

Companies use business models to describe how they create income by referring to the value chain structure and its relationship with the industry value system (Fisken and Rutherford 2002). Rambow et al. state that businesses striving for long-term commercial success are being pushed by digitisation to adapt business models to new market scenarios or build new business models when the old ones become outdated due to technological progress (2019). To take advantage of new technology and develop an innovation (as in Apple's case), a new model is frequently required.

Apple revolutionised portable entertainment, created a new industry and transformed the firm when it debuted the iPod and the iTunes store in 2003. In just 3 years, the iPod/iTunes combo became a roughly $10 billion product, accounting for over half of Apple's sales. Apple's market value soared from less than a billion dollars in early 2003 to more than $150 billion by late 2007 (Johnson et al. 2008).

Companies sometimes find themselves in a predicament that appears insurmountable. The core reason for all of these catastrophes is not that things are not done well. The seeming paradox arises from the fact that the assumptions upon which the organisation was founded and is still operating no longer hold. These are market-related assumptions. It's all about figuring out who your consumers and rivals are, as well as their beliefs and behaviours (Drucker 1994). All of these crisis processes may be managed by creating a proper business strategy. According to Masanell and Ricart, the success or failure of a company's business model, on the other hand, is mostly defined by how it interacts with the models of other industry participants (2011). For instance, one firm model may appear to be superior to others when analysed in isolation, but when interactions are taken into consideration, it provides less value than the others. Isolating models lead to inaccurate assessments of their strengths and faults, as well as bad decision-making. This is a big reason why so many innovative business concepts fail.

1.3.2   How to Create a Business Model?

The notion of a business model has grown significantly more popular as a result of digital developments. New applications, services, platforms, data and gadgets have created a crowded playground for all types of businesses looking to capitalise on new opportunities. New businesses have sprung on the scene, with varying degrees of success and exponential expansion. Both experienced and new players have the need to overcome similar obstacles in common. They need to generate value for connected consumers by developing new business models that function in the digital environment (Zott and Amit 2017).

The business model is a system of interrelated and interdependent activities that governs how a firm interacts with its stakeholders. How can a corporation improve its chances of establishing the best business model for its circumstances (Zott and Amit 2017)? Answers to questions seen in Fig. 1.5 and their repercussions make up a business strategy.

A business model consists of four interrelated components that work together to create and provide value. Every component of the model should be compatible with the rest of the model:

- Value proposition
- Targeted customer
- Value creation and value delivery
- Value capture and revenue model

1.3.3   Value Proposition (What Are They Offering?)

Developing a business model entails more than just finishing your business strategy and deciding which goods to pursue. It is also about figuring out how you'll provide your clients with continuous value. At the start of their business, every firm seeks to figure out if their business model will meet demand and be accepted by the market. It is not enough to know how to produce, create or supply anything; the product or service offered also has to be valuable to potential clients. It is a misconception to believe that "They will come" if you build it. Figure 1.6 depicts a value proposition canvas, which is useful for determining a new product or service's business model. The value proposition that will be built with the components on the canvas will aid in determining point of junction of the firms' products and the customer. Offering value propositions and marketing approaches feed customer awareness of firms (Myler 2013). Customers must find an advantage at least as much as the extra value they receive from the product they are now using to move from another to yours (Golub et al. 2000).

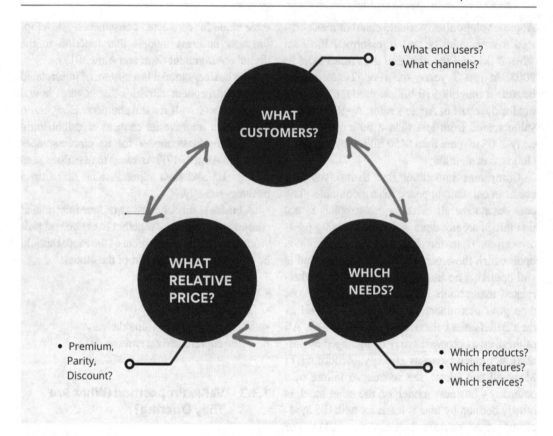

- What end users?
- What channels?

WHAT CUSTOMERS?

WHAT RELATIVE PRICE?

WHICH NEEDS?

- Premium, Parity, Discount?

- Which products?
- Which features?
- Which services?

Fig. 1.5 Business model creation diagram. (HBS 2019)

Fig. 1.6 Value proposition. (Thomson 2013)

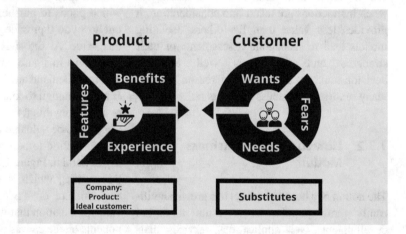

The purpose of the value proposition is to provide answers to the following questions:

- Why should customers buy your product?
- What is the benefit to them?

Firms make strategic decisions about how to position their innovations to disrupt rival firms. Disruptive innovations should be assessed in the context of a company's business strategy (Christensen et al. 2018, p. 1050). Accordingly, enterprises must propose and regularly renew new value propositions as part of their innovation operations and business strategy for a business model to disrupt other companies. In this way, companies can attract mainstream customers from their competitors (Schmidt and Scaringella 2020).

Companies create completely new consumer value propositions by solving a challenge that has never been solved before. One of these new propositions may be Apple's iPod and iTunes electronic entertainment distribution system (Johnson et al. 2008). As another example, when GE Aircraft Engines changed from selling airlines jet engines to selling them flying hours, they created a unique value proposition. This transferred the risk of downtime from the airline to GE, allowing the company to build a very successful service operation (Chesbrough 2007).

1.3.4 Targeted Customers (Who Are They Targeting?)

It is all about establishing your target market for marketing your company. A conceptual model for measuring the impact of marketing tactics on both targeted and untargeted clients is shown in Fig. 1.7. The individual you believe is most likely to buy your items is your targeted customer. The target customer base includes a certain age rather than a variety of ages, a specific income level rather than a wide range of income kinds and the possibility that these people will buy your items (Belcher 2019). This phase in the business model development process aids in the development of marketing strategies as well as the estimation of income and costs, taking into consideration the various types of business models and clients.

Recent techniques to engage with your consumer base faster and more accurately for targeting are available in the digital world. By utilising digital technologies to broaden conventional marketing channels, a wide range of channels may be supplied, going beyond the usual usage of established tools such as newsletters or email marketing (Schrock et al. 2016). In addition, digital technologies enable companies to collect new information about customers. In industrial marketing contexts, digital technologies enable communication between firms and their consumers in B2C environments and allow for the use of various channels to identify and target customers (Spieth et al. 2019).

Companies may appeal to various customer groups while retaining their current financial ratios by using a new value network. Nestlé, for example, created Nespresso, a coffee shop that has been compared to an upscale Starbucks, to appeal to young urban professionals. Nestlé's coffee company, which had historically offered instant coffee to the mass market through department and grocery shops, gained a new value network and a different customer base with Nespresso (Koen et al. 2011).

1.3.5 Value Creation/Value Delivery (How Are They Planning to Create and Deliver Their Service?)

After firms have decided on a value proposition, you need to ensure that it "echoes" across the business system, ensuring that every corporate activity reinforces the chosen value. A relevant framework for examining this echoing process is the value delivery system (Golub et al. 2000). Golub et al. state that customers base their purchase decisions on two factors: the advantages and pricing of a product or service. Generally, customers choose the item or service that provides the highest value among competing options. There is a significant point that the winning approach is usually the one that best executes the value proposition, not the one with the most appealing value. Figure 1.8 depicts the purchase decisions of customers and its relation to value creation and delivery.

In more conventional elections, managers split firms' systems of production into three segments: create value, develop a product and sell it. This method could be more useful for production-side issues like cost reduction. However, when developing a complex value offer, it is more beneficial to make the business system customer-centric:

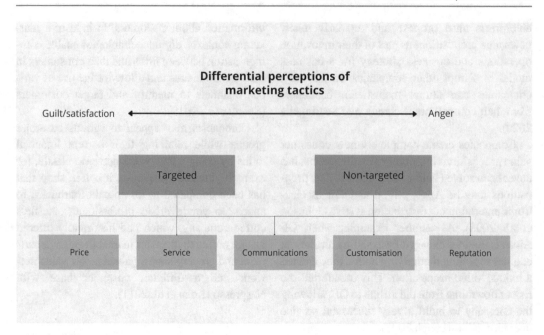

Fig. 1.7 Differential perceptions of marketing tactic. (Belcher 2019)

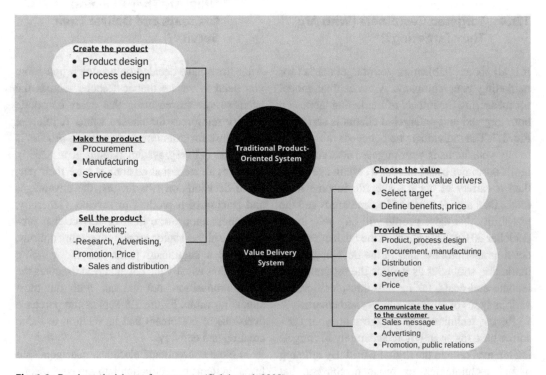

Fig. 1.8 Purchase decisions of customers. (Golub et al. 2000)

choosing the value, delivering the value and communicating the value to the customer. A value distribution system is a business system created in this manner. According to Daeyoup and Jaeyoung (2015), businesses plan how to distribute their value proposition to the value network and deliver to value users to have an effective business. The process of creating and delivering value has some advantages for businesses. Firms aim to enhance their methods for boosting customer satisfaction, understanding consumer preferences, lowering inventory, increasing inventory turnover, reducing stock-out situations and improving time to market as part of this process, which might lead to financial gains. Furthermore, the creation and delivery planning process provides insight into which values are more widely accepted by consumers, market participants and suppliers (2015).

1.3.6   Value Capture/Revenue Model (What Are the Sources of Their Expected Revenue, and How Are They Planning to Create This?)

The main goal of a business model is to produce money by extracting value from what it creates. According to Richardson (2008), it does not follow that a company that develops a compelling value proposition and effectively generates and distributes that value would earn higher returns or even be sustainable. It must also have a revenue-generating mechanism with a profit margin over its costs. The revenue model is made up of this part of the business model. It's important to distinguish between the income model and the economic model. The income model identifies the many sources of income or income that the company gets. On the other hand, the economic model is concerned with the firm's expenses, margins and different financial elements.

The business model should have well-defined strategies for monetising corporate value through value generation and capture (Daeyoup and Jaeyoung 2015). The revenue model consists of all strategy research to achieve this profit potential. Pricing the offer, service delivery and infrastructure improvements should all offer attractive profit potential for growth and innovation under a viable revenue model (Advantage 2021).

Different methods can be used when creating a revenue model – advertising models, subscrip-

tion models and sales tactics, to name a few. As an example of a revenue model, Spotify introduced its product to US audiences in 2011. Spotify's revenue model was noticeably different from other music streaming services. Unlike Pandora, which relied on advertising revenue, and Apple Music, which went with a premium approach, Spotify went with a freemium strategy. That is, it provided a free basic service and a paid premium service with additional material, functionality and a superior user experience (Tidhar and Eisenhardt 2020).

1.4   What Is Value?

1.4.1   What Do We Mean When We Say "Value"?

The word "value" is used in a variety of ways in today's commercial and economic environment. It often refers to the monetary worth that a person, company or the market assigns to a commodity, goods or service. In truth, in today's economy, most goods and services such as commodities, tangible properties, intangible properties, services and labour, lands and businesses are priced according to their monetary value. However, the term value has gained a different dimension in economy and business, especially with the emergence of the "triple bottom line" approach, first coined by John Elkington in 1994 as mentioned in "Enter the Triple Bottom Line" (Elkington 2004). In the following section, we will first analyse economic, monetary and business value, discuss the distinctions between them, examine the concepts of environmental and social value for businesses and analyse the triple bottom line.

1.4.2   Economic Value, Monetary Value and Business Value

1.4.2.1   Economic Value

A product or service's economic value is a metric used in economics to assess its benefit to a particular economic agent. It is often calculated using the people's willingness to pay for the product, which is usually expressed in monetary units. A person's preferences influence the economic value of an item or service, as well as the trade-offs they are prepared to make to get it. Suppose a person owns an apple, for example. In that case, the economic value of that fruit is the advantage that they obtain from using it. If they plan to eat the apple, the economic value is the pleasure they expect to gain from doing so. Economic value cannot be directly assessed since it is subjective and reliant on a person's preferences. However, several approaches for quantifying or estimating economic worth have been created, such as "willingness to pay (WTP)" or "hedonic pricing".

1.4.2.2 Monetary Value

The monetary worth of an item or service is the price that would be paid for it if it were transferred to a foreign party. The monetary worth of physical, intangible, labour and commodity assets, for example, is used to determine their prices. The financial effect of risk is estimated using the expected monetary value.

1.4.2.3 Business Value

In management, the phrase "business value" refers to all types of value that have a long-term impact on the health and viability of a company. The idea of a company's "business value" extends beyond its "economic worth" to encompass a variety of other types of value, such as the value of its employees, customers, suppliers, channel partners, alliance partners, managers and society at large. In many cases, various types of value can't be explicitly monetised. There is no agreed-upon definition of business value, yet there are some economists who argue that focusing just on financial metrics, such as profit and shareholder value, is not enough to help businesses make decisions.

1.4.3   Environmental Value

Environmental values and valuation of nature are difficult to understand in part because these terms are used in so many contexts and so many ways. Human and ecological values may be quantified as assistance to decision-making in certain cases. Some people are against environmental value because they believe it equates to "putting a price label on nature" or "lowering ethics to statistics and numbers". Yet environmental value remains a hot topic today that every company, organisation or community strives to integrate into their operations, plans and activities.

Economists who study the relationship between the economy and the environment confront an ever-increasing need for environmental values to be calculated now. There are several instances of how people profit from the use of natural resources, such as forests, rivers and other ecosystem services, which may be used as examples of how essential nature is to the economy. Because of this, environmental concerns are becoming a more prominent part of our daily lives and, as a direct result of this, in political and policymaking circles. Cost-benefit analyses (CBAs) commonly include environmental values as a consideration. This kind of research is used to examine and evaluate current and prospective programs and policies and to evaluate their impact on society.

The use of economic values to influence environmental decision-making has been embraced and implemented by many institutions and organisations that have developed environmental processes and standards for public sector projects that emphasise the need for environmental valuation and cost-benefit analysis. Environmental value is becoming more significant in policymaking, considering the stronger ties between natural capital and economies in emerging nations. Multilateral institutions, such as the World Bank, have begun incorporating environmental valuation methodologies and norms into their planning processes in response to the UN Sustainable Development Goals inception and adoption.

As stated above, there are many views and debates on environmental values, whether it can be evaluated through monetary evaluation techniques or whether it is ethical to value the environment within economic terms. As an expected result of these discussions, there are many views or criteria on valuation techniques of the environment. In this subsection, five basic categories for environmental valuation that are proposed by Harris and Roach (2017) are discussed (Harris and Roach 2017).

A. Market Valuation

Harris and Roach suggest that forests, fish populations, minerals and groundwater are just a few examples of environmental assets that are already being traded on the open market. They say that economists can assess the direct-use value of these resources by calculating the consumer and producer surplus (Harris and Roach 2017).

B. Cost of Illness Method

The cost-of-illness analysis is a method for estimating the financial toll that a particular sickness or condition has on a person. Harris and Roach define the cost-of-illness method as pollution's negative effects may be valued using the cost-of-illness method, which estimates the costs of treating illnesses caused by the pollutant (Harris and Roach 2017). Environmental impacts and operations of businesses often affect human life directly or indirectly.

C. Replacement Cost Methods

The cost of restoring or replacing a resource, such as fertilising the soil to restore fertility, is estimated using the replacement cost method. Harris and Roach state that these techniques consider the costs of acts that offer human-made replacements for lost ecological services. For instance, if a forest ecosystem were to disappear, a town may build a water treatment facility to compensate for the loss of water purifying advantages. To some degree, the pollination of plants by bees might be done manually or mechanically by some automatic system. They say that we can approximate society's willingness to pay for these environmental services by estimating the construction and labour costs of these alternative activities (Harris and Roach 2017).

D. Revealed Preference Method (RPM)

RPM indirectly infers market participants' values of environmental products and services. For instance, individuals' value on clean drinking water may be deduced from their expenditure on bottled water. This way, the environmental value of clean water can be estimated (Harris and Roach 2017).

E. Stated Preference Method

Surveys are used to understand market agents' preferences on environmental values. The main advantage of this method is that people can be surveyed on every type of environmental value such as carbon storage, nuclear energy, forest area for future generations, etc. But as a disadvantage, the validity of the results can be dubious (Harris and Roach 2017).

1.4.3.1 Planning and Creating Environmental Value

The key to effective planning is turning it from an intellectual exercise that culminates in another report on the shelf into a dynamic business integration process. Environmental specialists must enter the business, get a grasp of the larger organisation, listen to the requirements and objectives of other departments and find strategic options for resolving corporate difficulties. It entails transcending typical job descriptions to assume roles as strategists, entrepreneurs, sales agents and instructors. Therefore, successful planning may uncover value-creating possibilities while also providing essential insight into the most effective communication methods with important persons and groups.

A well-thought-out strategy has four key components: (1) knowing your business, (2) taking inventory of potential environmental impacts, (3) identifying value-creating opportunities and (4) prioritising activities (Global Environmental Management Initiative 2014).

A. Know Your Business

Evaluate the environmental actions' business perspective. Corporate environmental specialists are often uninformed of other divisions and employees' unique aims, priorities and requirements. Three things are required to provide value-added solutions. To begin, determine who the present and prospective environmental services clients are. Second, understand the business problems that your clients are attempting to address. Finally, be familiar with your company's long-term business strategies (Environment: Value to Business, Global Environmental Management Initiative 2014).

B. Evaluate Inventory Potential

Check the environmental impacts of the business. An environmental impact may be caused by the use of resources and production of wastes, the production and use of goods and the disposal or recycling of items. To begin compiling an inventory of environmental effects, identify the organisation's principal business operations and activities, including production and operational procedures. Finally, inquire about the environmental implications of each department's activities and products. How are the company's actions controlled in terms of their environmental impact? Whose earnings, development and public image are in danger due to environmental impacts (Environment: Value to Business, Global Environmental Management Initiative 2014)?

C. Identify Value-Creating Opportunities

Identifying value-creating possibilities at various levels of firm operations may be done after an inventory study on environmental effects has been completed. It is possible to cut expenses and increase profits by implementing environmental efforts beyond the minimum compliance standards. Search your organisation for areas and branches where value might be produced. According to corporate and government requirements, various environmental criteria and actions must be met. Even while these actions may be seen as a corporate expense, they have a basic environmental benefit. These may include things like obtaining a permit, avoiding environmental damage fines and so forth.

Finding creative methods to achieve more with fewer resources should be the objective of operations. Focusing on resources is essential. The costs of manufacturing, compliance and waste disposal and management may be reduced by lowering overall resource inputs, hazardous inputs or unwanted by-products. In addition, lowering environmental risk may avert major environmental damage as well. Among other things, ARCO recently obtained a new hazardous waste authorisation. In the future, an oil tank might be built on land once used to dispose of hazardous garbage. The tank's permit and construction were expedited. The approach was supported by long-term environmental data and a sturdy working connection with the local authorities. One million dollars in monitoring expenses were saved because of the company's innovations. In addition, they were able to recover the refinery-related landscape. Long-term monitoring costs have been reduced because of this project's successful repurposing of unused land and the subsequent savings (Environment: Value to Business, Global Environmental Management Initiative 2014).

Long-term costs connected with capital investment and design decisions cost businesses millions of dollars. Purchase of land, construction of facilities, start-up or redesign of manufacturing lines and new products may have significant financial repercussions. Environmental managers may provide value when it comes to capital budgeting and decision-making. For example, it is crucial for Duracell to have a supplier development program. To improve performance, Duracell engages with its providers continuously. There have been environmental measures included in supplier rankings for some time. A decrease in Duracell's greenhouse gas emissions and considerable cost reductions prompted the business to incorporate its energy management approach into its global supplier development program. During a meeting of major suppliers, Duracell invited them to join a cooperation agreement. Each business agreed to establish energy efficiency objectives, initiate actions to achieve those goals and adopt best practices throughout the group. Duracell vendors, in the majority, have agreed to participate in the project. Recognition will be given to the best performers. New cost-cutting initiatives for Duracell, new supplier contacts and assistance for suppliers in their cost-cutting efforts are some of the benefits of this arrangement (Environment: Value to Business, Global Environmental Management Initiative 2014).

D. Prioritising

There isn't enough time, personnel or money for corporate environmental professionals to explore all of the possible value-creating possibilities in the business. Environmental managers must prioritise environmental tasks and concentrate their efforts to efficiently use limited resources. Typical decision-making factors include the relevance of the project to the company's objectives, the project's magnitude in terms of money and resources needed and the project's level of complexity. Just because a job is simple and inexpensive, it is not implied that it should be done first. There is always a plan that considers the benefits that might be gained while also considering the financial and political resources needed (Environment: Value to Business, Global Environmental Management Initiative 2014).

1.4.3.2 Assessment and Measurement of Value Added

An essential yet difficult duty for environmental experts is determining the worth of environmental projects. The preceding Impact Assessment section presents an in-depth analysis of the environmental value discussion. Let us provide a quick overview of how to calculate the environmental value added. In this case, the right tools and methods are dependent on the questions being answered:

- Which of the following are you measuring: the effect of a single project, the value of environmental activity or the total value of all environment-related operations in the organisation?
- What are you doing? Do you evaluate the value that a proposed plan may generate, or are you looking to see whether an existing project has already created a value?

It is possible to utilise impact assessment and value measurement to check the results of environmental initiatives, giving useful input for future program adjustments as well as results that can be communicated to key stakeholders to keep their support. In addition to planning and prioritising environmental initiatives, impact assessment and value measurement are also really important. According to P&G's cost ratio analysis, health, safety and environmental (HSE) initiatives pay more than twice for themselves in terms of cost savings. Salary, employment, healthcare and facility activities such as dumping waste are included in the ratio as costs. Pollution prevention and eliminating the materials thrown away during manufacturing are included as utilities of the HSE initiative.

1.4.4 Social Value

1.4.4.1 What Is It?

A rising number of organisations look at their operations from a holistic perspective, including their activities' broader social, economic and environmental consequences. It's hard to define what exactly constitutes "social value", but companies that make a determined effort to guarantee that their outcomes are good might be considered to be benefiting the long-term well-being and resilience of individuals, organisations and humanity as a whole. The United Nations' Sustainable Development Goals are, in essence, a global social, moral charter. Including social value in the policy and spending decisions may help public sector organisations better serve their communities. Both what a company does and how it does it may positively impact society. By reporting on social value, corporations may externalise their programs by connecting them with precise, quantifiable results and doing so in an easy-to-understand manner for the benefit of consumers and other stakeholders.

Numerous entrepreneurs' principal objective is to establish a profitable firm. Apart from serving as a means of earning money, a business can also help to promote social ideas that benefit others. Social values can be incorporated into daily company plans or serve as the impetus for beginning a new firm.

Environmental stewardship is a value that can simply be included in your company procedures. Try to purchase recycled items, such as paper and printer ink cartridges, and correctly dispose of hazardous waste materials. Conserve energy by shutting off computers and lights when not in use and maintain proper operation of business vehicles to minimise hazardous emissions.

Additionally, businesses may utilise the company to assist humanitarian issues relevant to the corresponding industry. For instance, a food-related company, such as a restaurant or bakery, might consider donating a percentage of their profits to help feed the poor. Figure 1.9 summarises the benefits of social value.

Businesses may also express social ideals via their operations. Charging a fair price for products and services while putting a high emphasis on customer service is one possibility. Provide an equitable work environment and recruit workers that share the same values as the company or business. As can be seen from Fig. 1.9, there are so many easy ways to create social value, and the benefits of this approach are not limited to society. It comes back to the businesses themselves

too. The main strategy for generating this type of social value is to balance the desire to operate a profitable business and the desire to manage it ethically and bring benefits to society.

1.4.4.2 How Social Value Is Measured and Reported?

Although there are many ideas about what social value is, it is difficult to define and draw the boundaries of the concept of social value. For this reason, reconciling on a measurement criterion for social value is another complicated task. Geoff Mulgan (2010) argues that social value measurement was a hot topic among funders, NGO leaders and lawmakers. Unfortunately, they could not even agree on what it was, much less how to evaluate it. Geoff Mulgan suggests that the corresponding main problem was thinking that social value was absolute, set and stable. Yet it's easier to measure social worth when it's seen as a subjective, changeable and variable concept. Despite this, the assessment of social value is already becoming increasingly standardised. The National Social Value Measurement Framework, abbreviated National TOMs (TOMs stands for "Themes, Outcomes and Measures"), was created and released in 2017 by the Social Value

Jobs and Economic Growth	Health, Wellbeing and the Environment	Strength of Community
• Decent jobs for local people and hard to reach groups • Local people with the right skills for long-term employment • School leavers with aspirations of the industry • The local supply chain is supported and grown • Residents have comfortable homes which are affordable to operate • Thriving local businesses	• Good accessibility and sustainable transportation • Resilient buildings and infrastructure • High quality public and green spaces • Good mental health • Good physical health • Limit resource use and waste	• Strong local ownership of the development • Existing social fabric is protected from disruption • The new community is well integrated into the surrounding area • Thriving social networks • Vibrant diversity of building uses and tenures • Strong local identity and distinctive character

Fig. 1.9 Creating social value and its benefits. (UKGBC 2018)

Portal. It is a strategy for documenting and quantifying social value using a consistent metric. It serves as the "golden thread" connecting an organisation's broad strategy and vision to its execution. The TOMs serve as a "golden thread" connecting strategy and delivery of social impact in the following ways (Social Value Portal 2017):

- Themes: the elements that comprise an organisation's "vision" of social value.
- Outcomes: the intended positive consequences for the organisation.
- Measures: What types of measures will be utilised to determine whether these results are positive?

In addition to these three "golden threads", TOM provides more themes that can be counted as goals to accomplish social value. To achieve these goals, we need to encourage the development of responsible regional businesses and the creation of healthier, safer and more resilient communities (Social Value Portal 2017).

1.4.4.3 Seven Principles of Social Value

Social Value UK proposes seven principles for accountability and maximising the social value created by businesses. These principles serve as the foundation for anybody seeking to make decisions with a broader social impact (Social Value UK 2021b):

1. Involve Stakeholders: By incorporating stakeholders, influencing what is assessed and how it is quantified and assessed in a social value account.
2. Understand What Changes: Distinguish between good and bad changes and between those that are planned and those that aren't, using data and facts.
3. Value the Things That Matter: Stakeholders' values must be taken into consideration when deciding how to allocate resources since an outcome's significance is determined by its value.
4. Only Include What Material Is: Identify which data and information should be included in the reports to provide an accurate and fair picture

that enables stakeholders to make reasonable judgments about the effect.
5. Do Not Over-claim: Claim only the value that corresponding actions produce.
6. Be Transparent: Provide evidence of the foundation on which the analysis could be deemed reliable and honest and evidence of the fact that it will be presented to and discussed with key stakeholders.
7. Verify the Result: Ascertain that suitable independent assurance is in place to provide evidence.

Finally, Social Value UK states that the implementation of these principles will assist organisations in being more responsible for the outcomes of their work, which includes being accountable for more than whether the organisation met its goals (Social Value UK 2021b).

1.4.4.4 Creating Social Value: Business Approach

To grasp the idea of "social impact", it is a must to grasp the meaning of the term "social value". For example, efforts made by businesses and individuals to address social issues are referred to as "positive social effects". Business leaders must first establish the fundamental principles that drive their actions to effect social change. These values are often centred on critical social and environmental concerns affecting society, such as global warming, poverty, unemployment and other serious social and environmental difficulties. Purpose-driven executives might gain confidence to make a significant difference when their beliefs guide their decision-making and strategy. Matt Gavin (2019) from Harvard Business School Online suggests four business strategies to guide business leaders for their attempt at social change.

A. Conduct Business in an Ethical Manner and Encourage Ethical Business Operations

Gavin (2019) suggests that it is essential that firms examine their own practices to make sure that social responsibility is an integral part of what they do. It can be understood that businesses wishing to have social impact and create social value should explore how their procurement and production procedures may be more ethical and how they may be able to use social responsibility to modify their business practices.

B. Establish Strategic Alliances with Charity Organisations

Being a pioneer for systemic transformation is not an easy process. It demands a thorough awareness of society's issues and the tenacity necessary to overcome them. For this, Gavin (2019) thinks that increasing a company's social impact may be as simple as forming strategic alliances with non-profits that focus on the world's most critical issues. An example he gives is the partnership between Peet's Coffee and TechnoServe. Thanks to this partnership, farmers from Ethiopia, Guatemala and Rwanda have been educated about farming, business skills, sustainability practices and improving their quality of living.

C. Encourage Employees to Participate in Volunteer Activities

Being a purpose-driven organisation takes a commitment that transcends the corporate level. Employees must be convinced of the importance of their work and the organisation's transformational goals. Creating a work environment that encourages employees to give back may help companies solve important issues and foster a sense of belonging.

D. Motivate Action with Corporate Platforms

Gavin (2019) also believes that, in addition to creating programs and activities to address global issues, corporations may use platforms like blogs and other online channels as activism tools.

1.4.4.5 Ten Impact Questions to Maximise Social Impact

Considering effect entails making choices, such as between two strategies, one product or two attempts to enhance one product. Choices are made to have a greater influence than before. Constantly investigating new choices and alter-

ing your behaviour increases the likelihood of having the greatest influence possible. To make these decisions, businesses have to address several questions. "Social Value UK" proposes ten impact questions that are fundamental to maximising impact. These questions can function as a guide for businesses to evaluate and maximise their impact on social concerns and social value (Social Value UK 2021a):

- What issues are being attempted to resolve?
- What is the proposed solution to the corresponding issue?
- Whose lives are altered because of actions taken?
- Which outcomes or results are expected to be seen?
- What is the amount of change done by outcomes?
- How long should outcomes be measured for?
- How can the relative importance of the different changes be measured?
- How important is this distinction?
- How much of the change is done by the proposed solutions and actions?
- What are the most important changes?

1.4.5 Triple Bottom Line

Companies need to assess their social and environmental impact and financial performance as part of the "triple bottom line" concept. Financial, social and environmental accounting are all included in the three-part accounting paradigm of the triple bottom line. As Lim Mei said, using the term "triple bottom line" refers to reporting on the degree to which a company has helped society accomplish the three interrelated objectives of economic success, environmental preservation and social equality (Lim 2004).

Both governmental and commercial organisations are increasingly disclosing the environmental and social impacts of their actions, and environmental performance monitoring is becoming more and more commonplace around the globe. E. S. Woolard said in 1994 that industrial businesses are the only ones who can implement the green economy and culture of the

Fig. 1.10 Triple bottom line. (Lim 2004)

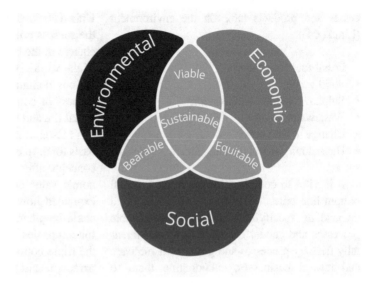

twenty-first century. Next-century environmental performance is the goal of the industry. A lot of companies are attempting to get there, but not everyone has yet. Because they won't be there in the long run, those who aren't trying aren't a long-term concern (Woolard and Global Environmental Management Initiative 1994). Companies are required to report environmental and social performance information, although there is a general lack of oversight. As a result of this lack of measurement, a business assessment method known as the triple bottom line (TBL) has grown in prominence, which incorporates economic, environmental and social factors. Figure 1.10 provides an overview of the TBL.

Three critical elements in triple bottom line, as can be seen in Fig. 1.10, are used as criteria to measure organisational performance. These are economic value (prosperity), social value (social justice) and environmental value (Lim 2004). As "what gets measured, gets managed", accurate measurement is critical for demonstrating how sustainable business practices may be utilised to improve a company's performance and analyse the organisation's progress. What can be derived from the triple bottom line is that for an organisation to be sustainable, it must be financially secure (i), behave in a way that minimises the negative environmental effect (ii) and act in harmony and convenience with societal expectations (iii) (Lim 2004).

1.4.5.1 Economic Bottom Line

The economic bottom line encompasses profit and the ideas that underpin a company's strategy or behaviour and the business' long-term sustainability. The following measures should be included while conducting an economic bottom line report (Lim 2004):

- Income and expenditures
- Taxes
- Annual operating performances
- Cash management and investor returns
- Job growth
- Credit rating

The economic bottom line part of TBL provides a guide to organisations or businesses to make more with what they have. Suppose an organisation wants to develop and grow within TBL. In that case, they need to determine the economic activity they want and examine how well other organisations are performing concerning sustainability criteria.

1.4.5.2 Environmental Bottom Line

The environmental value refers to the environmental effect of a company's goods or activities and the type of its emissions and waste and how the environment manages them. The following measures should be included to determine how much negative impact an organisation's processes and products have on the environment (Lim 2004):

- Fossil fuel consumption
- Solid waste
- Sulphur dioxide concentration
- Wastewater quality
- Change in land cover
- Hazardous gas-waste management

It is vital to comply with the environmental bottom line criteria because organisations build demand by specifying ecologically sustainable processes and goods. The need for environmentally friendly processes and goods will motivate and reward businesses, encouraging them to engage in research and development to enhance their processes and production that is less harmful to the environment. Furthermore, shareholders, investors and clients also get benefits from these new ecology-friendly processes and products (Lim 2004).

1.4.5.3 Social Bottom Line

The term "social bottom line" relates to an organisation's attitude towards gender and cultural diversity, work hours and compensation, employee safety and participation and social assistance or facilities. There is no consensus on whether an organisation meets its social responsibilities. However, these are some measures that could be used while analysing an organisation (Lim 2004):

- Employee safety
- Workplace stability
- Engagement with community and community support

1.4.6   Conclusion of Value Discussion

In today's world, especially in the business world, people think of the monetary value of a product or goods when we say value. In a world where the success of companies and businesses is measured in terms of realised profit, this is normal. However, especially with the emergence of the

United Nations' Sustainable Development Goals, the success criteria of companies have started to change in the business world along with the rest of the world. Nowadays, importance is given to various definitions and metrics to measure the success of companies, and their contribution to social life and the environment is increasing. For this reason, it is imperative to establish and set a basis for the environmental value and social value concepts after installing the definitions of economic value and monetary value. This section explained how environmental and social value could be planned, measured and applied briefly for companies and businesses. Finally, the role of the triple bottom line concept in this process of paradigm shift has been introduced.

In the following section, we explain in detail how the impact of companies and organisations on the environment and social life can be measured over the definitions made in this section.

1.5   Impact Assessment

When enterprises and institutions desire to evaluate their effects on the stakeholders and the system, they eventually need techniques for impact assessment (IA onwards). In scholarly discussions, the measurement and evaluation of the actors' impact are becoming increasingly influential (Simsa et al. 2014). This section will provide findings from academic literature and the business environment on IA to guide involved actors such as businesses and institutions. First, IA will be defined regarding different points of view. Second, the question of why actors desire to conduct IA will be investigated. Third, there will be a brief examination of the actors involved in IA. Fourth, we will inspect several methodologies for conducting IA. Then, we will share examples of IA use from both institution and business sides.

1.5.1   Definition

To assess impact accurately, a distinction between impact and outcome must be made. However, there is no unanimity on this subject (Simsa et al.

2014). Definitions differ significantly since distinct sectors and stakeholders require diverse viewpoints to assess the unique dynamics of their activities. To provide a general perspective, Stern examines the literature and divides IA definitions into two categories: content and methodological definitions. Content definitions seek to explore any effect, acknowledge that there can be constructive or adverse effects and address the long term (Stern 2015). OECD, for example, describes the impact as "the positive and negative, primary and secondary, long-term effects produced by a development intervention, directly or indirectly, intended or unintended" (2010, p. 24). On the other hand, methodological definitions are more tightly concentrated and employ experimental data, resulting in a shorter-term focal point (Stern 2015). For instance, according to Roche, the impact is the methodical investigation of a major change in people's lives that is caused by an activity or series of activities (1999, as cited by Stern 2015).

1.5.2 Reasons for Conducting IA

Although definitions vary substantially, reasons for desiring to conduct IA share commonalities. The first common reason is that actors put such a remarkable effort into receiving funds that they cannot risk losing credibility and being eliminated from further funding. IA is an immense way to demonstrate success and be accountable. Second, when they receive further funding, the need for improvements emerges, again, to show their accomplishments. To attain advancements, involved actors can also utilise the findings from IA to comprehend the effects of their efforts. Third, since public voice is an important part of the business process, actors should find advocates to continue their journey. They can build this support by exhibiting the results from IA. In a nutshell, O'Flynn lists the motivations for assessing impact as follows: (1) demonstrating success both to explain received funds and to obtain additional funding, (2) learning to understand the implications of initiatives to enhance effectiveness, (3) being accountable and (4) using IA results to advocate for changes (2010).

1.5.3 Who Is Involved in IA?

As indicated earlier, monetary, environmental and social values are assessed while managing a business. When we link the process with the outcome, we see that these values should also be looked at during the IA. First, the monetary value of the impact is mainly regarded for financing reasons. So, we can say that the actors in the funding mechanism, such as institutions, venture capitalists, banks, etc., play a significant role in the IA process. Second, businesses disclose their environmental value, to exemplify, showing their environmental dignity to the actors in the ecology, such as environmental non-governmental organisations, besides animals and plants. Lastly, there is a necessity to demonstrate social value. Social actors such as individuals and states can be said to be affected by the actions of the companies or institutions. Therefore, IA should be designed to cover those monetary, environmental and social stakeholders as well as the businesses and institutions as the main actors of the process.

1.5.4 How to Conduct IA

IA is studied in various contexts as an interdisciplinary topic. However, empirical and methodological literature is scarce on quantifying the impact at the macro-level (Simsa et al. 2014). The key, argued by Stern, is that evaluators should begin by imagining what they want to know about programs rather than focusing on a certain toolkit. Designing IAs necessitates making informed decisions on various issues, including the objective, required resources and skills, ethical requirements, data collection and analysis and methods for encouraging assessment adoption (Stern 2015). We will investigate some of the existing techniques for conducting IA in this subsection.

1.5.4.1 Approaches to IA
Simsa et al. emphasise the importance of the fields of evaluation research, social accounting, ecological and social IA, nonprofit organisation (NPO) research, social entrepreneurship, profit-oriented entrepreneurship, business ethics or cor-

porate social responsibility (CSR) (2014). First, in evaluation research, three categories of appraisals can be recognised: (1) analysing program theory, (2) assessing program process and (3) computing program impacts (Schober et al. 2013, as cited by Simsa et al.). Second, in the realm of accounting and accountability, IA incorporates non-monetary impacts in accounting, balancing and profit calculation. Third, in ecological IA, the natural environment was the focus initially, where social components were included later. Fourth, in NPO research, it is important to note that impact is not always synonymous with success. Countable and measurable outputs can be used as success criteria. Fifth, in entrepreneurship, social impact investors consider not only the financial but also the societal implications of their investments. As a result, indicator systems comparable to those used by traditional for-profit businesses are established. Lastly, the impact has been taken up by firms primarily in the context of CSR within the issue of business ethics. Initially, the emphasis was mostly on environmental sustainability; later, the social dimension gained prominence as in the case of ecological and social IA (Simsa et al. 2014).

In terms of technique, according to O'Flynn, there are three general approaches for impact assessment: (1) retrospective, (2) process-driven and (3) ex-post studies as described in Fig. 1.11 (2010).

1.5.4.2 Methodologies

IAs come in a variety of shapes and sizes. It is vital to understand and build the underlying assessment architecture to measure the impact of a project, an organisation or a sector (Simsa et al. 2014). The impact value chain or logic model is shown by Simsa et al. (2014). Arising from evaluation research, the logic model is a representation of the theoretical functioning of a program, an organisation or a sector that is used to appraise the intended goals. The model recognises and differentiates the input, activity, output, outcome and impact components. Let us define what those listed indicators mean. First, all resources invested in an organisation's activities are called input. Second, activities refer to specific acts, tasks and the organisation's achievements to fulfil its goals. Third, the term output refers to physical items and services that can be quantified directly as a result of an organ-

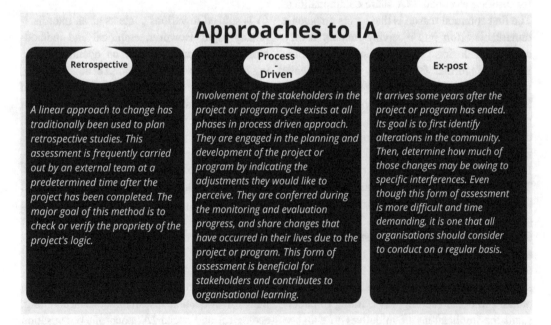

Fig. 1.11 Approaches to IA. (O'Flynn 2010)

isation's activity. Fourth, particular alterations in attitudes, behaviours, knowledge, abilities, etc. that occur due to an organisation's operations are hinted at as outcomes. Fifth, deadweight is the amount to which the outcomes would have happened anyhow and must be deducted from the outcome to spot the impact. Sixth, above and beyond the outcome that would have happened anyhow is named impact. Here, assessors should differentiate between performance measurement and IA. When it comes to performance evaluation, the activities or outcomes are the focus of attention. However, in the case of IA, identifying, measuring and valuing outcomes, including deadweight, is pivotal. The associated indicators, items and scales build the basis to measure impact empirically (Simsa et al. 2014).

Furthermore, Stern provides focal points to consider before starting to create an IA model. While there is no one-size-fits-all approach to making these decisions, some logical stages to follow might help guide decision-making. These steps are depicted as (1) evaluation questions, (2) program attributes and (3) available designs (2015). Various assessment commissioners will ask different types of impact questions, or even a different combination of such questions, such as how much of an effect can be attributed to the interference. When considering the program attributes, evaluators must generate a control group or comparator. There are designs and procedures such as instrumental variables that can help evaluators to discover specific impacts when an investigation is not possible. Furthermore, insisting on a precise measurement under all situations in many complicated program contexts is futile. Assessment commissioners frequently regard the advantages of combining approaches. Therefore, combined methods research that incorporates quantitative and qualitative techniques will increase the confidence of the results as they are formed on numerous different sources of information acquired in various ways. Few evaluations focus on a single subject; instead, they aim to measure impacts and explain what works where and when (Stern 2015).

Briefly, Stern argues that there are colossal differences in the shape, form, location, purpose, interrelationships and life cycle of programs. Then, it is not surprising that these attributes affect IA design. Determining the unit of analysis, creating theories of change and accounting for unpredictability are all the possible requirements of program features (Stern 2015).

Alternatively, according to Hailey and Sorgenfrei, the key is not the framework itself but how it is utilised – making the process as vital as the output is critical. It is also crucial to build breathing frameworks that demonstrate what actors attempt to assess the dynamic and multidimensional nature. Moreover, concerns such as power and authority, culture and context, as well as complexity and change, must be considered (Hailey and Sorgenfrei 2004).

1.5.4.3 Possible Failures and Solutions

Like all techniques, IA techniques also might come up with failures. According to O'Flynn (2010), eight possible reasons behind these failures are listed in Fig. 1.12.

O'Flynn (2010) also provides solutions to beat those challenges. Actors can overcome the first obstacle by devoting more time to figuring out how various processes are linked and complementing one another. To surpass the second one, it is suggested that assessors comprehend and explain the organisational zone of influence. The third challenge can be overcome by comprehending that measuring, evaluating and correlating facts have their own validity. Also, measurements should be done within their sphere of influence. To beat the fourth one, organisations should establish assessments for their purposes and then alter them regarding the demands of other stakeholders. To surmount the fifth one, evaluators can build rolling baselines and employ more quantitative data. The sixth possible reason for failure can be beaten through employing a small number of user-friendly tools. For the seventh one, assessors should devote effort to assuring that partners and stakeholders understand the assessment's purpose. For the last challenge, findings can be utilised to facilitate consensus and planning workshops, generate case studies and so on (O'Flynn 2010).

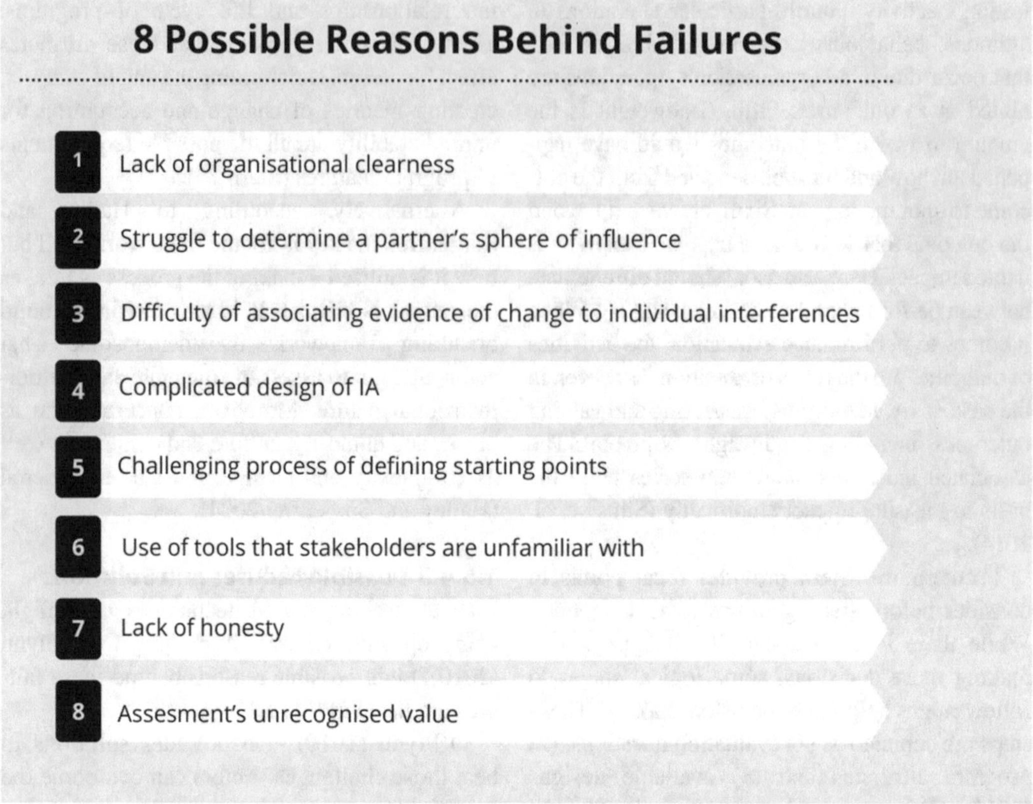

8 Possible Reasons Behind Failures

1 Lack of organisational clearness

2 Struggle to determine a partner's sphere of influence

3 Difficulty of associating evidence of change to individual interferences

4 Complicated design of IA

5 Challenging process of defining starting points

6 Use of tools that stakeholders are unfamiliar with

7 Lack of honesty

8 Assesment's unrecognised value

Fig. 1.12 Possible reasons behind failures. (O'Flynn 2010)

1.5.5 Examples of IA Methods

Now, it would be convenient to illustrate IA systems and principles through examples. First, the Anticipated Impact Measurement and Monitoring (AIMM) system of International Finance Corporation (IFC) will be investigated. Second, we will review the principles of EBRD for its IAs. The third example coming from the World Bank presents possible IA techniques and their comparison. Fourth, we will share our findings on IA by a business by examining Intel's corporate responsibility report.

1.5.5.1 International Finance Corporation (IFC)

IFC developed its own IA method, named the AIMM, to measure project-level and systemic outcomes of its actions. It is also an emerging model for impact investors and a tool to incentivise impact. Project and market outcomes are the two dimensions of this system. Project dimension looks into the influence in three different effect categories: (1) stakeholder effects, (2) economy-wide effects and (3) environmental effects. Second, the market outcomes dimension evaluates how well an intervention enhances market structure and function by encouraging competitiveness, resilience, integration, inclusivity and sustainability goals. IFC has created sector frameworks to evaluate projects in each of the IFC businesses. Sector frameworks assign ratings in four areas, as described in Fig. 1.13, to assist in IA (IFC 2021).

In the project outcomes dimension, the gap analysis evaluates the respective scale of the development problem that each intended effect is addressing. The intensity analysis assesses a project's contribution to reducing divergence in development. Evaluating intensity is based on normalised sector-specific criteria. When a gap evaluation and an intensity assessment are com-

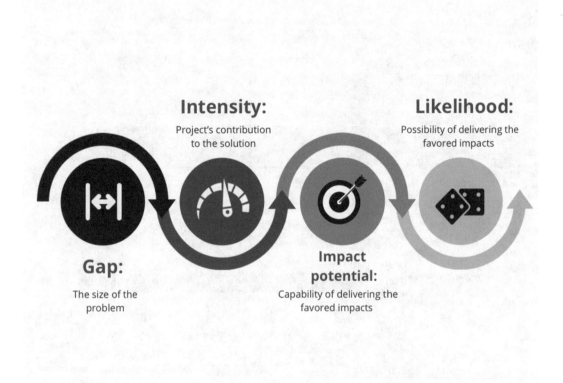

Fig. 1.13 Areas to be rated. (IFC 2021)

bined, as shown in Fig. 1.14, an overall potential impact rating is appointed. The potential for an effect to have an influence is determined by both the size of the problem (the size of the gap) and the effectiveness of the intervention (its intensity). This method prioritises projects that address larger development gaps and/or unique or constructed programs to generate results quickly (IFC 2021).

The market creation dimension appoints market stages to market development for each of the five market features. On the other hand, market movement is used to analyse a project's efforts to change markets' structure and functioning. As shown in Fig. 1.15, an overall potential impact rating is authorised after the market stage is combined with the market movement. The market's capacity for systemic change is determined by the market stage of development and the catalytic change that IFC anticipates its project to engender (IFC 2021).

The AIMM approach incorporates the uncertainty of realising and maintaining the desired impacts over time for both the project- and market-level dimensions. As shown in Fig. 1.16, the likelihood assessment is used to distinguish the potential results that a project could produce and the risks that could prevent them from being realised (IFC 2021).

The IFC uses AIMM evaluations to choose and compose interventions with the most significant impact possible. IFC administers its pipeline of interferences and develops plans to remedy inadequacies by aggregating AIMM ratings for different business areas. Portfolio ratings will also assist IFC in balancing strategic goals to pursue the portfolio strategy. They aim to maximise the development effect while making sustainable and risk-adjusted financial returns. Furthermore, IFC monitors and reports on ex-ante expectations, develops information feedback loops and compiles a lesson inventory (IFC 2021).

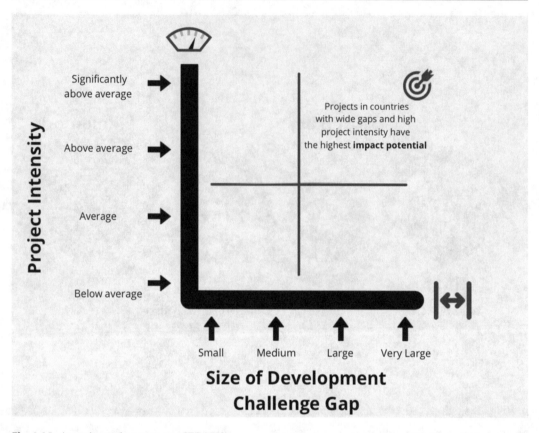

Fig. 1.14 Assessing project outcome. (IFC 2021)

1.5.5.2 European Bank for Reconstruction and Development (EBRD)

The EBRD has published a brochure to share its principles while assessing the impacts of its projects. There are nine principles, as listed in Fig. 1.17 (EBRD 2021).

Following the first principle, the manager must develop strategic impact goals for the portfolio or fund to create positive and measurable social or environmental impacts consistent with the Sustainable Development Goals (SDGs) of the UN or other broadly accepted objectives. The manager shall also ensure that the investment strategy has a plausible foundation for reaching the impact goals (EBRD 2021).

For the second principle, the manager must have a procedure in place to administrate IA on a portfolio basis. The principle's goal is to construct and track impact performance throughout the entire portfolio while considering that impact

might differ between individual investments. The manager shall also acknowledge connecting staff incentive schemes with impact achievement besides financial performance (EBRD 2021).

For the third principle, the manager must attempt to build and record a believable story on its input to each investment's impact. Contributions can be made in various ways, including financial and non-financial. The story should be told and backed up by proof whenever available (EBRD 2021).

The fourth principle requires that the manager examine and measure the concrete, constructive effect potential resulting from each investment beforehand. The evaluation should be conducted using a proper results framework that strives to address the following key questions: (1) What is the destined effect? (2) Who is influenced by this effect? (3) What is the magnitude of this effect's significance? The manager should also try to assess how likely the investment will have the

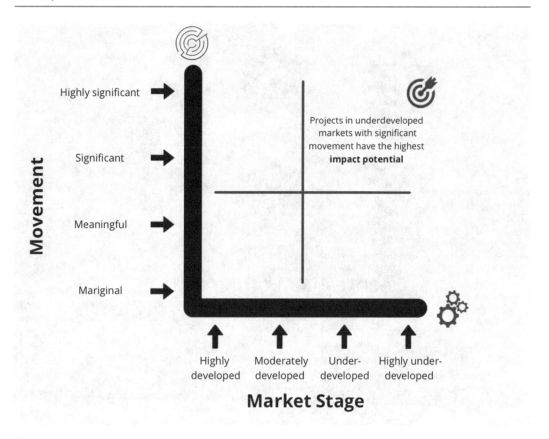

Fig. 1.15 Assessing market outcome. (IFC 2021)

desired effect. Moreover, substantial risk factors that could cause the impact to differ from ex-ante estimates should be identified. Furthermore, the evidence should be gathered to estimate the relative size of the difficulty addressed within the chosen geographical context. The manager must also discover ways to boost the investment's effect. Lastly, indicators must be adjusted with industry standards and best practices (EBRD 2021).

Within the fifth principle, the manager shall recognise and prevent. If avoidance is not practicable, alleviate and administer environmental, social and governance (ESG) risks for each investment as part of a systematic and documented approach. The manager must engage with the investor to obtain its promise to address gaps in current systems, procedures and standards, utilising best international industry practices. In addition, investees' ESG risk and performance should be audited. The manager

should engage with them, if needed, to avoid gaps and unforeseen events (EBRD 2021).

The sixth principle necessitates that the manager utilises the results framework in principle four to track the process to achieve positive impacts for each investment. A predetermined method to communicate performance data with the investee should be used to track progress. This should specify how frequently data will be gathered, the design for gathering data, data sources, data collection duties and how the data will be reported. Besides, the manager must strive to take necessary steps if monitoring reveals that the investment is no longer believed to reach its desired effects (EBRD 2021).

To comply with the seventh principle, when undertaking an exit, the manager must examine the effect of timing, design and exit procedure on the sustainability of influence (EBRD 2021).

For the eighth principle, the manager is responsible for revising and recording each

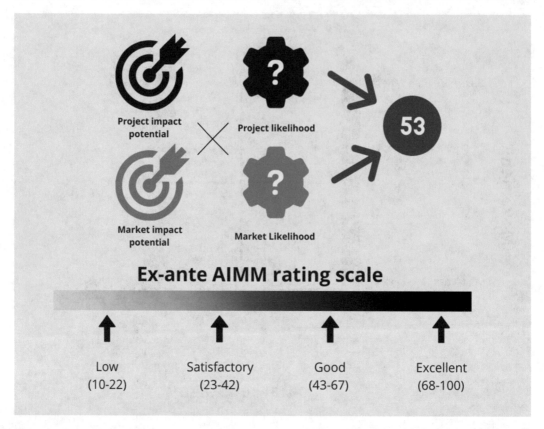

Fig. 1.16 Likelihood assessment. (IFC 2021)

investment's impact achievement and comparing the projected and actual impact as well as other consequences. Moreover, these conclusions should be used to enhance operational and strategic investment decisions and administrative procedures (EBRD 2021).

In the last principle, the manager must publicly report the consistency of its IA methods with the principles annually and arrange for separate proof of this alignment regularly. Furthermore, the findings of this verification report should be made public (EBRD 2021).

1.5.5.3 The World Bank

The World Bank desires to discover by using IA whether the changes in consumption and health could be linked to the program itself rather than to some separate aspect. It has issued a guidebook that explains the most common quantitative approaches used in ex-post IAs of programs and policies. The guidebook also goes through ways

to quantify distributional impacts and ex-ante approaches for predicting program consequences and methods (Khandker et al. 2010).

To assure that IA is functional, various measures should be followed, as stated by the guidebook. The importance and objectives of the evaluation, for example, must be explicitly stated throughout project identification and preparation. The nature and timeliness of evaluations are also a source of concern. To isolate the influence of the program on consequences, IAs should be planned ahead of time to assist program administrators in assessing and updating targeting throughout the intervention. The availability and quality of data are also valuable factors in dictating the capability of a program. Hiring and training fieldwork staff and establishing a consistent data management and access strategy are critical. From an administrative standpoint, the evaluation team should be carefully constituted during project implementation to include enough technical

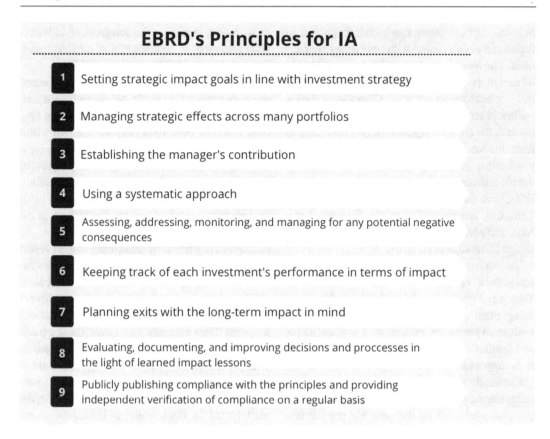

Fig. 1.17 Principles of EBRD for IA. (EBRD 2021)

and managerial knowledge to assure precise data and result reporting and transparency in execution to interpret the data properly (Khandker et al. 2010).

As the first approach in the book, randomised evaluations attempt to determine the effectiveness of a program by finding a group of participants with similar observed features and randomly assigning the treatment to a subset of this group. This strategy overcomes the problem of unobserved characteristics causing selection bias (Khandker et al. 2010).

The double-difference method is another method that allows the World Bank to see if consequences are traced for both participants and nonparticipants over a long enough period to capture any intervention effects. This strategy indicates that tracking results for both participants and nonparticipants over time will offer a solid foundation for determining the program's influence by utilising the double-difference method;

nonparticipants' observed changes over time yield the counterfactuals for participants (Khandker et al. 2010).

In addition, adding a third variable that affects just the treatment but not unobserved factors, an instrumental variable method reveals exogenous variation in treatment. These methods can be used with cross section or panel data. Instruments can be created via program design as well as other exogenous shocks that are unrelated to the desired consequences. Furthermore, regression discontinuity and pipeline techniques are the extensions of instrumental variable and experimental methods that use exogenous program rules to compare participants and nonparticipants in a narrow area (Khandker et al. 2010).

Although experimental methods are ideal for IA in theory, nonexperimental methods are commonly used practically, either because program managers are hesitant to exclude certain segments of the population from an intervention randomly or because a randomised approach is inappropriate for a rapid-action project with little time to experiment. The quality of IA, even with an experimental design, is determined by how the development and implementation are done. Compliance issues, spillovers and unobserved sample bias frequently obstruct the clean recognition of program impacts from randomisation. However, nonexperimental procedures, such as propensity score matching, double difference and the use of instrumental variables, have their own strengths and shortcomings. Thus, they are susceptible to bias for several reasons, including inaccurate evaluation framework design (Khandker et al. 2010).

According to the World Bank, no single assignment or evaluation method is flawless. Thus it is a good idea to double-check the results using other approaches. Ex-ante and ex-post evaluation methodologies, as well as quantitative and qualitative approaches, can all be integrated. It is important to utilise specific approaches. Understanding the planning and execution of an intervention, the aims and processes by which program goals can be met and the precise features of targeted and nontargeted areas are all key components in IA. One may also decide whether certain components of the program can be changed to make it better by directing pleasant IAs throughout the program and beginning early in the plan and execution stages of the project (Khandker et al. 2010).

1.5.5.4 Intel Corp.

To take a look at IA examples from the business side, let us investigate Intel's corporate responsibility report of 2020. Intel claims in this report that incorporating and expanding ethical business practices into their worldwide operations and supply chain reduces risks and promotes human rights respect. Intel's 2030 goals include taking steps to preserve and enhance their focus on maintaining and establishing a strong safety culture as their business evolves and grows, as well as expanding the global reach of their wellness programs. The goals also include a large increase in the number of suppliers covered by their engagement initiatives to increase human rights accountability throughout their global supply chain. Intel,

as it asserts, is also at the forefront of industry-wide initiatives to enhance ethical mineral sourcing and responsible mobility (Intel 2021).

As we can see above, Intel seeks to demonstrate its respect for people, the environment and, generally, the future well-being of all living species. But how does Intel evaluate the real-world impacts of their actions? IA commissioners present their beliefs and how they have progressed over the past year in their corporate responsibility reports. They track a comparable approach in all of their values while doing so. Let us look at the employee safety and well-being value as an example. They begin by describing in their report in 2021 that they want to ensure that more than 90% of their employees believe Intel has a solid safety culture and that 50% of their employees participate in their worldwide corporate wellness program. They continue with establishing a baseline. First, 37% of Intel employees engaged in Intel's EHS Safety Culture Survey, with a baseline average of 79% on "safety is a value" metrics; and second, 22% of Intel employees participated in Intel wellness initiatives at the start of 2020. Following that, Intel summarises its progress for the previous year. During 2020, their health and wellness teams worked to enhance employee knowledge and engagement in Intel's programs, emphasising preventative and early intervention programs and participation in the newly expanded virtual offerings of the Intel Vitality Program. They finish the assessment by looking ahead. Intel's safety culture aim will be to boost employee and management engagement in their safety programs, increase company-wide participation in their safety culture survey and expand the poll to 50% of employees by the end of 2021 (Intel 2021).

As a result, Intel, as a technology giant, uses these corporate responsibility reports to demonstrate its social and environmental dignity to the public. It uses several IA principles while doing so. First, it identifies the desired impact as ensuring more than 90% of their employees believe that Intel has a solid safety culture and that 50% of their employees participate in their worldwide corporate wellness program. Second, it creates baselines such as the Safety Culture Survey, with a baseline average of 79% on "safety is a value" metrics, etc., to compare improvements. Third, as we can see, it displays the progress using experimental data. Lastly, it learns lessons from its evaluation to set further goals, such as expanding the poll to 50% of employees by the end of 2021 (Intel 2021).

1.5.6   Summary of IA Discussion

In this section, we have provided our findings on IA from the academic literature and business environment. Since the definition of impact differs among businesses and institutions, distinctions are also present across existing IA techniques. It is crucial to start conducting IA by remembering this fact. To sum up the IA debate, it is vital to comprehend the design and implementation, the aims and processes by which action goals can be reached and the precise attributes of targeted and nontargeted areas of a business or institution's action. While forming the IA, evaluators should (1) decide on what is being assessed, (2) ensure the implementation of programs with impact in mind, (3) address the normative and ethical issues that activities raise and (4) differentiate between actors' "program theory" and the "theories of change" of how the program works in practice (Stern 2015). Furthermore, as stated by Hailey and Sorgenfrei, it is crucial to make the process as vital as the output (2005). In addition, it is not only necessary to identify impacts in IAs, but it is also necessary to comprehend how the various initiatives, programs and organisations operate (Simsa et al. 2014). When the businesses or institutions design and implement an effective IA, they consequently demonstrate success for the received fundings, making them eligible for further fundings, and understand the implications of initiatives to enhance effectiveness, be accountable and attain public advocacy for their actions.

1.6   The Innovation Journey

From the literature review and discussion above, Fig. 1.18 can summarise the innovation journey how ideas translate and transform into products and/or services as follows.

Before we explain the journey, we should stress that this is not a linear process; instead, it is a combination of interrelated complex processes. As the outcome-based approach or the LOTI approach suggests, Phase 0, envisaging, begins with looking for real-world outcomes and setting a vision. Then we proceed to the discover phase. This phase discovers and defines problems preventing the desired outcomes. After completing this, one should enable the necessary skills and competencies to solve the listed problems. The company can proceed to the develop phase when the skills are sufficient. In this phase, a product or service prototype is developed and tested. If the results are successful, we continue with the appraise phase. Assessing the impact and value of the innovation is essential in deciding whether the product/service will be scaled up for a more extensive commercial use or will stay as a prototype. If market conditions suggest scaling up, the innovation will penetrate into the market and reach a broader range of consumers and customers. Supporting mechanisms are vital to foster product/service development during this whole process.

Innovation districts and living labs are ideal testbeds for innovation development and trials. Regulatory sandboxes will alleviate the regulatory burden and pave the way to a vivid market environment, especially for start-ups. Funding and financing are other key issue for the innovation journey. Grants and loans could be a starting point for the process. As the journey proceeds to the develop and appraise phases, the public sector and private investors will see the opportunity and come into the picture as public-private partnerships or venture capitalists. The journey will also shape the business models of the innovations as innovators might change, adapt or update their value proposals, value creation and value capture as the product/service evolves. When the value is mentioned, we should highlight that revenue or the monetary value is not the only value. There is growing pressure on tackling climate change and achieving sustainability in the industry and businesses. Hence, the innovations should also provide environmental and social value to increase their overall impact, receive more funding and achieve better market diffusion.

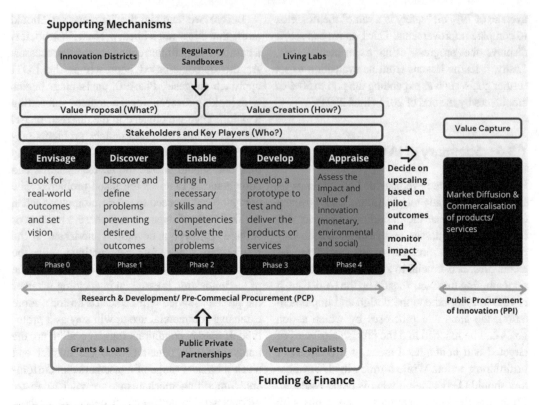

Fig. 1.18 The innovation journey

The remainder of the book is organised as follows. Chapter 2 presents brief descriptions and working principles of 34 emerging technologies. We solely focused on those in the "market diffusion and commercialisation of products/services" phase when deciding on which technologies should be included in this book. After this, we continue with 17 United Nations Sustainable Development Goals and 650 companies. We chose the companies and use cases with a comprehensive market scanning and by reviewing numerous sources that list start-up competitions and "best start-up" or "most innovative company" awards.[1] The use cases are presented in the "business model" approach by briefly mentioning value proposition (what?), value creation (how?) and value capture (revenue). When mentioning value capture, instead of the revenue model of the companies or how companies make money, we deliberately focused on how technologies capture economic and business, environmental and social and ethical value. Finally, we complete the book with a brief conclusion.

[1] See (Accessed Online – 2.1.2022):

https://www.forbes.com/innovative-companies/list/#tab:rank;   https://edisonawards.com/nomineegallery.php;   https://www.bonnchallenge.org/;   https://solarimpulse.com/topics;   https://www.valuer.ai/;   https://tech2impact.com/startups/;   http://www.businessfor2030.org/;   https://www.startus-insights.com/;   https://builtin.com/edtech/edtech-companies;   https://tracxn.com/explore/

Top-AR-VR-in-Education-Startups;   https://www.forbes.com/americas-best-startup-employers/#7d81a1556527;   https://www.cloudways.com/blog/best-startups-watch-out/#iot;   https://www.unwto.org/sdgs-global-startup-competition;   https://www.foundersbeta.com/tech-companies/20-tech-companies-to-watch-for-in-2021/;   https://www.businessinsider.com/recruitment-startup-technology-investment-workforce-covid-growth-talent-business-international-2021-7;   https://www.crn.com/slide-shows/storage/the-10-hottest-tech-startups-of-2021-so-far-

References

Advantage, Circular. "Innovative Business Models and Technologies to Create Value in a World without Limits to Growth 2014." URL: https://www.accenture.com/t20150523T053139__w__/il-en/_acnmedia/Accenture/ConversionAssets (2021).

L.M. Belcher, *Target Market vs. Target Customer* [WWW Document] (2019). https://smallbusiness.chron.com/target-market-vs-target-customer-20674.html. Accessed 5 Nov 2021

P. Belleflamme, T. Lambert, A. Schwienbacher, Crowdfunding: Tapping the right crowd. J. Bus. Ventur. Insights **29**(5), 585–609 (2014). https://doi.org/10.1016/j.jbusvent.2013.07.003

A. Bellucci, G. Gucciardi, D. Nepelski, *Venture Capital in Europe. Evidence-Based Insights About Venture Capitalists and Venture Capital-Backed Firms* (Publications Office of the European Union, Luxembourg, 2021)

L.M. Bettencourt, J. Lobo, D. Helbing, C. Kühnert, G.B. West, Growth, innovation, scaling, and the pace of life in cities. Proc. Natl. Acad. Sci. **104**(17), 7301–7306 (2007)

Brock, *Public-Private Partnerships* [online] (2021). https://www.investopedia.com/terms/p/public-private-partnerships.asp. Accessed 26 Feb 2021

Brookings (2014). https://www.brookings.edu/essay/rise-of-innovation-districts/ [online]. Accessed 26 Feb 2021

R. Casadesus-Masanell, J.E. Ricart, How to Design a Winning Business Model, in *Harvard Business Review* [online] (2011). Available at: https://hbr.org/2011/01/how-to-design-a-winning-business-model#. Accessed 23 Nov 2021

H. Chesbrough, Business Model Innovation: It Is Not Just About Technology Anymore, in *International Marketing Review* 35 (2007). Available at: https://l24.im/uzE

C.M. Christensen, R. McDonald, E.J. Altman, J.E. Palmer, Disruptive innovation: An intellectual history and directions for future research. J. Manag. Stud. **55**(7), 1043–1078 (2018) Available at: https://onlinelibrary.wiley.com/doi/10.1111/joms.12349 [online]. Accessed 23 Nov 2021

E. Cosgrave, K. Arbuthnot, T. Tryfonas, Living labs, innovation districts and information marketplaces: A systems approach for smart cities. Procedia Comput. Sci. **16**, 668–677 (2013)

D. Crisan, Buying with intent: Public procurement for innovation by provincial and municipal governments. School Public Policy Publ. **13**, 2–3 (2020)

K. Daeyoup, K. Jaeyoung, Business model innovation through value delivery differentiation: Multiple case studies. Indian J. Sci. Technol. 8 (2015). Available at https://citeseerx.ist.psu.edu/viewdoc/download?doi=10.1.1.887.1850&rep=rep1&type=pdf#:~:text=Value%20creation%20means%20to%20create,value%20in%20a%20value%20network. Accessed 23 Nov 2021

A. Dibrova, Business angel investments: Risks and opportunities. Procedia Soc. Behav. Sci. **207**, 280–289 (2015). https://doi.org/10.1016/j.sbspro.2015.10.097

P. Drucker, The theory of business, in *Harvard Business Review* (1994). Available at: https://publicpurpose.com.au/wp-content/uploads/2016/05/Theory-of-the-Business-HBR-Sept1994.pdf [online]. Accessed 23 Nov 2021

EBRD, *Operating Principles for Impact Management* [online] (2021). Available at: https://www.ebrd.com/who-we-are/operating-principles-for-impact-management.html. Accessed 21 Nov 2021

J. Elkington, Enter the Triple Bottom Line, in *The Triple Bottom Line Does It All Add Up? Assessing the Sustainability of Business and CSR*, ed. by A. Henriques, J. Richardson, (Earthscan, London, 2004), pp. 1–16

European Commission, *Shaping Europe's Digital Future* [online] (2020). Available at: https://ec.europa.eu/digital-single-market/en/innovation-procurement. Accessed 4 Mar 2021

European Commission, *Innovation Fund* [WWW Document] (2021). https://ec.europa.eu/clima/eu-action/innovation-fund_en. Accessed 5 Nov 2021

European Investment Bank, *InnovFin Energy Demo Projects*. [WWW Document] (2021). https://www.eib.org/en/products/mandates-partnerships/innovfin/products/energy-demo-projects.htm. Accessed 5 Nov 2021

J. Fisken, J. Rutherford, Business models and investment trends in the biotechnology industry in Europe. J. Commer. Biotechnol. **8**(3), 191 (2002)

M. Gavin, How to Create Social Change: 4 Business Strategies –Harvard Business School [online] (2019). Available at: https://online.hbs.edu/blog/post/how-can-business-drive-social-change. Accessed 9 Nov 2021

Global Environmental Management Initiative. "Environmental reporting in a total quality management framework." Washington, DC: Global Environmental Management Initiative (1994).

Global Environmental Management Initiative, *Environment: Value to Business* (2014). Available at: http://gemi.org/resources/EVTB_001.pdf

H. Golub, J. Henry, J.L. Forbis, N.T. Mehta, M.J. Lanning, E.G. Michaels, K. Ohmae, Delivering value to customers [online] (McKinsey & Company, 2000). Available at: https://www.mckinsey.com/business-functions/strategy-and-corporate-finance/our-insights/delivering-value-to-customers. Accessed 23 Nov 2021

Google Trends – Crowdfunding [WWW Document] (2021). Available at: https://trends.google.com/trends/explore?date=2010-01-01%202021-11-23&q=crowdfunding. Accessed 23 Nov 2021

Government of Canada, *Apply for funding under Sustainable Development Goals Funding Program – Grants* [WWW Document] (2020). https://www.canada.ca/en/employment-social-development/services/funding/sustainable-development-goals.html. Accessed 6 Nov 2021

J. Hailey, M. Sorgenfrei, Measuring Success: Issues in Performance Measurement [online] (Intrac, Oxford, 2004). Available at: https://www.intrac.org/wpcms/wp-content/uploads/2018/11/OPS44Final.pdf. Accessed 21 Nov 2021.

G. Hamel, *Leading the Revolution: How to Thrive in Turbulent Times by Making Innovation a Way of Life* (Harvard Business School Press, 2000)

M.J. Harris, B. Roach, Valuing the environment, in *Environmental and Natural Resource Economics: A Contemporary Approach*, ed. by M. J. Harris, B. Roach, (Routledge, 2017)

Harvard Business School, Unique Value Proposition – Institute for Strategy and Competitiveness – Harvard Business School [online] (2019). Hbs.edu. Available at: https://www.isc.hbs.edu/strategy/creating-a-successful-strategy/Pages/unique-value-proposition.aspx. Accessed 23 Nov 2021

J. Hemer, A snapshot on crowdfunding, Arbeitspapiere Unternehmen und Region, No. R2/2011, Fraunhofer-Institut für System-und Innovationsforschung ISI, Karlsruhe (2011)

D. Horgan, B. Dimitrijević, Social innovation systems for building resilient communities. Urban Sci. **2**(1), 13 (2018)

IFC, *How IFC Measures the Development Impact of Its Interventions* [online] (2021). Available at: https://www.ifc.org/wps/wcm/connect/af1377f3-4792-4bb0-ba83-a0664dda0e55/202012-IFC-AIMM-brochure.pdf?MOD=AJPERES&CVID=noLTBSi. Accessed 6 Nov 2021

Intel 2021. *2020–21 Corporate Responsibility Report* [online]. Available at: http://csrreportbuilder.intel.com/pdfbuilder/pdfs/CSR-2020-21-Full-Report.pdf. Accessed 21 Nov 2021

S. Javed, B.A.E. Aldalaien, U. Husain, Performance of venture capital firms in UK: Quantitative research approach of 20 UK venture capitals. Middle East J. Sci. Res., 432–438 (2019)

M.W. Johnson, C.M. Christensen, H. Kagermann, *Reinventing Your Business Model*. HBR [online], pp. 57–69 (2008). Available at: http://syv.pt/login/upload/userfiles/file/Reinventing%20Your%20business%20model%20HBR.pdf#page=57. Accessed 23 Nov 2021

R.O. Joseph Schumpeter, *The Theory of Economic Development*, 1st edn. (Harvard University Press, Cambridge, 1934)

S.R. Khandker, G.B. Koolwal, H.A. Samad, *Handbook on Impact Evaluation: Quantitative Methods and Practices* [online] (World Bank, Washington, DC, 2010). Available at: https://openknowledge.worldbank.org/bitstream/handle/10986/2693/520990PUB0EPI1101Official0Use0Only1.pdf. Accessed 21 Nov 2021

P.A. Koen, H.M.J. Bertelse, I.R. Elsum, The three faces of business model innovation: Challenges for established firms. Res. Technol. Manag, **54**, 52–59,

2011. Available at: https://doi.org/10.5437/089536
08X5403009. Accessed 23 Nov 2021

V. Lember, T. Kalvet, R. Kattel, Urban competitiveness and public procurement for innovation. Urban Stud. **48**(7), 1373–1395 (2011)

S. Leminen, M. Westerlund, A.-G. Nyström, *Living Labs as Open-Innovation Networks*. September (2012)

J. Lerner, R. Nanda, Venture capital's role in financing innovation: What we know and how much we still need to learn. J. Econ. Perspect. **34**(3), 237–261 (2020)

T. Liu, et al. "Emerging themes of public-private partnership application in developing smart city projects: a conceptual framework." Built Environment Project and Asset Management (2020).

M. Lim, Economic value + environmental value + social value = ? Interdiscip. Environ. Rev. **6**, 40–53 (2004)

'Loan' in Cambridge Dictionary, [WWW Document]. Available at: https://dictionary.cambridge.org/dictionary/english/loan. Accessed 23 Nov 2021

LOTI, *Our Approach: London Office of Technology and Innovation* [online] (2021). Available at: https://loti.london/about/approach/. Accessed 2 Mar 2021

J.H. Love, S. Roper, SME innovation, exporting and growth: A review of existing evidence. Int Small Bus. J. **33**(1), 28–48 (2015)

S. Manigart, K. Waele, K. Robbie, P. Desbrières, H. Sapienza, A. Beekman, Determinants of required return in venture capital investments: A five-country study. J. Bus Ventur. **17**, 291–312 (2002). https://doi.org/10.1016/S0883-9026(00)00067-7

C. Mason, M. Stark, What do investors look for in a business plan?: A comparison of the investment criteria of bankers, venture capitalists and business angels. Int. Small Bus. J. **22**(3), 227–248 (2004). https://doi.org/10.1177/0266242604042377

E. Mollick, The dynamics of crowdfunding: An exploratory study. J. Bus. Ventur. **29**(1), 1–16 (2014)

G. Mulgan, Measuring social value. Stanford Soc. Innov. Rev. **8**(3), 38–43 (2010) Available at: https://doi.org/10.48558/NQT0-DD24 Accessed 7 Nov 2021

R. Murray, J. Caulier-Grice, G. Mulgan, *The Open Book of Social Innovation*, vol 24 (Nesta, London, 2010)

Myler, L., *3 Steps To An Irresistible Business Model* [online] (Forbes, 2013). Available at: https://www.forbes.com/sites/larrymyler/2013/08/01/3-steps-to-an-irresistible-business-model-multiple-authentication-methods/?sh=459d5ac53c6c. Accessed 22 Nov 2021

Nesta, *Funding Innovation: A Practice Guide* [WWW Document] (2018). https://media.nesta.org.uk/documents/Funding-Innovation-Nov-18.pdf. Accessed 5 Nov 2021

O'Flynn, M., *Impact Assessment: Understanding and Assessing Our Contributions To Change* [online] (International NGO Training and Research Center, 2010). Available at: https://www.intrac.org/wpcms/wp-content/uploads/2016/09/Impact-Assessment-Understanding-and-Assessing-our-Contributions-to-Change-1.pdf. Accessed 21 Nov 2021

OECD, 2010. Glossary of Key Terms in Evaluation and Results Based Management [online]. Available at: https://unsdg.un.org/sites/default/files/OECD-Glossary-of-Key-Terms-in-Evaluation-and-Results-based-Management-Terminology.pdf. Accessed 21 Nov 2021

OECD and Statistical Office of the European Communities, *Oslo Manual*, 3rd edn. (OECD Publishing, Paris, 2005)

R. Parenti, Regulatory, s.l.: European Parliament (2020)

D. Politis, Business angels and value added: What do we know and where do we go? Venture Cap. **10**(2), 127–147 (2008). https://doi.org/10.1080/13691060801946147

Queensland Government, *Finding Grants and Business Support* [WWW Document] (2021). https://www.business.qld.gov.au/starting-business/advice-support/grants/finding. Accessed 5 Nov 2021

M.G.R.D. Radoslav Delina, Understanding the Determinants and Specifics of Pre-Commercial Procurement. J. Theor. Appl. Electr. Commer. Res. **16**(2), 80–100 (2021)

K. Rambow, N.G. Rambow, M.M. Hampel, D.K. Harrison, B.M.L. Wood, Creating a digital twin: Simulation of a business model design tool. Adv. Sci. Technol. Eng. Syst. J. **4**(6), 53–60 (2019) Available at: https://doi.org/10.25046/aj040607. Accessed 23 Nov 2021

J. Richardson, The business model: An integrative framework for strategy execution. Strateg. Change **17**(5-6), 133–144 (2008) Available at: https://doi.org/10.1002/jsc.821. Accessed 23 Nov 2021

R.B. Robinson, Emerging strategies in the venture capital industry. J. Bus. Ventur. **2**, 53–77 (1987)

M. Rolfstam, Understanding public procurement of innovation. SSRN Electr. J., 1–16 (2012)

K.W. Schilit, A comparative analysis of the performance of venture capital funds, stocks and bonds, and other investment opportunities. Int. Rev. Strateg. Manag. **4**, 301–320 (1993)

A. Schmidt, L. Scaringella, Uncovering disruptors' business model innovation activities: Evidencing the relationships between dynamic capabilities and value proposition innovation. J. Eng. Technol. Manag 57 (2020). Available at: https://doi.org/10.1016/j.jengtecman.2020.101589. Accessed 23 Nov 2021

W.A. Schrock, Y. Zhao, D.E. Hughes, K.A. Richards, JPSSM since the beginning: intellectual cornerstones, knowledge structure, and thematic developments. J. Pers. Selling Sales Manag. **36**(4), 321–343 (2016) Available at: https://l24.im/CTMW. Accessed 23 Nov 2021

J. Schumpeter, *Capitalism, Socialism and Democracy* (Harper & Brothers, United States, 1942)

R. Sharma, F. Pereira, N. Ramasubbu, M. Tan, F.T. Tschang, Assessing value creation and value capture in digital business ecosystems. SSRN (2013). Available at: https://ssrn.com/abstract=2286128. Accessed 30 Nov 2021

R. Simsa, O. Rauscher, C. Schober, C. Moder, *Methodological Guideline for Impact Assessment* [online] (2014). Available at: https://thirdsectorimpact.eu/site/assets/uploads/post/methodological-guideline-impact-assessment/TSI_WorkingPaper_012014_Impact.pdf. Accessed 21 Nov 2021.

Social Value Portal, *What is Social Value?* [WWW Document] (2017). Available at: https://socialvalue-portal.com/what-is-social-value/. Accessed 7 Nov 2021

Social Value UK, *Ten Impact Questions* [WWW Document] (2021a). Available at: https://socialvalueuk.org/ten-impact-questions/. Accessed 7 Nov 2021

Social Value UK, *What are the Principles of Social Value* [WWW Document] (2021b). Available at: https://socialvalueuk.org/what-is-social-value/the-principles-of-social-value/. Accessed 7 Nov 2021

P. Spieth, T. Roeth, S. Meissner, Reinventing a business model in industrial networks: Implications for customers' brand perceptions. Indus. Market. Manag. [online] **83**, 275–287 (2019) Available at: https://www.sciencedirect.com/science/article/pii/S0019850118305625. Accessed 23 Nov 2021

Startup Genome, *Global Startup Ecosystem Report* [WWW Document] (2019). https://startupgenome.com/reports/global-startup-ecosystem-report-2019. Accessed 5 Nov 2021

E. Stern, Impact Evaluation A Guide for Commissioners and Managers Impact Evaluation [online] (2015). Available at: https://www.betterevaluation.org/sites/default/files/60899_Impact_Evaluation_Guide_0515.pdf. Accessed 21 Nov 2021

P.J. Thomson, *Business Model Canvas Articles* [online] (2013). Available at: https://www.peterjthomson.com/tag/business-model-canvas/. Accessed 23 Nov 2021

R. Tidhar, K.M. Eisenhardt, Get rich or die trying… finding revenue model fit using machine learning and multiple cases. Strateg. Manag. J. **41**(7), 1245–1273 (2020) Available at: https://onlinelibrary.wiley.com/doi/10.1002/smj.3142. Accessed 23 Nov 2021

UKGBC, *Social Value in New Development: An Introductory Guide to Local Authorities and Development Teams* [online] (2018). Available at: https://www.ukgbc.org/wp-content/uploads/2018/03/Social-Value.pdf. Accessed 21 Nov 2021

VentureEU – the European Union venture capital megafund [online] (2020). Available at: https://ec.europa.eu/programmes/horizon2020/en/ventureeu. Accessed 2 Mar 2021

B. Zider, *How Venture Capital Works* [WWW Document] (1998). https://hbr.org/1998/11/how-venture-capital-works. Accessed 5 Nov 2021

C. Zott, R. Amit, Business model innovation: How to create value in a digital world. NIM Market. Intell. Rev. **9**(1), 18–23 (2017). https://doi.org/10.1515/gfkmir-2017-0003

Emerging Technologies

2

Abstract

This chapter presents brief descriptions and working principles of 34 emerging technologies which have market diffusion and are commercially available. Emerging technologies are the ones whose development and application areas are still expanding fast, and their technical and value potential is still largely unrealised. In alphabetical order, the emerging technologies that we list in this chapter are 3D printing, 5G, advanced materials, artificial intelligence, autonomous things, big data, biometrics, bioplastics, biotech and biomanufacturing, blockchain, carbon capture and storage, cellular agriculture, cloud computing, crowdfunding, cybersecurity, datahubs, digital twins, distributed computing, drones, edge computing, energy storage, flexible electronics and wearables, healthcare analytics, hydrogen, Internet of Behaviours, Internet of Things, natural language processing, quantum computing, recycling, robotic process automation, robotics, soilless farming, spatial computing and wireless power transfer.

Keywords

Emerging technologies · Use cases · Innovation · Sustainable development

2.1 3D Printing

3D printing, also known as additive manufacturing (AM), creates a three-dimensional product of any shape from a three-dimensional model or other electronic data sources by layering material under computer control (Dongkeon et al. 2006). In additive manufacturing, objects are built from the bottom up in layers. The layers are created in slicing software from a three-dimensional computational model of the object to be printed. These computational models are typically developed in computer-aided design (CAD) software and exported as .stl or .obj files for 3D printing. The description of the 3D printing process is shown in Fig. 2.1.

Like many other contemporary technologies, 3D printing has both positive and negative consequences. 3D printing is a widely accessible technology that allows consumers to create products in their own homes, using their own devices while also removing logistical and energy-related responsibilities from manufacturers. However, individual use of 3D printing technology may lead to unemployment among workers in the sub-production stages of manufacturing. Despite this, 3D printing technology offers overwhelming possibilities for innovation and efficient manufacturing.

3D printing technology has evolved since the first 3D printer was established in 1984, and

S. Küfeoğlu, *Emerging Technologies*, Sustainable Development Goals Series,
https://doi.org/10.1007/978-3-031-07127-0_2

41

Fig. 2.1 Process of 3D printing. (Campbell et al. 2011)

printers have gotten more functional as their price points have decreased. Rapid prototyping is used in a variety of industries, including research, engineering, the medical industry, the military, construction, architecture, fashion, education and the computer industry, among many others. The plastic extrusion technology most widely associated with the term "3D printing" was invented by the name "fused deposition modelling" (FDM) in 1990. The sale of 3D printing machines has increased significantly in the twenty-first century, and their cost has steadily decreased. By the early 2010s, 3D printing and additive manufacturing had evolved into alternate umbrella terms for AM technologies, one being used in popular vernacular by consumer-maker communities and the media and the other one being used officially by industrial AM end-use part producers, AM machine manufacturers and global technical standards organizations.

There are several 3D printing technologies, including stereolithography (SLA), digital light processing (DLP), fused deposition modelling (FDM) and selective laser sintering (SLS). However, the most commonly used techniques are FDM and SLA (Kamran and Saxena 2016). Scott Crump developed FDM in the late 1980s. Its wide use is due to its ease of manufacturing, relatively low cost and variety of applications. The FDM process has been applied in many areas such as biomedical, aerospace, automobile, pharmaceutical, textile and energy fields (Singh et al. 2020). FDM uses a stock material fed into a liquefier to shape the material in a liquid form easily. The material is heated to its melting temperature through various temperature treatment methods within the liquefier.

The melted material is then pushed through a nozzle to be extruded onto a Cartesian space (Campbell et al. 2011). While some printers allow the nozzle to move around in the Cartesian space, other printers build layers by moving the print bed under a stationary nozzle. The print bed

spans the x and y axes, and the layers of the object which are to be printed are added towards the z-axis. Other 3D manufacturing processes use fundamentally different methods to create the different layers which are needed to give form to the final object. One such process is stereolithography, where the object is "printed" by hardening layers in a pool of photosensitive polymer using an ultraviolet laser (Campbell et al. 2011). In this method, the energy of the laser is transferred to certain regions of the liquid polymer to harden it. When all the desired regions are hardened, the printed object can be taken out of the pool of polymer. A type of stereolithography is DLP invented by Larry Hornbeck in 1987. The difference between SLA and DLP is that DLP uses UV light to harden the shape of the object at once rather than hardening different selective sections of the resin over time. Another method called selective laser sintering, also developed in the late 1980s, uses lasers to melt layers of polymeric powder to obtain the final shape. The melted section hardens in time, and it can be removed from the powder once the hardening is complete (Campbell et al. 2011). Many other methods are in use or the phases of development.

3D printing simplifies the process of transforming ideas into products. The technology allows rapid and accurate production from various materials. 3D printing also streamlines the prototyping process by providing faster production, allowing businesses to stay one step ahead of the competition. The technology uses a simple interface, allowing more equitable and widespread use. 3D printing also helps product developers to produce low-cost prototypes early in the development process, resulting in better goods and fewer dead-ends. Materials science as a field is affected largely by the 3D printing applications, as the number of materials created by 3D printing has risen considerably in recent years. The possibility of 3D printing various materials has also allowed the technology to be used in different fields. Metals, polymers, ceramics, composites and smart materials have all been successfully used in 3D printing applications with varying costs. Different materials require customisations of the 3D printers due to different material properties such as melting temperature (Shahrubudin et al. 2019). An innovative material used in 3D printing applications is smart materials, which sense variations in their external environment and provide an effective reaction to fluctuations by modifying their material characteristics or geometries. In particular, energy connection or conversion between different physical fields, such as thermal energy conversion into mechanical work, is shown as a product of smart materials. Due to the potential of smart materials, 3D-printed components of such materials might change over time in a specified way. This leads to a new phenomenon known as 4D printing. 4D printing innovations are primarily accessible by recent progress that has been achieved in multimaterial printing. 3D printing of multi-smart materials or a mix of smart materials and conventional materials requires understanding the design and manufacturing processes (Khoo et al. 2015).

Benefits of 3D Printing:

- AM dramatically shortens the production period and process and provides great flexibility for the continuously changing market demand.
- Production run size can be maintained at low levels on a unit basis while manufacturing costs are not considered.
- AM reduces assembly mistakes and related expenses because pre-assembled components can be obtained through one subsequent operation, which is the quality control inspection.
- Tools are not included in the additive production process. It provides flexibility for the market adaptation and reduction or even elimination of related expenses such as toolmaking, stoppages due to referred changes, inspection and maintenance.
- Different production processes can be used in hybrid production with AM. In this situation, it can provide a combination of additive pro-

duction methods with traditional methods to make advantages offered by both.
- Optimum usage of materials can be provided, and it is possible to recycle any waste material through AM (Jiménez et al. 2019).

Studies attempt to predict the impacts of 3D printing on manufacturing, supply chains, business models, competition and intellectual property. A study by Jiang et al. makes economic and societal predictions for 2030 (2017). The study results predict a trend of decentralisation in supply chains across many fields since AM will allow cheaper and more accessible localised production capabilities. This is expected to decrease the environmental impact of manufacturing due to reduced transportation emissions. The study also predicts that more than 25% of applicable, final products will be sold digitally as files to be 3D printed instead of physical products. A distinction is made between complex and less complex parts where complex parts are made centrally in specialised manufacturing locations, and fewer complex parts are distributed digitally and produced locally (Jiang et al. 2017). The study makes additional predictions about consumer markets and business models changes by 2030. Businesses' competitive advantage will no longer depend on the efficiency of their production operations but their network of users and creators. Companies will seek employees with skills related to AM, and many jobs will be replaced in the manufacturing industry. This change is reportedly due to the expectation that more than 10% of all gains from manufactured products will be from 3D-printed products by 2030. These products are predicted to be made up of many materials and electronics since enhanced AM methods will allow for such products to be 3D printed at lower costs. The affordability of 3D printers will also induce a significant increase in 3D printer ownership of individuals, especially in industrialised countries. Thus, websites that feature 3D designs will gain more popularity and will allow designs to be sold or downloaded as open-source projects. This is expected to make it harder to detect violations of intellectual property

Fig. 2.2 3D Printing applications. (Mpofu et al. 2014, p. 2149)

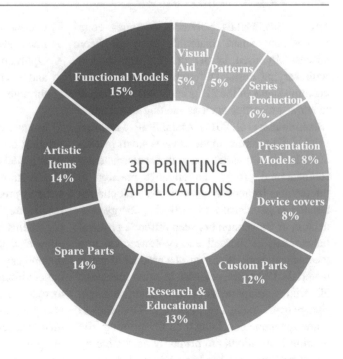

rights (Jiang et al. 2017). It is expected that 3D printing will become more affordable, refined, purposeful and widespread in the future. The words "create it" may soon become as ubiquitous as "print it". Examples include raw commodities, satellite networks, machinery, ships and factories. When the cost of manufacturing is reduced, as it is with 3D printers, to the point that virtually anybody can purchase the "means of production", everyone will say "make it." So, the future of 3D printing technology is promising. As the applications of this technology, shown in Fig. 2.2, surge in various areas while potential future applications arise. It is expected that the 3D printing manufacturing industry will grow by 18% each year and reach 8.4 billion dollars by 2025. Especially in the automobile and aerospace industries, the usage of 3D-printed parts will increase significantly in the upcoming years (Mpofu et al. 2014).

2.2   5G

5G is a contemporary technology that offers new interfaces to all end-user devices and network components. The quest for 5G stems from the rapidly developing desire to build a highly connected and globalised world in which information and data are easily and equitably accessible to everyone around the globe. Technologies that enhance access to information and data have gone through significant improvements, with new technologies constantly developing to address the shortcomings of previous iterations. 5G is expected to address the shortcomings of 4G technology and improve upon the promises of 4G. 5G technology promises higher capacity and data rate, lower latency, larger device connectivity, lower costs and more consistent quality compared to 4G (Gupta and Jha 2015). 5G can simultaneously connect more wireless technology users with smarter, faster predecessors. 5G technology allows for network connections using Internet technology that is specified to power, battery life, size and cost in the Internet of Things (IoT) applications. 5G technology provides for a revised technological solution in terms of tonnes of wireless technologies, and it opens up new possibilities for mobile connectivity that go well beyond what is now possible, allowing new applications to be utilised in a variety of different situations (Painuly et al. 2020).

From 1G to 5G, communication technology has evolved over time. Mobile connectivity technology began in 1979 with the first generation of mobile networks, also known as 1G. 1G was a fully analogue technology. Analogue technology and frequency division multiple access (FDMA) were used in 1G, as well as Nordic mobile systems (NMT) and advanced mobile phone system (AMPS) switching. The average speed of a 1G connection was 2.4Kbps. 2G was introduced in 1991 with more services and features, including enhanced coverage and capacity and superior voice quality to 1G. The speed of the 2G network was enhanced to 64 kbps, and the first digital protocols, such as code division multiple access (CDMA), time division multiple access (TDMA) and global system of mobile (GSM), were used. 2G technology was used for voice and data packet switching. When 3G was introduced in 2003, it represented a new mobile technology and services age. Using 3G technology, the speed was upgraded to 2000 kbps, and the first mobile broadband service was launched. A new age of mobile capabilities began with the rapid growth of smartphone Internet services after introducing 3G technology. Digital voice and web data are used separately in 3G, email and SMS. 4G was introduced in 2011 and is currently in use alongside 2G and 3G. 4G speed can reach 100,000 kbps. High-speed Internet and the next generation of transportation networks are required to meet this massive demand (Saqlain 2018). By the early 2000s, developers had realised that even the most advanced 4G networks would not be able to handle the demand. An academic team has begun work on 5G since 4G has a 40–60 ms latency, which is too high for real-time responses. NASA aided in developing the Machine-to-Machine Intelligence (M2Mi) Corp, which is tasked with developing M2M and IoT like the 5G infrastructure required to encourage it in 2008. In the same year, South Korea established a 5G Research and Development program, while New York University established the 5G-focused NYU WIRELESS in 2012 (ReinhardtHaverans 2021). Historical development of the 5G is represented in Fig. 2.3.

Like existing networks, 5G transmits encoded data between hotspots using a cell system that divides the territory into sectors and employs radio waves to do so. The spine of the network must be connected to each cell, either wirelessly or through a landline. With two different frequency bands below and above 6 GHz, 5G uses higher frequencies than 4G (Pisarov and Mester 2020). 5G's improved connection was expected to revolutionise everything from finance to healthcare. 5G opens the door to life-saving technologies like remote operations, education, medicine and more. Additionally, 5G technology creates opportunities for new capabilities and enterprises. Despite the power of 4G wireless network technology, fast speed, rapid response, high reliability and power efficiency, mobile services are not enough to sustain growing demand. Thus, these qualities have become important criteria for 5G services (Yu et al. 2017).

The 5G technology consists of three main types, which are enhanced mobile broadband (eMBB), ultra-reliable low-latency communication (URLLC) and massive machine-type communications (mMTC). eMBB enables improved customer experience in cell broadband. It requires high data rates across a large coverage region. URLLC's primary applications include industrial automation, automated driving and virtual surgery. Lastly, mMTC provides support for a variety of devices, such as remote controllers, actuators and system tracking within a small area (Noohani and Magsi 2020).

5G technology can also be divided into five main categories. As shown in Fig. 2.4, these categories include immersive 5G services, such as massive contents streaming and virtual reality/augmented reality; intelligent 5G services, such as crowded area services and user-centric computing; omnipresent 5G services, including Internet of Things; autonomous 5G services, including drones, robots and smart transportation; and public 5G services, such as emergency services, private security, disaster monitoring and public safety (Yu et al. 2017). Moreover, Fig. 2.5 represents the applications of 5G.

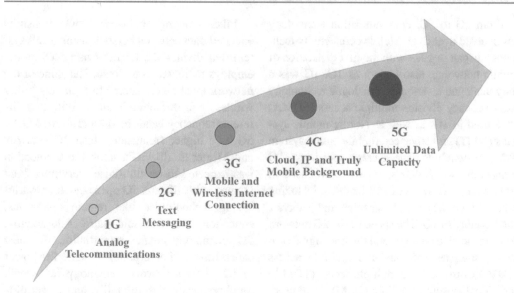

Fig. 2.3 Historical development of network technology. (Pisarov and Mester 2020)

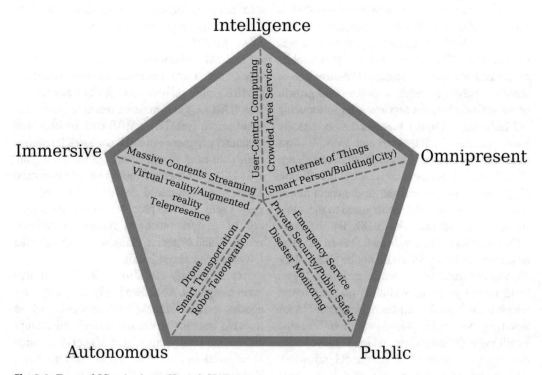

Fig. 2.4 Types of 5G technology. (Yu et al. 2017)

The widespread application of 5G is seen by many as inevitable given IoT requirements. Devices will require 5G capabilities to maintain continuous wireless connection and improve their speed and security. 5G offers significantly faster data rates compared to 4G networks. Furthermore, 5G has ultra-low latency (latency refers to the amount of time it takes for one device to deliver a data packet to another device). The latency rate in 4G is approximately 50 ms,

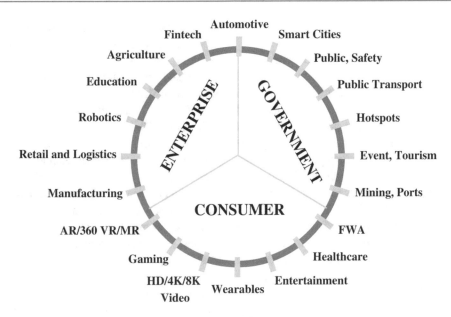

Fig. 2.5 Applications of 5G technology. (Juniper Networks 2021)

while in 5G, it will be under one millisecond. This is a critical rate for industrial usage and self-driving automobiles. 5G uses significantly less energy than previous technologies. Lower energy use facilitates the construction of battery-free IoT nodes, allowing IoT to operate as battery-free and maintenance-free endpoints.

Moreover, 5G consumes nearly five times less energy while being five times more cost-effective (Painuly et al. 2020). Also, 5G will allow a more connected world. Over the next 10 years, IoT is anticipated to develop tremendously, necessitating a network capable of supporting billions of connected objects. The capacity and bandwidth of 5G will be tailored to the user's demands (Lopa and Vora 2015). The vast majority of household devices in use, from routers to televisions, are powered by a single chip. With the advent of 2G and 3G technology, the world changed dramatically, and after the arrival of 4G technology, the world has changed even more. 5G technology is a major step forward, not only for the technology industry but for the entire globe. 5G is expected to create $12.3 trillion in world economic output and provide 22 million occupations by 2035. Moreover, it is estimated that 5G's overall contribution to the world's real gross domestic product (GDP) from 2020 to 2035 will reach the same size

as India's economy (Campbell et al. 2017). As mentioned, 5G will be used mainly in three areas: eMBB, massive Internet of Things (MIoT) and mission critical services (MCS) (Campbell et al. 2017). eMBB indicates extending cellular service and increasing capacity to include more structures, such as offices, industrial areas, shopping malls and major venues, and accommodating a much larger number of devices with high data volumes (Kavanagh 2021). This will allow for more cost-effective data transmission. Secondly, the costs associated with MIoT will be significantly reduced by the energy efficiency of 5G and its ability to function in both licensed and unlicensed spectrums and the potential to supply deeper and more flexible coverage. Lastly, the adoption of 5G will fulfil the application's high reliability and ultra-low latency connectivity needs. Thus, this technology will be frequently used in the operation of complex systems to eliminate the risk of failure. When all of 5G's components are fully deployed and functioning, no wire or cable will be required to supply communications. 5G has the potential to be the ultimate answer to the old "last mile" challenge of delivering a comprehensive digital connection from the carrier network's edge to the consumer without having to drill another hole through the wall.

2.3   Advanced Materials

The materials that developed and continue to evolve recently can be defined as advanced materials. Also, these materials show high strength, hardness and thermal, electrical and optical properties and have promising chemical properties and strength density ratios against conventional materials. Energy consumption value decreases by using advanced materials. Besides, higher performance and lower cost value can be obtained (Randall Curlee and Das 1991). The subgroups of the advanced materials can be classified as metallic materials, ceramics, polymers and composites, which are combined in terms of their nature. Besides, according to their properties and usage areas, advanced materials can be classified as biomedical, electronic, magnetic, optical materials, etc. Also, whereas advanced materials have superior properties

such as thermal, electrical, mechanical and a combination of these properties, they add value to the systems in which they are used (Randall Curlee and Das 1991). Also, there are a lot of advanced materials definitions. One of them is that advanced materials have potential usage in high value-added products. The other definition implies that enhanced processes improve the cost-performance efficiency of functional materials. Besides, advanced materials positively affect economic growth, life quality and environmental issues under enhanced processes and products. All new materials or modifications of existing materials with high properties have at least one aspect that can be classified as advanced materials. Furthermore, they can have completely new features (Kennedy et al. 2019). Figure 2.6 summarises some of the major advanced materials available for industrial and commercial use.

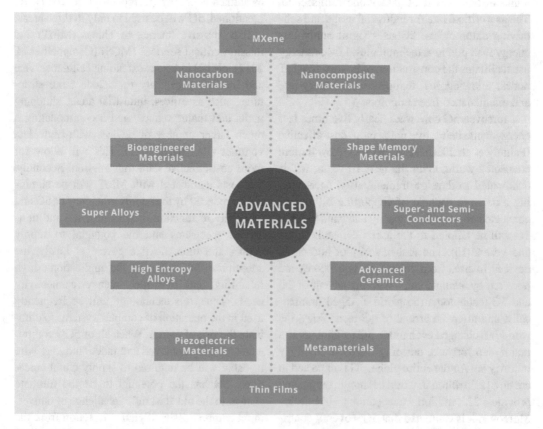

Fig. 2.6 List of advanced materials

(i) *Advanced Ceramics*

High-performance ceramics are designated advanced ceramics. They have a crystalline structure, and refined raw materials are used for the utilisation of advanced ceramics. Advanced ceramics can be carbides, nitrides, oxides and silicides such as zirconia, silicon nitride, silicon carbide, aluminium oxide, etc. They exhibit high mechanical properties like hardness, strength, modulus, etc. Also, they have high thermal and electrical conductivity, chemical resistance and low toxicity (Ayode Otitoju et al. 2020).

(ii) *Bioengineered Materials*

Bioengineered materials are mostly used for medical purposes. They are derived from natural structures, or they can be produced synthetically with different techniques. Materials such as collagen, gelatine and fibrin can be given as examples of naturally derived materials. Synthetic materials are not bioactive, unlike natural ones, and they need to undergo several processes to become compatible with biological environments (Sedlakova et al. 2019). Bioengineered materials can also be classified as biomedical materials and biomimetic materials where biomedical materials derive from natural structures and use their properties, and biomimetic materials are synthetic materials that imitate natural processes to function (Tirrell et al. 2002).

(iii) *High-Entropy Alloys*

The high mechanical, physical and chemical properties cannot be obtained by using pure metals. Because of this situation, other metals are inserted into the metal system to be used. Conventionally, alloy systems include at least one dominant metal atom and slightly alloying elements. However, the high-entropy alloys contain equiatomic or near equiatomic at least five principal metallic elements which have approximately 5–25% atomic percentage (Gludovatz et al. 2015). High-entropy alloys (HEA) have advanced properties like superior thermal stability, corrosion and oxidation resistance, high strength, hardness and wear resistance and so on. Besides, these promising properties are situated in the system thanks to four key effects of high-entropy alloys. These effects are named core effects, and one of these effects is the high-entropy effect that gives the name to the system. In addition to this, the others can be named as the cocktail effect, sluggish diffusion and severe lattice distortion effect (Tsai et al. 2013).

(iv) *Metamaterials*

Metamaterials are artificial materials with extraordinary properties. They are considered revolutionary as they can provide unusual optical and electromagnetic features (Adams and Barbante 2015). They apply in many different fields, from mechanics to acoustics. Several disciplines currently examine them since they promise a wide range of applications (Schürch and Philippe 2021). Recent work on metamaterials concentrates on the control of changing material properties (Adams and Barbante 2015).

(v) *MXene*

The compounds consisting of transition metal and nitride, carbide or carbonitride can be designated as MXene. These materials have a 2D structure and $M_{n+1}X_nT_x$ (for n = 1 to 3) formulation. M stands for transition metals like Sc, Ti, Cr, V, Nb, Hf, Zr and the like. Besides, X refers to carbon or nitrogen atoms, and T refers to hydroxyl, oxygen or fluorine. N + 1 layers of transition metals cover the N layers of carbon or nitrogen in this structure. $Ti_3C_3T_x$ is the first synthesised MXene, and in addition to this, MXene, including more than one M element, can be in two different structures, such as solid solution and ordered structure. Whereas random dispersion of two different transition metals is obtained in the solid solution structure, the one or two layers of a transition metal are covered by the layers of other transition metals in the ordered structure (Anasori et al. 2017).

(vi) *Nanocomposite Materials*

Polymer Nanocomposites

Polymer nanocomposites include polymeric matrices and nanofiller materials as additives (Abdulkadir et al. 2016). According to the types of polymer materials, the polymer nanocomposites are also divided into thermoset and thermoplastic nanocomposites (Zaferani 2018). Besides, reinforcement materials may be organic or inorganic filler. Thanks to a variety of polymer matrices and fillers, different kinds of properties can be obtained (Dhillon and Kumar 2018). Examples of the usage areas of polymer nanocomposites are drug delivery, energy storage, information storage, magnetic and electric applications and the like (Abdulkadir et al. 2016).

Metallic Nanocomposites

The nanosized additive materials are used to manufacture metal matrix composites. The metal matrix composites are produced to obtain high mechanical properties such as high strength, ductility, toughness, dimensional stability, hardness, etc. The obtaining of the high mechanical properties depends on the homogeneous dispersion of the additive in the metal matrix. If the agglomeration takes place, the mechanical properties decrease. Furthermore, the production of the metal matrix composites is divided into two subgroups – in situ and ex situ. In an ex situ process, the additives are produced before adding the metal matrix, while the addition of the reinforcement is a part of the composite production (Ceschini et al. 2017). Besides, the production of metallic nanocomposites can be classified as liquid-state, solid-state and semi-solid-state methods, respectively (Sajjadi et al. 2011).

Ceramic Nanocomposites

The ceramic nanocomposites include glass or ceramic matrix material and different types of nano additives such as nanoparticles, nanotubes, nanoplatelets and hybrids of these materials and so on. These types of nanomaterials are added to the ceramic matrix to improve the mechanical properties of thermal shock, wear resistance,

electrical and thermal conductivities and the like (Porwal and Saggar 2017). Also, there are types of ceramic composite materials that include nanocrystalline matrices. These ceramic nanocomposites are designated as nanoceramics, and the dimensions of the grain size of the matrix are smaller than 100 nm (Banerjee and Manna 2013).

(vii) *Nanocarbon Materials*

Graphene

Graphene is a single-layer 2D nanomaterial having carbon atoms in a honeycomb atomic arrangement. However, there are also two- and three-layered graphene structures. The graphene exhibits different properties than fullerenes and carbon nanotubes (Rao et al. 2009). For example, the properties of the graphene are given like promising quantum hall effect, superior young modulus, high thermal conductivity, large surface area, optical transparency and so on. There are a lot of production routes to obtain graphene as single or multi-layer. The production of graphene is classified in the two subgroups as bottom-up and top-down methods. Whereas the chemical vapour deposition, graphitisation, solvothermal and organic synthesis methods are bottom-up methods, liquid electrochemical and thermal exfoliation of graphite and liquid intercalation, reduction via chemical and photothermal ways graphene oxide are designated as a top-down method. In addition to the advantages of graphene, the graphene structures can be used as composite materials with polymers, organic and inorganic compounds, metal-organic frameworks and the like. These composite materials are utilised in distinct areas such as fuel cell and battery systems, photovoltaics, supercapacitors and sensing platforms (Huang et al. 2012).

Carbon Nanotubes (CNTs)

CNTs are cylindrical structures of graphite and can be classified into subgroups such as single-walled, double-walled and multi-walled carbon nanotubes. Whereas single-walled includes a single graphene sheet, the other two groups have more than one graphene sheet. These materials exhibit high surface area, large flexibility, low weight, high aspect ratio and the like (Mallakpour and Rashidimoghadam 2019).

Fullerene

The carbon atoms number can be 60, 70 and 80 in the structure of the fullerene. C60 has a canonical structure and exhibits icosahedral symmetry. Besides, the electronic structure of fullerenes and graphene is similar, and they can be soluble with toluene. Also, fullerene shows insulator properties like a diamond. The colours of the fullerenes are different. For example, while C60 has a violet colour, C70 has a reddish-brown. The carbon arc method is used for the production of fullerenes (Ramsden 2016).

Carbon Nanofibre

Carbon nanofibres have high mechanical properties, surface area, thermal and electrical conductivity and nanoscale diameter. These excellent properties are utilised in different application areas such as energy storage, composites as reinforcement and the chemistry industry. Also, various synthesis techniques like chemical vapour deposition, templating, drawing and electrospinning can be used to obtain these materials (Mohamed 2019). Different production routes cause a variety of morphologies; these are classified as herringbone, platelet and ribbon (Malandrino 2009).

(viii) *Piezoelectric Materials*

Piezoelectric materials produce electrical energy when the mechanical forces are applied, and a change of shape occurs when the electrical energy is given to the material. Ceramics, ceramic-polymer composites, films and crystals can be shown as subgroups of the piezoelectric materials. Most of these materials are ceramic, and the performance of the piezoceramic materials strongly depends on various properties such as elastic stiffness, thermal coefficient, dielectric constant and so on (Moskowitz 2014). The common examples of piezoelectric materials are PZT, BaTiO3, PVDF (polyvinylidene fluoride), ZnO, ZnS, GaN and so on (*Electronic Textiles* 2015).

The usage of lead-free piezoelectric materials has been increasing due to environmental issues. Therefore, the investigations focus on utilisation of the lead-free piezoelectric materials instead of PZT piezoelectric materials. For example, langasite, tungsten bronze structure, materials with perovskites can be shown as an example of these types of materials (Uchino 2010).

(ix) *Semiconductors*

Semiconductors are an essential component for the electronics and energy industries. The reason for them to be important is their chemical properties. Unlike other materials that act as either a conductor or an insulator, semiconductors do not have a fixed value for conductivity. Thus, they can be manipulated by external stimuli to work as a conductor while they are insulators under natural circumstances. Also, their ability to carry electrical current by positively charged matter, called "holes", in addition to electrons, enables the production of electronic parts, such as transistors and solar cells (Neville 1995).

(x) *Shape Memory Materials*

Shape memory materials (SMMs) can return to their original shape after their shape is changed by another subject or impact (Huang et al. 2010). They are primarily used in medical applications, but R&D studies are on using these materials in industries such as aerospace and automotive (Bogue 2009).

(xi) *Superalloys*

The superalloys having surface stability and mechanical strength are materials that consist of VIIIA base elements. Thanks to these superior properties, these materials are utilised at high temperatures above 650 °C. The superalloys are divided into three subgroups depending on the base elements of superalloy, which are nickel, cobalt and iron-based alloys. Besides, powder microstructure, cast and wrought are other subdivisions of the superalloy. High mechanical properties come from precipitation hardening and solid solution strengthening mechanisms. Turbine blades and aero-engine discs are examples of usage areas of the superalloys (Liu et al. 2020).

(xii) *Superconductors*

Superconductivity can be defined with an instant decrease of electrical resistance to zero at a transition temperature named critical temperature (Tc) (Bardeen et al. 1957). The superconductors can be divided into low-temperature and high-temperature superconductors. There are different types of novel superconductors. Some examples are lithium, boron or transition metals like uranium and transfer salts' complexes. For example, C60 fullerenes are promising candidates for novel superconductors because this material has advanced properties such as high critical current and magnetic field. Magnesium diboride MgB2 is also given as another example of novel superconductors. The critical temperature of this material is 39 Kelvin, and it is thought of as a high-temperature superconductor. It also has dual-band superconductivity. The other examples of this type of advanced materials are alkali oxide fullerenes, RNi2B2C, RNi2B2C, CeMIn5 (M = Co, Rh, Ir), CePt3Si, CePt3Si, Sr2RuO4 and so on. Besides, boron-doped diamond, $Na_xCoO_2H_2O$ and CaC_6 materials can be shown as the other types of new superconductor materials (Shi et al. 2015).

(xiii) *Thin Films*

Thin films have a thickness between nanometres and micrometres, and they have different properties from their thicker equivalents. They serve as surface coatings in several fields. Biomedical, mechanical, electric and thermal industries utilise thin films as protective surface coatings (Mylvaganam et al. 2015). These advanced materials can be used in different application areas such as aerospace, energy storage, refrigeration, etc. With the increasing utilisation in new energy applications like electric vehicles and fuel cells, the advanced materials give high energy and power density and flexibility. The usage of nano-enhanced materials increases the life cycle and capacity of the components used in the energy storage devices (Liu et al. 2010). Also, nanomaterials provide mechanical and electrical advantages in the systems. Thus, it is thought that these materials will become next-generation materials (Shearer et al. 2014). New types of metallic materials such as high-entropy alloys and superalloys exhibit high promising mechanical properties such as ductility, high-temperature properties and fracture toughness (He et al. 2016; Liu et al. 2020). Thanks to these advantages, the usage of new metallic materials in the medical, turbine blades and other application areas have been increasing (Anupam et al. 2019; Ma et al. 2020).

2.4 Artificial Intelligence

Artificial intelligence (AI) is defined as a system that can collect data, learn, decide and take rational actions using appropriate methods such as machine learning, deep learning and reinforcement learning. Alan Turing put forwards the first question that led to the development of the term. In his article "Computing Machinery and Intelligence", he wondered if machines could think someday as humans do (Turing 2009). Thus, research in this field started thanks to Alan Turing and then accelerated when John McCarthy coined the term "artificial intelligence" for the first time in 1955. However, many people previously thought of the term "intelligence" as a concept that only humans can have (McCarthy 1989). So, if machines could 1 day have intelligence, according to McCarthy, then the word "artificial" intelligence is more appropriate for the description.

Nevertheless, artificial intelligence as a term should be considered a system that can think and act logically, unlike human intelligence. It is necessary to model humans as rational beings instead of emotional ones (Guo 2015). So, human thinking needs to be modelled, and this can be reduced to four steps. First, the data needs to be collected, then learned and a decision will be made as a

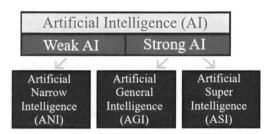

Fig. 2.7 Types of artificial intelligence

result of this learning. And in the last step, the work will be done in line with the decision made (Küfeoğlu 2021). As artificial intelligence is an emerging technology, its boundaries keep enlarging, and it continues to evolve. There are two types of artificial intelligence that human beings can produce now and hope to have in the future. Types of artificial intelligence can be seen in Fig. 2.7.

The first type is generally called "weak AI", or, namely, narrow AI or artificial narrow intelligence (ANI). This type of AI can be trained and then perform specific objectives. Unfortunately, most of the "artificial intelligence" products or services today are weak AI, and their ability is limited. Some of the good example products and services are digital assistants such as Alexa and autonomous vehicles.

The second one is called "strong AI" which consists of artificial general intelligence (AGI) and artificial super intelligence (ASI). AGI can learn from past experiences and solve problems. Also, it can plan any task. This type of artificial intelligence is in use in some buildings. However, it is still a technology that needs to be developed, and its usage area is limited because it is not especially useful.

On the other hand, ASI, also called superintelligence, is a type of artificial intelligence expected to exceed the intelligence and ability of the human brain. Also, it is the type of "artificial intelligence" that people mostly come across as an idea in movies. This type of artificial intelligence exists only theoretically, and no usage area can be given as examples from our daily lives. In addition, machine learning and deep learning come into play at this point in artificial intelligence. Artificial intelligence is achieved through

machine learning (ML) and deep learning (DL). The relationship between all these can be observed in Fig. 2.8.

Machine learning is a sub-branch of AI and is based on statistics. Hence, input is given in the system, and a permanent estimate determines the output. Not only is manual coding done, but also it is aimed to develop a system that re-codes itself according to the ever-increasing data and increases its accuracy (Mohammed et al. 2016). In addition, machine learning has constantly been changing until today and has evolved with steps such as supervised learning, unsupervised learning, reinforced learning, deep learning and deep reinforced learning. As stated in Fig. 2.1, deep learning is a sub-branch of machine learning and, therefore, artificial intelligence. It is a system that simulates human speech and thinking with neural networks. It is used in many fields, such as convolutional neural networks (CNNs) and recurrent neural networks (RNNs) (Küfeoğlu 2021).

Artificial intelligence (AI) is an important technology that facilitates, accelerates and even saves human life from time to time. It was invented to be useful in daily life. For example, AI devices doing housework are easier than doing it manually, and it takes more time. AI shortens the information processing process and increases efficiency (Küfeoğlu 2021). For example, in the field of medicine, early detection of diseases has become easier. As a result of the data provided by AI, which performs morphological evaluation, it reduces the workload of healthcare professionals and facilitates the diagnosis of the disease (Mintz and Brodie 2019). Furthermore, thanks to the use of AI, the patients' medical data are stored and analysed to improve the healthcare system (Hamet and Tremblay 2017). Thus, the workflow accelerates, and patients can get the necessary treatment faster and easier.

People use many examples from daily life and do not even know that they are using AI technology. Machine translations are vastly used on the Internet and social networks, improving day by day. The computer has learned to recognise both spoken and written speech. The other example is computer games. AI is used to create a game universe that controls

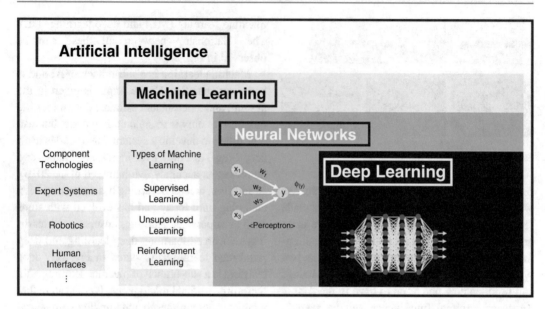

Fig. 2.8 Relationship between AI, ML and DL

bots – characters that people do not play. AI is used to create game strategies. Also, AI technologies are used to create smart houses. A special program controls everything that happens in the house – electricity, heating, ventilation and household appliances. Robot vacuums scan their surroundings to determine if they need to get started. Another example of the usage of artificial intelligence is in agriculture. It is imperative to use artificial intelligence in agriculture because it is difficult to achieve quality food in the modern world (Küfeoğlu 2021). With the use of AI, farmers can achieve a better harvest by providing their crops with more optimal conditions (Sharma 2021).

As can be seen from the examples given, nowadays, people are surrounded by AI technologies, and it is in every sphere. Thus, the main goal of these technologies is to make life easier and faster. The next thing related to the previous is that these technologies save time to spend on other things they want to do. Therefore, this is the marketing approach too. There can be an example of social media like Facebook and Instagram, where users tag their friends on the pictures, and now AI can do it automatically for them. So, the users save their time by using the media more in other ways.

There are some reasons for using AI technologies. Firstly, AI can store and process huge amounts of data. With deep learning, new data is added to the previous data, so people reach more accurate data every time artificial intelligence is used. In addition, humans are constantly affected by their emotions when making decisions. A machine with AI is not influenced by its emotions when making decisions, as it has no emotions. Therefore, the decisions taken are more objective and logical (Khanzode and Sarode 2020). As a result, AI makes the device to which it is added faster, smarter and more efficient. According to the research, artificial intelligence technology has improved data science by 9.6%, health services by 6.3%, defence systems by 5.3% and natural language processing by 5.1% (Shabbir and Anwer 2018).

If emerging technologies are considered, predicting the future is essential too. AI has been a popular, prominent research and application area recently, but the question is, will artificial intelligence evolve more in the future? AI technology can reach a wide number of applications because of the ability of machines to work with humans, collaborate digitally without any limits, make sensible decisions with the results it analyses from the data at critical points, bring various

Fig. 2.9 Usage of
artificial intelligence

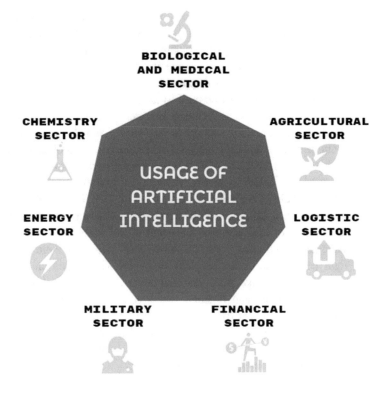

ideas together and integrate them to produce
physical or digital prototype properties. These
abilities make AI a perfect technology that can
continue to grow in the future. The most up-and-
coming sectors that will use AI in the future are
shown in Fig. 2.9.

In traditional systems, medical and biological
systems do not work efficiently due to the complexity, a large amount of data and human errors.
However, efficiency in new generation biological
and medical systems has increased thanks to artificial intelligence algorithms. As stated by
Shabbir and Anwer (2018), future applications of
AI for various sectors are provided below:

- Artificial intelligence tools in the financial
  sector will be used to prevent market manipulation, fraud, reduction in trading costs and
  market volatility. It is expected that there will
  be systems that warn or directly intervene to
  solve the problem of system failures and other
  risks.
- AI enables the establishment and refinement
  of supply chains in manufacturing sectors,

reliable forecasts due to data analysis, demand
regulation, inventory accuracy and optimisation of programs. Therefore, the applications
of AI are faster, smarter and environmentally
efficient.

- The use of artificial intelligence in the agricultural sector provides smart production, storage
  and distribution solutions. It will also play a
  role in deciding on the fertilisers and chemicals to be used for the crop by instantly receiving data. Strengthening the decision-making
  mechanisms by showing the buying-selling
  balance through to its data analysis feature, by
  chatbot application usage preparing customer
  support texts to retain the customer, speeding
  up the support time, reducing the number of
  responses to the request and increasing its
  quality and machine learning, and this application will help people for their diet, organising their daily habits, etc. It will play a role in
  increasing the price-performance ratio for the
  consumer.
- It is expected that students will develop methods that increase the rate of study and learning

by analysing their working styles, working hours and learning styles. In addition, thanks to machine learning, analysing students' physical or psychological conditions and increasing their success rate are made possible.

- Artificial intelligence will be used to identify new drugs in large-scale genome research, providing the necessary support to find new genetic problems and efficiency. AI will help decide or prescribe drugs, whilst being aware of the patients' health problems.
- Within the field of logistics, the use of artificial intelligence will improve the making and management of delivery schedules and efficient channel vehicles.
- When people use online services or social media, AI will make interface adaptations, personal assistants and chatbots more consumer centric and user friendly. The required product quantities will be determined by the analysis specific to customers.
- AI can be used to integrate and control renewable energy sources, enable self-healing networks and harness power system flexibility, especially to encourage the use of renewable energy in developing economies.
- Potentially, military balance and future warfare will heavily be affected by the future of AI technology because of the developments in robotics and automatisation (Allen and Chan 2017). Some reconnaissance and attack missions will be planned to be done using AI-supported unmanned weapon systems that will be deployed soon. Moreover, it is expected that clandestinely designed AI systems might be used to penetrate advanced air defences. Lastly, the development and use of AI-augmented systems are expected to reduce the vulnerability of an army to cyberattacks (Johnson 2019).

2.5 Autonomous Vehicles

Autonomous vehicles are vehicles that can perform their functions with artificial intelligence algorithms defined in their content, sense their environment and operate without the need for human intervention. With the astounding growing speed of technological developments, significant improvements have been made in the development of autonomous vehicles. The automotive industry, especially, has achieved significant advances in the mechanical and electrical characteristics of vehicles since the 1920s. Autonomous vehicles have also been envisaged as the most popular objective in this respect. Various automotive firms and universities made numerous attempts to pioneer autonomous cars between 1920 and 1980. In the 1920s, a radio-controlled driverless automobile was one of the earliest demonstrations (Davidson and Spinoulas 2015). The fact that there was a lot of development in the field of science and technology in this time period was the most important factor affecting the situation of autonomous vehicles from the 1980s to nowadays. Although the dream of autonomous vehicles dates back to old times, it took the 2010s to meet the technical requirements and take realistic steps.

In the working mechanism of autonomous vehicles, many components are widely used, such as complicated artificial intelligence algorithms and devices with high processing power, sensors and actuators (Gowda et al. 2019). GPS (global positioning system), LIDAR (light detection and ranging), RADAR (radio detection and ranging) and video camera technologies are also integrated with these components (Ondruš et al. 2020). In the context of GPS, the users of vehicles, municipalities and technology-driven businesses get help in the field of transportation planning from the mapping and power of data functions of GPS (Bayyou 2019). Therefore, the inclusion of GPS inside of autonomous vehicles can increase the efficiency of vehicles by applying smart route optimisation plans and gathering information about the environment. Secondly, lidar is defined as a remote sensing technology that detects the distance between a target by making it visible with light particles and works by detecting the returning light (Ondruš et al. 2020). Radar and lidar have similar properties in terms of working principles, except for the transmission source used, such as light and what it is intended to measure. In radar, which works with the principle of signal give and take, a change in the frequency of the signal occurs during the return phase from the receiver while measuring and this change is used to determine the speed of the vehicle (Sarkan et al. 2017). Lastly, detecting randomised human factors and physical elements that cannot be identified in the system but are present in the traffic is not easy with radio waves and light without the contribution of video cameras (Yun et al. 2019). All of these technologies and technical components are the factors that developed the performance level of autonomy in the vehicles to provide users with a well-prepared and safe experience while driving. In this direction, it has been claimed that AVs permit "drivers" to free up the time customarily spent checking the roadways, empowering them to utilise their time more successfully by resting, eating, unwinding or working during the time customarily spent driving (Haboucha et al. 2017). Figure 2.10 shows how AVs work briefly.

Considering the effects that autonomous vehicles can offer when integrated into human life, it is obvious that it is a very critical technological revolution. According to Beiker and Calo, by eliminating the driver from the equation and relying on cars to manoeuvre themselves through traffic, this technology has the potential to enhance safety significantly, efficiency and mobility for humans (2010). Over the years, via the increase in connection speed with technologies such as 5G, advances in the Internet of Things and the strengthening of the interconnectivity of mobile devices, feasible and applicable solutions have emerged in autonomous technologies. These advancements have prepared the path for autonomous vehicle (AV) technology, which promises to minimise collisions, energy consumption, pollution and traffic congestion while also boosting transportation accessibility (Bagloee et al. 2016). Consequently, it can be

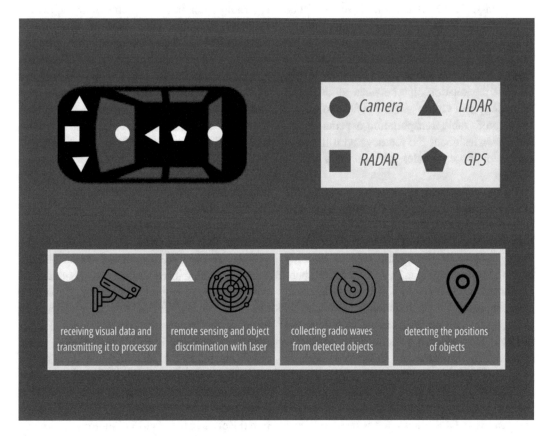

Fig. 2.10 How automated vehicles work

said that the use of autonomous technologies on vehicles, such as collective, individual and scientific research, has been helping the processes become more streamlined, adaptable and efficient for the user due to advancements in emerging technologies.

As has been stated earlier, autonomous vehicles have become significant for a large number of areas in life. Chan highlighted autonomous cars' beneficial contributions and impacts on several levels, including users, infrastructure and sustainable cities and societies (Chan 2017). Firstly, for individual users, crashes occurring with vehicles due to lack of attention can be prevented by providing a more reliable driving experience by software and hardware components included in autonomous vehicles. It would not be unfeasible to have more secure and quicker transportation in cities, which save time with the help of self-driving cars. Such junctions are expected to have an important impact on the road system of each city. The travel and waiting times will be considerably shorter (Zohdy et al. 2013). Additionally, it is thought to have an effect that can prevent 9 out of 10 accidents that occur under normal conditions (Chehri and Mouftah 2019). To sum the relationships between autonomous technology and citizens, it is possible to have a more comfortable transportation experience and fewer worries about the journeys with the integration of those vehicles into the daily life of humans.

On the other hand, in city road planning, the problems which are faced under normal circumstances can be decreased by autonomous technologies. More controllable vehicles provide a clear structure of roads, low cost of building parking lots and roads and more accessible public transportation services which can help urban planning, easier public and mobility services, an incentive for private investors on their business models. This emerging technology conserves resources for infrastructure in the city, such as parking and road development, while vehicle technology also decreases traffic and eliminates possible parking problems. The use of advanced and real-time GPS allows for a more efficient navigation experience, resulting in more accessi-

ble, dependable and adaptable routes. Besides, the sensors implanted nowadays within the autonomous vehicle are "intelligent" since they do not as it gave an information estimation but are sent with a coordinated computer program brick competent to perform, to begin with, a stage of preparing this data (Chehri and Mouftah 2019). Consequently, more efficient infrastructure due to improved vehicle control is provided by GPS and sensors.

The last promise of autonomous vehicles, sustainable cities and high-level comfort of societies can be achieved. Most governments used to develop additional roads and streets to address the rising urban environment demands. Due to a lack of public funding and physical space, the transportation network, which has a lower capacity than the population, has been overburdened, causing additional congestion, CO_2 emissions and significant disruptions to people (Dameri 2014). The previously mentioned developments regarding autonomous vehicles offer impressive solutions for sustainable cities in response to these problems. Autonomous vehicles play an important role in reducing physical and environmental noise pollution, reaching the desired level of city traffic flow, eliminating the security concerns of the city's people, speeding up regional procurement processes and reducing procurement costs (Seuwou et al. 2020). As a result, it is not impossible to reach smart, sustainable, green and information cities with the integration of AVs into urban life. Figure 2.11 summarises potential use areas of AVs.

AVs are used in many different areas of industry according to the level of automation of the vehicle. To clarify the unique function of each vehicle, defining its capabilities and complexities with their autonomy level is a must. Building a classification analysis of AVs is significant in terms of their capability to make tasks autonomously (Ilková and Ilka 2017). According to the model created by SAE International, it presents a taxonomy with precise definitions for six levels of driving automation, ranging from no driving automation (level 0) to complete driving automation (level 5), in the context of motor vehicles and their operation on routes (2018). The classifica-

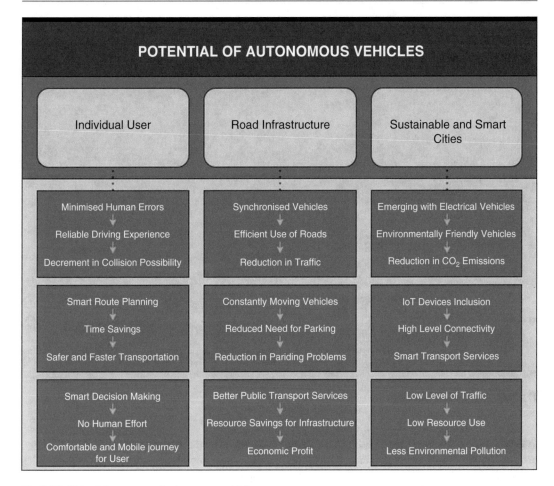

Fig. 2.11 Potential use areas of autonomous vehicles

tion system is based on how steering and braking are managed, how much human control is required when driving and whether the AV can operate without it in all scenarios (Alawadhi et al. 2020). For each level, elements represent the low processing capabilities. The difference between levels 2 and 3, where the human driver does part of the dynamic driving task and level 3 when the automated driving system performs the full dynamic driving work, is important (Ilková and Ilka 2017). Fig. 2.12 explains the automation levels in detail.

When everything is taken into consideration, it would not be an exaggeration to claim that autonomous vehicles will be very influential for the future trends of the technology world and automotive industry. When the future situation is examined, it is obvious that minor and major developments in autonomous vehicles are in interaction with each other. For user-oriented improvements, research states that the adaptation of users into autonomous vehicles will be possible with the modified after-sales mechanism, easy solutions to technical problems and flexible supply chain systems (Bertoncello and Wee 2015). Furthermore, the most used ways of transportation will be AVs by saving drivers more than half an hour per day, making quite a lot of parking space suitable for use and providing life and property safety by offering a lower error rate in driving experiences by 2050 (Bertoncello and Wee 2015). When considering the future state of autonomous vehicles, there is no significant obstacle to the increase in usage rates. Up to 15% of new automobiles produced in 2030 might be fully driverless after technology, and regulatory

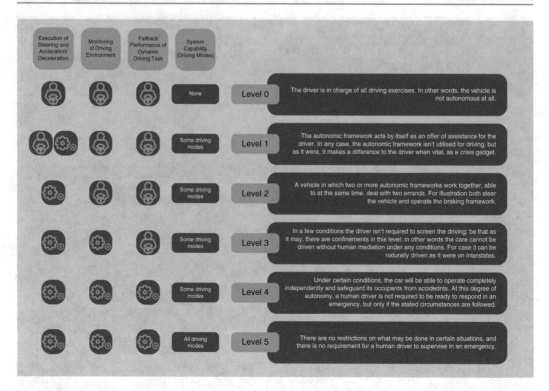

Fig. 2.12 Six levels of autonomy

concerns are overcome (Gao et al. 2016). There are serious studies of many research companies related to this subject. BCG predicted that worldwide sales would level at about 100 million per year by 2030 and that by 2035, 30% of the fleet would be electric, and 25% would be autonomous (Jones and Bishop 2020). Additionally, according to KPMG, by 2030, linked cars will account for 75% of the UK motor-park (used vehicles), with about 40% being partially automated and fewer than 10% being totally autonomous (Jones and Bishop 2020). A great deal of work falls on manufacturers and technology decision-makers to make these predictions feasible and to truly feel the impact of autonomous vehicles in the future. Considering the cumulative progress of technology, it is obvious that the development of autonomous vehicles will gain momentum with the production of prototype projects. Millions of lines of code are already embedded in the latest vehicles rolling off European manufacturing floors; the next phase of autonomous driving plainly requires both engi-

neering acumen and digital smarts to develop, build and distribute successful automobiles (Gupta 2021).

2.6   Big Data

Big data research is at the forefront of modern business and science. It mainly includes data from online transactions, videos, images, audios, emails, logs, clickstreams, postings, social networking interactions, science data, health records, sensors, search queries, mobile phones and associated apps. These data are stored in databases, which became highly complicated to capture, store, form, distribute, manage, analyse and visualise using standard database software (Sagiroglu and Sinanc 2013). At the end of 2016, it was stated that 90% of the world's data had been produced in just 2 years, at a rate of 2.5 quintillion bytes per day. Furthermore, data is growing at an exponential pace, with estimates of more than 16 zettabytes (16 trillion GB) of useful data by 2020.

The advent of the Internet of Things, as well as the global proliferation of mobile devices – technologies, not just humans, are producing data – and the growth of social media, which has transformed everyone into a broadcaster and hence a data producer, is also adding to the quickly growing volume of data. The vast bulk of the information is no longer numerical and poorly organised. As a result, the majority of data is in the form of unstructured data, such as text, video, audio and images, which are becoming increasingly widespread (Suoniemi et al. 2020).

To put it simply, big data refers to massive volumes of information. However, size is not the only factor to consider (Oliveira et al. 2019). Although there is no singular definition for big data, there are some relevant definitions in the literature. Big data comprises structured data found in organisational databases and unstructured data created by new communication technologies (e.g. Internet of Things), such as images, videos and audio (Sestino et al. 2020). Big data also refers to a collection of enormous, complicated datasets that are too vast for traditional data processing tools and other relational database management technologies to analyse, manage and record in the timescale required. Big data also implies the diversity and velocity of data and its volume. The three Vs of big data are shown in Fig. 2.13.

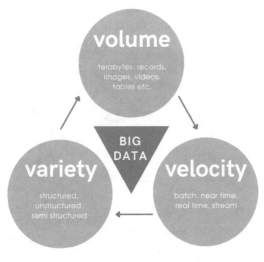

Fig. 2.13 Three Vs of big data

Volume refers to the amount of data generated from various sources. The structural variability of a dataset, which might be structured, semi-structured, or unstructured, is referred to as variety. Data that has been arranged in a way that makes it easier to analyse is known as structured data. On the other hand, unstructured data is data that is difficult to analyse and includes movies, photos and audio files. Although semi-structured data is not subject to as stringent standards as tabular data, it can be saved in XML (extensible markup language) format. Finally, velocity refers to how quickly data is produced and processed from various sources such as social media and the Internet (Oliveira et al. 2019).

According to the TDWI (transforming data with intelligence) Big Data Analytics survey, the benefits of using big data include better-targeted marketing, more direct business insights, automated decision-making, client segmentation, sales and market potential recognition, risk quantification and market trends, more lucrative investments, understanding of company transformation, better planning and forecasting (Sagiroglu and Sinanc 2013). According to current research, big data facilitates corporate decision-making through technology, systems, techniques, practices and applications related to gathering, storage, analysis, integration and deployment of large amounts of structured and unstructured data. From \$3.2 billion in 2010 to \$16.9 billion in 2015, the vendor market for big data technology has grown over 40% annually (Suoniemi et al. 2020).

Non-expert staff, cost, the difficulty of designing analytical systems, poverty of database software, scalability issues, inability to make big data usable for end-users, incompetence in reaching enough data load speed in current database software and lack of compelling business case are some of the disadvantages of big data mentioned by TDWI (Sagiroglu and Sinanc 2013). According to McKinsey Global Institute's Report, the value potential of big data is mostly unexplored and underused by businesses today (McKinsey Global Institute 2011). Three key problems that are preventing businesses from getting larger benefits from big data are (1) organisational structure and procedures; (2) strategy, leadership and talent; and (3) information technology (IT) infrastructure. Many businesses are unclear on how to integrate big data and afraid to spend on new information technology, or they just consider big data analytics to be arduous.

We live in a big data era, defined by the rapid accumulation of omnipresent information. Big data contains an infinite amount of information. It is expanding in various industries, giving a method to enhance and simplify operations (Lv et al. 2017). By becoming a necessity of our age, big data is almost everywhere. Any industry that accumulates a large amount of data, such as e-commerce, geography and transportation, research and technology, health, manufacturing and agriculture, can benefit from big data analytics. According to Andreas Weigend, professor at Stanford University and Amazon's former chief scientist, "Big Data is when your datasets become so large that you have to start innovating how to collect, store, organize, analyse and share it"

(Backaitis 2012, cited in Gobble 2013). This large amount of data collected by companies and institutions is processed, providing them with an opportunity to evaluate their performance and gain insights by creating "information", which has now become a valuable resource like money (Vassakis et al. 2018). Some of the key application areas of "big data" in different sectors are shown in Fig. 2.14, and those areas can be summarised as follows (Zellner et al. 2016; Memon et al. 2017)

- Agriculture: With the latest developments in agriculture, the use of big data has increased gradually with a better understanding of its importance. With the help of the data collected from plants, it is possible to follow them in real-time, and the obstacles in front of them to grow most healthily and efficiently are greatly alleviated.
- Banking sector: Today's financial firms have been transformed into online and mobile banking. The increased usage of online and

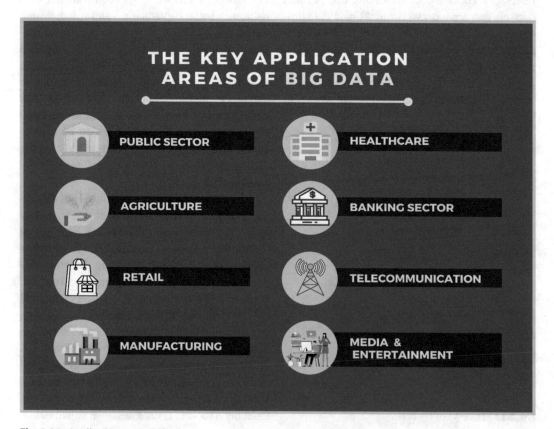

Fig. 2.14 Application areas of big data

mobile channels has resulted in fewer face-to-face contacts between consumers and banks while boosting virtual interactions and the volume of client data. Banks now have far more data about their clients than ever before regarding both volume and variety. However, only a small part of them is used to achieve successful commercial outcomes. Like most e-commerce firms, big data technology can make efficient use of consumer data, assisting in the development of customised products and services. Main application areas of big data within the banking sector include credit scoring and risk management. Automated procedures based on big data technology, such as machine learning algorithms, make loan and credit decisions in seconds. Moreover, the need for better risk monitoring, risk coverage and increased predictive capability in risk models has never been greater. Big data technology, along with hundreds of risk indicators, can help banks, asset managers and insurance companies detect possible hazards earlier, respond more quickly and make more informed choices. Big data may be tailored to an organisation's specific needs and used to improve several risk categories.

- Healthcare: Recent issues such as increasing healthcare costs and a higher need for healthcare coverage can trigger the demand for big data technology. As high-quality health services necessitate analysing large datasets, big data analytics help categorise patients into certain groups to take into consideration the differences between different patient groups. This enables healthcare providers to focus on more specific questions of concern to certain patient groups, thus significantly enhancing the quality and effectiveness of care.
- Manufacturing: With the better integration of IT technologies, the manufacturing industry has been undergoing significant changes. As the connectivity in every stage of the production process is growing, data management is highly engaged in making the existing data more manageable, standardised and integrated. The use of big data technologies plays a key role.

- Media and entertainment: Because of the impact of digitalisation, anybody can create, share and publish material. Media companies are more and more linked to their consumers and competitors. This implies that the use of big data technology to process a wide range of data sources, and if necessary, in real-time, is a significant asset that corporations are willing to invest in to make informed decisions.
- Public sector: Governments generate and collect vast quantities of data through everyday activities. The applications of big data in the public sector include the generation of data-driven insights to identify patterns and generate forecasts. Fraud and threat detection, planning of public services, supervision of private sector-based activities and prioritisation of public services are included in big data analytics in the public sector (Mckinsey Global Institute 2011; Yiu 2012 cited in Zillner et al. 2016). Moreover, improvements in effectiveness by increasing transparency through the free flow of information, creation of innovative and novel services to citizens and personalisation of services that better fit the needs of citizens can be considered as earnings of the public sector from big data (McKinsey Global Institute 2011; Ojo et al. 2015, cited in Zillner et al. 2016).
- Retail: The collection of in-store, product and customer data plays a key role in the retail sector. By enabling accurate information extraction from huge data, big data technologies provide this sector with important benefits and opportunities, such as understanding consumer behaviour and generating more context-sensitive and consumer-oriented tools.
- Telecommunication: The achievement of operational excellence for telecom players can be summarised as a combination of benefits in the management of marketing and customer relationship, service deployment and operations, which all depend on big data technologies to make sense of large amounts of internal data and data from huge numbers of users.

The increased production and availability of digital data in many areas, along with improved analytical skills due to improvements in computer sciences, has resulted in new findings utilised to improve results in many fields. In parallel with these developments, organisations are also undergoing a systemic transformation in the knowledge-based economy. Information management and big data analysis are concerned with strategies for maintaining a shared foundation of corporate knowledge, allowing different organisational units and functions to coordinate their efforts, exchange knowledge to support decisions and generate competitive advantages. In this sense, businesses effectively leverage big data to streamline processes, create efficiencies and improve services provided to customers, especially online shopping platforms such as Amazon (Madden 2012, cited in Gobble 2013). Through big data applications, corporate knowledge may spread globally and be kept in several formats, including skills and expertise in the minds of researchers and workers and organised information in databases and other big data corporate resources (de Vasconcelos and Rocha 2019), which offers opportunities for fostering innovation.

Economics, management and business data analytics are all changing in the age of big data. The emphasis on economic and management science has shifted to empirical studies and the systematic use of information technology and computer systems. The digital revolution and the global big data phenomenon are anticipated to have more impact in economic research. Researchers and corporate executives and consultants are increasingly relying on large-scale business data obtained through partnerships and corporate networking. Thanks to the Internet, corporate intranets have grown into sociability and knowledge sharing centres. There is a higher reliance on making sense of big data in today's highly connected organisational environment. This implies the need for software solutions that enable the efficient and methodical assessment of massive amounts of company data. This cause-and-effect connection leads to predictive analysis in knowledge-intensive enterprises, such as data mining methods based on machine learning and artificial intelligence. Making sense of often unstructured data may be a time-consuming effort. To successfully solve technological, skill-based and organisational difficulties, businesses must acquire a diverse assortment of big data-related IT resources (Suoniemi et al. 2020).

Today, in parallel with the increase in the number of users and devices connected to the Internet, a large amount of data has been collected in the databases of companies and technologies that will enable this data to turn into commercial value have come to the fore. With the increasing importance of data, many companies have transferred a significant amount of workload to the departments of these companies. To survive and compete, businesses must integrate industry 4.0 techniques into their activities. They must modify their management, organisation and production practices to achieve that. The best way to achieve this aim is by "reengineering": Its origins in the realm of information technology have now broadened to include the wide process of revamping key business operations to improve organisational performance. Reengineering methods give conceptual references targeted at rethinking and rebuilding corporate processes through digitalisation. The industry 4.0 revolution has stressed a collaborative link between business process digitalisation and IT since it began to develop more flexible, coordinated, group-oriented and real-time communication capabilities (Sestino et al. 2020).

Nowadays, big data is at the first stage of its evolution. Most of the businesses in different sectors still have not implemented the big data concept. However, many of them continue to work in this direction, as described before. Even in this early stage, the positive effects on the enterprises cannot be ignored. Many individuals envision big data as the planet's core nervous system, with individuals serving as its sensors. However, it is obvious that the concept of big data will lead to another revolution in the concept of business shortly, based on the competency it offers to interpret and analyse even the most variable and unrelated data (Chauhan and Sood 2021).

To put it shortly, current tools and approaches perform data processing inefficiently. The objective of all present analytical techniques and data

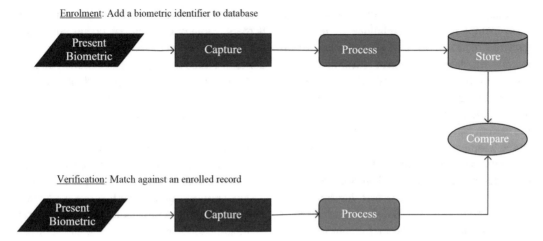

Fig. 2.15 Biometric system. (National Biometric Security Project 2008)

processing technology is to process a small amount of data. Existing technologies for large data processing reduce efficiency and generate a slew of complications. As a result, present technologies cannot entirely resolve large data challenges. Cloud computing, artificial intelligence, parallel computing, grid computing, stream computing, bio-inspired computing, quantum computing, semantic web and software-defined storage are critical research issues that need to be studied for the big data concept to be understood and applied properly (Yaqoob et al. 2016).

2.7 Biometrics

Globalisation has allowed humanity to become more interconnected, with communication between individuals increasingly mediated through technological platforms and transactions increasingly frequently conducted remotely (Fairhurst 2019). The term biometrics is derived from the Greek term *bio*, meaning "life", and *metric*, meaning "measurement" (Gillis 2020a, b). Biometrics refers to the authentication of a person's identity through chemical, physical and behavioural characteristics. Biometric technology offers a safe and convenient identification system; users do not have to remember complex passwords or carry identification documents, easily lost or stolen. Biometric identity verification

systems compare an individual's live-captured unique characteristics to a biometric template stored in a database to determine their resemblance. The system authenticates the information acquired in real-time against the reference model of biometric data to verify an individual's identity. Biometric verification has a high industrial acceptance rate worldwide due to the introduction of digitalisation and computerised databases, which ensure security through fast personal identification. Figure 2.15 demonstrates the working structure of a biometric system.

Biometric technology, at its most basic level, consists of pattern recognition systems that collect biometric patterns or characteristics utilising either image acquisition devices or a combination of both. In the case of fingerprint and iris recognition systems, such as scanners or cameras, and voice and signature recognition systems, movement acquisition devices like microphones are used (National Biometric Security Project 2008). An individual must be enrolled in the system before a biometric system can be used to determine identity, as shown in Fig. 2.15. The registration process involves the collection of measurement of the individual's characteristic(s) and the storage of this data as a biometric template within the system. The template is matched to live-captured biometric data in subsequent usage. Biometric systems typically work in one of two modes after the enrolment process: biometric authentication or a one-to-many comparison. The technique of matching gathered biometric data to an individual's biometric template to confirm identification is known as biometric authentication. Rather, to identify an unknown individual using biometric identification. The individual is acknowledged if the system can match the biometric sample to a stored template within an acceptable threshold.

Biometrics have a long history, with the first examples present in the ancient Mesopotamian civilisation of Babylon. The first descriptions of biometrics are from the Babylonian civilisation around 500 BC, while the first record of a biometric identifying system dates from the 1800s. Biometrics has been around as today's technology since the 1960s. Biometric technology has continued to evolve over the years and has developed into many forms by 2021. Contemporary biometric devices collect a variety of identifying information and use it for diverse purposes across sectors. Some devices can authenticate identity without any interference or direct contact with the person whose information is being collected; this includes voice recognition, walking gait and other specific behaviours. These are considered behavioural biometrics, which detects unique distinguishing features depending on how individuals interact with their systems. Behavioural biometric systems are especially useful for cybersecurity and online fraud protection. Many behavioural biometrics applications, unlike physiological solutions, do not require an apparatus for data gathering.

Unlike behavioural biometrics, physical biometrics are based on an individual's unique and quantifiable physical characteristics. Fingerprints, retinae and DNA sequences collected from blood, saliva and other bodily fluids are key in biometric technology widely used in forensics, medicine and criminal justice cases. These biometrics require a device that links these unique properties to an existing database. The objective is to match the individual's unique characteristics to an existing record or file to identify them. Biometric travel documents are required to cross most international borders, and they provide an elevated level of security to the countries attempting to regulate who comes in and out and secure their borders. Biometric voting documents could also provide a new level of security in elections (DHS 2021). Detailed descriptions of common biometric technology are listed below, and the historical development of these two biometric types is shown in Figs. 2.16 and 2.17.

1. Fingerprint: There are two types of fingerprint identification systems, automated fingerprint identification systems (AFIS) and fingerprint recognition systems. The first one is typically used only by the provision of law. Fingerprint identification provides a unique template based on the properties of the fingerprint without preserving or even allowing for image reconstruction. The first image is obtained by scanning the finger in real-time while it is in direct touch with a reader device, checking for confirming features like temperature and pulse.

2. Hand geometry: Hand geometry is defined by the relative dimensions of fingers and joint placements. In the late 1960s, the Shearson-Hamill investment bank on Wall Street applied Indentimat, one of the earliest automated biometric systems. It utilised hand geometry for nearly two decades. Some systems are capable of taking simple two-dimensional measurements of the hand's palm. Others want to take a basic three-dimensional picture from which to extract template characteristics.

3. Face recognition: Face recognition is still in its infancy, with the majority of research and applications taking place on tiny databases. The face of the individual must be exposed to a video camera for biometric identification purposes. The possibility of tricking or confusing some systems with cosmetics is an obvious flaw in several present approaches.

4. DNA: Human DNA is a genetic structure that can be obtained in many ways, such as human hair, nails, saliva and blood. This structure is found in every cell in the human body, and it contains a lot of genetic information. Also, every person's DNA is unique, except for identical twins.

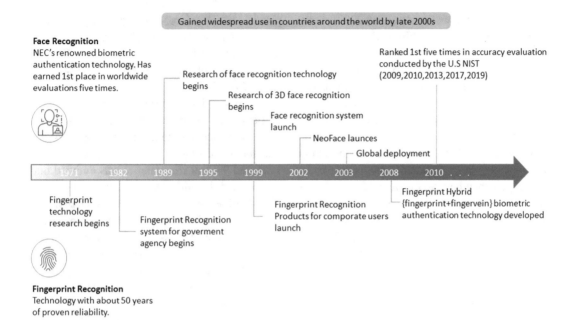

Fig. 2.16 Historical development. (RecFaces 2020)

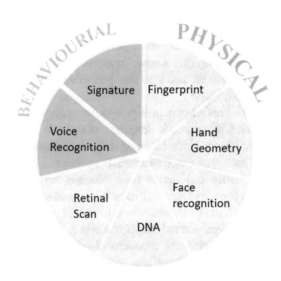

Fig. 2.17 Types of biometrics. (RecFaces 2020)

5. Retinal scan: Retinal identification is an eye signature. What allows this definition to be made is the vascular structure of the retina. A retinal scan is taken by looking at a specific target with a lens, and a retina scan is a reliable feature.

6. Voice recognition: There are two voice recognition techniques, automatic speaker verifica-

tion (ASV) and the other is automatic speaker identification (ASI). The main data used in in-person identification using these methods is the person's voice. After this data is received, the system detects the voice of the person whose identity is desired to be determined by making comparisons with the templates registered in the database.

7. Signature: Identification with signature data is made with some criteria created by experts over the years. Computer systems developed for this identification method can now be as successful as an expert at detecting distinctive features in signatures. In addition, signature identification systems do not only focus on the shape of the signature but also can detect the "speed of using the pen" of the signer or the "pressure applied to the surface" while signing (Phadke 2013).

These different forms of biometrics have applications across a variety of sectors and applications which are rapidly increasing as the technology develops. As a result, biometric technology has progressed substantially in recent years, with increasing performance, faster transaction rates and lower costs (Xiao 2007). Some experts believe that new biometric technologies and applications, such as brainwave biometrics, vascular pattern recognition, body salinity identification, infrared fingertip imaging and pattern recognition, may emerge soon (Asha and Chellappan 2012). Because a biometric sensor will never capture the same data twice, comparing biometric characteristics is an inaccurate comparison; computational intelligence-based techniques may be able to solve this problem in the future. In recent years, various approaches, such as neural networks, fuzzy logic and the evolutionary algorithm, have increasingly addressed complicated biometric authentication and identification issues. Due to rising security demands, technological advancements and decreasing prices, we can expect the development of more biometric applications in the future (Xiao 2007). Although convenient and widely applicable, biometrics may also come with numerous challenges. Therefore, the following requirements should be fulfilled to execute the technology on large scales: high levels of accuracy and performance under varied operating conditions and user composition; sensor compatibility; a fast collection of biometric data in difficult operation settings; low failure-to-enrol rate, high degrees of privacy and template protection; and protecting and securing supporting information systems (Jain and Kumar 2010).

2.8   Bioplastics

Plastics constitute a huge part of people's lives because they are used everywhere. Moreover, they are employed in a variety of industrial sectors, from chemical to car. Plastics' chemical structure may be altered to create a variety of strengths and forms to get a larger molecular weight, low reactivity and long-lasting material; hence, synthetic polymers are advantageous. Bioplastics are simply plastics that are created from plants or other biological sources rather than petroleum. These can be categorised into biobased plastics, which are created at least partly from biological stuff, and biodegradable

plastics, which microbes can partially or totally break down in an acceptable time scale under particular conditions.

Properties of common types of plastics are explained in Table 2.1.

The properties of these plastics are explained in Table 2.1. While they are useful, they pose a threat to the environment. Unfortunately, 34 million tonnes of plastic waste were produced per year, and 93% of the material was dumped into oceans and landfills (Mekonnen et al. 2013). Plastics dissolve in nature extremely slowly. This causes the products to pollute the environment throughout the years. Biobased plastics, which is organic material from animals or plants, is a better alternative than petroleum-based plastics. Biobased plastics can dissolve in nature faster when suitable conditions are provided. Its process depends on environmental conditions such as temperature, materials and application. Biodegradation is when microorganisms found in nature convert materials into natural substances with a chemical reaction (Kerry and Butler 2008). These are important features for plastics to prevent negative environmental impacts. In Fig. 2.18, plastics are divided into four according to these characteristics:

The shaded part in Fig. 2.18 represents bioplastics. Bioplastic is described as a plastic substance that is either biobased or biodegradable or has both qualities (European Bioplastics 2020). Bioplastics can derive from biomass such as sugar cane or cellulose; this process is called biobased (Gill 2014). Renewable biomass resources derived from natural biopolymers (e.g. carbohydrates, proteins and recovered food waste) are used to manufacture bioplastics (Brodin et al. 2017; Yamada et al. 2020). Besides the biopolymers, microalgae is a strong biomass source to produce bioplastic. *Chlorella* and *Spirulina* were the most common algae species used in manufacturing biopolymers and plastic blends. Bioplastic can be made from by-products of high-value chemical manufacturing from microalgae, as per a biorefinery approach (Onen Cinar et al. 2020).

The increasing demand for plastic use day by day, and due to the degradation time of some plastics in nature being longer than 100 years

Table 2.1 Types of plastics

Plastic	Properties
PE	Polyethene has a long chain of carbon atoms. For instance, CH_2
Biobased-PE	Bio-ethylene is converted to bio-PE using a traditional catalyst-driven polymerisation process
PET	The polyester family of polymers includes polyethene terephthalate, which is a general-purpose thermoplastic polymer
PA	Polyamides are known as nylons. Such kinds of plastics are stiff; therefore, they resist bending and abrasion
PTT	Polytrimethylene terephthalate is a synthetic material. Condensation polymerisation or transesterification is the procedure used to make it
PLA	Polylactic acid is polyester made from renewable biomass, most commonly fermented plant matter
PHA	Polyhydroxyalkanoate is a kind of polyester created by bacterial fermentation in nature
PBS	Polybutylene succinate is a biodegradable plastic that dissolves into water and carbon dioxide when it interacts with microorganisms in the soil
PBAT	Polybutylene adipate terephthalate is a biodegradable polymer that decomposes due to the action of naturally occurring microorganisms
PCL	Polycaprolactone is a biodegradable polyester that has a very low glass transition temperature
PP	Polypropylene is a synthetic polymer made from natural resources

Di Bartolo et al. (2021)

Fig. 2.18 Classification of plastics. (European Bioplastics 2020)

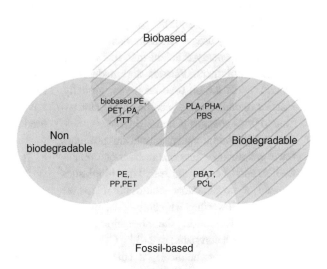

(only an estimate due to lack of time), this emphasises the need to either reduce the use of plastics (which should be legally encouraged) or replace non-degradable plastics with plastics which are degradable and are produced in a sustainable way (Karan et al. 2019). According to 2019 data, bioplastics constitute approximately 1% of the 360 million tonnes of plastic produced annually. But as more complicated materials, applications and products emerge, the demand for bioplastics is increasing and the market is growing rapidly. The bioplastic industry has become a young and innovative sector, both economically and ecologically, in the name of a sustainable bio-economy, as it uses its existing resources more efficiently and produces lower carbon emissions (European Bioplastics 2020). Biodegradable plastics have a bright future ahead of them. The following are some of the benefits of bioplastics (Bezirhan Arikan and Ozsoy 2015):

- Carbon footprints are reduced. It should be emphasised that the carbon footprint of a bioplastic is highly reliant on the plastic's ability to permanently store the carbon that the growing plant extracts from the air. Plastic generated from a biological source sequesters CO_2 emitted by the plant's photosynthetic process. This sequestration is reversed if the resultant bioplastic degrades back to CO_2 and water. However, a permanent bioplastic that is meant to be like polyethene or other common polymers permanently retains CO_2. Even if the plastic is recycled numerous times, the CO_2 collected from the atmosphere remains retained (Chen 2014).
- Independence. Bioplastics are made from renewable resources such as maize, sugar cane, soy and other plants, as opposed to conventional plastics, which are made from petroleum (Yu and Chen 2008).
- Efficient energy usage. The manufacturing process is more energy-efficient than that of conventional polymers. Plastics are made from around 4% of the oil used each year globally. Plastics manufacturing becomes more susceptible to price changes (Chen 2014).
- Eco-safety. Additionally, bioplastic produces fewer greenhouse emissions and is contaminant-free. Yu and Chen demonstrated that bioplastics provide a considerable contribution to the objective of lowering GHG emissions, emitting just 0.49 kg CO_2 during the fabrication of 1 kg of resin. When compared to 2–3 kg CO_2 equivalents from petrochemicals, it equates to an approximately 80% reduction in global warming potential (Yu and Chen 2008).

Today, bioplastic is used extensively in four industries. Figure 2.19 summarises the use areas as follows (Bezirhan Arikan and Ozsoy 2015):

Recent increases in crude oil costs and the potential market for agricultural resources in the area of bioplastics give a push to utilise ecologically acceptable alternatives to materials generated from fossil fuel sources. Thus, bioplastics

have established a new study topic for scientists by providing a viable option for global sustainable development ("Bioplastics – are they truly better for the environment?" 2018). The environmental effects caused by normal plastics, which are not biodegradable for a long time, have pushed scientists to develop materials that are produced from natural sources such as plants, bacteria and biomass and can dissolve in nature in a short time. New developments may lead to increased productivity in production and new opportunities in the field of bioplastics in the future. In addition, since microorganism biology can be applied and commercialised in different sectors such as agriculture, medicine and pharmacy, it also provides an opportunity for bioplastic production. Therefore, current guidelines for the manufacture, usage and disposal of bioplastics should be defined. Labelling regulations should be designed following the emission values of the items, the raw material used and the energy used. Recent advances in technology, global support and continuous innovation are important for promoting and commercialising bioplastics. Here, instead of competing with traditional materials, bioplastics should aim to increase their usage rates over time (Sidek et al. 2019). Bioplastics have some challenges which should be considered for future implementations:

- It is accepted that normal plastics' costs are lower than bioplastics. Nevertheless, when the production of bioplastics increases, the cost is expected to decrease (Bezirhan Arikan and Ozsoy 2015).
- If bioplastic and normal plastic are not distinguished from each other, confusion occurs in the recycling process (Bezirhan Arikan and Ozsoy 2015).
- Bioplastics are produced from renewable sources, which means that sources could be reused repeatedly (Lagaron and Lopez-Rubio 2011).
- Some terms can be misunderstood. For example, bioplastics are known as compostable. This feature confuses people's minds. They think that they can produce bioplastics from food, but the truth is that the production of

Packaging Industry	Packaging products such as shopping bags, garbage bags, bottles, labels, packaging films, cushioning packaging materials have a serious importance in the plastics industry.
Textile Industry	Unlike plastic fibres, which have disadvantages such as static electricity and poor air permeability, fiber made from bioplastic has a better feel and air permeability.
Manufacturing Industry	The bioplastic can be used in the manufacturing industry, for children's toys and home interior decorations. Children always put toys in their mouths; however, chemical plastics contain toxic substances. In this respect, bioplastic is much safer than general plastic.
Medical Industry	Currently, bioplastic is used in the medical industry for medical bone nails and tissue skeletons to avoid multiple surgeries for patients in a convenient and acceptable way.

Fig. 2.19 Bioplastics industries

bioplastics is an industrial implementation (Warren 2011). Manufacturers are responsible for that misunderstanding because they try to make their products more attractive on the market (Bezirhan Arikan and Ozsoy 2015).

- Due to the lack of legislation, many countries produce bioplastics despite the absence of any law and legislation. The number of bioplastics produced is expected to increase from day to day; however, the lack of legislation makes it difficult to accurately monitor (Bezirhan Arikan and Ozsoy 2015).

A variety of assays may be performed to evaluate how bioplastics degrade. It is critical to establish worldwide standard techniques that are comparable. Regrettably, existing standards have not been equated and are mostly applied in the nations where they were developed. All details must be standardised as soon as possible. For the manufacture, use and management of bioplastic waste, a new guide and standard should be developed specifically for bioplastics. In addition, labelling laws might be modified depending on a product's raw material usage, energy consumption, manufacturing and use emissions (Bezirhan Arikan and Ozsoy 2015).

2.9 Biotechnology and Biomanufacturing

In the most general sense, biotechnology can be defined as the synthesis of modern technology and naturally existing biological processes. Biotechnology, or biotech, uses biological systems and/or living organisms to develop new technological instruments, products and machines across a wide range of fields and disciplines. There are three main branches of biotechnology: genetic engineering, protein engineering and metabolic engineering (Gavrilescu and Chisti 2005). Biomanufacturing is a type of manufacturing that uses biological systems (such as living microorganisms, resting cells, animal cells, plant cells, tissues, enzymes or in vitro synthetic – enzymatic – systems) to make commercially feasible biomolecules for use in the agricultural, food, material, energy and pharmaceutical industries (Zhang et al. 2017). Despite its connotation as some of the most advanced technology in the contemporary world, biotechnology has existed for centuries and even millennia, albeit in simpler forms than those we know today. Two often-cited examples of early biotechnology are bread bak-

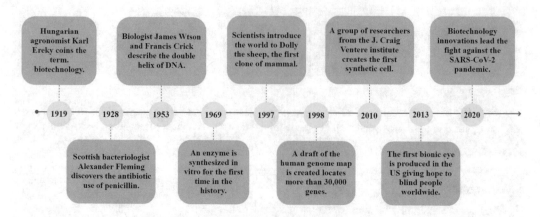

Fig. 2.20 Other milestones of biotechnology. (What is Biotechnology? Types and Applications – Iberdrola 2021)

ing and beer fermentation. Both of these processes combine the naturally occurring biological properties of wheat and yeast and use human intervention and technology to create the desired product. As scientific knowledge expanded, so did humanity's ability to understand more complex biological processes and eventually intervene in them to create new technologies.

Modern biotechnology and biomanufacturing began with the discovery of proteins in 1830. In the twentieth century, revolutionary knowledge on the structure and process of DNA allowed for rapid advancement in biotechnology. The discovery of DNA's structure helped to understand human genetic code while contributing to the foundation of genetic engineering and recombinant DNA technology (Khan 2011). By 1976, scientists developed the first working synthetic gene. This development was followed by recombinant human insulin and human growth hormone production in the late 1970s. By 1994, the susceptibility gene for breast cancer was discovered. The year 1997 marked another significant milestone when the first clone of a mammal, a sheep named Dolly, was created. Additionally, a remarkable development was achieved in 2008 when a blue rose was developed through genetic modification. Over the past few decades, there have been several other discoveries in biotechnology that could be classified as milestones, such as the invention of antibiotics and the application of selective breeding in plants and ani-

mals, which led to better crop and livestock production (Khan 2011). Other milestones of biotechnology are represented in Fig. 2.20.

As biotechnology developed and its applications increased, the field was divided into seven branches coded by seven colours, shown in the list below (Iberdrola 2021).

1. *Red:* Biotechnology in the medical field. Stem cells and gene therapy are clear examples of red biotechnology. DNA fingerprint technique is another prominent application.
2. *White:* Biotechnology applied to industry. Examples include the production of insulin to treat diabetes and the development of new enzymes. Pharmacogenomics, regenerative medicine, nanobiotechnology and biopharmaceuticals are other emerging fields of biotechnology that could be classified as both red and/or white.
3. *Green:* Biotechnology in the agricultural sector. Examples include genetically modified crops, bacteria-based plant fertilisers and pest-resistant grains.
4. *Grey:* Biotechnology is used to protect the environment. It has emerged to prevent environmental contamination and sustain the ecosystem (Khan 2011). One example is microbes that digest oil.
5. *Blue:* A defunct category for biotechnology applications in the ocean. An example application of this area is wound treatment.

6. *Gold:* Biotechnology is responsible for gathering, storing, analysing and separating biological information, particularly that linked to DNA and amino acid sequences. It is also known as bioinformatics.
7. *Yellow:* Biotechnology focused on food production. It is now being used to lower saturated fat levels in cooking oils, for example.

The types of biotechnology can also be more simply classified as animal, agricultural, medical, industrial and environmental biotechnology. Figure 2.21 represents the types of biotechnologies, and applications of biotechnology are shown in Fig. 2.22.

Biotechnology has allowed for unprecedented possibilities and potential for cures and treatments. For instance, preventative therapies are executed using medical biotechnology, which generally refers to harnessing living cells to develop pharmaceutical treatments and cures for a range of diseases and conditions. These types of technologies have been transformative in the prevention and treatment of several types of cancer (Pham 2018). Medical biotech has also played a significant role in reducing the impact of infectious diseases. The mRNA vaccines developed in 2020 to protect humans against the COVID-19 virus provide an excellent example of biotechnology developed in response to a threatening infectious disease (Jackson et al. 2020). Biotechnology has also allowed for earlier and more accurate diagnoses of diseases that are caused by genetic factors as technology now allows for genomics to analyse patients' genetic sequencing and look for risk factors and/or existing conditions based on DNA (Pham 2018).

Another revolutionising biotechnological development is genetically modified crops. Many of these genetic interventions were developed to increase food production and profits for the companies developing them. Genetically modified crops may quickly become a worldwide necessity as climate change transforms the farmland needed for agriculture and demands crops with the ability to survive droughts, increased heat, increased storms, etc.

There are numerous other products in the biotechnology area:

- Cosmetics and personal care items: Biotechnologically textured products are highly pure, non-irritating, smoother, less greasy and environmentally sustainable.
- Bread: Genetically engineered microorganisms allow for longer shelf life for higher quality bread while helping to remove potassium-carcinogenic bromate.
- Vitamin B2: Genetically improved microorganisms generate this fermentation in a single

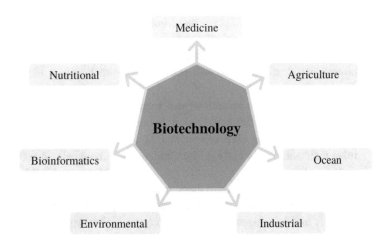

Fig. 2.21 Types of biotechnology. (Khan 2011)

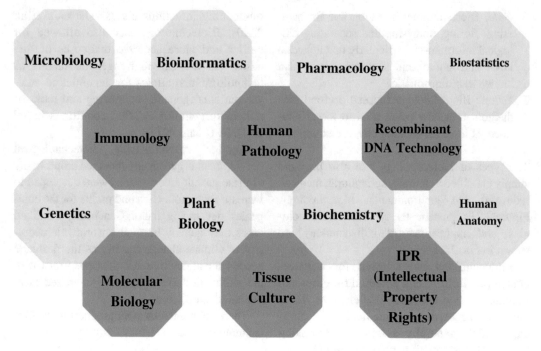

Fig. 2.22 Fields working with biotechnology. (Khan 2011)

step, which reduces CO_2 emissions and energy consumption by 33%.

- Diapers: A biodegradable polymer is created using *Bacillus*, which is environmentally and skin-friendly.
- Detergent: Proteases, amylases and lipases, which are within the biotech enzymes, assist in saving energy and cleaning clothes luminously.
- Tissue paper: With wood bleaching enzymes, the pulping process is faster, more environmentally friendly and cost-effective.
- Textiles and stonewashed jeans: The usage of biotech cellulose generates soft textiles with decent colours.
- Foam and nylon: The biotech-treated foam is utilised to create furniture and nylon to manufacture softer grade fibre tapestries that are more resistant to stain and UV light than ordinary nylon.
- Plastics: Biodegradable polymers produced with *Bacillus* microbes are highly beneficial in foods and beverages and are environmentally benign in the production of food services and containers.

- Synthetic rubber: Natural compounds are polymerised into synthetic rubber and elastomers, which are highly pure and cost-effective (Saxena 2020).

Biotechnology has always played a significant role in human life for millennia, with its importance only becoming greater over time. Medical biotechnology is already serving more than 350 million patients around the world through the treatment and prevention of everyday and chronic ailments. New fields of study emerge as technology and knowledge develop, including nanobiotechnology, bioinformatics, pharmacogenomics, regenerative medicine and therapeutic proteins (Khan 2011). Additionally, the use of recombinant organisms will have a wide range of applications, including new vaccines, solvents and chemicals. Another area that will gain importance is biochips, which are relatively more energy efficient compared to silicon chips. These products could have an impact on hormone secretion and heart rate. Gene therapy is another field that will have a rapidly increasing demand. With the help of this application, genetic diseases or

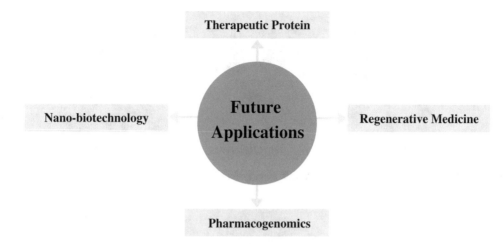

Fig. 2.23 Future Applications of biotechnology

disorders can be prevented by implementing a healthy, mutated gene (Khan 2011). Furthermore, with their environmental enhancement impacts, biotechnology and biomanufacturing will have key roles in the adoption of sustainability on a long-term scale. In this context, biological production systems, as one of the applications of biomanufacturing, are appealing because they employ fundamental renewable resources to generate a wide range of compounds using low-energy processes (Gavrilescu and Chisti 2005). Instead of finite and volatile fossil fuels, industrial biotech can assist in the fight against climate change by providing an alternative and safer source of global energy. It has resulted in significant reductions in greenhouse gas emissions and aims to reduce 2.5 billion tonnes of carbon dioxide emissions per year by 2030 (*Report on Industrial Biotechnology and Climate Change: Opportunities and Challenges – OECD* 2011). Figure 2.23 shows numerous future biotechnology application areas.

Another promising area of biotechnology is synthetic food. The effective integration of food science and synthetic biology is a key technology for addressing current food safety and nutrition issues and a key approach for overcoming the sustainability concerns associated with traditional food technology. It may be possible to eliminate the disadvantages of traditional agriculture while boosting resource conversion effi-

ciency by incorporating synthetic biological technologies into future food. In general, the synthetic biology-driven food sector has the potential to address future food supply issues. A future food revolution powered by synthetic biology is possible in three stages. Firstly, synthetic biology can enhance traditional food production and processing. Secondly, synthetic biology can improve food nutrition or provide new functions. Thirdly, using created microbial communities in synthetic biology can change the conventional fermentation food production method (Lv et al. 2021). As a result, biotechnology is promising. We can imagine a society free of cancer, AIDS and Alzheimer's disease, as well as a world with sustainable development that addresses the energy, food and environmental demands of an ever-growing population without jeopardising either Earth's resources or future.

2.10   Blockchain

Blockchain can be defined as a technology protocol that enables data sharing with trust-based transactions such as identification and authorisation in a decentralised-distributed network environment without the need for approval or control of a central authority. In Fig. 2.24, the decentralised structure of the blockchain system is compared with other systems.

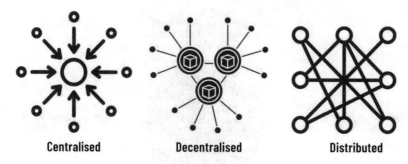

Fig. 2.24 Comparison of the blockchain system with other systems

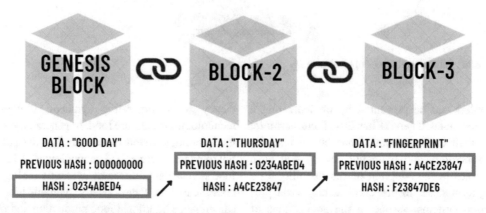

Fig. 2.25 Blockchain technology working principle

Blockchain is also defined as a database system that provides decentralised and intermediary transactions, allowing the blocks in which data is stored to be processed, stored and arranged in temporal and linear order. Produced blockchains are traded on a decentralised network structure. Due to this dispersed network structure, it becomes impossible to make theoretical changes to the data. This also helps to build confidence in the data. The distributed network structure makes the system more secure. These scattered data can be open source or closed source.

Blockchain technology has exciting implications for the future potential of decentralised structures, which includes a trustworthy and transparent trading infrastructure for peer-to-peer energy trading (Küfeoğlu 2020). Figure 2.25 shows how blockchain technology works.

Blockchain does not offer editing on legacy data. Therefore, it works not like traditional databases but as a digital ledger where transactions are listed one after the other. In the blockchain,

transactions are stored in blocks, and each newly created block refers to the previous block with a unique identification number called a "hash". Because each block is cryptically linked to the previous one, changing any of them changes all subsequent blocks.

Within the global and fast-paced system, it is seen that both companies and countries are trying to catch up with digital transformation with their efforts to expand their field of activity. The most important feature that ensures its reliability is that a saved database cannot be changed or corrected again; with this aspect, it would not be wrong to say that blockchain is not a database but a data recording system. In this way, users can connect to the network, perform new transactions, verify transactions and create new blocks without intermediaries. Blockchain technology proposes several advantages such as increased security, transparency, high speed, low cost and decentralised nature.

(i) *More Secure*

Blockchain technology provides a layer of security by using cryptography to the data saved on the network. The decentralised aspect of blockchain provides superior security because it is combined with encryption. Blockchain is a decentralised and cyberattack-resistant database. It is not possible to change the history of the ledger or send the same transaction twice (i.e. double-spend) as every transaction ever made on the network is recorded and stored permanently. This certainty creates mutual trust. Congestion management using electric vehicles in grid services; energy data registration in a secure medium as an open ledger; and billing, switching providers, swapping capacities and so on are just a few examples of how blockchain technology can be employed (Dena 2019).

(ii) *Transparency*

Everyone, not just its users, has access to the blockchain database. As a result, the control is visible. A block's transactions, wallet addresses, transaction ID (shipping code) and quantities may all be viewed by anyone. Open blockchain networks are truly "open". For example, in the Bitcoin network, it is possible to see all the blocks created to date and the money transfers in them. All transaction information can be accessed via Blockchain.com.

(iii) *High Speed and Low Cost*

Due to its fast and low cost in health, food, forensic cases and keeping records of international companies, blockchain technology leaves traditional methods behind one by one. The most important reason for this is the transfer of data directly from one user to another quickly and cost-effectively. Transactions, especially international transactions, take seconds rather than weeks to complete.

(iv) *Decentralised*

The revolutionary feature behind blockchain is that transactions are completed not one by one but by many computers at the same time. All computers reside on the same network called a peer-to-peer network (P2P). This model is often referred to as the "distributed trust model". Figure 2.26 shows the advantages of blockchain in general.

Three features of blockchain technology come to the fore. They are a distributed architecture that ensures that a copy of each data is kept on thousands of nodes in the network. The transparency allows for the tracking of all transactions made on the network. Immutability that prevents processing of the data produced to the blockchain. With these features, blockchain is a candidate to be the backbone of the new Internet structure.

Blockchain offers groundbreaking technologies that have the potential to change the Internet and even the world for many reasons. Participants can share excess energy and purchase or sell carbon credits using blockchain's revolutionary P2P energy trading platforms (Küfeoğlu 2020). One thing is certain: blockchain technology is ushering in a new era of digital information sharing, as

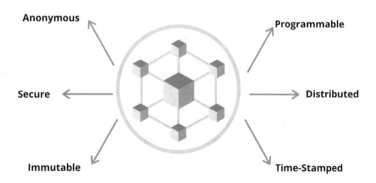

Fig. 2.26 Advantages of blockchain

well as a novel means of data storage and transaction representation. Blockchain technology offers alternative methods to solve the heavy paperwork process, delays, incorrect transactions, high costs, fraud and many more problems on the logistics side. It is possible to design functional systems by integrating international trade with blockchain technology. Over $1 billion was invested by venture capitalists in 215 blockchain-based firm deals in 2017 (CBINSIGHTS 2018). The technology is highly promising in delivering a secure and trustless transaction and data storage medium. By 2027, it is predicted that blockchain would have stored roughly 10% of the global gross domestic product (World Economic Forum 2015). The technology can be utilised in a wide range of areas and businesses such as banking and payments, data security, voting and elections and energy and distribution networks.

(i) *Banking and Payments*

There are still many obstacles to a perfect financial sector, whether it is identification or fraud difficulties in developed countries or security issues in areas where technology is not widely used. Blockchain has the ability to tackle these issues in a revolutionary way, benefiting every element of the industry.

(ii) *Data Security*

Messages and data are encrypted with a cryptocurrency that uses a public key infrastructure (PKI). Personal information is less likely to be revealed and replicated via this way.

(iii) *Voting and Elections*

Blockchain has the potential to play a significant role in digital transformation, allowing citizens to vote from the comfort of their own homes or from anyplace else. Vote stealing may be prevented at every level with fast and precise verification and accurate vote counting. It will be impossible for a person to vote with personal identification numbers more than once. Systematic breaches will be detected quickly because of the distributed ledger.

(iv) *Energy and Distribution Networks*

Blockchain enables people to share extra energy and buy or sell carbon credits through revolutionary peer-to-peer energy trading platforms. This paradigm shift towards decentralised local energy exchange via peer-to-peer (P2P) will drastically minimise transmission losses while also deferring costly network upgrades. Unlike centralised architectures, the blockchain distributed ledger does not require the intervention of third parties to preserve the system's integrity and security. Blockchain is a distributed ledger that employs automated technology to create smart contracts that improve cybersecurity and optimise energy operations, lowering transaction costs considerably (Küfeoğlu 2020). Individual customers may be able to swap electricity and make payments in a frictionless manner thanks to blockchain's decentralised transaction verification. Better network and congestion management and the challenge of renewable generation intermittency can all be aided by digitalisation, allowing for more efficient network operation and more effective network monitoring (Küfeoğlu et al. 2019).

Blockchain technology, as a unique technology that creates a new consensus process to eliminate a single central authority, could be very valuable in energy trade (Küfeoğlu et al. 2019). Some peer-to-peer energy marketplaces are built on the blockchain platform, which allows all transactions to be authenticated and stored permanently without the need for a central authority (Küfeoğlu et al. 2019).

2.11   Carbon Capture and Storage

The concentration of uncontrolled dispersed CO_2 may cause crucial climate change as an increase in average global temperatures. It is estimated that carbon dioxide emissions around the globe will be higher in 2050 than they were in 2018. Forecasts indicate that CO_2 emissions will increase gradually every 5 years and will exceed 40 billion metric tonnes from 2045 (Statista 2019). Carbon capture and storage (CCS) tech-

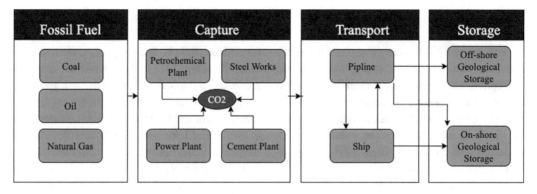

Fig. 2.27 Carbon capture and storage phases

nology has emerged to mitigate climate change by reducing CO_2 emissions (Singh 2013). The first large-scale project on CO_2 capturing was planned to enhance produced oil by injecting CO_2 and increasing the pressure on the oil reservoirs in the 1970s (IAE 2016). The increase in oil and gas production through injected captured CO_2 has revolutionised the energy sector, and the importance of carbon storage has also expanded. Although CCS is used to increase production, it has become one of the agenda items of many countries since it is known that the long-term effects of CO_2 released after this production on global climate change will be substantial. While CO_2 emissions dropped dramatically by 5.8% in 2020, the demand for fossil fuels in 2021 seems to reverse this situation. A 4.8% increase in CO_2 emissions is expected in 2021, with demand for coal, oil and gas recovering after the temporary impact of the pandemic (IAE 2016). Most of the CO_2 emissions are caused by the use of fossil resources. The intensive use of coal, natural gas and oil in power plants, industrial facilities, vehicles, residences and workplaces to meet energy and heating needs are the biggest cause of the emission problem. The use of these resources releases large amounts of greenhouse gases into the atmosphere. Increasing livestock farming, cutting down trees, etc. are other causes of increased CO_2 emissions. However, their impacts are not as notable as fossil fuels and applying CCS for these causes ensures emission reduction in the longer term. In this regard, CSS methods are mostly focused on fossil fuels. Figure 2.27 illustrates the CCS stages in detail.

Many research and development projects for CCS have been established among the presented cases in many countries. Lawmakers primarily aim to reduce carbon emissions for measures related to climate change. Governments and companies have adopted carbon-neutral methods to reduce this impact of fossil fuels, which have an equal balance between emitting carbon and absorbing carbon from the atmosphere in carbon sinks (European Parliament 2019). In this direction, incentives and orientations towards carbon-neutral technologies are increasing, together with regulatory methods such as carbon tax and carbon footprint monitoring. However, carbon-neutral methods are still not enough in line with climate targets. Countries that have committed to zero carbon emissions until 2050 at the Paris Climate Summit have started to use carbon-negative methods and carbon-neutral methods to achieve this goal. Carbon-negative methods aim to reabsorb more carbon dioxide released into the atmosphere than disseminated (IEA 2020). Although it is difficult to withdraw more carbon than is emitted today, carbon-negative methods have a significant role in reducing the amount of carbon in the atmosphere. Among the most common carbon-negative methods are carbon capture and storage technology.

(i) *Carbon-Capturing Methods*

The main CO_2 capturing technologies are chemical and/or physical absorption, physical adsorption and membrane separation. Carbon capture is a technology with multiple methodologies, but its basic logic is based on the separation of free CO_2 from the air. The capture of CO_2 could be classified into three types, i.e. post-combustion capture, pre-capture and direct capture (Kuckshinrichs and Hake 2015). The fuel is not directly burned in pre-combustion capture but transformed into synthesised gas at the appropriate temperature and pressure. Afterwards, CO_2 is transformed into carbon dioxide and H_2, and CO_2 is collected for H_2 as the fuel (Rackley 2017). CO_2 is captured from the industrial process waste into a nearly pure CO_2 stream in post-combustion. (e.g. cement plant flue gases). In direct capture, pure CO_2 is captured directly from the air (e.g. mineralisation of steel slag) (Goel et al. 2015). The separated CO_2 could be used for different purposes such as soda ash production, oil drilling and alternative energy sources production. Thus, the storage of captured carbons has great significance. Figure 2.28 articulates these carbon capture methods.

Geological storage is the most used method to store carbon. First of all, carbon is captured in different ways. Then it is injected into different geological forms like oil and gas reservoirs, saline forms, unmineable coal seams and basalt formations. CO_2 is stored by impermeable cap-rock (Rackley 2017). By significant developments of technologies in the injection industry, geological storage is in oil and gas reservoirs preferred by companies to store carbon. Stored CO_2 is also utilised for unmineable coal because its molecules could easily interact with the coal surface. The main problem for geological storage is transportation cost. Although it is not necessary to construct storage facilities, it is important to invest in transportation infrastructure to deliver CO_2 to these facilities. Ocean storage is one of the greatest storage since oceans are major candidates to store captured CO_2. However, storage must be in-depth not to release CO_2 into the atmosphere. Direct dissolution is sent by ships with pipes to deep waters at supercritical fluid, and it creates CO_2 lakes in the depth of the oceans. Creating CO_2 lakes in the oceans is an efficient method for long-term storage (Rackley 2017). Mineral storage is suitable with the law of thermodynamics so it can occur in nature without any application by humans. CO_2 reacts with metal oxides and produces stable carbonates. This process takes over a long timescale in nature. When this process operates at higher temperatures and pressure, it can be accelerated, creating energy costs (Rhodes 2012).

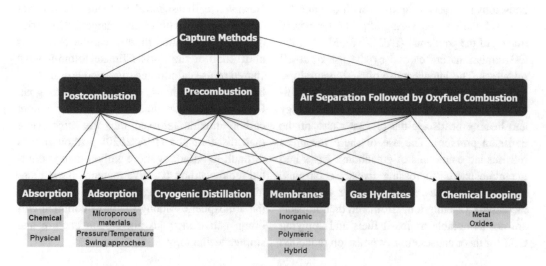

Fig. 2.28 Carbon capture methods. (Creamer and Gao 2015)

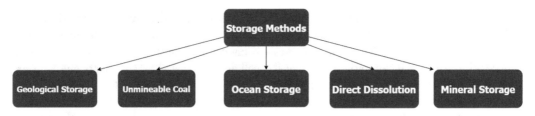

Fig. 2.29 Carbon storage methods

Although CCS is a useful technology, it has not received the necessary response in the industry due to its cost. The high unit and technology costs are among the main reasons why the private sector does not invest in this field. Incentives from governments, reduction in technology costs and private sector investments in this field play an important role in the future of CCS technology. In sectors where carbon capture is easy, and CO_2 is heavily emitted, CCS technologies perform more efficiently and more economically. Examples of these are the cement, iron and steel and refinery sectors. In addition, decomposed CO_2 has great importance, and it needs to be used or stored without being released back. The most suitable areas for storage are oil reservoirs. In addition, this storage method is used to increase oil production by pumping it into the oil deposits stuck between the layers. Figure 2.29 illustrates the abovementioned storage methods.

The efficiency of carbon-capturing technology may be as high as 90% when it comes to capturing the CO_2 that is released from industrial processes. These industrial processes are in a broad spectrum, from the energy sector to manufacturing and construction. Among all, the energy sector is covering the highest percentage when it comes to carbon-emitting, by almost 40%, when examined globally (Leung et al. 2014). Steps of capturing, storing and utilising carbon are crucial to the environmental act against global warming by reducing and preventing the negative effects at some level. Overall carbon emissions may be negative with a combined strategy of capturing carbon and utilising biomass. In addition to the contribution of the global warming fight mentioned above, calculations indicate that CCS's contribution to emission reduction attempts may be needed to reach a percentile of 14 to keep the globe's temperature at a certain level of 2 °C or less by the year of 2050.

Renewable energy, which has become increasingly important, cannot meet the rising energy demand on its own, and it seems that fossil fuels will still play an important role. According to Jackson, although renewable energy is now considered the most cost-effective source of power generation worldwide, the growth in energy demand and the growth needs of governments indicates that fossil fuels will continue to play an active role (Jackson 2020). In the future, CCS technology appears to grow more and become crucial in terms of maintaining the environmental balance by absorbing CO_2 while supplying energy demand from fossil fuels. Aiming to benefit from the ecological and economic advantages of CCS, the UK government will provide £1 billion in funding to support the development of four CCS centres and cluster projects across the UK by the end of the decade (Kelly 2020). Furthermore, the Norwegian climate solution will heavily rely on CCS with developed versions of the technology. Equinor, Shell and Total are investing in the Northern Lights project, which is Norway's first CO_2 storage licence and a key component of the Norwegian government's "Longship" strategy (Equinor 2020). Canada is another country investing in Direct Air Capturing (DAC) technology. *Carbon Engineering* states that DAC technology can be scaled up to remove up to 1 million tonnes of CO_2 per year from the atmosphere in Canada (Jackson 2020). In addition to recent developments mentioned above, Climeworks, one of the CSS companies, expands its direct air capture and storage technology, making a permanent CO_2 removal solution more readily available. It includes the infrastructure and foundation of the next generation of Climeworks CO_2 collectors.

Their solution includes the installation of plants and machinery in Iceland and is expected to be completed in spring 2021. What makes Climeworks' use of DAC so intriguing is that it can capture pollutants straight from the air, rather than merely eliminating emissions connected with electricity generation, and also this is the company's largest facility to date, with a CO_2 collection capacity of about 4000 tonnes per year (Jackson 2020). As a result, CCS technology will gain even more importance and develop in the coming years. CCS seems like a viable solution to address the climate change problem.

2.12   Cellular Agriculture

Throughout the years of extortion of natural resources, possible new solutions to neutralise the effects started to emerge. First, a decrease in consumption was suggested. Other suggestions included using less water to preserve the resources, less plastic and more biodegradable resources to prevent pollution and so on. Among many topics, the consumption of meat has always been on the agenda. Setting aside the ethical arguments, meat consumption is not sustainable in many ways and causes substantial global warming. At this point, cellular agriculture steps forth as a remedy. This technology focuses on the production of an animal product-like substance that does not harm the environment in the process of production or consumption. This substance is produced from cell structures rather than animals. Using advanced biotechnology, tissue-based products like meat (of fish, cattle, sheep, goats, chicken, turkey, etc.) and protein-containing foods like eggs and dairy products become available alternatives. Focusing on both sustainable food production and the safety of food, cellular agriculture uses many techniques to imitate and meet the standards of nutritional values while being almost carbon neutral.

A. *Cell Cultivation*

Cell culture takes cells from a plant or animal and grows them in a controlled environment. The cells can be extracted directly from the tissue and crushed enzymatically or mechanically before being cultured, or they can be replicated from a cell line or cell strain. The primary culture phase begins after the cells have been separated from the tissue and multiplied under optimal circumstances until all substrates have been occupied. To provide more vacancies for growth, the cells must be subcultured by transferring them to a new case containing fresh growth. The primary culture is referred to as a cell line or subcloned after the first subculture. Cell lines are reproduced from primary cultures which have a limited lifetime, and when they are crossed, cells with the highest growth capacity dominate. As a result, genotype and phenotype will be uniform (Invitrogen and Gibco 2021). This cell line becomes a cell strain when a subclone's subpopulation is positively selected from the culture through cloning. Following the commencement of the parent line, a cell strain frequently acquires further genetic changes. Culture conditions are variable for each cell type, but the environment generally contains essential nutrients, hormones, enzymes and gases that are necessary for substrate (Invitrogen and Gibco 2021). Figure 2.30 presents the stages of cellular agriculture with cell cultivation.

B. *Precision Fermentation*

The second method in cellular agriculture is precision fermentation. This method uses microorganisms to obtain the proteins in animal products. This method can be explained in four steps:

1. Introduce: Firstly, a gene from a farm animal is introduced, for microorganism production, to a host cell optimised to produce an animal protein kind of yeast. Integrating this gene from farm animals into the host cell will instruct the host cell on how to create specific animal proteins.
2. Feed: Secondly, the production of animal proteins at the cellular level starts after the implementation of those farm animals' genes to host microorganisms' cells. Host cells need nutrients to ensure that the cellular agriculture

Fig. 2.30 The method
of cellular agriculture
with cell cultivation.
(Cellular Agriculture
Society 2021)

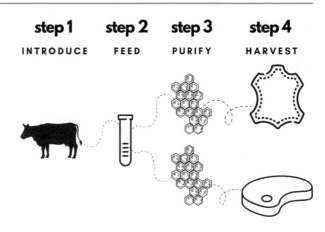

process takes place properly and for its continuation. You need to feed the host cell
nutrients in a controlled environment called a
fermentation cultivator.
3. Purify: After the production of animal proteins is completed, the host production cells
must be removed. Thus, only animal proteins
remain in the final product. Only the final
purified animal proteins remain after the host
cells are separated from the produced
proteins.
4. Harvest: In this final step, animal proteins go
through the post-processing steps. After rigorous testing for safety and quality, a real animal
product is ready to be harvested (Cellular
Agriculture Society 2021).

Figure 2.31 presents the stages of cellular
agriculture with precision fermentation.

It is possible with cellular agriculture to benefit from animal meat without the need for animals. The current system can produce animal
meat that is enough to cover the existing consumption rate, but it will be unable to do so in the
future due to factors such as increasing population. Experts estimate that the human population
will be 9–11 billion in 2050 (UN 2017).
Increasing population means increasing food
needs. Thus it seems that animals alone will not
be enough for humans. Rather than encouraging
consumers to choose plant-based diets more, the
other ideal solution is the innovative improvements in meat production, which stands out as
the task of cellular agriculture. There are three

main benefits of cellular agriculture: benefits for
the environment, benefits for the animals and
benefits for human health. Figure 2.32 demonstrates some benefits of cellular agriculture.

The huge quantity of resources required by
livestock farming has a wide environmental
impact. It is astounding how much water, land
and power livestock use. Khan states that in addition to the 1.6 kg of feed necessary to make a
0.23-kilogram steak, the manufacturing process
necessitates 3515 litres of water and just as much
energy is needed to charge a laptop as much as 60
times. Moreover, different greenhouse gases are
released into the atmosphere containing a total of
4.54 kg of CO_2, which is equal to 2 litres of gasoline. This data represents what is needed to make
an 8-ounce steak and not the entire animal.
Roughly 25% of the surface of the world is dedicated to livestock. This is approximately 70% of
all agricultural land (Khan 2017). Furthermore,
animals consume around 30% of the world's
freshwater. Livestock account for 14.5% of all
emissions of greenhouse gases. Cellular agriculture promises to minimise global greenhouse gas
emissions and to encourage more ethical usages
of natural resources. Cellular farming is an eco-
friendly and sustainable option in comparison
with animal production. Fleece made from cellular farming requires less than 1/10 of the land
and water (Khan 2017). The greenhouse gas
emissions in this beef will also be considerably
reduced. The number of animals required in the
production process is reduced due to cellular
agriculture, eliminating livestock. With the

Fig. 2.31 The method of cellular agriculture with precision fermentation. (Cellular Agriculture Society 2021)

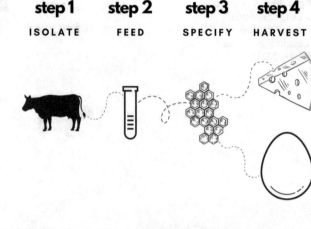

Fig. 2.32 Benefits of cellular agriculture

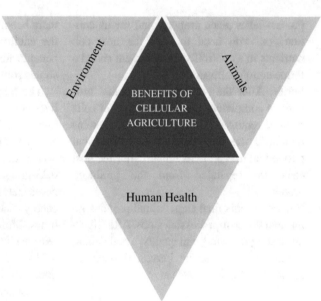

increasing population, the demand for meat is increasing even more. On the other hand, meat production facilities quickly raise animals under unsanitary conditions to produce more meat without due concern given to animal welfare (Khan 2017). Cellular agriculture is among the most effective solutions to fulfil the increasing meat demand while ensuring animal welfare. About 80% of all antibiotics sold in the USA are used in animal agriculture. This situation increases antibiotic resistance in humans and causes various health problems. For instance, most bacterial contaminants, *Salmonella* and *E. coli*, that cause food-borne diseases commonly interact with contaminated animal excrement

(Röös et al. 2017). Bacterial contaminants like *Salmonella* or *E. coli* will not be the case in cellular agriculture since no livestock will contaminate the meat or other goods.

Meat production has long been the subject of controversy. Animal rights activists think that obtaining meat from animals is a massacre, and they look for other ways to meet their protein needs without eating meat. For example, they turn to a vegan and vegetarian lifestyle. However, cultured meat, the most popular topic of cellular farming, could change this. Cultured meat comes as an alternative for its ecological and animal welfare benefits and to feed humanity's growing population. Besides all the environmental bene-

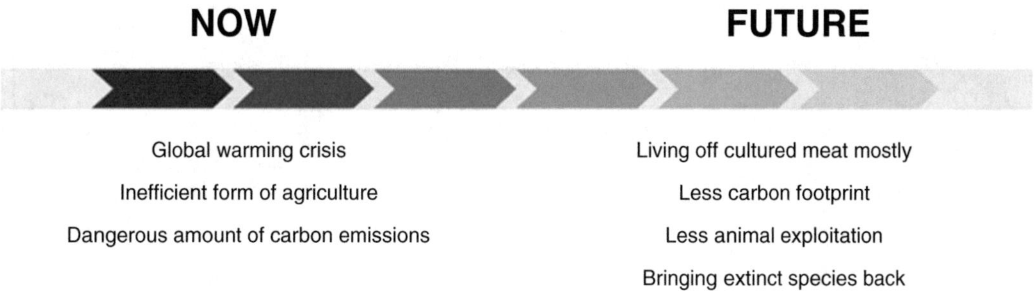

NOW   **FUTURE**

Global warming crisis   Living off cultured meat mostly

Inefficient form of agriculture   Less carbon footprint

Dangerous amount of carbon emissions   Less animal exploitation

                                               Bringing extinct species back

Fig. 2.33 Motivation for cellular agriculture

fits, cultured meat is also healthier than normal meat because it is produced under laboratory conditions in a controlled manner. Cultured meat does not contain bacteria and other disease-causing agents. While plastics/microplastics are seen in normal meats, they are not seen in cultured meats (Gasteratos 2019).

Humanity continues its search for life on another planet. People have long kept their eyes on the Moon and Mars. The long-term missions have been targeted, and there is a basic problem in meeting the protein needs of astronauts. Experts have invested in cultured meat for a long time to solve this problem. If colonies are established on the Moon and Mars in the possible future, keeping human beings alive is planned by using cultured meat. Israel-based company Aleph Farms aims to accelerate cell-based meat production in space. Aleph Farms says that "We want to make sure that when people live on Mars, we'll be there too" (Morrison 2020). In this respect, cellular agriculture is most likely to be on the main agenda in the future.

Gareth Sullivan, deputy director of the University of Oslo's Hybrid Technology Hub, is experimenting with technology to generate stem cells from endangered species such as the northern white rhino. According to Labiotech, working with Ian Wilmut, one of the scientists who cloned Dolly the sheep in 1996, the researcher obtained a fundamental grasp of stem cells. Stem cells from endangered animals are being stored for a future in which technology can bring extinct species back to life. The project received an investment of €220,000 from the Good Food Institute, which conducts vegan and cultured

meat R&D (Smith 2021). Figure 2.33 demonstrates the present and future motivations for cellular agriculture.

2.13   Cloud Computing

Cloud computing is a technology that provides elastic and scalable computing techniques to fulfil information technology capabilities delivered in varying service models through the Internet. Moreover, it is an easy way to share the folders with other people and work by collaborating with them via the Internet from the personal computer or network servers. This sharing can be private or public. Cloud computing technology generally includes many clouds, and these clouds communicate with each other through application programming interfaces and using web services (Mirashe and Kalyankar 2010). Cloud computing is quite popular among researchers, citizens and governments nowadays. One of the reasons for this is that when the memory of personal computers is full, it indirectly inhibits speed and performance. To prevent this, a personal account is created by transferring personal data to computers with a lot of memory via the Internet. Thus, both storage space and computer resources can be acquired without sacrificing the personal computer's memory. Since the main consideration behind cloud computing is to minimise the burden on the terminals of the user, cloud storage, as one of the subdisciplines of cloud computing, comes forwards in line with this purpose. Cloud storage services can provide both data storage and business access. It consists of necessary stor-

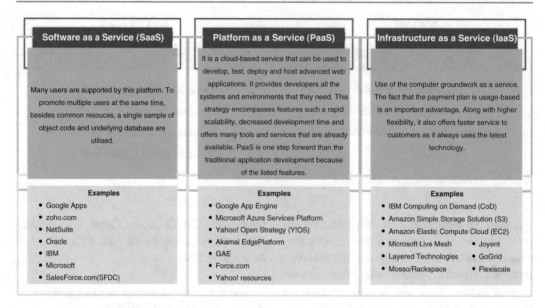

Software as a Service (SaaS)	Platform as a Service (PaaS)	Infrastructure as a Service (IaaS)
Many users are supported by this platform. To promote multiple users at the same time, besides common resouces, a single sample of object code and underlying database are utilised.	It is a cloud-based service that can be used to develop, test, deploy and host advanced web applications. It provides developers all the systems and environments that they need. This strategy encompasses features such a rapid scalability, decreased development time and offers many tools and services that are already available. PaaS is one step forward than the traditional application development because of the listed features.	Use of the computer groundwork as a service. The fact that the payment plan is usage-based is an important advantage. Along with higher flexibility, it also offers faster service to customers as it always uses the latest technology.
Examples	**Examples**	**Examples**
• Google Apps	• Google App Engine	• IBM Computing on Demand (CoD)
• zoho.com	• Microsoft Azure Services Platform	• Amazon Simple Storage Solution (S3)
• NetSuite	• Yahoo! Open Strategy (Y!OS)	• Amazon Elastic Compute Cloud (EC2)
• Oracle	• Akamai EdgePlatform	• Microsoft Live Mesh • Joyent
• IBM	• GAE	• Layered Technologies • GoGrid
• Microsoft	• Force.com	• Mosso/Rackspace • Flexiscale
• SalesForce.com(SFDC)	• Yahoo! resources	

Fig. 2.34 Cloud providers and some example establishments. (Dillon et al. 2010)

age devices, and it huddles all of them together in the application software for usage. As a result, it can be thought of as one of the cloud computing systems responsible for large capacity storage (Liu and Dong 2012). Many companies, such as Microsoft Azure, Amazon Web Services (AWS), Rackspace and GoGrid, provide particular cloud computing services (Chopra 2017). Anyone can benefit from these services for a monthly subscription fee. In other words, this service can be thought of as renting a computer with very high memory and performance far away from the users themselves. It is especially useful because it does not require physical hardware to perform computation or storage. Since the data is stored and the other resources are available on the Internet, the data is always accessible anywhere and anytime as long as there is an Internet connection (Huth and Cebula 2011).

Although cloud computing provides ease of use to users, it is a modular technology with a very different operating system. Therefore, to understand what cloud computing is exactly, it is also important to understand how to choose the cloud providers. However, first of all, it is necessary to talk about what cloud providers are. Each provider offers specific functionality that gives users more or less control, depending on its type.

Therefore, choosing the right provider becomes important. There are three service providers in total. These can be listed in Fig. 2.34 as software as a service (SaaS), platform as a service (PaaS) and infrastructure as a service (IaaS).

According to the user, there are different types of clouds present. The public cloud is accessible to any user that has an Internet connection. Private cloud is created for a group and/or organisation. Only that group and/or organisation can access this type of cloud. Community clouds can be shared between more than two organisations with requirements similar to one another. Lastly, hybrid clouds are formed when more than two clouds with various types or the same type are merged (Abualkibash and Elleithy 2012).

There are a lot of advantages and disadvantages to using cloud computing as all technologies which we use. The disadvantages of cloud computing are: it cannot work without the Internet or in low-speed Internet environments; sometimes the features may not be enough; the stored data may not be reliable because it is opened to the Internet if your password or cloud account is lost (Mirashe and Kalyankar 2010). On the other hand, cloud computing offers various advantages to its users. It makes the idea of connection via the Internet quite appealing over a connection through immovable physical hardware. According to Lewis (2010) and Hurwitz et al. (2012), listed below are some features of cloud computing that describe its prominence for organisations, companies and general users:

- *Availability:* Users are allowed to reach their documents and resources at any time via the Internet.
- *Collaboration:* Users can work at the same time and share the necessary and common information that they need.
- *Elasticity:* Resource utilisation of a user, which can be shaped according to the changing needs, can be transparently managed by the service provider.
- *Lower infrastructure costs:* Public cloud users can choose whichever resource that they want to use, and they pay per use. This results in no payment for the maintenance or upgrade of the hardware and software of the corresponding resource.
- *Mobility:* Networked access to the resources and information makes the attainability of the cloud by the user globally, from the places where there is an Internet connection possible.
- *Risk reduction:* Some check tests and demonstrations applied in the cloud can help organisations to examine their ideas and concepts, deciding whether to invest in that technology or not without incurring big upfront expenses.
- *Scalability:* A variety of resources supplied by cloud providers can be scaled according to the demands of the corresponding cloud users.
- *Virtualisation:* Regardless of how users utilise a resource's installed software and hardware, the provider can serve way more users with less physical hardware installed. The hardware of a certain resource can serve many users at the same time.
- *Security:* In an unlikely event, such as the loss of a file on the device when the corresponding device is damaged, it will remain in the account and safe in the cloud system.
- *Device performance:* The speed and performance of the computer used, depending on the memory, are not adversely affected since all the necessary sources are kept in the cloud.

- *Quick entry to market:* In contrast to a traditional strategy that includes purchasing hardware and software, cloud infrastructure and services may be installed fast. As a result of the preceding argument, a business can bring new items to market faster than competitors who rely on their infrastructure.

Apart from the general advantages of cloud computing, it can also be seen as an auxiliary technology to other emerging technologies. The importance of cloud computing is seen in spheres like IoT, serverless computing and quantum computing technologies.

IoT Technologies Smartphones, televisions, smartwatches, household appliances, sensors for systems such as "smart home", "smart city" and more are among the gadgets ("things") connected to the Internet. All of these devices communicate with one another or with the control software, often without the need for human participation. The more IoT turns into an autonomous system, the Internet within the Internet, the more this will accumulate huge amounts of data in real-time.

Serverless Computing Platform as a service (PaaS) is becoming more common in the software development world. The customer does not need to worry about server hardware or operating system administration because computer resources are automatically scaled as the load increases or falls. AWS Lambda, for example, is a platform that works on the premise of event-driven computing, in which the appropriate pool of resources is assigned in milliseconds in reaction to an event, such as the addition of a new code module.

Quantum Computing Quantum computers are the next step in the evolution of traditional supercomputers. Such computers are planned to be used primarily for working with large amounts of data. At the same time, both data and computing resources (qubits) can be placed in the cloud.

Since the cloud providers supply the cloud service, reliability is an issue in cloud computing.

Providers should offer decent performance with reliability. One cloud service that provides reliability and resilience with decent performance is called "reliability as a service (RaaS)". Deep and machine learning is expected to be used in RaaS for failure prediction. Failure datasets will be used to characterise failures, leading to the development of a failure prediction model. This will be an opportunity to provide a failure-aware resource guaranteeing reliability and decent performance (Buyya et al. 2018). Another possible application of cloud computing will be based on SDN. It seems that the capability of SDN to shape and optimise network traffic will affect the studies on cloud computing (Vahdat et al. 2015). Due to workload/resource fluctuations or features, the renting fee of the cloud resource cannot be predicted by the user beforehand, which creates a necessity for tools to resolve this problem. With demand from the big data community, it is understood that the cloud environment requires new visualisation tools to be explored (Buyya et al. 2018). Other future research expectations and future-oriented cloud computing applications are summarised in Fig. 2.35.

2.14　Crowdfunding

Crowdfunding has been around for a century, but a British rock band launched the first successful crowdfunding campaign for a tour with the help of online donations from fans in 1997. Crowdfunding is the collaborative method to raise cash through friends, families, consumers and individual investors. This method uses people's collaborative efforts – mostly online through social and crowdfunding media – to leverage their networks to reach them more widely. It is initiated by developing the accurate product to be funded and providing a solid history of its development, producing a unique video for a crowdfunding platform and establishing a monetary objective. Figure 2.36 shows the successful execution of a typical crowdfunding campaign, and Fig. 2.37 illustrates the components and outline of the crowdfunding concept.

There are four main types of crowdfunding: first, equity-based crowdfunding is the real capital exchange of private corporate shares. In this type, businesses are allowed to set investment ceilings, minimum pledge amounts, etc. and accept or reject investors interested in viewing their business paperwork. Second, reward-based crowdfunding is the most popular and useful kind of crowdfunding. This form of crowdsourcing requires the determination of several levels of awards that match the commitments. Third, peer-to-peer lending-based crowdfunding allows businesses to collect cash in the form of loans to repay lenders on a predefined schedule with a specified rate of interest. With crowdsourcing donations, campaigns accumulate donations without providing value in return. This kind of marketing is great for social reasons and charity. Lastly, donation-based crowdfunding is to raise money from other individuals for charitable causes. Campaigns are often 1–3 months in length and work well for amounts under $10,000 (Hossain and Oparaocha 2017). Table 2.2 summarises the characteristics of equity crowdfunding, rewards-based crowdfunding and peer-to-peer lending.

The investment of crowdfunding is an alternative money source for companies that ask numerous investors to invest for a small amount. Then, investors receive the company's equity shares (Chen 2021). The 2015 Jumpstart Our Business Startups Act (Jobs Act) declared that it is allowed for diverse types of investors to invest with crowdfunding when the investment infrastructure is better (Securities and Exchange Commission 2015). It has been said in the crowdfunding industry report for 2019 that the food and beverage, health and beauty categories are the fundraisers' leaders. Figure 2.38 demonstrates the share of industries where crowdfunding is applied.

Also, investments through crowdfunding may open an opportunity to avoid debt. A large group of backers invest in the company knowing the purpose of the loan and interest rate. A higher interest rate is received than the market rate by lenders (Chen 2021). Companies choose to crowdfund when the other debt instruments are too costly or cannot get credit from financial institutions because of credit default swaps.

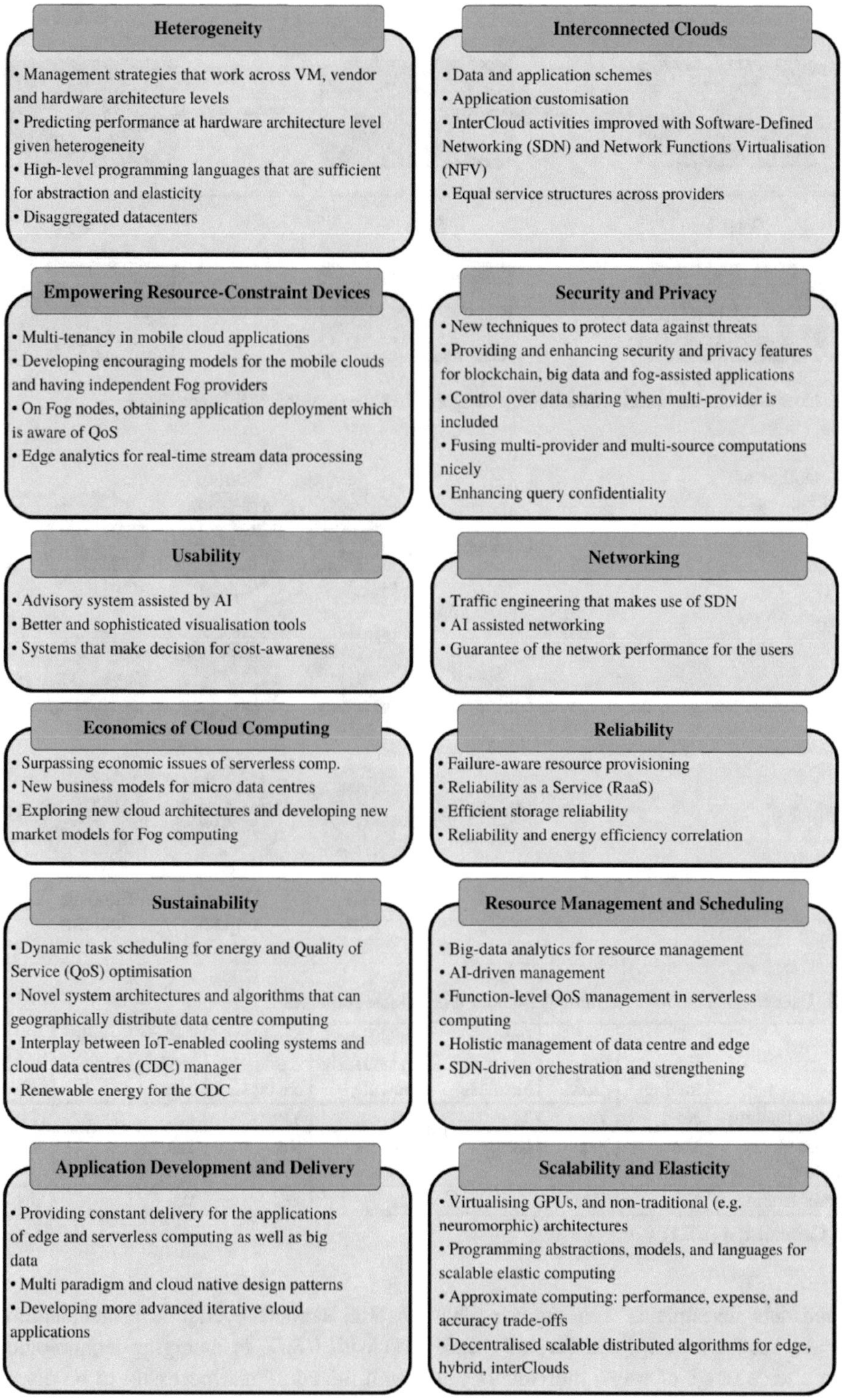

Heterogeneity

• Management strategies that work across VM, vendor and hardware architecture levels
• Predicting performance at hardware architecture level given heterogeneity
• High-level programming languages that are sufficient for abstraction and elasticity
• Disaggregated datacenters

Interconnected Clouds

• Data and application schemes
• Application customisation
• InterCloud activities improved with Software-Defined Networking (SDN) and Network Functions Virtualisation (NFV)
• Equal service structures across providers

Empowering Resource-Constraint Devices

• Multi-tenancy in mobile cloud applications
• Developing encouraging models for the mobile clouds and having independent Fog providers
• On Fog nodes, obtaining application deployment which is aware of QoS
• Edge analytics for real-time stream data processing

Security and Privacy

• New techniques to protect data against threats
• Providing and enhancing security and privacy features for blockchain, big data and fog-assisted applications
• Control over data sharing when multi-provider is included
• Fusing multi-provider and multi-source computations nicely
• Enhancing query confidentiality

Usability

• Advisory system assisted by AI
• Better and sophisticated visualisation tools
• Systems that make decision for cost-awareness

Networking

• Traffic engineering that makes use of SDN
• AI assisted networking
• Guarantee of the network performance for the users

Economics of Cloud Computing

• Surpassing economic issues of serverless comp.
• New business models for micro data centres
• Exploring new cloud architectures and developing new market models for Fog computing

Reliability

• Failure-aware resource provisioning
• Reliability as a Service (RaaS)
• Efficient storage reliability
• Reliability and energy efficiency correlation

Sustainability

• Dynamic task scheduling for energy and Quality of Service (QoS) optimisation
• Novel system architectures and algorithms that can geographically distribute data centre computing
• Interplay between IoT-enabled cooling systems and cloud data centres (CDC) manager
• Renewable energy for the CDC

Resource Management and Scheduling

• Big-data analytics for resource management
• AI-driven management
• Function-level QoS management in serverless computing
• Holistic management of data centre and edge
• SDN-driven orchestration and strengthening

Application Development and Delivery

• Providing constant delivery for the applications of edge and serverless computing as well as big data
• Multi paradigm and cloud native design patterns
• Developing more advanced iterative cloud applications

Scalability and Elasticity

• Virtualising GPUs, and non-traditional (e.g. neuromorphic) architectures
• Programming abstractions, models, and languages for scalable elastic computing
• Approximate computing: performance, expense, and accuracy trade-offs
• Decentralised scalable distributed algorithms for edge, hybrid, interClouds

Fig. 2.35 Future directions of cloud computing. (Buyya et al. 2018)

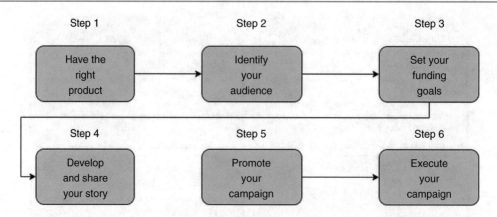

Fig. 2.36 How to run a successful crowdfunding campaign in six easy steps. (MindSea 2021)

Fig. 2.37 Outline of crowdfunding. (Dashurov 2021)

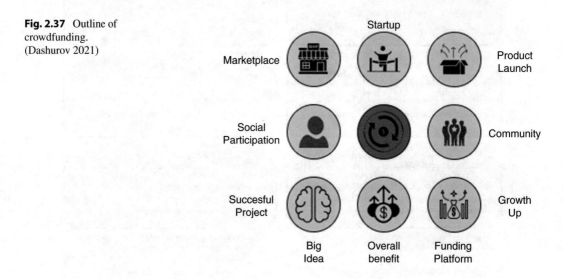

Table 2.2 User check table for determining whether it is suitable for her/him

	Pre-trading	Pre-profit	Profitable, growing business	Established and steadily growing	Established stable business	Launching new product/service/brand	Making acquisitions
Equity crowdfunding	No	Yes	Yes	Yes	Yes	Yes	No
Rewards-based crowdfunding	Yes	Yes	No	No	No	Yes	No
Peer-to-peer lending	No	No	Yes	Yes	Yes	Yes	Yes

European Commission (2021)

Equity and debt investments' funding are risky investments, but the investor can diversify their capital in a wide range of ways. Individuals can directly support the companies they feel connected to by crowdfunding investment infrastructure (Chen 2021). In the path of developing a project or launching a business, having access to capital is crucial even if the project has nothing to do with R&D. In emerging technologies' case, with the additional necessity to be possessed of enough material to research and develop truly, crowdfunding becomes even more than a way out. Funding by investors and VCs is not an easy method, especially when the project is not finan-

Technology, 6.9%
Food and Beverage, 18.2%
Medical, 13.0%
Health and Beauty, 7.4%
Sport and Fitness, 6.2%
Transportation, 5.3%
Media and Entertainment, 5.2%
Banking and Investment, 4.3%
Mobile App, 3.9%
Blockchain, 3.2%
Other, 26.3%

Fig. 2.38 Crowdfunding industries. (Okhrimenko 2020)

ADVANTAGES
• Marketing: Recognition amongst competitors due to clear online presence
• Evaluation and Consultancy: Free beneficial briefing by professionals
• Public Response: Surveying the response of the public and collecting data
• Investors: Breaking away from traditional investors and their trite patterns
• Finance: A modern choice to raise funds
DISADVANTAGES
• Hard Work: Attracting attention prolonged is necessary way before launching the project
• Sensitive Reputation: The risk of failure is worrisome due to possible loss of reputation
• Idea Theft: The risk of losing the unique idea if the copyright is not set right

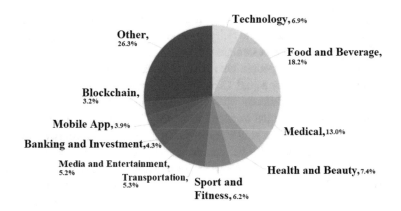

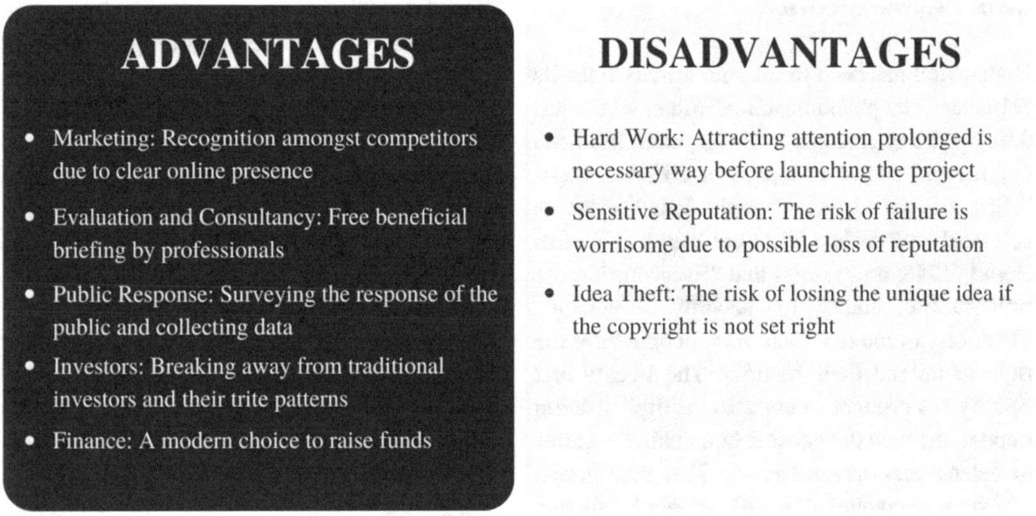

Fig. 2.39 Advantages and disadvantages of crowdfunding. (nibusinessinfo.co.uk 2021)

cially beneficial right away. Crowdfunding is a functional solution to this very problem. Surely, this does not mean crowdfunding does not have its criteria or standards to evaluate within but means that its broad spectrum of funders enlarges the criteria extent. Figure 2.39 summarises several prominent benefits and drawbacks of crowdfunding.

Started by a British rock band in 1997, crowdfunding has become a multibillion-dollar business in less than two decades for artists, filmmakers, organisations, people and now companies. The World Bank report predicts worldwide crowdfunding investments to reach 93 billion dollars by 2025 (Fridman 2016). Although the first crowdfunding portal, ArtistShare, launched in 2000, it is expected that there will be a return to specialised platforms in future years (Fridman 2016). With so many crowdfunding choices, platforms will see value in narrowing their emphasis and attracting a specialised audience. Niches such as gaming, education, music, charities, researchers and local initiatives are creating their platforms to better serve supporters and make themselves noticeable. Rewards-based platforms such as Indiegogo and Kickstarter have garnered a lot of attention in the USA. Equity crowdfunding was the next great wave that began in 2016. Titles III and IV of the JOBS Act have made it feasible for unaccredited (i.e. not wealthy) investors to readily purchase ownership shares in new firms they care about (Fridman 2016). Investing in a start-up's shares was once restricted to rich people. Because of the new rules, businesses are trying to attract investors as well as people who wish to back a company's concept or a goal. Crowdfunding provides firms with a different option for obtaining cash than venture capital or angel investors. The new investment class, a bigger group, made up of Middle Americans, will be the driving force behind investing and hence decision-making (Fridman 2016).

2.15   Cybersecurity

Humankind has been faced with crimes since its existence. The phenomenon of crime, which has developed and changed over the years, has been defined and posted in tonnes of different ways. When it is asked, most people define crime as acts that break the law. The Declaration of Human Rights (1948, art.3) states that "Everyone has the right to life, liberty and security of person". Therefore, in international laws, people have the right to defend their security. The twenty-first century has brought an era with the digitalisation trend to the new danger that humankind requires to defend themselves against. This new threat, which is encountered in all areas of life and whose importance increases day by day, is cybercrime. Various governments and companies are taking many measures to prevent these cybercrimes. Aside from the variety of measures, cybersecurity is still a huge concern for many. Cybersecurity is the practice of protecting systems, networks and programs from digital attacks. Cyber resilience, on the other hand, is the measure of an individual's or enterprise's ability to continue working as normal while it attempts to identify, protect, detect, respond and recover from threats against its data and information technology infrastructure.

This section mainly focuses on the challenges faced by cybersecurity on the latest technologies. It also focuses on the latest developments in cybersecurity techniques, ethics and the future of cybersecurity. For clarity, the cybersecurity concept will be examined in subsets. These subsets include types of attacks, types of cybersecurity threats and sectoral risk management. These will be examined as defence mechanisms against these cybercrime types that endanger individuals and organisations. Nowadays, with the increasing impact of technology in people's lives, the security of information is exceedingly important. Cybersecurity plays an important role in the field of information technology. Information security has become one of the biggest challenges of our time. That is why one of the most valuable things is knowledge of cybersecurity. With the increasing use of technology, a large amount of digital information emerges individually and institutionally. The protection of digital information requires just as much importance. Multiple cyber methods compromise digital information. In the continuation of the chapter, these vulnerabilities and methods that endanger cybersecurity will be examined, and which systems are at risk and what this developing technology promises will be mentioned. Cyberspace can be defined as a worldwide territory in the data ecosystem, the interconnected IT infrastructures including the Internet, telecommunication systems, computer networks and embedded processors and controllers (Committee on National Security Systems (CNSS) 2015).

It is seen in Fig. 2.40; various motivations lead the attackers, from personal gain to drawing attention to social problems. To receive social admiration, cybercrimes might cause many problems by taking large masses behind them. There are many attacks and types of attacks in cyberspace that will require cybersecurity experts to protect their computers. The most known attack types are backdoor, phishing, social engineering and malware.

(i) *Backdoor*

In cybersecurity, a backdoor refers to any approach that enables authorised and unauthorised users to overcome ordinary safeguards and obtain access to high levels through a computer, a network or a software program. Backdoor attacks are a form of adversarial attacks on deep networks. The attacker provides poisoned data to the

Fig. 2.40 Attackers' motivation

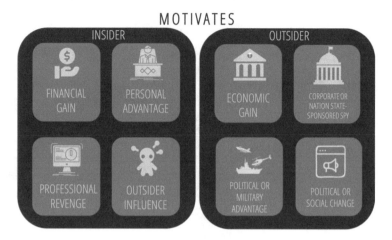

victim to train the model and then activates the attack by showing a specific small trigger pattern at the test time. Most state-of-the-art backdoor attacks either provide mislabelled poisoning data that is possible to identify by visual inspection, reveal the trigger in the poisoned data or use noise to hide the trigger (Saha et al. 2020). When cyber thieves are in the system, personal and financial information can be stolen; other software and hijack equipment can be installed.

(ii) *Phishing*

Phishing is a scam to share confidential info, such as passwords and credit card details. Victims get a harmful email or text message that is like a workman, a bank or a government agency, imitating a person or institution they trust. If the user clicks the bait, a legally valid website counterfeit will be provided. Suppose people enter their username and password to log in. In that case, the attacker who uses it to steal identity sees this information and can thus trade their bank balances and black-market personal information. At this point, different methods are recommended to users and service providers to protect themselves from phishing attacks. The NCSC (National Cybersecurity Center) recommends an innovative layered method called the 4-layer approach for large institutions and cybersecurity professionals ("Small Business Guide" 2018; "Phishing attacks" 2018). According to the research, financial institutions are the target of phishing attacks,

as seen in Fig. 2.41, in the first quarter of 2021, followed by social media and SaaS/webmail. Due to increased digital banking transactions during the pandemic period, business e-mail compromise (BEC) is even more costly (APWG 2021).

(iii) *Social Engineering*

In computer science, social engineering relates to the tactics used by thieves to get the victims to do a kind of dubious measure that usually involves security breaches, money sent out or private information. As it is colloquially known, social engineering hacks human behaviour and influences the modelling of societies in a devious way. Social media has produced a culture where sharing everything about everyone is normal and even encouraged by some (Hadnagy 2018). If it is said that cyber thieves hack people's systems using malware and viruses, social engineering is just the same by doing it with people's thoughts. On the other hand, social engineering could be done with information shared online by users themselves. As the Cambridge Analytica case has demonstrated that standard privacy laws may not be enough to safeguard customer data, another branch of the privacy protection debate may be worth investigating (Sun 2020). Victims who solved the quizzes gave their personal data and information to the harmful social engineering companies. Emotions that can be utilised are love (in its various forms), hate or anger (us versus

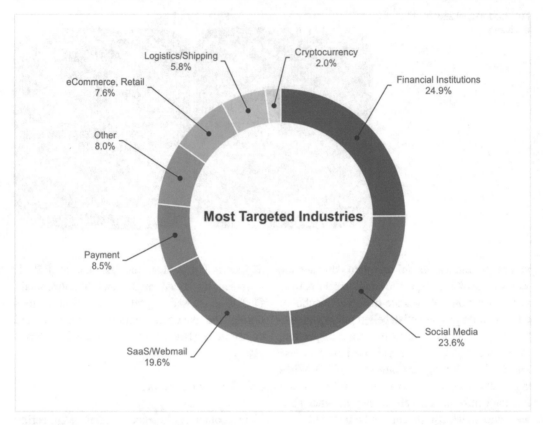

Fig. 2.41 Most targeted industries

them), pride (in themselves or their organisation) and futility (in themselves or their organisation) (there is no other option). Picking the correct emotion is simpler in person because we can read body language or via phone to assess the tone of voice and alter the approach based on the scenario. This approach aims to influence the target's emotions so that they overcome their natural cognitive replies (Winterfeld and Andress 2013). Then by researching and analysing the flip side of this coin, US elections were affected in a roundabout way by this kind of harmful social engineering by social media companies.

(iv) *Malware*

Malware is an inclusive word describing any malicious software or code damaging to computers. Malware has become a significant tool for unlawful economic activity, and its developers go to great lengths to avoid detection by anti-

malware programs (Rieck et al. 2008). Malicious software (malware), hostile, intruder and purposely bland seeks the use of computers, computer networks, tabling and mobile devices to enter, harm or disable, frequently gaining partial control of activities of a device. It interrupts regular operations, like copying, encrypting, destroying or modifying essential computer functions without knowledge or permission and monitors computer activities. Commonly encountered types of malware are viruses, keyloggers, worms/trojans and ransomware. This type of malware is placed on your devices by attackers to threaten users. Except for the ransomware attacks, most types of cyberattacks aim to convey data/information to the attackers, but the ransomware attacks are more complicated. The attacker encrypts the data with his attack and demands a ransom from the victim. In such cases, the crucial point is that when the attacker gets what he wants, he can access and damage the information in the system again since he has encrypted software in the system. This type of malware is placed on your devices by attackers to threaten users by spying on users' personal data. Spyware, malicious software, regularly collects data from targeted victims' devices. Spyware, *Pegasus*, caused a major international scandal. As data from Forbidden Stories collaboration with Amnesty Security Laboratory states, approximately 180 journalists were affected by this spyware (Rueckert 2021).

Unlike companies, individual users could lose their data in unfortunate situations because of cybercriminals. On the other hand, this threat can cost the data of numerous users. Therefore, companies may lose their reputation and reliability. Companies should care about cybersecurity as a corporate culture to prevent these threats and attacks from outside. Creating a cybersecurity culture requires altering how everyone works, how leadership is involved, how processes are implemented and how challenges are addressed. At the core of a culture of cybersecurity, each employee can carry out their daily tasks in the most secure way imaginable. To create this culture, companies need to take some external and internal actions. There are some influencing factors such as external regulations, national and social cybersecurity culture. Another factor affecting this culture in the company are activities such as managerial communication plans, training and performance measurements. In this way, values and beliefs will change in terms of both the individual and the group, and a cultural environment on cybersecurity will be created. As these beliefs become established, new behaviours will emerge that are both work-related and company-related.

A. *Systems at Risk*

(a) *Utilities and Industrial Equipment*

Computerised systems provide many services such as opening-closing valves in electricity networks, nuclear power plants, water, gas and treatment plants. Even when devices are not connected to the Internet, they can be vulnerable and potentially under attack. Devices that are not connected to the Internet are threatened by a worm called Stuxnet. Great damage can be done by individuals and nations. National goals in cyberattacks might include harvesting information, reducing the target nation's war-making capabilities, threatening other nations by showing their capability making the target nation feel weak and demoralised and/or creating a national distraction in the target nation. Since no one is directly killed in a cyberattack on the power grid, nations have been willing to conduct them without a declaration of war (Ahern 2017).

(b) *Aviation*

As in many sectors, the aviation sector works with complex systems. For this reason, everywhere, from airports to cockpits, is at the risk of cyberattacks. Interconnectivity of systems and dependency on technology created the optimum premises for new risks to emerge. The aviation industry uses a wide computer-based interconnected system, spanning from air navigation systems, onboard aircraft control and communication systems, airport ground systems, flight information systems, security screening and others used daily and for all aviation-related operations. The trend of the aviation industry is to become increasingly digitised. Digitalisation brings along new hazards as the interactions between people and systems make the risk harder to predict (Civil Aviation Cybersecurity 2021).

(c) *Consumer Devices*

In the past decade, with technological developments, people started to frequently use everyday devices such as smartphones, tablets, smartwatches, e-readers, etc. in their daily lives. Although personal computers and laptops have existed in our lives for a long time, the number of devices that can connect to the Internet has increased with more recent technologies. These kinds of daily use devices are commonly vulnerable to the threat of cyberattacks.

(d) *Large-Scale Corporations*

Large-scale corporations have always been the biggest targets of cybercrime. The crime incidents collect information such as financial, registered credit card, demographics, or address information, and then this information is sold in environments such as the deep web in exchange for money. In addition, attacks on large companies are not only aimed at financial gain but also to harm companies, damage or embarrass their brand image.

(e) *Autonomous Vehicles*

Vehicles are increasingly computerised. For example, almost every part of a car, from the complex engine to the door handle, now contains electronic circuits. Door locks, cruise controls and many vehicles use mobile phone networks, Bluetooth and Wi-Fi. Recently, with the increase in driverless and electric vehicles, new cyber risks have emerged. There are many ways to initiate an AV cyberattack. An attack can target the software that manages visual information and road infrastructure, or it could be a physical attack on the vehicle's hardware (Alves de Lima and Correa Victorino 2016). There are many risks, such as brake-accelerator pedals, door locks and motors out of the driver's control. Even driverless vehicles receive software updates over the Internet and require many security policies.

(f) *Governmental Institutions*

Activists attack the state system, military, police and intelligence systems. Public institutions have become potential targets as they are now digitised. Compliance with various safety certifications and quality systems is required.

(g) *Internet of Things and Physical Vulnerabilities*

The Internet of things, or IoT, is a system of interrelated computing devices, mechanical and digital machines, objects, animals or people that are provided with unique identifiers (UIDs) and the ability to transfer data over a network without requiring human-to-human or human-to-computer interaction (Gillis 2020a, b). One of the biggest concerns in the spread of IoT devices and technologies is the security problem. People are worried about undesired control of technological devices in all areas of life, and even more, personal and unconventional devices such as IoT are afraid of being controlled by unwanted people. In the future, although IoT devices are an indispensable part of life, they will be exposed to more cyber threats.

B. *Protection*

(a) *Attacker Types, Motivations and Gains*

Cyberattacks are becoming more common day by day. The main reason for this is that cyberattacks provide a lot of benefits to attackers. Cyberattacks are categorised according to where the attack came from. Threats may occur in the internal or external sources. Internal attacks could be more harmful because the attacker has been directly accessing the users' information.

(b) *Computer Protection (Countermeasures)*

Raban and Hauptman (2018) indicate that some measurements and protections enable threats to be absorbed and recovered and company activities restored to normality quickly. It involves several strategies, such as adaptive reaction, variety, redundancy, disappointment and proactive resistance. In recent decades, computer and network innovation have quickly developed and have been used rapidly in different areas, bringing enormous effects on human civilisation and promoting the economy, society, science and engineering of society as a whole considerably (Chunli and DongHui 2012). Computer technology and network technology are utilised in current civilisation and are still continually deepened in many aspects of culture. The article now suggests the suitable methods and technological solutions that may rapidly develop in computer networks to address the abovementioned security concerns:

(c) *Control Security of Password and Authentication and Authorisation*

The organisational leaders should regularly identify an information security weakness to help them identify all sorts of variables and diversity of safety security issues to detect and process any flaws in a timely manner to rectify any security flaws and verify the results immediately. The identification of weakness generally involves the scan of the network's security, data security scan and the scan of the database server.

(d) *Applying Firewalls*

A firewall is a computer system application that enables people to filter attackers, viruses and malware which try to enter the Internet on the computer (Reddy and Reddy 2014). Any received data via the firewall must be tested to see if to accept, refuse or divert, verify and regulate all inbound and outbound network services and visitors, assure data protection and safeguard the computer networks as far as possible against malicious assaults. As a result, the growing complexity, openness of the network, complicates the issue of security. Also, it demands the development of advanced security technologies in the interface of between varieties of networks for security domains, for instance, Intranet, Internet and Extranet (Abie 2000).

(e) *Data Authentication and Encryption*

The data security contained in the database can be assured by encoding some critical data. After encrypting, do not be concerned about the loss of data even when the existing network is destroyed. The files received must be authentically verified before download if they came from a trustworthy and dependable provider and are not modified.

(f) *Computer Users and Managerial Sensitivity Upgrading Training*

Individual Internet users select separate passwords, data to apply for legal operations, keep other users from illegal network connectivity and use cybersecurity sources for network security training following their duties and authority. At the same time, the usage of antivirus software updates, which are the front end of the network, should be considered when the virus strengthening is in use. Improve information security awareness management, employment morals, sense of commitment development, build, perfect safety management system, steadily strengthen computer system network security centralisation management, enhance information system safety build, security and provide a reliable guarantee.

The market growth of cybersecurity is quite vivid as there is a growing concern for security and cyber resilience in enterprises. Figure 2.42 shows the market share in cybersecurity applications in 2020 and 2021.

2.16   Data Hubs

Data hubs are structures that store, analyse, classify and organise the data obtained from various sources as a central model while maintaining the hierarchical structure of the data and providing access to all partners to the content (Küfeoglu and Üçler 2021). Data hubs can also be defined as a solution that utilises different technologies. These technologies are data warehouses and data science (Christianlauer 2021). By integrating a system or component with a data centre over data hubs, all data related to this system or component is shared. Data can be easily transformed and distributed to various cloud data warehouses and various business intelligence (BI) tools thanks to data hubs (Choudhuri 2019).

Many businesses are looking at numerous options on the market to develop their data hubs to handle their core vital company data, and data hubs are becoming more popular. However, this technology is commonly misunderstood as a substitute for data warehouses or data lakes. Data hubs serve as hubs of intermediation and data interchange, whereas data warehouses (DWH) and data lakes are thought to be endpoints for data collecting that exist to assist an organisation's analytics. A summary of each solution's

Market Segment	2020	2021	Growth (%)
Application Security	3,333	3,738	12.2%
Cloud Security	595	841	41.3%
Data Security	2,981	3,505	17.6%
Identity Access Management	12,036	13,917	15.6%
Infrastructure Protection	20,462	23,903	16.8%
Integrated Risk Management	4,859	5,473	12.6%
Network Security Equipment	15,626	17,020	8.9%
Other Information Security Software	2,306	2,527	9.6%
Security Services	65,070	72,497	11.4%
Consumer Security Software	6,507	6,990	7.4%
Total	133,775	150,411	12.4%

Fig. 2.42 Growth of market segments (in USD)

properties can be seen in Fig. 2.43 (Christianlauer 2021).

Data permanence is only one aspect of a modern data hub. The goal of previous data hub generations was to centralise data into a single location and store it for a limited number of sectoral use cases. Today's data hubs must fulfil a growing variety of operational and analytical use cases and centralise data. Some characteristics of a modern data hub are listed below.

1. A modernised data hub is not a permanent platform. On the other hand, the modernised data hub is a virtual or physical gateway through which data flows.
2. Data is represented in a modernised data hub without being physically persistent.
3. A modernised data hub has a corporate scope, even in today's complicated, multiplatform and hybrid data landscapes.
4. Modernised data hubs differ significantly from traditional ones. A single data domain or use case is limited in most traditional hubs, such as a customer master or a staging area for incoming transactions. Typically, a modernised hub is multi-tenant, serving many

business units and including all data domains and use cases.
5. A modernised data hub isn't the same thing as a silo. A hub cannot be a silo if it integrates data widely, provides physical and virtual viewpoints, reflects all data regardless of physical location and is properly governed. A contemporary data hub with these capabilities is an antidote to silos (Russom 2019).

A data hub collects information from a variety of sources, including data warehouses, data lakes, operational datastores, SaaS applications and streaming data sources. One or more business apps can access the information in the hub. For years, data hubs have been used in applications like master data management which aggregates consumer data from several systems to detect missing data and correct inconsistencies and errors across all data sources (Ivanov 2020). The enterprise data hub (EDH) is a solution to big data challenges. EDH is a data management solution that includes storage, processing and analytics applications for both new and old use cases. New open-source technologies, machine learning (ML), artificial intelligence (AI) and cloud-based

	DATAHUB	DWH	DATA LAKE
STORAGE OF DATA	YES	YES	YES
INDEX	YES	YES	NO
LATENCY OF DATA	SMALLER	BIGGER	SMALLER
ALL KINDS OF DATA	YES	NO	YES
INNATE ANALYTICS	YES	NO	YES
MACHINE LEARNING OPTIMIZED	YES	NO	YES

Fig. 2.43 The properties of data hub, DWH and data lake. (Christianlauer 2021)

architectures all necessitate a flexible EDH that offers faster data access and cheaper costs than traditional data storage systems. The partnership between business and information technology (IT) to create an EDH will result in a faster time to market, more product variety and more profits (Mukherjee et al. 2021).

A data hub is a contemporary, data-centric storage infrastructure that enables enterprises to aggregate and exchange data to fuel analytics and AI applications (PURESTORAGE 2021). Although it is a technology, this is an approach to arbitrate more effectively, share, connect and/or determine where, when and for whom target data should be sustained. Endpoints, which might be programs, processes, people or algorithms, interact with the hub in real-time to send or receive data from it (Lauer 2021).

A data hub sets up a connection to each system or component that needs to be integrated and ensures that the connection is shared with all other systems that must interact. Data services can be exposed and posted consistently, allowing for better integration of system-wide data and the

need for data replication to support business processes between systems. For example, any change made by anyone to their credentials takes place within this data hub, and all subscription applications can continue to use the connection. The data hub simplifies data governance requirements by keeping data in a central location. Data can be easily transformed and distributed to other endpoints such as cloud data warehouses and analytic BI engines (Choudhuri 2019).

A data hub provides an opportunity for data custodians and data users to collaborate on determining whether data is critical for distribution to the user community. This is a paradigm change from data warehousing systems where data custodians made all decisions about which data was made available to consumers. These benefit both parties in this equation: Data custodians may focus their resources on what is recognised as having the most demand/need while collecting input on datasets' quality and usefulness. Data consumers can thus obtain more data and spend less time negotiating access to datasets maintained within organisations. With these aims in mind, a data hub is for people as much as it is for data (Delaney and Pettit 2014).

Master data management (MDM) focuses on mastering collections of business data based on programmed (hard-coded) rules to enforce pre-set rules and synchronise (bi-directional) operational systems on one set of "golden rules" thanks to its well-documented definition and purpose. This was a much-needed guideline for data management and governance activities. Many businesses, however, have failed to implement MDM initiatives due to their complexity and cost, as well as the hazardous and ambitious nature of attaining the objective of having a single, agreed-upon set of data semantics provided across the organisation (Semarchy 2021). It is quite difficult to form a clear idea with the available data from organisations running multiple and independent systems (Precisely Editor 2021). Simultaneously, data analytics require mastered data and lineage to establish a data hub with accurate data attributions. As a result, analytics-driven companies began to migrate away from operational system connections and towards smaller, localised data hubs with agreed-upon analytics and application semantics. This is not to suggest that MDM is no longer necessary; rather, it demonstrates how businesses understood they needed to be more capable, flexible and nimble. It is noted in this context that there are a variety of scenarios linked to the utilisation of data hubs (Semarchy 2021). Data hubs have been developed as a solution in the presence of complex and constantly updated data sources, in cases where they actively benefit from the data at hand, when real-time and operational data are desired to be used in contrast to past snapshots within the enterprise, and a reliable integration system is needed (Marklogic 2021). Figure 2.44 represents the type of communication between multiple peers and the data exchange structure of business before the data hub (Küfeoğlu and Üçler 2021).

The benefits of data hubs, both about the mentioned scenarios and in general scope, are listed below:

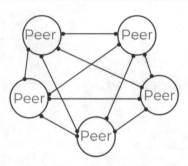

Fig. 2.44 Peer-to-peer data exchange and market structure before data hub. (Küfeoğlu and Üçler 2021)

- Provides visibility to all data.
- Controls data consumption, ownership and sharing from a single location.
- Demands complex qualities that cannot be developed on their own.
- High-performance data pipelining moves data at the optimum latency.
- Provides data operations with fine-grained control via rules and processes.
- Creates a linked architecture out of what would otherwise be a collection of silos (Russom 2019).

Due to the scenarios mentioned above and the various benefits it offers, data hub technology emerges as a solution that can help to easily overcome many difficulties that may arise in various operational processes and help the emergence of technology-supported business processes that users need (Choudhuri 2019). Figure 2.45 demonstrates the market communication system after the establishment of the data hub (Küfeoğlu and Üçler 2021).

A data hub is a digital environment that allows business and data teams to share data and access and deliver rich data services highly secure. It provides the flexibility and worldliness required for designing and developing unique use cases (Dawex 2021). Considering the future of data hubs, one of the most important issues in business life, "marketing", cannot be overlooked. Data hubs are the next step in the marketing stacks'

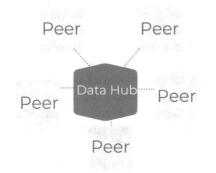

Fig. 2.45 Peer-to-peer data exchange and market structure after data hub. (Küfeoglu and Üçler 2021)

development, and they're likely to become more popular in the future. The hub idea is essential because it can assist intelligently in organising campaigns across the digital and physical domains and because a data hub serves as an information aggregator rather than another marketing solution. The main distinction between a data hub and a marketing stack is that data hubs are completely back-end solutions that allow for cross-channel insights without being limited to a certain set of solutions. As a result, it is predicted that data hub utilisation would increase over time. More data involvement every day in marketing, such as cross-channel editing and real-time personalisation, may provide significant long-term effects. The data hub is a central location where consistent, tailored messages may be provided through a single interface via email, online domains, display advertisements and mobile applications and then tracked as part of the decision-making process. Companies that capture this future are likely to succeed where others falter (Zisk 2016). Marketing operations that work successfully create, sustain and expand demand for goods and services in society (Chand 2021). Considering the importance of marketing activities in economic development, data hub technology, which is one of the marketing strategies, contributes to the economy.

2.17 Digital Twins

Digital twins have become more popular with the Internet of Things (IoT) technologies which enable monitoring physical twins in real-time at high spatial resolutions. The monitoring process takes place by using both miniature devices and remote sensing that produce ever-growing data streams (Pylianidis et al. 2021). The main goal of the digital twin is to create highly accurate virtual models of every physical entity of the original model to mimic their states and behaviours for further optimisation, evaluation and prediction (Semeraro et al. 2021). Before the industrial revolution, artisans primarily made physical artefacts, resulting in one-of-a-kind examples of a given template. However, when the notion of interchangeable components was introduced in the eighteenth century, the way things were designed and made changed dramatically as firms mass-produced products' replicas. The mass customisation paradigm has recently emerged, which attempts to combine these two well-established manufacturing techniques to attain low unit costs for personalised items. Even though such paradigms of manufacturing allow for the mass production of vast amounts of comparable, i.e. tailored components or products, the created instances are not duplicates that are related. On the other hand, building a twin is creating a duplicate of a component or a product and using the duplicate to consider various other conditions of the same component or product, therefore building a relationship between various copies. This concept is believed to have come from NASA's Apollo program, when "at least two identical space vehicles were created to allow mirroring of the space vehicle's circumstances during the trip" (Rosen et al. 2015). While the terminology has evolved since its beginning in 2002, the underlying principle of the digital twin model has stayed relatively constant. It is mainly linked to the idea that the creation of a digital informational construct of a physical system can be independent of that particular physical system. This digital information then could be a "twin" of the information that belongs to the initial physical system and would be connected to the original system itself during the system's lifespan. The basic working principle of digital twin technology is shown in Fig. 2.46.

The term digital twin dates back to 2002 when the University of Michigan held a presentation

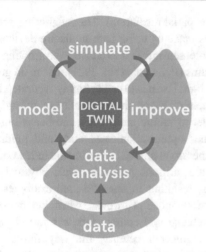

Fig. 2.46 Basic working principle of a digital twin

for the industry about the formation of a product life cycle management (PLM) centre. The presentation's slide in which the digital twin was introduced, shown in Fig. 2.47, was called "Conceptual Ideal for PLM". The slide included all the components of the digital twin, such as real space, virtual space, the link for information flow from virtual space, the link for data flow from real space to virtual space and virtual sub-spaces. The model's proposition was composed of two different systems. The first system was about the physical system that has always existed. The second one was a new virtual system that held all the information and conditions about the physical system. This meant that the real spaces were mirrored to its virtual space model or vice versa. The term PLM, or product life cycle management, indicated that it was not a static representation but rather the two systems would be linked over the system's lifespan. As the system moves through the four phases of creation, production, operation and disposal, the virtual and the real systems are linked together to create a more efficient way of working (Grieves and Vickers 2017).

The rise of the Internet has permitted the creation of more complex virtual models of various physical objects and the integration of such models into systems engineering during the last few decades. These models are utilised as the master product model, which includes the model-based description of needed product features and design verification and validation. The advancement of "microchip, sensor and IT technologies" cleared the path for the creation of smart products that track and transmit their operational conditions, allowing them to contribute data regarding their status into their product models. Advanced sensing procedures, which go beyond mathematics and scanning, enable the collection of huge quantities of data from physical objects in a simple, fast and reliable manner. The significant advancements in simulation technology, along with the expanding capabilities for obtaining and transmitting data from goods, allowed virtual twins of actual products to be created. As a result, the current concept of the "digital twin" idea has emerged (Schleich et al. 2017). Some general questions and answers about the digital twin are listed in Table 2.3.

Digital twin technology helps us to see how efficient and effective the system is in the operations and support/sustainment phase. Moreover, by using digital twin technology, companies can prevent undesirable behaviours, both predicted and unpredicted, to avoid the costs of unanticipated "normal accidents". In addition, by using a digital twin, we can significantly reduce the cost of loss of life by testing more conditions that a system can face in a real-world environment (Grieves and Vickers 2017). As the manufacturing process steps become more digitised, new potential for increased productivity emerges. Additionally, as the number of applications for digital twins grows, the cost of storage and computing decreases (Parrott and Warshaw 2017). Today, the technology exists to construct the foundations of a digital twin to aid in the care and management of people with various chronic illnesses. The next step is for forward-thinking companies and institutions with high-quality technologies and high expertise in subject matter to begin field testing such systems in real-world settings, to assess the impact of the constantly improving design on engagement, health outcomes and service utilisation (Schwartz et al. 2020).

Fig. 2.47 Conceptual ideal for PLM. (Grieves and Vickers 2017)

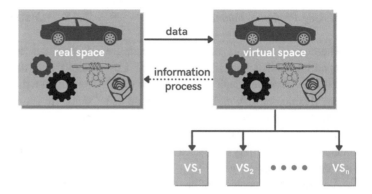

Table 2.3 Questions and answers on digital twins

Question	Answer	
What is a digital twin?	A collection of processes that simulate the behaviour of a physical system in a virtual system that receives real-time input to update itself throughout its lifespan.	
	The digital twin duplicates the physical system to detect failures and modify opportunities, suggest real-time measures for optimising unpredictable situations and monitor and evaluate the operational profile system.	
Where is the digital twin used?	• Healthcare	• City management
	• Maritime and shipping	• Aerospace
	• Manufacturing	• AR/VR
Why should a digital twin be used?	Digital twins can help businesses enhance their data-driven decision-making processes substantially. Businesses utilise digital twins to evaluate the capabilities of physical assets, adapt to changes, enhance operations and add value to systems by connecting them to their real-world versions at the edge.	
Who is doing digital twins?	• Microsoft azure	• Akselos
	• Ansys twin builder	• GE Predix
	• Siemens PLM	• Aveva

The reasons for using "digital twins" to achieve business goals can be gathered around five headings as (Arnautova 2020):

1. *Risk Evaluation and Manufacturing Time Are Both Accelerated*

Companies may test and evaluate a product through digital twin technology before the product comes into existence in the real world. It allows engineers to realise process-related problems before manufacturing through duplication of the intended production process. Engineers can disrupt the system to create unexpected circumstances, analyse the response of the system and come up with mitigation methods. This new capacity improves risk assessment, speeds up the creation of new goods and increases the dependability of the manufacturing line.

2. *Accurate Predictive Maintenances*

Businesses may examine their data to detect proactively any faults inside the system because a digital twin system's IoT sensors create large data in real time. This capability enables organisations to facilitate more precise predictive maintenance, which leads to an increase in production line efficiency and a decrease in costs of maintenance.

3. *Synchronised Monitoring Remotely*

Getting a real-time, detailed perspective of a huge physical system is typically challenging, if not unachievable. On the other hand, a digital twin may be accessed from anywhere, allowing users to monitor and adjust the system's performance remotely.

4. *Enhanced Association*

The automation of processes and reach to system information 24 hours a day, 7 days a week enables technicians to get deeper into communication between teams, resulting in increased productivity and operational efficiency.

5. *Making Profitable Financial Choices*

The cost of materials and labour, which are grouped and called financial data, can be integrated into a virtual depiction of a real-world object. Businesses may make more accurate and fast decisions on whether or not changes to a manufacturing value chain are financially viable, thanks to the availability of a vast amount of real-time data and powerful analytics.

A digital twin consists of a user interface, monitoring and analytics components. The components that are mentioned are the initial stage in enabling a digital twin to monitor, analyse and evaluate agricultural systems while also providing a continuous stream of operations. A more complex version of a digital twin could include actuator parts to control fans and windows in a greenhouse. If needed, the monitoring and control operations would be performed constantly and can report relevant information to different stakeholders. The more improved digital twin model needs simulation components to decide based on past and future predicted conditions of the physical twin (Pylianidis et al. 2021). It may utilise considerably less expensive resources in designing, producing and running systems because information replaces wasted physical resources. It can better comprehend systems' emergent forms and behaviours by modelling and simulating them in virtual space, reducing the accidental errors mainly made by humans (Grieves and Vickers 2017). Future advancements could be expected since computer technologies show no signs of slowing down. Finally, because the digital twin reflects the physical system, we may be able to use the virtual system while the actual system is in use. Capturing and utilising in-use data, as well as system front running, are two possible applications. The digital twin idea has the potential to alter how we think about system design, production and operation as well as minimise the number of UUs (unpredictable and undesirable circumstances) in complex systems and supplement systems engineering (Grieves and Vickers 2017).

2.18 Distributed Computing

Distributed computing is a field of computer systems theory that investigates theoretical concerns relating to the organisation of distributed systems. In a more limited sense, distributed computing is described as the use of distributed systems to tackle time-consuming computational problems. In short, it is the simultaneous solution of various parts of one computational task by several imaging devices (Косяков 2014). DARPA established the first distributed system in the 1960s under the name "ARPANET". Ethernet is the first widespread distributed system that was invented in the 1970s. Although the goal is to program a single piece of hardware to run multiple computers, these computers work as a single system. The aim is to create a network connected with different computers. Figure 2.48 shows how distributed computing works.

Distributed computing is a technique for solving time-consuming computational problems by combining several computers into a parallel computing system (Косяков 2014). Multiple software runs on different computers as a single system affects a distributed computer system. It is possible for the components of the distributed computer system to be either close to each other, connected by a local network, or physically remote, connected by a wide area network. Personal computers and other components such as mainframes, workstations, minicomputers and so forth can be grouped to form a distributed system (IBM 2017).

There are many definitions on this topic, but the most original one belongs to Leslie Lamport. According to him, distributed computing is the name given to the cooperation of two or more machines interacting with each other on the network for a purpose. By machine, it means a wide spectrum ranging from supercomputers or personal computers. By network means close areas such as the same campus or intercontinental, and it has a wide range of machine types. If it is needed to separate and analyse distributed computing in the literal sense, "distributed" means

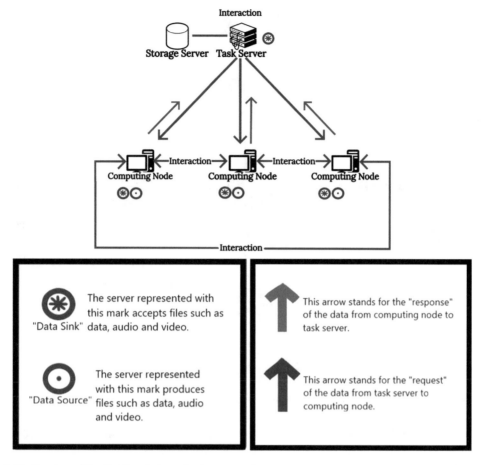

Fig. 2.48   Structure of the distributed computing

spread out in space (Lamport and Lynch 1990). However, well-known Russian professor Andrew S. Tanenbaum defines this system as a set of *independent computers*, which presents the users of the system as a *single* united system. The independent computers by themselves cannot be a single united system. This is possible only if the special programs are used, which is the *middleware* (Косяков 2014). Thus, there is one single system to which connected nodes and all of them together solve a problem. To understand distributed systems and distributed computing, examples of application areas can be seen in Fig. 2.49.

Communication networks called *closely coupled,* and *loosely coupled* are a characterisation factor of the distributed system. The location of the processors relative to each other indicates the communication network; consequently, interpro-

cessor communication's speed and reliability can be roughly defined considering the communication network. Components of the closely coupled network are spatially close to each other, and generally, communication is said to be fast and reliable. On the other hand, a system consisting of physically distant components is called a loosely coupled network where generally reliability and the speed of the communication are less than that of a closely coupled network (Bal et al. 1989).

Also, according to Gibb (2019), distributed computing systems have three characteristic features, the first one works on all parts in the system simultaneously, the second the clock concept is not global and the last one does not affect the other parts in the system when one part fails. The main connecting link of distributed computing

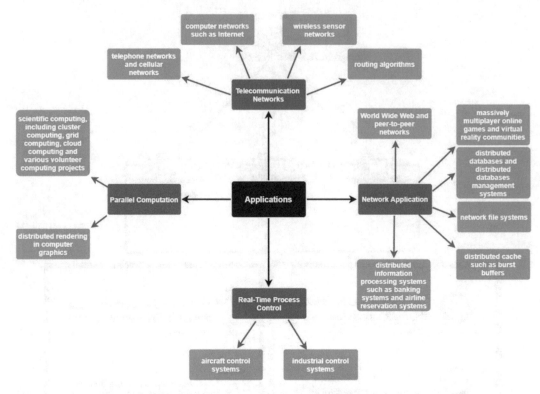

Fig. 2.49 Examples of distributed computing's application areas. (Jahejo 2020)

systems is software. A distributed computing system is a software and hardware system dedicated to solving a certain problem. On the one hand, each computing node is a self-contained unit. The software component of the DCS, on the other hand, should give users visibility into working with a unified computing system. In this way, the DCS is distinguished by the following significant characteristics:

- The ability to work with a variety of devices.
- Including those from diverse vendors.
- Compatible with a variety of OS systems.
- On a variety of different hardware platforms (Г.И. Радченко 2012).

There are several classifications of distributed control systems:

- Resource discovery methods.
- Resource availability.
- Resource interaction approaches.

Many different technologies provide search and discovery of resources in the WAN (e.g. resource discovery services such as DNS, Jini Lookup and UDDI). An example of a centralised resource discovery method is DNS (domain name system). This service works on principles that are extremely like the principle of the phone book (Г.И. Радченко 2012). Also, there are four types of distributed systems (Gibb 2019). These are shown in Fig. 2.50.

We are living in a technology era, and research, explorations, inventions and the development of useful applications are usually done with the help of computers. Because of the abundance of information, the run time of the simulations and the required memory scale-up, we will eventually require computers with high performance and many computation resources to use time efficiently. Therefore, distributed computing has become a trend for high-performance computation for complex applications (Lim et al. 2011). The fact that the organisations that use the programs are scattered is one of the main reasons

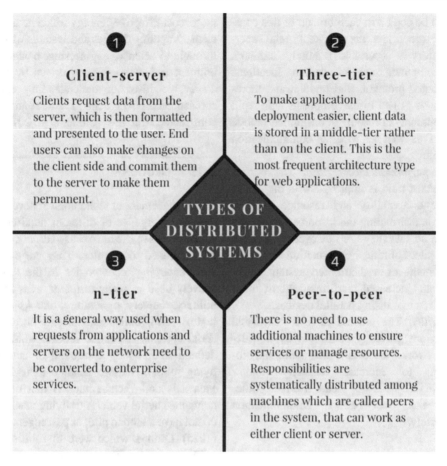

Fig. 2.50 Types of distributed systems. (Gibb 2019)

why distributed programs are useful. A company, for example, is divided into multiple divisions. Each division has its own set of tasks and uses internal data to carry out operations. However, so the divisions supply services to one another, there must be some communication between them. Physically tight divisions or geographically distant divisions are also possible (Liskov 1979). In addition, distributed computing and distributed systems offer many other advantages and useful features for the users. Some of which are listed (Kshemkalyani and Singhal 2007) as follows:

* *Enhanced reliability:* Reliability can be said to be one of the critical features of the distributed system. This is because the corresponding application or simulation can be replicated and checked several times in a distributed

structure. Besides, it is quite unlikely for the components of a distributed system to malfunction simultaneously, which ultimately increases reliability.
* *Inherently distributed applications:* A considerable number of applications should be distributed. Particularly for such applications, distributed computing becomes imperative and indispensable.
* *Physically distant data and resource accessibility:* It is often not reasonable to copy all the data to all the components of a distributed system to achieve a distributed computing application because of the sensitivity or capacity of the corresponding data. For example, payroll data of a worldwide company is kept in a common single server accessible by all its branches remotely because of the unsuitability of this

data to be copied in each branch of this company. Furthermore, the data kept in the supercomputers is accessible remotely, although supercomputers are in certain locations. Distributed protocols and middleware, therefore, have gained further emphasis with rapid developments in resource-constrained mobile devices as well as wireless communication technology.

- *Better performance/cost ratio:* The performance/cost ratio is improved with the use of remote accessibility and resource sharing, which is distributing the resources across the system so that they are not overloaded on a single site. Splitting a computation across the components of a distributed system might yield an increased performance/cost ratio compared to utilising parallel devices.
- *Scalability:* The processors in a distributed system are generally connected by a wide-area network. Hence, attaching additional processors to increase capacity and/or performance does not create a direct bottleneck that can affect communication immensely.

Previously, economic concerns supported sharing a single computer among many users, and effort in operating systems was focused on supporting and controlling such sharing. However, there is no longer any need to share a single, costly resource. Processors and memory have become much more affordable thanks to advancements in hardware technology (Liskov 1979). With the rapid development of microprocessors, distributed computing systems have become economically attractive for many computer applications. The calculation is the most important thing in computer science. After this technology, there will be some contribution to other emerging technology areas. For example, datasets grow day by day in every field, requiring more parallelism. So, machine learning algorithms will need more performance and drive skills. Deep learning models are among the highest computational applications available today, and they frequently work with large datasets or search multiple purpose spaces. The demand for

more cost-effective, energy-efficient and proficient computing devices and systems will expand throughout science, engineering, business, government and entertainment, driven by society's achievable ideas of understanding extremely sophisticated phenomena in natural and human-constructed structures (Stoller et al. 2019).

2.19 Drones

The advancement of technological developments and the emergence of different requirements in various areas of life have caused drones to start to be mentioned more often. They are one of the most emerging technologies in the last years, actively used in many different areas including military, delivery, agriculture, etc. As a technological term, unmanned aircraft or unmanned aerial systems are called drones, which can be defined as remotely controlled or autonomous flying robots. A more detailed definition from Valavanis and Vachtsevanos is that drones are unmanned aerial vehicles or flying machines that do not have a human pilot or passengers on board (2015). Drones, which were first utilised in the military in the nineteenth century, have since become more widespread in all parts of daily life (Dalamagkidis et al. 2012). Drones are most significantly related to the military, although they are also utilised for rescue operations, surveillance, route planning and weather forecasting (Udeanu et al. 2016). Drones come in a variety of types due to their wide range of applications. The technical characteristics of drones should be discussed to have a better understanding of them. Vergouw et al. state that drones can be categorised according to their kind, fixed-wing or multirotor, autonomy, weight and shape and energy source. These dimensions are significant for the drone's cruise range, maximum flying time and payload capacity, among other factors. As stated previously, drones are an important technical feature of drones. Two of the most common drone types are "fixed-wing drones" and "rotary-wing drones". The vast majority of current drones belong to one of these two categories (2016). These two broad drone categories have their own

The Technical Characteristics of Drones			
Size	**Range**	**Wing**	**Power**
Nano <30 mm	Close-range < 0.5 miles	*Rotary-wing*	Electric
Micro 30-100 mm		Single Dual Rotors	
		Multi-Rotor	
Mini 100-300 mm	Mid-range 0.5–5 miles	Tricopter	Gas
		Quadcopter	
		Hexacopter	
Small 300-500 mm		Octocopter	
		Fixed-wing	Nitro
Medium 500 mm - 2 m	Long-range 5 > miles	Low-wing	
		Mid-wing	
		High-wing	
Large >2 m		Delta-wing	Solar
		Hybrid	

Fig. 2.51 Technical characteristics of drones. (Jiménez López and Mulero-Pázmány 2019)

set of positive and negative attributes. Fixed-wing drones, to give an example, have a higher maximum speed and a greater capacity, but they ought to sustain constant forwards mobility to stay above, making them unsuitable for applications that require stability, such as close inspection. On the other hand, rotary-wing drones can travel freely and stay fixed in the air, despite their mobility and payload limitations (Zeng et al. 2016). Hybrid drones are outside of these two categories. Hybrid drones are expected to become more common in the future as manufacturing and design improvements and expenses descend (Saeed et al. 2018). Figure 2.51 demonstrates the abovementioned characteristics of drones in better detail.

Drones, whose capabilities are improving, and their areas of application are expanding, are becoming more important and popular each day. Drone technology is on the rise because it has the potential to disrupt large industries. According to Giones and Brem, drones are anticipated to become as normal as smartphones are now. They have autonomous functionalities due to advances in artificial intelligence, image processing and robotics, which have increased their revolutionary potential (2017). Another reason drones, whose capabilities have increased in parallel with technology advancements, are significant is that

they may be used to tackle global issues. According to Kitonsa and Kruglikov, drones may be a big force for good since they have an immense opportunity for being utilised to achieve the sustainable development goals (SDGs) of the United Nations. Hunger, diseases, poverty and other issues plague developing countries; drone technology is important since it can help solve many of these issues (2018). The role of drones in resolving these issues is discussed in further detail in Fig. 2.52 with application areas.

The new era of drones promises the autonomous system of flying for robots. Drones can be associated generally with applications of defence. They can also greatly impact civilian duties such as agriculture, transportation, protection of the environment, communication and disaster affect minimisation (Floreano and Wood 2015). Drones can make a difference in distorted areas for light package supplies transportation which can be more important after a disaster occurs (United Nations 2021). A new vision with a wide range and perspective, locations that are hard to reach, static images, video records and detection of objects will come in with drone technology developments. Drone-based datasets will be used in different fields, such as visions of computers and related areas of them (Zhu et al. 2018). Also, passive or active

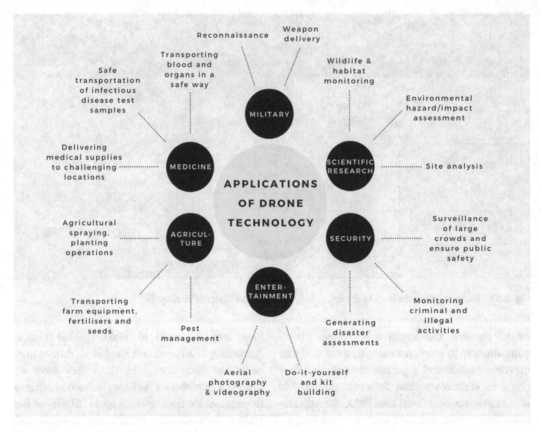

Fig. 2.52 Applications of drone technology

sensors will become more important in drone development, and they will be designed specifically. 3D modelling for landscapes will be easier with drone technology development. Drones will displace the kites, balloons and blimps, helicity, which are used for inexpensive low-level aerial photography. Increasing drone instrumentation, such as GPS, can have effects on cost, payload, range of flight and drone structure, and lower-cost platforms should be improved; drones can be more automated (Campana 2017). Additionally, in our day, drones are used in a large scale of civilian activities, such as photographing intense moments in extreme sports, construction surveillance, racing and agriculture, and their use is expected to expand in the upcoming years. The Federal Aviation Administration (FAA) of the USA estimates that the registered drone numbers in its database will reach 3.8 million by 2022 (Tezza and Andujar 2019).

(a) *Military*

Drones have a vital role in the military sector, which is where they were first utilised. Military forces use unmanned aerial vehicles (UAVs), known as drones, to attack high-value stationary targets. Unmanned ground vehicles (UGVs) can have explosives and supplies for forces on the ground, for example, heavy weapons or more ammunition, but also provide real-time video monitoring capabilities. Ground forces' combat power is increased by minimising their physical load (Fernández Gil-Delgado 2021). According to Pobkrut et al. a "survey drone" is a kind of drone designed for military purposes that use sensors such as an infrared camera and a motion detector to detect threatening targets. This means the drone is expected to possess the ability to visually detect the objectives. It is extremely difficult for a visual survey drone to detect hidden or invisible targets, so installing a system that imitates a nose to a fixed-wing drone to sense and categorise chemical volatiles or odours is a very useful method for locating hidden targets containing threats such as bombs and chemical weapons. The rationale of this method is that usually, the explosive parts of mass destruction weapons leak some gases that can be identified. Such technology will increase the survey drone's productivity and considerably benefit security services (2016).

(b) *Scientific Research*

Drone research was started for the military, and after that, it has developed in different fields of science. As electronic technology has become smaller and cheaper, camera and sensor costs have fallen, and the battery power has increased. Previously, scientists could only examine the globe from above using manned planes or satellites; nowadays, they can expand, enhance and refine their studies thanks to drones. Drones are also used to monitor rivers to predict floods. They can locate places in which trees are illegally cut down. They can detect the growth of algae as well as the trespassing of saltwater into water bodies. Plant species are determined, and diseases in forest trees are detected. In the field of energy, drones are used to detect methane leaks in the production process of oil and gas and to monitor the effect of pipes and solar and wind installations (Cho 2021).

(c) *Security*

Drones are quite popular for delivery services. UAVs which are used to transport packages, food, medical equipment and other commodities are known as "delivery drones". To accomplish a delivery, a drone has to specify the personal information of the customer and any data exchanged between the drone and the customer's site, such as a landing area, needs to be shielded from eavesdropping and drone capture. Available operating systems of drones, on the other hand, lack security support and depend solely on security measures at the link level (e.g. Wi-Fi protected access). As a result, they are subject to

common malicious attacks such as impersonation, manipulation, interception and hacking. Drones are also vulnerable to physical capture attacks because they are mobile and may pass through hazardous places. Outside landing locations that are not protected are especially vulnerable to physical capture by attackers. The security-related concerns for delivery services, such as authentication, non-repudiation and secrecy, must be addressed. Security measures to combat physical capture assaults are also necessary for delivery drones or outdoor landing places. A flexible system design is necessary to meet these security concerns for a broad range of parties and applications (Seo et al. 2016).

(d) *Entertainment*

In the entertainment industry, drones are frequently used. According to a survey, drones have a great potential for use in the entertainment industry with the help of AR (augmented reality) and VR (virtual reality) technology and will become more popular in visual arts, interactive tourism and live entertainment (Kim et al. 2018). Drone hobbyists have a wonderful experience building their own drones and competing in drone contests. In addition, drones have made aerial photography, which is usually highly expensive, more affordable.

(e) *Agriculture*

Drones are used extensively in agricultural operations. Drone technology offers significant benefits, such as precise monitoring of regions challenging to access by man, tracing illegal transactions, wildfire observations and crop yield surveillance on agriculture farms. Farmers may use drones to examine farm conditions at the beginning of any crop year. They also create 3D maps for soil testing. Drone-based soil and field studies also offer irrigation and nitrogen management data in fields for improved crop development (Puri et al. 2017). Drones in agriculture and smart farming are more effective than satellite technology since they can provide farmers with an overview of their fields while keeping close to

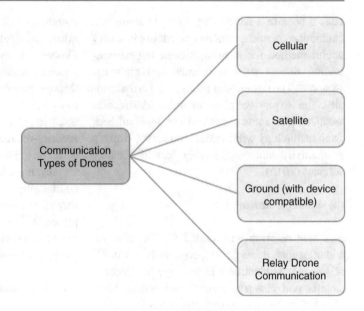

Fig. 2.53 Communication types of drones. (Yaacoub et al. 2020)

the land and therefore delivering more exact evaluations (Tripicchio et al. 2015).

(f) *Medicine*

Medicine is another field where drones are actively used and where their application is becoming increasingly prevalent. Providing catastrophe assessments when access routes are heavily limited; delivering first-aid packages, medicines, vaccines and blood to remote areas; and supplying safe transport of test samples and kits in areas with high contagion risk are commonly used applications of drones in the healthcare industry, and despite certain regulatory restrictions, drones have the potential to revolutionise medicine in the twenty-first century (Balasingam 2017). Drones, which have shown to be effective in the field of medical and health, appear to be promising for future advances in this sector.

(g) *Transportation*

Drone delivery is being considered a potential answer to future last-mile delivery issues. Meantime, the autonomous mobility trend provides flexible transportation within a city, reducing future traffic congestion (Yoo and Chankov 2021).

The four main categories of drone communications are drone-to-drone (D2D), drone-to-ground station (D2GS), drone-to-network (D2N) and drone-to-satellite (D2S). The diagram of communication of drones is shown in Fig. 2.53 (Yaacoub et al. 2020). Figure 2.54, on the other hand, illustrates the future application areas of drones.

The future of drone applications is evolving in parallel with the development of emerging technologies and is also considered as the maturation of their current usage areas. In the military, this is the situation. Conducting short-range surveillance is already mature and used, yet long-range surveillance and image capture are not at the maturity level. They are expected to be in 2–5 years. It is predicted that offering multimedia bandwidth by emitting signal/video/sound will mature in 1–3 years. In addition, it is expected that human transportation and cargo delivery via drones will reach advanced levels (Cohn et al. 2017). Also, the advancement of artificial intelligence for smartphones which are capable of recognising human users, understanding their behaviours and constructing representations of their surroundings, will continue to drive rapid advances in cognitive autonomy. Without the use of wearable devices, face recognition and gesture-based interaction will become largely available

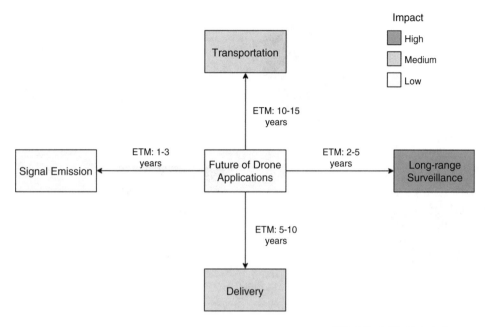

ETM: Estimated time to maturity

Fig. 2.54 Future of drone applications. (Cohn et al. 2017)

for hobby and toy drones over the next few years, to give an example, by attaching small drones with gaming-industry-developed human motion-detecting sensors (Floreano and Wood 2015).

2.20   Edge Computing

Edge computing is one of the trending and promising technologies that attract the attention of many users and researchers. Computational services and needs are satisfied better when using edge computing. Edge computing services allow the collection of data or perform a specific action in real time. Edge computing can be considered an alternative approach to the cloud environment, as real-time data processing takes place near the data source, which is considered the "edge" of the network. This is because applications that run with edge computing physically run on the site where the data was generated, rather than in the central cloud system or storage centre (Jevtic 2019). Until the development of edge computing, there were four waves in the history of computing, including edge computing: monolithic sys-

tems, the technology of the web, cloud computing and edge computing (Mannanuddin et al. 2020). Figure 2.55 illustrates the history of computing with these four main waves.

To begin with, it is needed to understand cloud computing to understand better what edge computing is. In brief, it is the storage where one's database is in. So, the idea of edge computing is to push the cloud services closer to the edge of the network. It gathers the data from the beginning, and the data processes at the very machines that gathered the data from the beginning. Thus, edge computing can be called a decentralised cloud. Also, edge computing is remarkably close to IoT technologies too. At this point, as IoT technologies gather the data, edge computing is the right service for it.

Cloud computing's centralised processing mode is insufficient to manage the data generated by the edge. The centralised processing paradigm transfers all data across the network to the cloud data centre, which then uses its supercomputing capability to solve computing and storage issues, allowing cloud services to generate economic

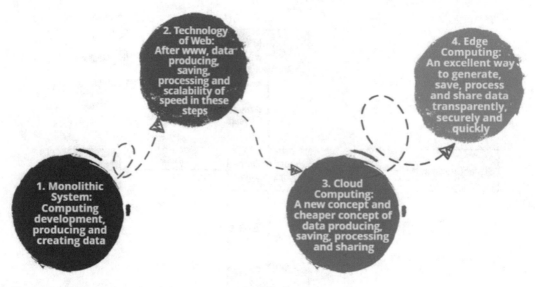

Fig. 2.55 History of computing

benefits. Traditional cloud computing, on the other hand, has significant drawbacks in the context of IoT, such as:

- *Latency:* Latency becomes more obvious in traditional cloud computing due to additional data transmission and reception time to the central cloud and end device, respectively. If the tolerable latency is surpassed, its consequences might be horrendous, as in the case of latency problems in high-speed autonomous vehicles.
- *Bandwidth:* The real-time transmission of massive amounts of data generated by edge devices to the cloud will put a lot of strain on network bandwidth.
- *Availability:* As more Internet services migrate to the cloud, the availability of these services has become a necessity in everyday life.
- *Energy:* Data centres consume a lot of energy.
- *Security and Privacy:* Leaking or examining information about private life in the system is among the possible vulnerabilities. For example, when the camera or photo images are transmitted to the cloud, the recordings/ images that are not wanted to be seen will be in the cloud (Shi et al. 2019).

Edge computing is caused by the need for overloading, latency and inability to perform real-time analysis when using cloud computing. In edge computing, the data that comes from cloud servers is transmitted directly to a network edge. This brings users and services together. If the performance features of edge computing are considered, it can be seen that bandwidth is high, latency is extra-low and real-time access in the network is faster. As a result, it reduces the load of the cloud and offers low-latency processes. Cloud technology has a centralised structure, while edge computing has distributed servers and has a decentralised system (Khan et al. 2019).

There are two kinds of edge computing: edge server, which is a piece of IT equipment. The other one is edge devices, and this is a piece of equipment that was built for some purposes. For example, suppose a vehicle automatically calculates fuel consumption. In that case, sensors based on data received directly from the sensors, the computer performing that action is called an edge computing device or simply "edge device" (El-Sayed et al. 2017). Nowadays, all modern electronic devices can compute. Thus, this means that people can work everywhere, even where they did not consider before. To sum up, edge computing allows reaching the devices that we want and doing our job. So, these emerging or modern technologies are pieces of IT equipment, and they have servers. Figure 2.56 summarises the components of edge computing.

Centralised Cloud:
The main place of data where it can be stored and shared. Centralized cloud is the outermost branch of the complete system. On the contrary of the edge of the complete system, it is capable of providing better computing performance and holding vast amount of data, networking resources.
Edge Infrastructure: Distributed data places that connect centralized cloud and edge devices. This section hosts plenty of resources
Edge Devices:
The data is processed simultaneously in edge devices according to the corresponding application of edge computing. However, it suffers from processing limitations.
Edge Sensors and Chips:
The section of data colleted and created

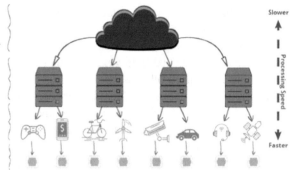

Fig. 2.56 Infographic of edge computing. (CBINSIGHTS 2019)

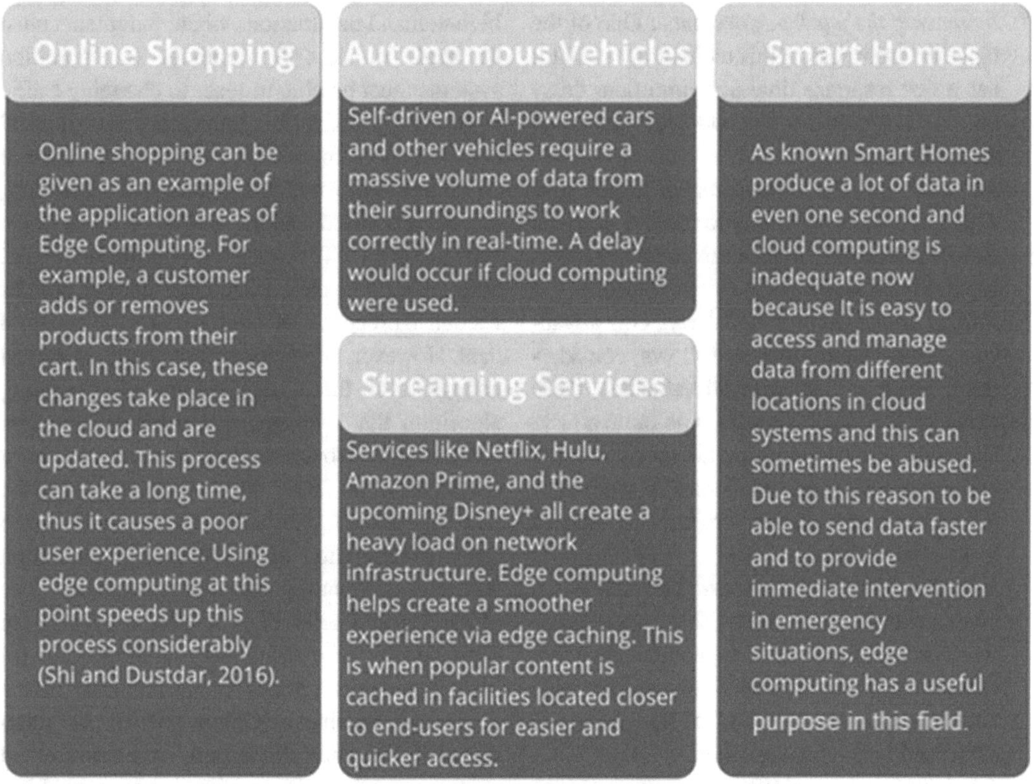

Online Shopping

Online shopping can be given as an example of the application areas of Edge Computing. For example, a customer adds or removes products from their cart. In this case, these changes take place in the cloud and are updated. This process can take a long time, thus it causes a poor user experience. Using edge computing at this point speeds up this process considerably (Shi and Dustdar, 2016).

Autonomous Vehicles

Self-driven or AI-powered cars and other vehicles require a massive volume of data from their surroundings to work correctly in real-time. A delay would occur if cloud computing were used.

Streaming Services

Services like Netflix, Hulu, Amazon Prime, and the upcoming Disney+ all create a heavy load on network infrastructure. Edge computing helps create a smoother experience via edge caching. This is when popular content is cached in facilities located closer to end-users for easier and quicker access.

Smart Homes

As known Smart Homes produce a lot of data in even one second and cloud computing is inadequate now because It is easy to access and manage data from different locations in cloud systems and this can sometimes be abused. Due to this reason to be able to send data faster and to provide immediate intervention in emergency situations, edge computing has a useful purpose in this field.

Fig. 2.57 Applications areas of edge computing. (Khan et al. 2019)

When a new technology emerges, it is quite essential to understand where it can be utilised. The best way to demonstrate the use of this method is through some key edge computing examples. Figure 2.57 compiles some application areas and examples of edge computing used for various scenarios to comprehend and internalise it better.

Edge computing aims to bring computing resources from a hyper-scale cloud data centre, which may be located a bit farther away (at the network's "core"), closer to the user or device at the network's "edge". This method lowers network latency and gathers computing power to process data close to its base. Mobile applications could leverage artificial intelligence and machine learning techniques more effectively over the edge network since they are currently fully reliant on the computing capacity of mobile processors (Стельмах 2020). Edge computing is everywhere as we need vehicles, houses, planes, buildings, etc. Low latency, high bandwidth, device processing, data offload and trustworthy computing and storage are the major advantages of edge technologies (Ekudden 2021). Edge computing offers some advantages to its users. The following are the advantages of edge computing listed by Mannanuddin et al. (2020):

- *Increasing the work performance:* One of the biggest advantages of this technology is that it has a fast response time and minimum delay because it is closer to the end-user than cloud technology.
- *No limit for the dimension of expansion:* Again, due to its proximity to the end-user and real-time analysis, it does not need any memory limit and is always open to expansion.
- *Protect your data at a high level:* This advantage can be exemplified as follows, considering a single cabinet with all valuables, these items will be under threat in a possible attack. However, if these items are allocated to small but numerous closets, the loss that will occur if a locker is unlocked is very small compared to all existing items.
- *Reducing infrastructure costs:* The data is different, and the demand for data is not exact. Some of the important or expandable data that comes from cloud systems are used by edge nodes. In this way, the cost of the process is decreased by edge computing.
- *Efficiency and reliability in business:* Since edge computing only stores the required amount of information on its nodes, there will be no loss of data and no delay in case of an interruption.

The data that has been taken from cloud services are transmitted easily and fast; the transmitted data and the velocity increase. In IoT, pulling data from sensors and going to process will be yielding and safe because wireless communication modules spend a lot of energy. However, edge computing does not. Normally, data is produced and presented to the consumer, but nowadays, data must also be obtained from consumers thanks to social media. Therefore, the cloud network cannot be located in one place. Processing and storing data at the edge provides better protection than transferring that data to a cloud (Shi et al. 2016).

Edge computing, day by day, is affecting more and more areas. According to Techjury, the total data volume will be around 40 trillion gigabytes by the end of 2021, with a generation rate of 1.7 megabytes per person each second (Tadviser 2019). First and foremost, edge computing is in high demand in situations where judgments must be made quickly. Autonomous transportation systems must be able to react to changing traffic conditions quickly, changing speed, direction and even the entire route. It is believed that they will be connected to the central cloud in some way, but operational decisions will have to be made "on board". IoT systems are developing and attracting more data. A reliable source will be needed to process, store and optimise the accessed data. However, edge computing is not developed enough to do these yet. An efficient scheduling algorithm that manages and controls this edge computing is also developed for energy efficiency (El-Sayed et al. 2017). One more example is the "intelligent" IoT sphere. When it is compared to the last generation of IoT, the new generation with edge computing will be more reliable. The efficiency and reliability of such a system are improved by processing data at the border (in local data centres, micro-clouds and even on the devices themselves) (Орлов 2019). As mentioned in the smart home part, sometimes cloud computing is not safe, and data is accessible from everywhere, and space problems can happen, so, in the future, this technology will develop about cloud offloading. In this direction, navigation systems and real-time applications games, augmented reality will develop. After the latest developments in social media, with mobile phones' smartness, many video analytics technologies are insufficient. Therefore, it will be used in the future, especially to increase security, for example, to catch a criminal, be quick and intervene immediately (El-Sayed et al. 2017).

2.21 Energy Storage

Energy systems are critical for gathering energy from different sources and transforming it into the energy forms necessary for use in various industries, including utility, manufacturing, construction and transportation. Energy sources can be used to meet consumer demand since they are easily storable while not in use. Early societies used rocks and water to store thermal energy for later use. Flywheels have been employed in pottery manufacturing for thousands of years. With the industrial revolution, new energy storage systems began to be used by people in many areas. Thermal, mechanical, chemical, electrical and magnetic energy may now be stored, converted and used thanks to many different methods. Modern energy storage devices have a wide range of uses in everyday life, for example, battery-operated portable devices such as computers, power banks, tablets, phones and smartwatches. Grid energy storage systems are necessary to maximise the introduction of energy efficiency. The electrochemical energy storage system, known as a battery system, has huge potential for grid energy storage. Energy storage methods can be used in a variety of applications. The form of transformed energy mostly determines the categorisation of energy storage systems. As can be seen from Fig. 2.58, energy storage systems are grouped under five main headings: mechanical, thermal, chemical, electrochemical and electrical energy storage. The following topics include energy storage systems and technologies, their use areas and potential future.

(A) *Mechanical Energy Storage*

Five different storage systems for mechanical energy storage systems are examined in this section.

(a) *Pumped Thermal Energy Storage (PTES)*

The heat pump system is utilised to transform the electrical energy and store it as thermal energy in this system. This technology is cur-rently an emerging technology. This system consists of four different system elements. The system consists of two solid-filled storage tanks plus a thermal engine that can perform both the functions of a heat pump and a heat engine. While electricity is utilised, the machine in the system works as a heat pump and gas is produced at high pressure and temperature. While the hot gas produced here is transferred to the hot storage tank, cold gas is injected into the cold storage tank. In this way, the gases are pumped into hot and cold stores and diffused into the solids that fill the tanks. In the discharge cycle, the machine used in the system works as a heat engine. It uses two storage temperature differences to operate the electric generator here. While the gas in the high-temperature storage tank has high-pressure values, the pressure value in the tank is kept at ambient pressure in low-temperature storage. The pressure difference between these two storage tanks is determined by the temperature difference, the solid material used and the working fluid (Barbour 2013). While these systems have a storage capacity from kilowatt to a megawatt, they can perform this storage process at 70–80% efficiency values (Ruer et al. 2010).

(b) *Compressed Air Energy Storage (CAES)*

Compressed air energy storage is an old technique to store energy as the first CAES plant was built in 1978. Large CAES facilities utilise underground places such as salt mines and rock caverns as storage locations (Gallo et al. 2016). In CAES, electricity using compressors catches and compresses the air. Then, the electrical energy used by compressors converts into the potential energy of compressed air. When energy is demanded, stored air is released and goes through gas turbines, where turbines convert the energy into electrical energy. The air compression process generates heat. Different sub-methods of CAES, such as D-CAES (diabatic), A-CAES (adiabatic) and I-CAES (isothermal), are named based on their approach to waste heat (Budt et al. 2016).

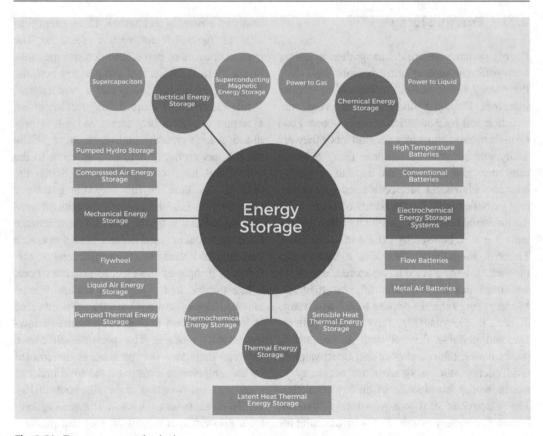

Fig. 2.58 Energy storage technologies

(c) *Flywheel*

The stored energy is in the rotational kinetic energy form in flywheel systems. The motor is used to charge this system. This motor utilises electricity for the rotation of the flywheel rotor. While the discharge process is carried out in the system, the same engine performs as a generator and produces electricity by reducing the engine speed. In these systems, the power ratio is determined by the characteristics of the power electronics and the engine-generator set in the system. The storage capacity also depends on the rotor speed, material and shape (Chen et al. 2009). There are two different flywheel configurations. These configurations depend on the maximum rotor speed. High-velocity flywheels could get 100,000 rpm, and composite materials are preferred in these systems. In low-speed flywheel systems, they operate at less than 10,000 rpm. Steel is commonly pre-

ferred as the rotor material in these systems. Due to the engines and materials preferred in high-speed flywheel systems, system costs are higher than low-speed flywheel systems (Lund et al. 2015). When they are modularised, the storage value can reach MW levels. These systems are capable of operating with 75–85% efficiency rates, and they are important storage alternative systems as they are fast-acting and long-lasting systems (Hadjipaschalis et al. 2009).

(d) *Liquid Air Energy Storage (LAES)*

Liquid air energy storage is an emerging technology that enables the storage of energy by storing liquified air in tanks. Unlike PHS and CAES, LAES does not require broad land to build a storage facility, and there is no significant environmental issue caused by LAES (Gallo et al. 2016). Therefore, LAES can be considered geographically independent and environmentally safe. LAES takes thermal and electrical energy as inputs and outputs thermal and electrical energy. Air liquefaction and power generation are two different cycles that act as the opposite of the other. Gaseous air is captured by a compressor and liquified by a condenser at the liquefaction cycle. This liquid air can be stored in a tank. When energy is demanded, the power generation cycle becomes active, and liquid air is released from the tank and pumped into a heater. Heated air moves through a turbine and spins it (O'Callaghan and Donnellan 2021).

(e) *Pumped Hydro Storage (PHS)*

The PHS technique stores energy as water's potential energy by creating two water reservoirs where one is placed lower. Water is pumped from the lower reservoir to the upper one when stored energy is needed. Once power is required, blockings between the reservoirs are removed, and turbines produce energy while water travels to the lower reservoir (Rehman et al. 2015).

(B) *Electrochemical Energy Storage Systems*

(a) *Conventional Batteries*

• *Lead-Acid*

This technology has been used for 150 years, especially in the automotive industry. The components of this technology are lead metal as well as oxide electrodes, also a solution of sulphuric acid (Chen et al. 2009). Whereas lead and lead oxide transform to lead sulphate, as the concentration of the electrolyte decreases in the discharge cycle, the deposition of the lead and increasing concentration of the electrolyte that take place on the anode occurs in the charge reaction. High efficiency and low cost are advantages of lead-acid batteries; nevertheless, low cycle life, low specific energy and the toxicity of the lead are disadvantages of this type of battery (Beaudin et al. 2010).

• *Lithium-Ion*

Lithium-ion (Li-ion) batteries are a high energy density, rechargeable energy storage technique that employs lithium ions as the main component. Lithium atoms in the anode are ionised during a discharge cycle and recombined with their electrons in the charge cycle. The electrolyte provides a transfer medium between the anode and the cathode, but this electrolyte is highly flammable. High energy density, extremely low self-discharge rate and relatively low price make Li-ion batteries popular in various areas.

• *Nickel-Cadmium*

Nickel hydroxide is the cathode, and cadmium hydroxide is the anode in this battery. The anode reaction is the conversion of the cadmium hydroxide to cadmium metal, whereas the cathode reaction is the transformation of nickel hydroxide to nickel oxyhydroxide (Gallo et al. 2016). The properties of this type of battery are long cycle life (1000–1500 cycles), elevated specific energy (55–75 W h/kg) and so on. However, the toxic property of cadmium is the negative side of this technology. Also, nickel-metal hydride batteries can be obtained by removing the cadmium in the system. This battery is an environmental version of nickel-cadmium batteries. When the energy density increases, the self-discharge and reduced durability occur in the nickel-metal hydride batteries (Gallo et al. 2016).

(b) *High-Temperature Batteries*

• *Sodium-Sulphur Batteries (NaS)*

Sodium sulphur (NaS) batteries started to commercialise in 1984. This battery technology contains molten sulphur and sodium ions at cathode and anode, respectively. A high temperature is required for this technology to keep sodium and sulphur in the liquid phase. Due to this situation, this technology can be designated as high-temperature batteries. Besides, the electrodes are made from alumina, which only gives sodium transportation in the electrolyte. To keep this battery at a high temperature, electrical heaters are used in this type of battery. Electrical heaters give rise to discharge capacity losses (Schlumberger Energy Institute (SBC 2013).

• *Sodium-Nickel Chloride (Na-NiCl$_2$)*

Sodium-nickel chloride batteries have molten materials at their electrodes. The first investigation started in the 1970s, and General Electric has been searching for this battery since 2007. An Italian company FIAMM produces commercial sodium-nickel chloride battery systems (Gallo et al. 2016). NiCl$_2$, or the mixture of NiCl$_2$ and FeCl$_2$, behaves as active material in the cathode. Also, sodium-nickel chloride batteries include beta-alumina electrolyte and molten sodium chloroaluminate (NaAlCl$_4$). This type of battery system has high specific energy, efficiency and cycle life; however, the disadvantage of this system is that the heating up from the frozen state of the battery takes a long time, nearly 15 hours (Gallo et al. 2016). Na-NiCl is safer than Na-S batteries. Because when a failure occurs in the electrolyte, molten sodium initially gives the reaction of solid chloroaluminate. As a result of this reaction, non-dangerous products also inhibit any further reactions (Gallo et al. 2016).

(c) *Flow Batteries*

To generate electricity, flow batteries use chemical reduction-oxidation processes. The anolyte and the catholyte, two chemical solutions, are held in tanks separated by a membrane. The differential charge levels on either side of the membrane that is used as potential are referred to as redox (Ferrari 2020). Ion exchange occurs when liquids are pumped over the membrane, causing an electric current to be generated, with the charge being supplied or withdrawn via two electrodes. The energy capacity is solely determined by tank size, while the power is determined by anode surface area. The most popular electrolytes are vanadium and iron solutions (Ferrari 2020).

(d) *Metal-Air Batteries*

This type of technology can be designated as emerging technology due to promising concepts for the future. This technology utilises the oxygen from the atmosphere in the porous cathode and metal electrodes as an anode – for instance, sodium air, lithium-air, zinc-air and magnesium air. High specific energy can be obtained with this technology. Although this technology has not reached its potential, sodium air batteries have great interest due to the abundance of sodium in the world and easiness of reaction (Hartmann et al. 2013). Two properties need to be developed: cost and life cycle. The EU determined the 3000 life cycles as an objective (Gallo et al. 2016).

(C) *Electrical Energy Storage Technologies*

(a) *Supercapacitors*

Capacitors store electric charges; however, they are different from batteries. Capacitors can be charged much faster than batteries and stabilise the circuit's electric supply. Supercapacitors are types of capacitors that have higher capacitance values than the other capacitors. But they have lower voltage limits. They are also called ultracapacitors in some resources. Electrochemical capacitors known as supercapacitors are used for fast power delivery and recharging (Simon et al. 2014).

(b) *Superconducting Magnetic Energy Storage (SMES)*

Energy storage via decreasing temperature below a critical temperature principle is utilised for this type of storage system. This technology can directly store electricity. These storage systems induce a dynamic electric field or generate a magnetic field by passing a current through a superconducting coil. Since the coil is made of a superconducting material, the current can flow through it almost without a loss (Luo et al. 2015).

(D) *Chemical Energy Storage Technologies*

(a) *Power-to-Gas (PtG)*

Power-to-gas technology depends on the conversion of energy. The working principle of this technology is that electricity is taken into the system and transferred to the electrolysis machine. The electrolysis process is carried out here, and two outputs, hydrogen and oxygen, are obtained. These products are then converted to methane by a process called mentation. After this process, the product that comes out is a kind of synthetic natural gas or substitute natural gas. This created product has the same properties as natural gas and can be transferred, used and stored just like natural gas. This system functions as an efficient energy storage system. In this system, if renewable energy sources are used as an energy source, and carbon capture technology is used in the mentation process, the resulting gas turns into a carbon-neutral gas (MAN 2021).

(b) *Power-to-Liquids (PtL)*

As the name suggests, power-to-liquid technology is a technology that converts energy into various liquids. The main sources used in this technology are electricity which is produced by renewable energy sources, water and carbon dioxide. Here, electrolysis is carried out with the help of electricity and water, and hydrogen is produced. The hydrogen obtained as a result of this production and the ready-made carbon dioxide are used for the production of liquid hydrocarbons. It is refined according to the type of hydrocarbon produced. There are two main production routes to realise liquid production in this technology, the Fischer-Tropsch (FT) pathway and the methanol (MeOH) pathway. With these production methods, the desired liquid hydrocarbons are produced (Schmidt et al. 2016).

(E) *Thermal Energy Storage (TES) Technologies*

One type of energy storage system is thermal storage of energy. In these systems, thermal energy is stored by heating or cooling the material or environment. The energy to be stored can come from waste cold, waste heat or thermal solar energy. In addition, electrical energy can be converted into a storage source for these systems after it is converted into heat energy (Gallo et al. 2016). The energy here can be stored daily, weekly or even seasonally, then stored energy could be utilised to heat, cool or generate power. Thermal energy storage systems are divided into three in themselves. This distinction is as follows.

• Sensible heat thermal energy storage (SH-TES)
• Latent heat thermal energy storage (LH-TES)
• Thermochemical energy storage (TCES)

The sensible heat thermal energy storage method is the most commonly utilised method for thermal energy storage systems. The liquid or solid used in such systems is heated, increasing its temperature. Then, the energy stored here is released when needed by lowering the temperature of the material. The heat capacity of the material used here is essential. In these systems, materials with high heat capacity are used, so the amount of material used is kept as low as possible. Material thermal properties, material storage capacity and operational temperature values are factors that affect material selection (Hauer 2012). Although these systems are less efficient (50–90%) than other thermal storage systems, they are preferred because of their simple structure and low cost (Connor 2019).

Latent heat thermal energy storage systems utilise the change of phase to store thermal energy. In these systems, phase-changing materials (PCM) are preferred as storage materials. These systems eliminate the two disadvantages of SH-TES systems. The first of these is about specific energy. The obtained specific energy increases between five and fourteen by using this system than the usage of SH-TES. Secondly, while the discharge temperature remains constant in LH-TES systems, the discharge temperature changes in SH-TES systems. The efficiency of these systems is around 75–90%. The PCM used in these storage systems is incorporated into building walls. The temperature of this material decreases due to the decrease in air temperature at night, and it solidifies. With the increase in temperatures during the day, the temperature of the material also increases and the material melts and becomes liquid. During the phase change that takes place here, the wall temperature remains constant and reduces the heat input to the interior. In this way, it can reduce or eliminate the need for air conditioning to cool the environment. This cooling process using these systems is called passive cooling (Abele et al. 2011).

Thermochemical energy storage systems are energy storage systems that perform chemical reactions using thermal energy and convert thermal energy into chemical energy. The purpose of these systems is not to synthesise new products to be used later. In these systems, reversible processes such as hydration-dehydration, adsorption-desorption and redox are used to store thermal energy to be utilised. These systems have a denser storage capacity than other thermal storage systems, and thus the material used for storage is much less. The efficiency of thermochemical storage systems is 75–100%, and these systems are a good alternative for long-term storage. These systems lose almost no energy during the storage period, which makes these systems suitable for long-term storage needs. Storage in these systems is usually carried out at ambient temperature (Abedin 2011).

The energy storage enables the reduction of energy costs, increases energy system reliability and flexibility and integrates different energy systems into the system. In addition, storage systems also contain an environmentalist approach. Energy storage contributes to reducing energy costs both on the producer side and on the consumer side. While the operational costs of energy production companies in frequency regulation or providing spinning reserve services will decrease, the consumption costs of consumers will decrease thanks to the use from the warehouse, which they will make at peak times of energy consumption, especially thanks to storage. In addition, both producers and consumers will not be adversely affected by power cuts that may occur, and it will

be possible to mitigate the economic losses due to interruptions to some degree (Energy Storage Association 2021).

The grid's flexibility and reliability increase by the development and integration of energy storage systems. The reliability of the grid is a very important issue for both producers and consumers. Thanks to the storage, in any negative situation that may occur in the grid, consumers are not adversely affected by power cuts because the energy in the warehouse is activated and can be given to the grid. In this way, the costs caused by the negativities will be prevented (Energy Storage Association 2021). In addition, energy storage facilitates the connection of renewable energy sources to the grid. Today, the biggest disadvantage of renewable energy sources is the instability of energy production, and this is one of the biggest obstacles to the choice of renewable energy sources. However, thanks to energy storage, the stored energy is transferred to the grid at the point where renewable energy sources are insufficient to meet the grid needs. In addition, energy storage is important for existing energy production systems. Especially in cases where it becomes difficult to meet the required production due to the sudden increase in demand in the network, these storage systems come into play and ensure that the demand is met. Thanks to the storage, flexibility is provided to the system (Energy Storage Association 2021). The extensive use of energy storage plays an important role in carbon emissions reduction through the common usage of renewable energy sources. In addition, since the networks can operate much more efficiently thanks to storage, the consumed energy will be used more efficiently. Accordingly, there will be a decrease in the amount of carbon released per unit of energy. Since storage systems contribute to decreasing carbon emissions, they also contribute positively to the environment (Energy Storage Association 2021).

There has been a sharp growth in the use of renewable energy sources in recent years, and this trend is projected to continue. Suppose nations maintain their current and previously declared policies. In that case, the worldwide capacity of solar photovoltaic (PV) is expected to reach 3142 GW by 2040, surpassing coal and gas to become the world's greatest energy source. Similarly, wind's proportion in electricity generation will rise from 5% in 2018 to 13% in 2040, with a capacity of around 1856 GW. As a result, the total wind and solar capacity will be 4998 GW. In addition, when hydro and other renewable sources are considered, the overall percentage of energy generation will rise from 26% in 2018 to 41% in 2040 (Hossain et al. 2020). It has become a necessity to meet the energy needs from renewable energy sources since the negative changes in the climate. However, renewable energy production systems cause various problems in continuous energy production, and fluctuations may occur in the production. Storing the produced energy becomes important at this point. In this period of transformation in energy systems, one of the keys to facilitating the transition to renewable energy is the development of new storage systems and the increase in their number. The storage capacity available worldwide in 2014 was around 140 GW (Xylia et al. 2021). By 2020, this amount has increased to 170 GW, and this corresponds to approximately 3–4% of the energy produced today (Kamiya et al. 2021; Xylia et al. 2021). It is aimed to increase the storage level to 450 GW by 2050 (Xylia et al. 2021).

The use of electric vehicles is increasing day by day. This increase brings with it new opportunities. It is possible to use electric vehicle batteries in two different ways for energy storage. One of these is to utilise batteries that have completed their life for energy storage (Renault Group 2021). Batteries in electric vehicles need to be replaced when the capacity percentage drops to 60–70%, but they are still usable (Cagatay 2021). In this case, the batteries taken from the vehicles can be combined and converted into fixed energy storage systems. Another possible usage method is the vehicle-to-grid (V2G) technology. In this technology, vehicles store the electricity they receive from the grid and transfer the stored energy back to the grid when there is a lack of energy in the grid. In this way, mobile energy storage can be provided (Renault Group 2021).

2.22   Flexible Electronics and Wearables

Flexible electronics and wearables (FEAWs) are technologies that support each other and need to be looked at together. The development of one improves the other so that they will be covered together in this chapter. Conventional electronic systems have an inherently rigid and unalterable form. Re-developing these systems by adding features such as flexibility and stretchability allows electronics to be added to a broader range of applications and products where flexibility is required. New form factors and new products can be developed using technologies such as printed electronics. According to experts, the market share of flexible electronics will increase soon. For example, by 2024, the worldwide flexible electronics market is estimated to reach USD 87.21 billion (Grand View Research 2016).

Flexible electronics present many innovative technological developments in the electronics field, such as the flex circuit board. A flex circuit board is a type of printed circuit board with at least one readable feature of the board. Flexible circuits are called FFC (flexible flat cable) and are used by replacing cable wires and connectors. In this case, the flex circuit is designed without electronic components. Another common use includes parts mounted on flexible circuitry such as LED strip and LCD panel interface. The electronic circuit is designed on a flexible plastic substrate, usually polyimide film, which is resistant to high heat to make it suitable for soldering assembly components. Flexible electronics offer low-cost solutions to a wide range of applications such as foldable displays and TVs, e-paper, smart sensors and transparent RFIDs. The main advantages of flexible electronics over existing silicon technologies are low-cost manufacturing methods and inexpensive, flexible substrates. The fact that flexible electronics are light, bendable and portable and require low-cost electronics is also becoming a very interesting material for next-generation consumer products (Cheng and Huang 2009).

Wearable technology is a phrase for items that have acceptable electronic functions and aesthetic qualities, consisting of a simple interface to provide specific activities to meet the demands of individuals. Since wearable technology provides many conveniences to individuals in terms of usage, portability and data utilisation, a considerable amount of attention has been paid to it in recent years, and the market share of wearable technology has increased like flexible electronics. In 2019, the worldwide wearable technology market was estimated at USD 32.63 billion, and it is expected to increase at a compound annual growth rate (CAGR) of 15.9% from 2020 to 2027 (IEA 2021). Flexible electronics and wearables have a mutualistic relationship with each other, as mentioned earlier. These emerging technologies show themselves in many different areas, various sectors such as fitness, finance, entertainment, education, medical and textiles (Wright and Keith 2014). Figure 2.59 demonstrates vari-

ous sectors where FEAW applications can be seen.

1. Fitness: One of the areas where FEAW is used is health monitoring in sports, for example, pedometer, heart rate monitor that tracks calories burned, distance taken, activity duration, etc., during sports. Moreover, recording these data can provide reference data for both athletic training and health management. In addition to the sports analytic tasks, wearable equipment may be utilised for both physical and mental health controls (Borowski-Beszta and Polasik 2020).

2. Finance: Another area where FEAW is gaining popularity is the financial sector, with its ease of payment. Wearable equipment can conduct financial transactions and save cus-

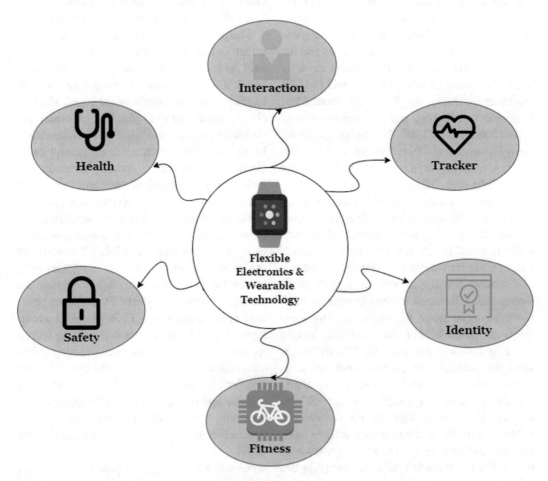

Fig. 2.59 Flexible electronics and wearable technology sectors

tomers' time at payment points by integrating NFC, a wireless communication technology enabling data transfer at an approximate distance of 10 cm (Leong et al. 2013). They may also count as a bitcoin wallet, and it is claimed to be one of the most secure wallets out there with this feature.

3. Entertainment: There are many FEAW integrated into the entertainment industry. CGI technology (computer-generated imagery) added to the costumes allows all effects prepared on the computer to be transferred to movies. In this way, the imaginary world of people can be transferred to real life. Wearable technologies that integrate virtual reality to all of the digital and physical elements to increase the gaming sensation in the real world are also used in the game industry. They are also included in the music industry, allowing individuals to listen to music freely, such as wireless headphones.

4. Education: FEAW could be regarded as "the most important tool utilised in education" in a way that enables remote students to watch and listen to the lectures without physical presence but with a sense of being present in class. It enables students to study differently and without the danger associated. It may present students with actual scenarios and bring them the locations that are difficult or occasionally inaccessible, such as space studies, archaeological courses, medical education, chemical engineering and aviation training (Attallah and Ilagure 2018).

5. Medical: Wearable bioelectronics provides clinicians with real-time monitoring of a patient's physiological parameters, which has interested many researchers and technology companies. Wearable bioelectronics contains three types of sensors: pressure, temperature and biochemical. Tech companies use those sensors to monitor, measure and inject hormones and enzymes into the skin (Parlak et al. 2020). A tech company, Omnipod 5, designed an insulin pump that sticks into the skin and controls blood sugar levels of type 1 diabetes patients (Henderson 2021). Also, many companies try to improve the conditions of indi-

viduals with wearable technologies. Sony designed a wearable air conditioner that can fit into a pocket and declares that the air conditioner can decrease body temperature by 13 °C (Byford 2020).

6. Textiles: The textile industry tries to adapt to many developments in flexible electronics. The companies insert the electronics that are sensors, communication modules into textiles. In recent designs, all the components of textiles are electronics which are smart textiles. With the integration of electronics in textiles, clothes transform into electronic devices. That integration inspires tech companies to produce various products (Paret and Crégo 2018).

Flexible electronics are becoming increasingly popular as a result of the numerous advantages they offer. Flex circuits will continue to be employed for a range of applications as more organisations discover their potential for increased customizability, affordability and portability. Flexible electronics are used by a wide range of sectors and professionals in their equipment and products because they provide a variety of benefits such as, but not limited to, being affordable, flexible, customisable, innovative and portable. From a different standpoint, flexible displays and flexible sensors will revolutionise wearable technology by allowing devices to conform to our bodies and clothing while providing increased utility. For instance, the utilisation of FEAW technology has become more medically oriented, an increasing number of wearable devices are being programmed to interact with humans' bodies and collect data that can be used to inform sports science and health research. It will be critical to ensure that these devices are shatterproof, unobtrusive and "unawareable" to design successful healthcare applications. Figure 2.60 summarises certain benefits and drawbacks of FEAW.

Flexible and wearable technologies will become more popular and cheaper to create as a result of advancements in the industry. In addition to having a larger market cap, it will be evaluated to be utilised in other industries in the future (Skilskyj 2018, 2019).

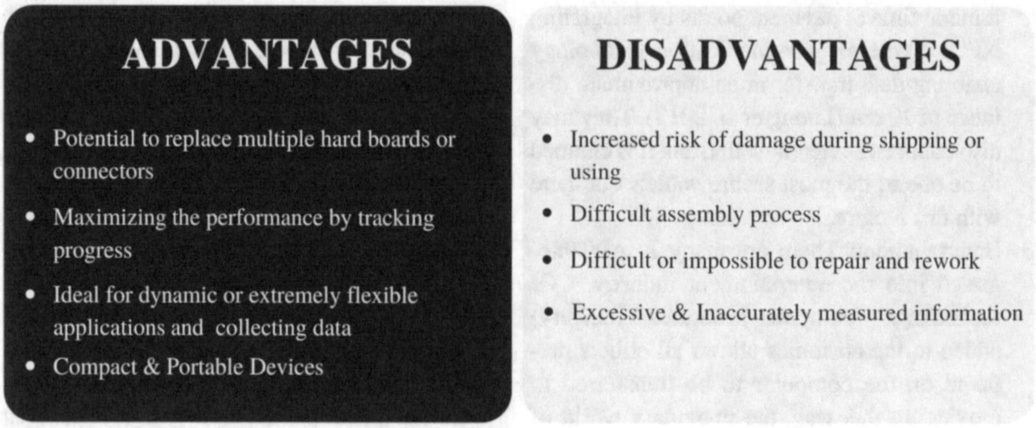

ADVANTAGES	DISADVANTAGES
• Potential to replace multiple hard boards or connectors • Maximizing the performance by tracking progress • Ideal for dynamic or extremely flexible applications and collecting data • Compact & Portable Devices	• Increased risk of damage during shipping or using • Difficult assembly process • Difficult or impossible to repair and rework • Excessive & Inaccurately measured information

Fig. 2.60 Advantages and disadvantages of flexible and wearable technology. (Lee et al. 2016)

Flexible electronics and wearable technologies have high potential. Soon, they will be able to do things that people cannot even think of right now. Major developments in medical, energy and electronics are coming. Studies in the medical field enabled innovative solutions. One of the featured solutions is electronic skins. Patients will be able to wear a plastic sheet with organic circuits embedded in it if they choose to use electronic skin. The skin can monitor the pulse, oxygen levels and temperature of the user's blood. Users can become more aware of potential health issues and provide invaluable data to their healthcare providers by utilising electronic skin. The UV radiation monitoring patch will also be available in the future (Moser et al. 2016). Users will be able to track how much dangerous UV light they receive using a UV radiation tracking patch. The patch may provide the data directly to smartphones so that people may notice whether they have to search for melanoma and other potentially harmful consequences of UV radiation more vigilantly. Flexible circuits in contact lenses, another development that we will encounter in the field of health, is another technology that we will encounter in the future. Contact lenses with electrically conducting polymers are an intriguing opportunity for the future of flexible electronics. A user may transfer a picture from a TV or computer screen right into their contact lenses with these contact lenses (Watkins et al. 2021). Additionally, sensors integrated inside the lens could potentially detect a user's glucose level and then project that level onto the lens, allowing the user to see it.

Studies in the field of energy continue day by day. One of the most exciting innovations in the energy industry is printable solar panels. Organic photovoltaics (OPVs) are a type of flexible electronics that are expanding solar energy possibilities. An OPV may print solar energy technology into the fabric of your drapes or laminate it onto your window panes rather than requiring solar panels to be placed on a roof. Consumers may see the technology integrated into solar-powered clothes, which could be used to recharge cell phone and laptop batteries (Global Electronic Services 2021). OPV cells are thin and flexible because they are placed on a flexible substrate. Sunlight is absorbed through technology, which then transmits that energy to devices. These OPVs are not only offered to make houses simpler and meet specific demands but may also be considerably cheaper than standard solar panels. The lower cost is primarily since OPVs employ a polymer-based semiconductor layer, whereas most solar panels on the market use more expensive semiconducting materials like silicon. Researchers and business professionals are developing techniques to increase the availability of OPVs and incorporate them into a wide range of products. In the future, experts expect to see a lot more of these OPVs (Watkins et al. 2021).

The last area to investigate in FEAW is electronics. The gadgets and devices that employ flexible electronics will be increased for consumers in the future. For example, foldable smartphones can be given as an example. While some of these modern phones are struggling for durability, flexible circuits help change this as technology progresses. As organic light-emitting polymers (OLEDs) develop and become more widespread, expect more durable foldable cell phones. In the future, individuals will be able to see televisions that can wrap around the walls of their houses, in addition to cell phones (Watkins et al. 2021; Delta Impact 2021, Global Electronic Services 2021). OLED TVs will grow more popular, offering consumers a smaller and more versatile alternative to traditional LCD TVs.

2.23   Healthcare Analytics

The technique of examining past and current industry data to forecast tendencies, increase accessibility or even effectively control disease spread is known as healthcare analytics. The area covers various sectors and offers both international and micro viewpoints. It can point the way to better patient care, clinical data, diagnostics and corporate management. When paired with marketing intelligence suites and visual analytics, healthcare analytics help managers make more informed decisions by providing real-time information that can support choices and provide valuable insights. Healthcare analytics is a compilation of administrative and financial data that may help hospitals and healthcare managers improve patient care, provide better services and modernise existing processes (Sisense 2021). Wise decisions based on accessible data could help alleviate problems that can occur in traditional healthcare systems and ease the transition to value-based management. In their management systems, healthcare facilities are incorporating information technology. This system collects a significant amount of data on a constant schedule. Analytics supplies skills and strategies for extracting information from this complicated and extensive data and converting it into data that can be used to aid healthcare decision-making (Islam et al. 2018). Figure 2.61 repre-

Fig. 2.61 Applications of healthcare analytics

sents some of the usage areas and applications of healthcare analytics technology.

Healthcare data refers to any data about a person's or a population's health. To gather this information, healthcare providers, insurance companies and governmental organisations employ a variety of health information systems (HIS) and various modern tools, the combination of which can demonstrate a comprehensive picture of each patient as well as trends related to geography, socioeconomic status, race and propensity. The data gathered can be separated into distinct datasets, which can subsequently be examined. A number of tools are used to collect, store, distribute and analyse health data. Some of these methods are mentioned below:

1. Electronic health records (EHRs)
2. Personal health records (PHRs)
3. Electronic prescription services (E-prescribing)
4. Patient portals
5. Master patient indexes (MPI)
6. Smartphone apps

Every second, more and more healthcare data are being evaluated thanks to digital data collecting. A substantial amount of data is being collected in real-time as electronic record keeping, applications and other electronic means of data collecting and storage become more prevalent. There is a need for a centralised, systematic method of gathering, storing and analysing data so that it may be used to its full potential. In recent times, data collecting in health situations has become more efficient. The information might be utilised to enhance day-to-day operations and patient safety and predictive modelling. Both datasets can be used to track trends and generate predictions instead of merely looking at historical or present data. Preventative steps can be taken, and the results can be tracked in this manner (University of Pittsburgh 2021). Four main types of healthcare analytics are stated below:

- *Descriptive analytics* commonly takes the form of a dashboard, leverages historical data to provide insights into trends or benchmarks. While descriptive analytics can help under-

stand what happened in the past, it can't give any substantial insight into how to affect future health outcomes or predict what might happen in the future.

- *Predictive analytics* employs modelling and forecasting to predict what will happen next. Although forecasting the future is beneficial, the conclusions are based on the premise that all conditions remain constant. Predictive analytics cannot be used to forecast what will happen after a specific intervention or modification.
- Machine learning technology is used in *prescriptive analytics* to recommend a course of action or strategy based on a variety of criteria. Prescriptive analytics can help you comprehend the consequences of a certain action; nevertheless, the inherent ambiguity and lack of maturity of this sort of analytics might lead to suboptimal decisions.
- Machine learning technology is used to examine raw data to find interconnections, patterns and outliers in *discovery analytics*. While discovery analytics aids to determine what is needed to investigate further, raw data can be incomplete or erroneous, limiting its utility (ArborMetrix 2020).

In recent years, there has been a significant trend towards predictive and preventative approaches in public health due to a growing need for patient-centric or value-based medical treatment. This is made feasible through the use of data. Rather than just treating the symptoms as they occur, clinicians can detect individuals at significant risk of acquiring chronic diseases and intervene before they become a problem. Preventive care may help to avert long-term problems and expensive hospitalisations, saving costs for practitioners, insurance companies and patients. If hospitalisation is required, data analytics can assist clinicians in predicting infection, worsening and readmission risks. This, too, can assist in lowering expenses and improving patient outcomes. In terms of epidemics, healthcare analytics analyses collected data in real-time to better understand the consequences of epidemics and predict future trends so that the spread can be slowed and future epidemics can be avoided (University of Pittsburgh 2021).

Pressures on healthcare institutions throughout the world to save costs, enhance coordination and results, deliver more with less and be more patient-centric are mounting. The growth of medical data systems, digital healthcare data and connected health equipment has resulted in an unparalleled information explosion, which has increased the industry's unpredictability. Nonetheless, the proof is growing that unsatisfactory medical outcomes and inefficiency progressively beset the sector. Developing analytic skills may assist such companies in using "big data" to provide meaningful insights, define their future vision, increase performance and save the critical time necessary to examine healthcare data. New analytics approaches may be leveraged to generate clinical and operational improvements to tackle business problems. Analytics in healthcare will progress from a conventional base point of transaction tracking utilising basic reporting techniques, spreadsheet applications and software reporting components to a prototype that will ultimately integrate predictive analytics, allowing companies to "see the future", provide more individualised health services, predict patient behaviour and allow for dynamic service (Cortada et al. 2012).

Healthcare providers are obliged to report on a variety of key performance measures. Proper data analytics are now a critical role for modernising digital healthcare software. It can increase productivity, manage daily operations and prepare for the future through trend analysis. An essential aspect is how healthcare analytics may benefit many stakeholders in the healthcare business (SourceFuse 2021).

As the ordinary adult lifetime rises along with the population of the world, data analytics in healthcare are prepared to make a significant impact in current treatment. The application of healthcare analytics can cut treatment costs, forecast disease outbreaks, avoid preventable diseases and enhance overall care for patients and standards of living. Data analytics simply digitises enormous amounts of data and then unifies and analyses it using particular technologies.

With healthcare expenses surpassing expectations, the sector needs data-driven solutions. These solutions are beneficial to both healthcare experts and the industry. As more providers are paid depending on medical results, health companies have an economic incentive to reduce expenses while simultaneously enhancing patients' lives. Furthermore, because physicians' judgments are increasingly supported by evidence, studies and medical information offered by healthcare analytics are in great demand. Figure 2.62 shows the reasons for the importance of healthcare analytics technology (Kent State University 2021).

The healthcare industry has a history of being sluggish to react, yet it is in a unique position to benefit from data and analytics insights. The COVID-19 pandemic has emphasised the value of leveraging technology to improve efficiency in remote patient care and telemedicine. Given the popularity of virtual health, it is apparent that the healthcare sector will increasingly rely on AI and big data to fill in the holes in conventional healthcare systems. Instead of remaining just huge storage, the move to electronic health records in clinics has opened up the option of using data models to utilise this information to deliver proactive healthcare actively. As a result, the idea of a "data-driven physician" is gaining momentum (Tabata 2021). Healthcare analytics in the future will contain increasingly bigger datasets for healthcare companies to interpret and manage. As new technologies develop and customer desire for personal control grows, it will become increasingly vital to understand how to navigate the competitive landscape and scale patient's data to stay relevant. Leading hospitals are attempting to guarantee that data-generating technologies are utilised to produce the greatest outcomes for patients as data-generating technologies have spread throughout society and industry. The Internet of Things includes sensors that monitor patient health and machine status and wearables and patients' mobile phones. Because of the network of this equipment, physicians get a complete picture of what is going on in the hospital and may be informed in real-time if a data anomaly reveals changes that require immediate attention.

Fig. 2.62 Why healthcare analytics. (Kent State University 2021)

This drastic move towards data can help doctors make better judgments and, in turn, improve patient outcomes. Artificial intelligence and clever algorithms will evaluate healthcare data to improve medical practitioners' abilities, from picking which individuals to cure to the effective methods to guide them throughout assessment and treatments. These breakthroughs are changing the way the community views healthcare, culminating in healthier communities with extended lifespans (Huiskens 2020). The use of AI inside the forms of natural language processing (NLP) and machine learning (ML) can add a lot of value to delivering better outcomes across the existing healthcare continuum. The application of these technologies in healthcare will also aid in the development of new "value-based care" models, and with the rise of big data, it may be used to drive greater personalisation and transformation in healthcare analytics for patients. Aside from digital disruption, inventive start-ups have a unique opportunity arise and develop solutions that address specific challenges in the healthcare ecosystem (Saxena 2019). Also, the healthcare analytics market is expected to grow at a CAGR of 28.9% over the anticipated period, from an estimated USD 21.1 billion in 2021 to USD 75.1 billion in 2026 (Healthcare Analytics Market 2021).

2.24   Hydrogen

Today, increasing economic and environmental concerns increase the interest in alternative fuels. Hydrogen is one of the important alternative energy carriers' sources that are emphasised (Sazali 2020). Hydrogen is the simplest member of the chemical element family, which is flammable, tasteless, odourless and colourless and represented by the symbol H. This element is normally found in nature as a hydrogen molecule in pairs. Although hydrogen is widely found in nature, this element makes up only 14% of the earth's crust by weight. Hydrogen is not found alone in nature but as a part of the water in lakes, seas, glaciers and similar structures (Jolly 2020). Since hydrogen is not found in a pure form in nature, it must be produced.

Today, there are various methods for hydrogen production, but the four most well-known of these methods are given in Fig. 2.63. Electrolysis, steam methane reforming, direct solar water splitting and biological methods are some of the known methods.

In addition to the shared methods for hydrogen production, there are many more methods, and various classifications are made according to the environmental sensitivity of these production methods and the resources used for production. Three of these methods are widely known (Sazali 2020) and used. These are grey, blue and green hydrogen (Boykin 2021). In the grey hydrogen production method, natural gas (CH4) is the main material, and hydrogen production is carried out using auto thermal reforming (ATR) or steam methane reforming (SMR) methods. As a result of this production, CO_2 is released into nature. Due

to this emission, this method is not environmentally friendly, and for this reason, it is called grey hydrogen production. In the production of blue hydrogen, the same methods as in grey hydrogen are used, and raw material (natural gas) is used.

However, in this method, the carbon dioxide produced due to the decomposition of natural gas is captured and stored. In this way, this gas that causes the greenhouse effect is not released into nature, and the harmful gases that come out during hydrogen production are largely eliminated. Although this method is more environmentally friendly than the grey hydrogen production method, it still has various effects on the environment since it is not possible to capture all the carbon dioxide that occurs during hydrogen production. Hydrogen that is produced by adopting and using the most environmentally friendly production methods in hydrogen production is called green hydrogen. The production steps performed in this production method are given in Fig. 2.64. The method used in this production method is electrolysis. The raw material used is water. Water is broken down into oxygen and hydrogen by electrolysis. The hydrogen obtained as a result of the separation is stored, and the oxygen, which is harmless to nature, is released into nature. Electricity is required for electrolysis, which is the method used in this production method. In this method, electricity produced by renewable energy sources such as the sun and wind is used to produce environmentally friendly hydrogen (Petrofac 2021).

Hydrogen can be used in different fields. These usage methods are shared in Fig. 2.65. Petroleum refining, chemical, stationary fuel cell, transportation and energy storage are the usage

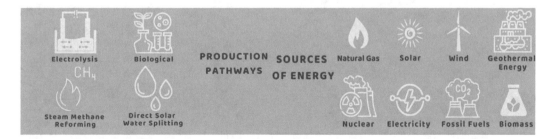

Fig. 2.63 Hydrogen production methods and energy sources used in production

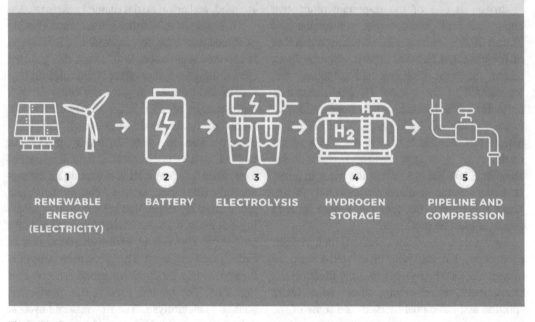

Fig. 2.64 Green hydrogen production

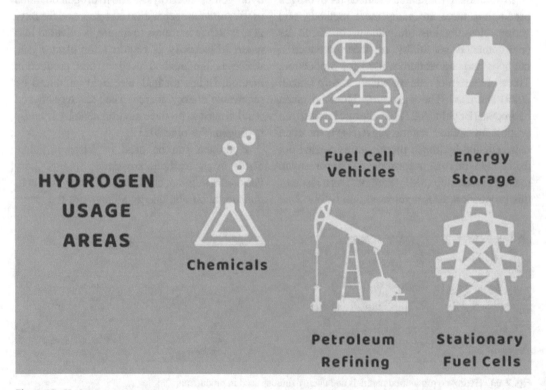

Fig. 2.65 Hydrogen usage areas

areas of hydrogen. Developments in transportation and energy storage appear as important application areas with the potential to increase the widespread use of hydrogen.

Hydrogen storage brings with it various advantages compared to other storage systems. For example, batteries can store certain amounts of energy for a certain time, and long-term energy storage is not possible in these systems. Here, the long-term storage feature of hydrogen storage systems comes to the fore. Depending on the size of the hydrogen storage facility, it can be stored and can keep the amount of hydrogen stored for days or even weeks after the storage process so that there is no storage-related loss during energy production. Although another storage system is pumped water storage systems that do not have various problems in terms of storage time or amount of stored energy, they require large lands and large constructions and are required to be formed in desired geographical conditions. In this direction, hydrogen storage systems also precede such storage systems (FCHEA 2021). Long-term storage is an important feature. These storage systems stand out as important candidates in emergency generators or other critical energy applications. The widespread use of renewable energy sources can bring along various problems. With this diversity in the sources of energy production, various problems can be experienced in ensuring the supply-demand balance in energy systems. These production imbalances experienced here cause the energy produced to be wasted if the production is more than needed or the system to be supported with fossil fuels due to the insufficient amount of energy being produced when the demand is high. In this case, energy storage systems gain importance. Hydrogen storage systems are becoming an important alternative at this point. These systems provide the needed power to the electrolysis machines when the energy need is low, hydrogen production is carried out and electrolysis machines can produce green hydrogen. This hydrogen produced can be stored, transferred to stationary fuel cells for power generation, transferred to fuel cell vehicles for use in transportation, transferred to these pipelines to reduce the

carbon density in natural gas pipelines, or for later use as a cryogenic liquid, compressed gas (FCHEA 2021). When an energy system needs the energy, stored hydrogen can be used for electricity production.

Another area where hydrogen can be used is in transportation systems. Here, in these hydrogen fuel cell vehicles, the vehicles still work with electric motors, but at this point, unlike the battery electric vehicles used today, they are systems that produce their electricity. In other words, these vehicles do not meet their energy needs from an internal battery that can be charged externally, as in electric vehicles. Thanks to the fuel cells they contain, these vehicles effectively have their efficient power plants. By using this fuel cell technology in vehicles, the reverse of the electrolysis process used to produce hydrogen is applied. Thanks to this reverse electrolysis, the hydrogen taken from the tanks react with the oxygen taken from the environment. As a result of this reaction, water vapour comes out of the exhaust and electricity is produced. These vehicles are vehicles that operate without causing emission problems, just like electric vehicles. In addition, while long charging times in electric vehicles are a problematic issue, these vehicles can be charged in a short time, and this problem is significantly eliminated. Therefore, this technology constitutes an important alternative for the future (BMW 2020).

As it can be understood from these two usage areas, fuel cells gain importance at the point of conversion of stored hydrogen into electricity. Fuel cells are devices that produce electricity via an electrochemical way. The components situated in fuel cells are anode, cathode, electrolyte and circuit. Chemical reactions occur in anode and cathode; moreover, hydrogen is used as a fuel while oxygen is used as an oxidant. The working procedure of the fuel cells are very similar to a battery; however, if the fuel is available, the heat and electricity can be obtained without problems such as recharging against battery systems. The fuel and oxygen are inserted into the system at the anode and cathode sites, respectively. Electric energy and heat are produced during fuel cell activities. The electrons go through the circuit, whereas the protons are transmitted through the electrolyte membrane (Behling 2012). The electron movement can be designated as current. At the anode, the splitting of the hydrogen into electrons and protons takes place. As a by-product, water is emitted by the reaction that is carried out by combining protons, electrons and oxygen at the cathode site (Behling 2012).

Hydrogen can be used as a stationary and portable energy transfer source. This source is an energy carrier with great potential for clean and efficient power. Hydrogen energy is an important candidate for energy security, reducing oil and natural gas dependency, reducing greenhouse gas and air pollution. In addition, it is an important alternative fuel source for land or air transportation (Smith 2016). It contributes to the solution of the problem of supply-demand balance, which is in front of the storage of hydrogen and the transfer of energy thanks to hydrogen and the use of renewable energy sources. Since these systems can provide long and efficient storage, they become more advantageous systems than normal short-term and low-efficiency storage systems. Thanks to the storage of green hydrogen, which is produced using renewable energy sources such as the sun and wind, energy storage can be realised. In this way, when excess supply occurs in the supply-demand balance, the excess energy that has been produced can be stored. Hydrogen storage systems provide long-lasting energy storage. In this way, when energy is needed, hydrogen is converted back into electrical energy and an energy supply is provided. By using this storage method, the problem of not being able to reach the energy when needed, which is one of the negative aspects of renewable energy sources, is eliminated. In this way, when energy is needed, instead of preferring the use of fossil fuels to ensure energy balance, the use of stored green hydrogen can be used to prevent CO_2 emissions that may arise from energy production. In this way, energy is produced and used more cleanly. Hydrogen technology is emerging as an alternative that can be used in the field of transportation, and with its contributions in this field, it can contribute to reducing the use of fossil fuels and reducing CO_2 emission values. Today, electric

vehicles are increasing day by day, but charging time is still an important problem in these vehicles. Since hydrogen vehicles provide an important alternative for this existing problem thanks to their short filling times, they stand out as an important option for the dissemination of such environmental-friendly vehicles in transportation. In this way, it can be an important solution for the environmental pollution problem caused by the transportation sector by increasing the environmental transportation options.

The reason why hydrogen attracts so much attention today is because of this low carbon emission. If hydrogen is used in energy transfer, it does not cause carbon dioxide emissions as in fossil fuels. When hydrogen is consumed, only heat and water are produced (Clark 2012). Although hydrogen seems to be such an attractive source, its usage areas are limited today and it is used in a limited way, especially in projects that can contribute to the environment. Today, the amount of hydrogen widely used in the petroleum and chemical industries is approximately 80 million tonnes. The amount of hydrogen produced is expected to increase to 100 million at the end of this decade and 500 million by 2050, especially with the development of technologies such as energy storage and hydrogen transportation. At this point, it becomes an important economy with the expected growth in the field of hydrogen. Today, 95% of the hydrogen produced is produced as grey hydrogen. In other words, fossil sources are used in the production of hydrogen (Schnettler 2020). The main reason for this is that this method is much cheaper than green hydrogen production. Grey hydrogen production costs are 1–2 Euros per kilogram, while green hydrogen production costs are 3–8 Euros per kilogram in Europe and 3–5 Euros in regions such as the Middle East, Russia, the USA and Africa. For this reason, grey hydrogen is a much-preferred source. However, with the increasing awareness about the environment, the demands and policies for green technologies are increasing. Efforts are also being made to reduce green hydrogen production costs. In the future projections, it is predicted that the green hydrogen production costs will decrease by half by 2030 and the kilogram cost of green hydrogen will decrease to 1–1.5 Euros by 2050. With the arrangements to be made and the technological developments to be experienced, it is foreseen that the hydrogen economy will turn into a green hydrogen economy and hydrogen production will increase rapidly, especially after 2030 (Van Hoof et al. 2021). At this point, although hydrogen is an important energy-transportation alternative for the future, it also has important economic potential.

2.25   Internet of Behaviours

The increase in the accuracy of analysis studies that can be done from existing sources and the expansion of their scope has allowed many different technical disciplines to emerge thanks to the development of technology. One of the technologies that can be mentioned in this context is the Internet of Behaviours (IoB). IoB can be defined as studies on data to provide information on user behaviours, interests and various possible preferences (Techvice Company 2021). While the Internet of Things (IoT) technology deals with the interaction of electronic devices with each other, IoB combines location, face and vari-

ous preference information obtained from users and tries to match this information with various behaviours (International Banker 2021).

The starting point of IoB is to create an optimum level of choice with the integration of data obtained by electronic devices providing data exchange over the Internet with IoT (Todaro 2021). To achieve the integration mentioned here, data can be collected and processed from various sources such as customer information, social media, location tracking and citizenship data provided by government agencies (Techvice Company 2021). In this context, transactions related to IoB can generally be considered as a combination of behavioural science, data analytics and technology (Tech The Day 2021). Today, the concept of IoB is used effectively in several areas. These areas are listed in Fig. 2.66 (Pal 2021).

Consider Uber and its Internet of Things (IoT) application. It is used to keep tabs on drivers and passengers. A survey is done at the end of each ride to assess the passenger experience. They can go further by using IoB instead of IoT to collect data without needing to evaluate the experience through a survey. It is conceivable to monitor the driver's actions and then analyse the passenger

Fig. 2.66 Some usage areas of IoB technology. (Pal 2021)

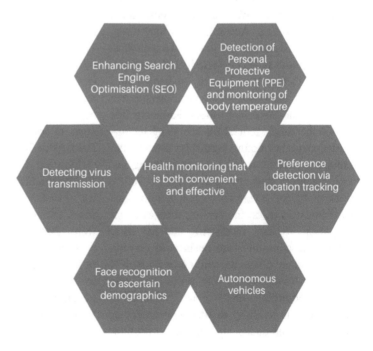

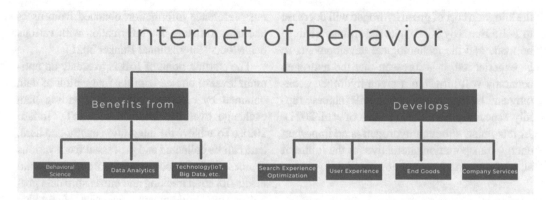

Fig. 2.67 Features of the IoB technology

experience to work on. Figure 2.67 demonstrates the features of the IoB technology.

The incorporation of IoT devices into many aspects of our lives does more than merely assist us in optimising and automating various procedures. It is profoundly altering sectors and ways of operation, including digital marketing. The importance of IoT technology and the IoB cannot be underestimated today, as they will have an impact on consumer behaviour and the marketing platforms used to capture their attention. So, it is critical to begin incorporating the IoB technology into the digital marketing plan of the business world as soon as possible to profit and obtain the greatest number of pleased consumers (Kolomiiets 2021). According to McKinsey, behavioural insights are key to unlocking an 85% boost in revenue and a 25% rise in gross margin. Businesses can utilise data to do behavioural analysis and establish where customers come from and what happens to them. Knowing this data via the IoB is crucial for analysing recommendations, constructing predictive models and developing effective methods to improve engagement, retention and conversion (Singh 2021). Furthermore, IoB integrates current technologies that directly focus on the person, such as facial recognition, location monitoring and big data. While some consumers are hesitant to provide their data, many others are willing to do so if it offers data-driven value. For businesses, this includes modifying their image, advertising items more successfully to their consumers or improving a product's or service's customer

experience (CX). In theory, data on all aspects of a user's life may be collected with the ultimate objective of increasing efficiency and quality (Vector ITC 2021).

The biggest challenge that IoB technology will face is cybersecurity issues. Analyses made on people's data can be of great interest to cybercriminals. In addition, the sharing of this data by various companies due to commercial concerns is also a problem for personal security. The legal process from the past related to this situation should also update itself and make users feel safer (Techvice Company 2021). Apart from these problems, the free use of users' data is another issue that makes people prejudiced against IoB. This issue can be overcome by providing customised prices and services to individuals based on data obtained from individuals. The leading sectors that can do this are the banking and insurance sectors (personal interest rate, insurance pricing, etc.) (Kidd 2019). The advantages of IoB are schematised in Fig. 2.68 (IoT Desing Pro 2021).

One of Gartner's technological emerging trends is the IoB, which will enable this "plasticity" – or flexibility – and allow organisations to react, endure and even prosper during a crisis (Sinu 2020). IoB is still in its early phases, according to Gartner, but by 2025, more than half of the world's population will have been exposed to at least one IoB program, whether from the government or a private organisation. Furthermore, according to Gartner, by 2023, 40% of the global population's digital actions will be

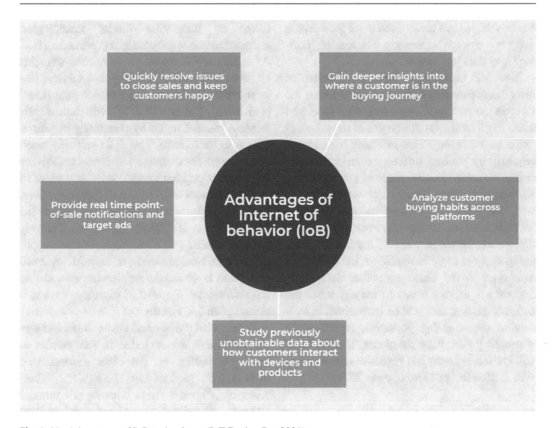

Fig. 2.68 Advantages of IoB technology. (IoT Desing Pro 2021)

tracked in order to affect human behaviour (Axios 2020). However, for these developments to take place, this technology must be accompanied by a transparency policy that is as user-friendly as possible, protecting privacy rights. In addition, it is very important to protect the cybersecurity of this data as highly as possible (International Banker 2021).

To minimise negative customer reactions, it will be critical to keeping a balance between customised offerings and interference. Any firm that adopts an IoB strategy must ensure that it has robust cybersecurity in place to secure all of that sensitive data. In the hands of the right people and with the proper data protection regulations, IoB will play a significant role in the near future (Sayol 2021). In personnel management, it will be feasible to measure the quality of workers' work and analyse how much each department, division or particular employee contributes to the organisation's overall success. Managers in

Europe and the USA are already preparing to chip their staff. CNN Business noted that CEOs in the west are optimistic about implants on their employees. A Citrix survey reports that these implants can replace cards, keys and more to increase productivity. Respondents believe that such technologies will be widely available by 2030 and will be incorporated into the IoB.

Intelligent VoIP systems can find keywords in conversations. There is a strong belief that it will be possible to analyse even more to gather information about one's verbal, paraverbal and non-verbal characteristics. These feats would facilitate the management of atmosphere among employees, which is especially important for organisations in which contact hygiene is essential, such as insurance, audit and law firms. These teams are highly vulnerable to negative effects that can be caused by a single toxic or depressive behaviour of one specialist. Using IoB technology can potentially revolutionise the fashion business. It can be selected as a customised outfit for a person using this modern technology. The idea of "fashion" may then just vanish.

New IoB technologies have the potential to revolutionise medicine. For example, during the Coronavirus pandemic, many individuals acquired a new term: saturation (the degree of oxygen saturation in the blood). This indicator is measured regularly by patients suffering from severe diseases. Wearable devices, such as smartwatches, will almost certainly learn to monitor them fast as well. The same may be said about other appliances. Subcutaneous chips, for instance, will arise that record the temperature of the human body, the quantity of sugar in the blood, the number of leukocytes in the blood and other data. Also, humanity is moving closer to the day when the patient's medical data will be transmitted in real-time to the attending physician, regardless of where the patient is on the planet. The treatment will get more distant and objective (Sannacode | Web & Mobile App Development 2021).

2.26   Internet of Things

The Internet of Things (IoT) connects the virtual world with real-world physical activity. The fundamental idea behind the Internet of Things is to create an independent and secure connection that allows data to be shared between physical objects and real-world applications (Chopra et al. 2019a). The Internet as we know it today has expanded into the real world, embracing ordinary things that characterise the IoT concept. As we know it today, the Internet has expanded into the real world, embracing ordinary devices that constitute the Internet of Things vision. The components of the IoT vision include steady improvements in information technologies, microelectronics and communication technologies that we have seen so far, as well as their potential to continue into the foreseeable future (Mattern and Floerkemeier 2010). As the number of IoT applications grows, intelligent cars, smart cities, smart factories, smart homes, agriculture and energy as components of a broad IoT ecosystem are gaining greater consideration. More potential applications may be pre-

dicted by integrating similar technologies, technical methods and concepts (Vermesan 2013). As a result, IoT is envisioned as an ecosystem that grows to integrate surroundings and services better to satisfy better human life expectations (Lutui et al. 2018). The growth of IoT will dramatically increase Internet traffic by connecting numerous objects to the Internet. This will eventually result in increased data storage requirements. Privacy and security problems would arise as a result of such a huge network. To fulfil all these objectives, a suitable architecture is required. Although the IoT architecture is still being developed, it already has several characteristics. Business layer, application layer, middleware layer, network layer and perception layer are the five levels that make up this architecture. Figure 2.69 illustrates the architecture of an IoT system.

Artefacts of the physical environment and sensor equipment are included in the perception layer. Depending on the object's identifying method, these sensors might be RFID tags, barcodes or infrared sensors. This layer's primary role is to categorise things and collect information about them using sensor tools. This information on the object's position, inclination, humidity, velocity and orientation may depend on the type of sensor. The network layer's primary responsibility is to transfer information from the sensors to the processing unit reliably. This information can be delivered both wired and wirelessly. The middleware layer is largely in charge of managing service between IoT devices and apps. It can also communicate with the local datahub. It transfers information from the network layer to the database. This layer analyses the data and conducts computations regularly. Finally, it makes automatic judgments based on the results of the tests. Based on the object information processed in the middleware layer, the application layer offers global management of the whole program. IoT may be used to implement various applications, including smart homes, intelligent manufacturing and more. The business layer connects with the IoT network, which includes various apps and services. The business layer generates business models using information from the application layer (Chopra et al. 2019b).

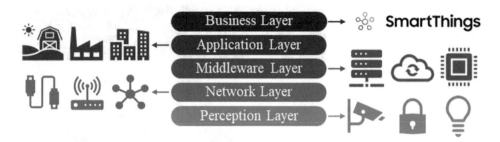

Fig. 2.69 Architecture of IoT

Fig. 2.70 Smart object functions. (Miragliotta et al. 2012)

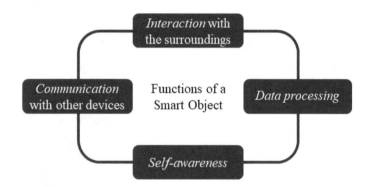

As Fig. 2.70 articulates, an object is called "smart" if it achieves one or more functions: interacting with the surroundings, measuring or sensing environmental variables, communicating with other objects and central systems, processing the obtained data and performing self-awareness and localising.

Smart objects are devices that provide intelligence to IoT implementations and an interconnecting network and allow intercommunication of appliances. Properties of a smart network are depicted in Fig. 2.71.

Smart houses are a hot topic in IoT research. In other words, IoT and home automation are inextricably linked because all passive home electronics are becoming digital, wirelessly linked and capable of communicating over local networks or the Internet (Raj and Raman 2017). More capable and smarter appliances are the beginning of a smarter home environment. Real-time energy usage monitoring for all smart home equipment may considerably cut energy expenditures. In terms of smart security, combinations of alarms and sensors at the range of the property may immediately notify the police or fire depart-

ment and homeowners. Second, by dividing the utilisation of IoT technologies in manufacturing into two categories: manufacturing and user experience, the use of IoT in the industry can be attested. Manufacturers utilise IoT to automate operations, track machinery and determine whether machines require repair. Furthermore, IoT devices in Internet networks are utilised to collect data from customers, which improves the user experience. Then it looks through the information it has acquired to see how it might improve the outcome and possibly solve problems. Finally, IoT-enabled remote patient monitoring (RPM) might enable patients to be tracked while at home. Caregivers can perform the first-step medical check-up without considering the distance, using data of the wearables that collect data of subject's vitals (e.g. blood pressure) (Greengard 2015; Hassan et al. 2018). Fourth, the Internet of Things (IoT) is an initiative to introduce smart city applications that use prevalent sensors in municipal infrastructures to provide real-time data (Latre et al. 2016). To meet the needs of urbanisation, energy management, transportation, health, governance and other

Fig. 2.71 Smart
network properties.
(Miragliotta et al. 2012)

Fig. 2.72 Applications
of IoT

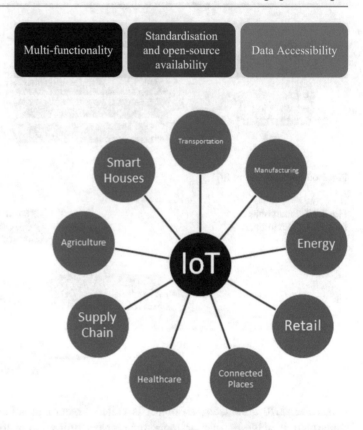

aspects of modern smart cities are all centred on sustainable and effective solutions (Ejaz et al. 2017). Finally, agriculture is one of the examples of IoT applications that are worth highlighting. Farmers and producers would benefit from incorporating the Internet of Things into their operations, as well as a variety of equipment or gadgets that obtain and convey data through cloud services, because it would give them more choices to settle on, cuts costs and increase production efficiency (Namani and Gonen 2020). IoT applications may be found in a variety of settings, as shown in Fig. 2.72.

The demand for efficient systems to regulate and optimise energy and resource use in homes, industrial and agricultural facilities is increasing. With the deployment of the IoT, potential opportunities for reducing resource consumption and boosting efficiency may be identified. In recent years, IoT technologies have presented an intriguing potential to develop sophisticated solutions for sensor devices, and Internet-based information systems are used to monitor and control energy consumption. With an adequate power management system, energy hotspots may be detected and energy production can be raised in real-time in proportion to their process (Chen et al. 2018).

Chen (2018) also proposed an energy monitoring and management framework for a machining workshop. Depending on the tasks of the machinery, unused auxiliary equipment can be turned off. To improve energy efficiency, real-time energy-focused planning can be used. Moreover, by tracking power consumption frequency in real-time, effective load balancing can be used to reduce energy consumption during peak times. The Internet of Things attempts to increase energy use visibility and comprehension by including smart sensors and smart metres into the system. Consequently, data on real-time energy use from complex systems may be easily gathered and analysed to help people or home automation make more energy-conscious decisions. By acquiring vast quantities of energy-related data in near-real-time, these types of solutions generate a great deal of knowledge. As a result, it is critical to plan ahead of time how to integrate these IoT energy monitoring devices into the overall energy management system for the environment (Shrouf and Miragliotta 2015).

Also, IoT is seen as a feasible technical option for tracking, evaluating and making quick decisions about infrastructure provision before it fails. A smart water management network could be imagined by integrating IoT across the whole infrastructure architecture. When IoT is fully integrated, the water tracking system becomes a smart entity capable of detecting and treating problems without the need for constant third-party monitoring (Ramakala et al. 2017). The control application can also provide a real-time summary of the recorded data and make recommendations for corrective action (Geetha and Gouthami 2016).

Conventional approaches may be combined with cutting-edge technology like the Internet of Things and wireless sensor networks (WSNs) to enable a wide range of uses in contemporary agriculture for long-term sustainable food production. New IoT technologies are addressing agricultural issues by boosting farm production efficiency, quantity, productivity and cost-effectiveness. This rapid adoption of IoT in agriculture and smart farming technologies is gaining momentum, intending to have 24/7 visibility into the health of the land and crops, equipment in use, storage conditions, animal behaviour, energy consumption and water usage rates.

The concept of collecting and storing all relevant data, analysing it and giving meaningful outcomes enables the Internet of Things to become a significant modulator of existing urban and rural life. We might argue that robots and people would become increasingly intertwined after the first industrial revolution due to the gradual blurring of their boundaries between them (Greengard 2015). The disappearance of these boundaries allows for the formation of greater direct or indirect ties between humans and nature. Air is a direct method to engage with the environment, and pollution is a major issue nowadays. Air quality data analysis may be used to monitor toxic gases and harmful particles such as carbon

dioxide and soot generated by factories and farmlands (Patel et al. 2016). Following that, necessary measures may be implemented right away. The energy derived from nature is the most extensive sphere of human connection. Because of the massive amount of data and traditional home power networks, measuring, administering, pricing and monitoring the energy that appears in every aspect of a person's daily life is extremely challenging. IoT appliances have the potential to occupy a large field of home appliances on their own, but in a network with a lot of data and continuous processing, the system grows wiser over time and develops its intelligence. The expansion of the Internet's area and depth above and beyond human contact will affect people's lives in the future (Greengard 2015).

2.27 Natural Language Processing

Natural language processing (NLP) is a multidisciplinary field consisting of computer science, artificial intelligence (AI) and linguistics subfields. The study of NLP is concerned with computers' capacity to apprehend words from texts or speech as well as humans. Through the combination of language modelling, computational linguistics, statistical, machine learning and deep learning models, NLP aims to enable computers to comprehensively process language, recognising the intent and sentiments of speakers or writers (IBM 2020). Many languages and text-oriented studies such as translating between languages, building a database, extracting summaries or understanding text content can be performed using this technology (Allen 2003).

The goal of NLP researchers is to learn how humans comprehend and utilise language so that suitable tools and techniques may be developed to help computers understand and modify natural languages to execute the tasks they are programmed to do (Chowdhury 2003). During NLP, the sentence's grammatical structure and word meanings are analysed by breaking down the sentences at hand. This way ensures that the computer understands and reads both spoken and written text at the human level (Wolff 2021). Only if the source text is a speech is a phonological analysis used. This analysis has to do with the internal and between word translation of tone of language, which can convey much about a word or sentence's meaning (Banerjee 2020).

If the source text is written, tokenisation is used. Simply put, tokenisation is the division of large amounts of text into smaller pieces. This process breaks down the raw text into tokens to assist in understanding the context or developing the NLP model (Chakravarthy and Nagaraj 2020). Morphological analysis is a method that concerns the internal structure of words, often used for NLP; it refers to the process of decoding words based on their smallest meaningful units known as morphemes. Banerjee (2020) explains this process through an example phrase, namely, "unhappiness". In his words: "It can be broken down into three morphemes (prefix, stem and suffix), with each conveying some form of meaning; the prefix un- refers to "not being", while the suffix -ness refers to "a state of being". The stem happy is considered as a free morpheme since it is a "word" in its own right. Bound morphemes (prefixes and suffixes) require a free morpheme to which it can be attached to and can therefore not appear as a "word" on their own". An additional method used for NLP is lexical analysis, which refers to the process of determining and examining the structure of words through separation into smaller pieces such as paragraphs, phrases and words. A language's lexicon is a collection of words and phrases that make it up and when working with lexical analysis, lexicon normalisation is considered to be essential. Among others, stemming and lemmatization are considered to be the most prevalent approach to lexical analysis. Stemming is a rule-based approach that regards the elimination of suffixes of a word (e.g. "ing", "ly", "es", "s" and so on). Additionally, lemmatisation is a method that combines vocabulary (the prominence of terms in dictionaries) and morphological analysis and is used to define the root form of a word. Syntactical analysis refers to the investigation of words in sentences to discover the sentence's grammatical structure. The grammatical evaluation and relative arrangement

of words in sentences are used to perform syntactic analysis parsing. The semantic analysis focuses on the interconnections between word-level meanings in a phrase to find probable meanings. Some individuals feel that the step determines the meaning, but all of the stages ultimately decide the meaning. In contrast, discourse integration regards texts as a whole, examining aspects that transmit meaning by linking component phrases. For instance, sentences are linked together or dependent on prior context. This can be illustrated by the word "it" in the sentence "It wasn't that difficult". According to pragmatic analysis, without being encoded in the text, extra meaning is read into it. This demands a comprehensive awareness of the world that includes understanding intentions, plans and goals (Banerjee 2020).

Figure 2.73 is a sample word cloud technique that is used to visualise the result of processed text data. The text data from this section of the book is used to generate the word cloud given in Fig. 2.73.

According to Jusoh (2018), Natural language user interfaces, automated text summarization, information extraction, translation software, questions answering platforms, speech recognition, text mining and document retrieval are all examples of areas where NLP is applied. Figure 2.74 compiles the applications of NLP that can be encountered in daily life (Tableu 2021).

NLP is a strong technology with numerous advantages, even though there are several limitations and issues on that. Table 2.4 demonstrates the advantages and challenges of NLP technology (MonkeyLearn 2021).

NLP is a set of techniques used to create a grammatical structure and semantic relationship, produce natural language and create an output that conforms to the rules of the target language and the data at hand (Reshamwala et al. 2013).

NLP has found great use across various fields. It may not be difficult to imagine the usefulness of this technology in business contexts, as it was shown in multiple cases. For instance, NLP can be used in business process management (BPM) to significantly reduce the amount of effort that is

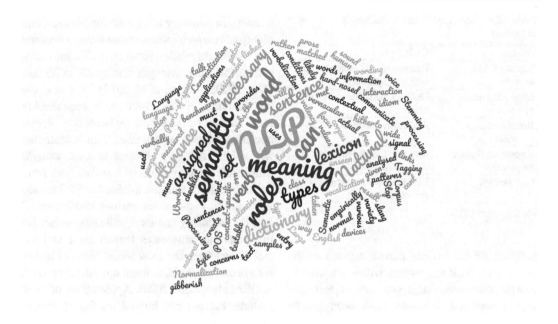

Fig. 2.73 A sample use of word cloud technique

Fig. 2.74 Some applications of NLP technology. (Tableu 2021)

required to ensure these processes work smoothly by enabling automation (van der Aa et al. 2018; Delicado et al. 2017). Furthermore, NLP can be used in human resources (HR) practices such as selection and recruitment. Analysing and screening applicants' resumes, responses and various other types of data with NLP allow for more efficient, less time consuming and less biased hiring processes (Talview 2018). However, NLP is an interdisciplinary technology, and its application has a great range, as demonstrated by its success across fields. For instance, Demner-Fushman and

Table 2.4 Advantages and challenges of NLP technology

Advantages	Challenges
Improved data analysis	Homonyms and contextual words and phrases
More efficient procedures	Faults in the written or spoken word
Better customer experience	Ambiguity and irony
Employees that are empowered	Research and development are lacking
Cost savings	Low-resource and domain-specific languages

MonkeyLearn (2021)

Simpson (2012) explain that, compared to the recent past, NLP techniques, which are used to analyse immensely large amounts of text data from biomedical literature, have been greatly improved and its promising results are a source of excitement in the field of biomedicine.

Furthermore, there has been increasing interest in NLP's application and use for sustainability. A recent report by Chakroun et al. (2019) underlines the importance of NLP in the context of sustainable development, which provides an exemplary use of NLP that contributes to the goal of reaching quality education. The example is of a chatbot that can successfully reproduce everything that a good teacher should do. In other words, it provides students with direction and support that is tailored for each student with the use of NLP. Another interesting application of NLP was demonstrated by analysing sustainability reports using NLP (Luccioni et al. 2020). Suggested by its developers, a custom model of NLP named ClimateQA was specifically designed to analyse, otherwise demanding, amounts of text from financial reports to detect segments that are of interest to climate change.

Additionally, NLP has found use in sustainable, responsible and impact investing (SRI), which emphasises the sustainability of organisations in three main categories called ESGs. Letter E stands for environmental factors such as carbon emissions, S stands for social indicators such as diversity and G stands for corporate governance factors such as bribery and corruption policies (Mukherjee 2020). These elements indicate the overall sustainability of a given organisation, and plenty of research indicates that there is a strong and significant relation between ESG's and financial performance (Morgan Stanley 2021; Whelan et al. 2020; O'Brien et al. 2018). To assess the sustainability of organisations, a considerably large amount of data from sustainability reports and articles has to be examined. This is where the use of NLP is highly efficient as it can analyse immensely large amounts of text data both from existing and live sources, mining for ESG related insights. NLP addresses various challenges of SRI by producing results significantly easier for interpretation, eliminates human error and the feat of including the most recent changes makes for a better, more efficient and up to date approach to SRI (Mukherjee 2020). Applications of NLP include, but are not limited to, the aforementioned examples across numerous fields such as biomedicine, psychology, business, finance and economics, all underline its usefulness and effectiveness in the automation of analytic processes.

NLP is a challenging problem of AI today. The biggest reason for this difficulty is that human language always contains semantic breadth and uncertainty. In studies conducted in this context, difficulties are usually encountered at the lexical and structural levels (Jusoh 2018). However, the increase in the amount of text data available every day and the potential to be used in other applications will make this technology even more important. Large volumes of unstructured, text-heavy data must be analysed efficiently by businesses. A great majority of the data created online and stored in databases is made up of natural human language. Until recently, businesses have not been able to utilise it properly. However, NLP can assist them in making efficient use of the largest available data (Lutkevich 2021). The mere fact that NLP is used even in the digital marketing industry is an indication that this technology will take a more important place in our lives (Lee 2019).

One of the greatest challenges to NLP stands to be code-mixed language, which refers to altered forms of language possibly unique to certain locations such as urban areas (Markets and Markets 2021). This limits the accuracy and efficiency of NLP processes with possible changes to normal meanings of sentences; thus, its application among various fields is also limited.

There has been a growing interest in automation and the facilitation of hiring processes among organisations and researchers. Personality is one of the most significant indicators, even more so than cognitive ability, of many work-related outcomes such as job performance, as established by a growing body of research (Barrick et al. 2001; Judge et al. 2013; Chiaburu et al. 2011; Salgado 2002). NLP methods have a highly promising future in its application to assess personality through numerous types of data, which can come from various sources such as job interviews, resumes and more, which in turn can be used to predict important job-related outcomes (Campion et al. 2016; Andrew 2021; Hickman et al. 2021).

Finally, regarding its financial future, there is a great expectancy of exponential growth in the market of NLP. Forecasts predict that its market share will exceed over 43 billion US dollars in 2025, an enormous increase of 14 times the size in 2017, estimated to be around 3 billion US dollars (Liu 2020). Additionally, the costs of commercial NLP solutions are considered to be high and may not be very tempting for smaller businesses and instead appeal to advanced programmers (Nadkarni et al. 2011). However, costs are expected to decrease in the future due to the increase in demand. This would facilitate the commodification of NLP (Markets and Markets 2021).

2.28   Quantum Computing

Quantum computing is a recent and emerging technology that uses subatomic particles during the computation process through a specific device called a quantum computer. The first question in this technology began when Feynman asked, "What kind of computer are we going to use to simulate physics?" He saw that classical computers are impossible to simulate the physical world. Because of the simulating time, the computers were not making a simulation. They were making

only imitations. He asked at this point "Is there a way of simulating it, rather than imitating it?" in the space-time view. The answer to his question is superposition and entanglement in twenty-first-century technology, although this was unknown when Feynman was considering these questions. Therefore, nowadays, these questions give rise to the starting steps of quantum computing. It is understood that quantum mechanics can simulate the physical world with quantum machines (Feynman 1982). A "bit" is formed in classical computers to process and transfer data. It can be either "1" or "0". Bit 1 corresponds to an electrical signal in the wire, whereas bit 0 does not correspond to any electrical signal. In quantum computing, characteristics of quantum mechanics are utilised in expressing and processing the information as quantum bits. Quantum bits are so small that they work with the physical properties of atomic particles like classical bits. Although quantum bits, aka qubits, are similar to classical bits, there is a major difference: qubits can be in both 1 and 0 positions at the same time. The calculation process, which is also called "superposition", also measures all possibilities, namely, positions, according to the size of the problem. Graphical representation of the bits and qubits can be seen in Fig. 2.75.

While analysing the probabilities of a problem simultaneously, it chooses the correct answer according to some mathematical operators. Thus, quantum computers are designed to solve the

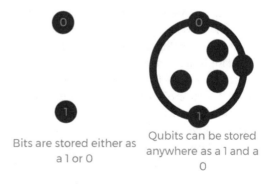

Fig. 2.75 Comparison of classical bits and quantum bits

Fig. 2.76 Comparison
between classical and
quantum computers

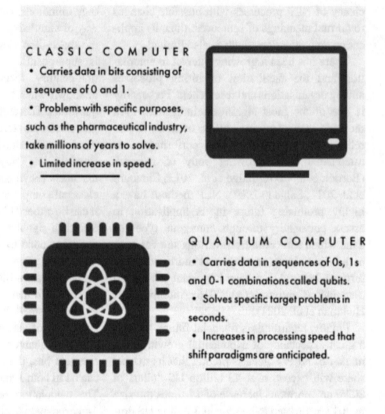

CLASSIC COMPUTER
• Carries data in bits consisting of
a sequence of 0 and 1.
• Problems with specific purposes,
such as the pharmaceutical industry,
take millions of years to solve.
• Limited increase in speed.

QUANTUM COMPUTER
• Carries data in sequences of 0s, 1s
and 0-1 combinations called qubits.
• Solves specific target problems in
seconds.
• Increases in processing speed that
shift paradigms are anticipated.

problems prepared for them in seconds. This makes the computation process much faster because 0 and 1 are not to be performed separately (Hughes et al. 2021). Figure 2.76 shows the comparison clearly. According to this, classical computational approaches, algorithms run on quantum computers can even reach exponential speedups. A quantum computer can operate at a cost that scales polynomially while operating on an exponentially large computational space thanks to certain features of quantum physics. Hence, quantum computing algorithms have the potential to make hard and time-consuming problems yielding and quickly solvable (Martonosi and Roetteler 2019).

Nature is surrounded by many quantum phenomena, such as Bose-Einstein condensation, which is a state of matter, superconductors and magnetic materials. As Feynman argues, it is often difficult for scientists to understand and simulate these materials because of the number of parameters needed to characterise a many-particle quantum system. It is mentioned before this part, a quantum computer can be used as a

quantum simulator to simulate quantum systems way more efficiently than a classical computer (Eisert and Wolf 2013). According to Moore's famous law, the number of transistors located per square inch of an integrated circuit increases exponentially year by year (Moore 1998). If this trend somehow keeps continuing, the quantum effects will dominate in the computer components which are at the atomic scale (Eisert and Wolf 2013). Then, quantum computing will become nothing but a must ineluctably.

After this technology started to develop, some simulation problems can be solved as explained the following:

• While designing sensitive materials such as superconductors provide a better understanding of the material structure and its effects on the human body at room temperature.
• When performing the high-precision chemistry calculations needed to find alternative routes to the Haber process used in fertiliser synthesis in agriculture.

- When performing real simulations of dynamic molecules for increased accuracy.
- When examining problems with drug and molecule design (Maslov et al. 2019).
- Quantum computing offers better solutions for power grid optimisation for a country, more foreseeable environmental modelling development and discovering new materials.
- According to research, it is found that quantum computing could remarkably shorten the time spent performing calculations such as option-pricing and risk assessment (Sara Castellanos 2019).
- Daimler Mercedes-Benz is trying to develop a new battery for electric vehicles using quantum computing (Kanamori and Yoo 2020).

Based on the applications and examples given above, it is seen that the use of this technology is inevitable. Many big companies such as IBM, NASA and Google are quite interested and invest in this technology, which has promising future potential (Kanamori and Yoo 2020).

Researchers who deal with machine learning continuously use principal component analysis, vector quantisation, Gaussian models, regression and classification methods (Ho et al. 2018). It is assumed that quantum computing technology can be utilised to surmount vast amounts of data to yield better scalability and performance in machine learning algorithms (Perdomo-Ortiz et al. 2018). Since quantum computing offers reduced computational time, it may cause many applications of classical computing to evolve quantum computing applications. Robots that are used in drug discovery, logistics, cryptography and finance need to deal with large amounts of data. This creates a necessity for faster computation; as a result, quantum computing can be utilised to perform intensive computational tasks with less time required compared to the required time of classical computing (Buyya et al. 2018). In robotics, solving the problems related to kinematics, such as mechanical movement or unexpected behaviours against a command is challenging. It is expected that quantum neural networks or the other quantum computing algorithms will be able to handle these problems (Gill

et al. 2020). Quantum computing is also expected to be useful in weather forecasting in the future. The computational power of the supercomputers that are used today is limited for some applications such as flood forecasting, urban modelling, sub-surface flow modelling and allied complex tasks. So, quantum computing algorithms can be adapted to solve such problems and achieve better Earth system models in the future (Frolov 2017). Quantum computing in the future in biochemistry and nanotechnology is expected to play an active role. With quantum computing, the results of biochemical processes are calculated faster than normal computers, and it is thought to play a role in the developments in the field of biochemistry in the future. Near future goals are listed in Fig. 2.77.

2.29 Recycling

Contrary to popular belief, products that have lost their function are not wasted. With the loss of product functions, the EoL stage begins. The end-of-life (EoL) phase can be defined in a variety of ways, such as when a consumer or operator disposes of a product without making any structural modifications (Gebremariam et al. 2020); making a non-functional product a reusable form of remanufacturing (Wang and Hazen 2016); recycling, which is the collecting and processing of discarded items as raw materials to create comparable products; with and without energy recovery, incineration and conversion of combustible wastes into gases; burying garbage or throwing it in a landfill ("EEA Glossary – European Environment Agency" 2021); otherwise, it simply results in a leak into the environment (Duque Ciceri et al. 2009). All items having an EoL date become trash. All stages of a product's lifespan are shown in Fig. 2.78.

In Fig. 2.78, the life cycle assessment (LCA) of products is shown. LCA is the most important instrument for determining a product's environmental impact. It is feasible to account for all of a product's environmental consequences using LCA, which covers all stages of the product's life

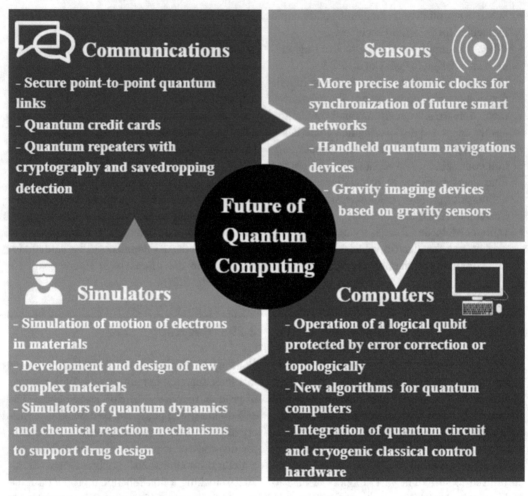

Fig. 2.77 Future of quantum computing

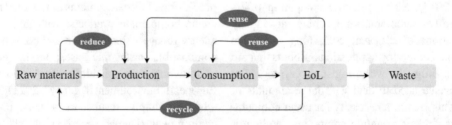

Fig. 2.78 General life cycle of products. (Spilka et al. 2008)

cycle, from resource extraction to waste disposal. The main principle of this cycle is based on the 3R principle. In Fig. 2.79, the main principles of waste management are explained.

According to these principles, preventing waste formation is the most effective and long-term solution to the waste problem. Secondly,

products should be used in other possible means even after they complete their lifetime. After all this, recycling is a necessity. Recycling can be categorised according to the type of material released. Closed-loop recycling, in which the product produced from the material obtained as a result of the recycling of the EoL product, is the

Fig. 2.79 Principles of waste treatment

RECYCLING STAGES	
INITIAL PROCESSES	**SECONDARY PROCESSES**
• Sorting and Seperation • Washing • Size Reduction • Pressing • Briquetting and Granulation	• Composting of Solid Waste • Combustion and Pyrolysis • Biogas Generation • Electrochemical Metals and Recovery

Fig. 2.80 General recycling stages. (Spilka et al. 2008)

same. The utilisation of recovered resources to create a new product is referred to as open-loop recycling. This results in either open-loop upcycling, which refers to the conversion of waste materials into something of more value and/or durability, or open-loop downcycling, in which the quality and usefulness of the resource is decreased or the capture of the material for future use (Dorn and MacWhirter 2016).

In Fig. 2.80, the recycling process is separated into two main stages:

Figure 2.3 shows recycling stages separated by the recycling stages as primary and secondary. Primary processes provide the appropriate situations for the actual transformation process that will begin in the secondary processes. Computer vision is used by artificial intelligence to map intricate material streams by analysing millions of photos. AMP neuron uses deep learning to improve the precise identification and categorisa-

tion of metals, plastics and paper based on their physical features, brand and other factors, as well as contextualising and data storage about each item it observes ("AMP Robotics" 2021). Waste may be used in a variety of ways in terms of technology in processing technology. In general, the initial stages are common, whereas the second stages vary depending on the EoL material. Five categories of secondary processes (Spilka et al. 2008; Vindis et al. 2008):

Chemical recycling: The process which alters the wastes' chemical structure, such as fuel oil production from plastics.

Resources recycling is the degradation of the macromolecule; for example, hydrolysis and alcoholysis.

Biological recycling: The method delivers oxygen treatment, including solid waste composting and waste treatment without oxygen.

Thermal recycling: This method is used to describe waste items that are in the last stages of disposal.

Material recycling: This process aims to acquire material that is a high-value raw material for further processing.

The following are new technologies that have been created to improve the recycling process stages:

- Concrete: Advanced dry recovery (ADR) and heating air classification system (HAS) are technical methods of producing coarse-grained, fine-grained and ultrafine-grained fragments by recycling end-of-life concrete material. The ability to move these methods works favouring them to get installed near precast concrete manufacture facilities or directly into the area of demolishment. Generally, the important features of these methods have been increasing the greenhouse gases (like CO_2 and NOx) produced by transportation, helping to conserve the resources and utilise the sources more efficiently. These technologies offer a promising future to close the material loop completely and execute a circular economy in the construction sector (Gebremariam et al. 2020).
- Metal: A method to sort out the discarded metal by using a laser called laser-induced breakdown spectroscopy (LIBS) has been established by the Fraunhofer Institute for Laser Technology and Cronimet Ferroleg. The recycling process improved efficiency compared to earlier examples thanks to the recently developed detector. The method is created to handle high-speed steel alongside different materials too. High-speed steel objects could have cobalt in them, which is a highly valued alloy, and one could come across them in nearly every hardware shop, for instance, in gimlets or milling tools. LIBS is a method that can be used to differentiate at least 20 alloying metal types in tiny metal objects rapidly, automated and deprived of contact ("Laser-based sensor technology for recycling metals" 2021).

- Plastics: Today's technology like incineration, pyrolysis and mechanical recycling are technologies with high costs and not adequate in environmental and economic terms. The solution for this issue is collecting and processing plastic waste and producing more valuable outputs much more affordable and desirable. The new method is named low-pressure hydrothermal processing. It offers a process to make gasoline, diesel fuel and other valued matters out of the most frequent type of plastic, polyolefin, securely in the environmental and economic regards. A newly developed alteration method works on the combination of hydrothermal liquefaction with effective departure. When the alteration of plastics into oil or naphtha happens, it is proper to work as a reserve for different synthetics or to be torn into monomers, different types of solvents or various outputs ("pulp and paper mill waste becomes fish feed, energy and more" 2019).
- Textiles: Recycling and recovering old textiles and waste fibre materials through various procedures result in recycled textiles. Old or used textiles, tires, footwear, carpets, furniture and non-durable products such as sheets and towels are the most common sources of recyclable textiles in municipal waste. Recycled textiles minimise the demand for virgin resources like wool and cotton, pollution and water and energy use.
- Compared to virgin products, the cheap cost of recycled items is predicted to help the recycled textiles business grow overall. In 2019, the global recycled textiles market was worth $5.6 billion, and by 2027, it is expected to be worth $7.6 billion. Between 2020 and 2027, the market is expected to develop at a CAGR of 3.6% (Parihar and Prasad 2021).
- Paper: A group of Swedish scientists are using waste sludge from wastewater effluent treatment at pulp and paper mills to develop new biobased goods. Bioplastics, bio-hydrogen gas and fish feed may all be made from that organic waste. Over 20 experts are working together on processes that link material and economic flows important to fish farmers, Swedish pulp and paper mills and Paper

Province, the world's most important bio-economy cluster. The goal is to provide advantageous market conditions for innovative biobased products and services (Brunn 2021).

- Lithium-ion batteries: Electronics, toys, wireless headphones, portable power tools, small and big appliances, electric cars and electrical energy storage devices all utilise lithium-ion (Li-ion) batteries. They can affect human health or the environment if they are not properly handled at the end of their useful life. Cobalt, graphite and lithium, which are all considered essential minerals, are used in Li-ion batteries. Critical minerals are raw resources that are economically and strategically vital to the USA, with a high risk of supply disruption and no viable replacements. We lose these important resources when these batteries are thrown away in the garbage (US EPA 2019).

The massive population growth in recent decades and the desire for people to embrace better living circumstances have resulted in a huge increase in polymer consumption, mainly plastics. The continued use of plastics has resulted in a rise in the number of plastics ending up in the waste stream, sparking a surge of interest in plastics recycling and reuse (Francis 2016).

The effect on the economy is explained through the plastics industry. The progress of the plastics sector reflected the so-called linear economy model, which emphasised the beneficial life of plastic items. This was true until the last decades. Growing environmental recognition at social and legislative levels has aided the adoption of the worldwide circular economy model (CEM) in the plastic sector in recent years. This model proposes that the plastic waste created after its useful life be recycled effectively and efficiently (Guran et al. 2020).

There are several advantages of recycling for an individual. First of all, recycling can reduce pollution. It can reduce the demand for new resources by recycling materials. This process may minimise the number of pollutants generated when creating new materials by using recy-

cling methods. Secondly, it leads to lower costs. Recycling has an environmental as well as a financial point for companies. Working with recycled material is substantially less expensive than working with brand new material. For this reason, firms may save money by employing recycled materials. On the other hand, recycling is related to saving energy; since recycling materials saves a lot of energy compared to making new ones. With the increasing environmental awareness, it is beneficial for businesses to be seen as sensitive to the environment ("Advantages of recycling" 2021). These advantages are shown in Fig. 2.81.

Recycling can help eliminate the problem of large volumes of waste simply thrown into plants and still need to be addressed. Through recycling, in most cases, these incinerated wastes are recycled to combat global warming and air pollution. On the other hand, landfills and incineration sites have a very harmful effect on the environment that surrounds them. These sites can cause irreparable damage to animals' habitats. Through recycling, the need for these harmful landfills can be reduced by decreasing the amount of waste sent to them. In this way, especially animals in danger of extinction are protected. Moreover, recycling offers us a more environmentally friendly alternative to extracting raw materials from the soil. This helps conserve resources. So, recycling can help protect the world's natural resources for future generations ("Advantages of recycling" 2021).

Every year, humanity produces 2 billion metric tonnes of trash. According to global waste statistics, waste production would increase by 70% to 3.1 billion by 2050 if the current trend continues (Kaza et al. 2018). Also, the immensely increasing world population may speed up this period. Therefore, the 3Rs (reduce, reuse and recycle) have a significant role in preventing this problem and preserving the environment and natural sources. The success of the 3R implementation depends on the balance between the programs and policies at the local level. The essential points of action relate to governance issues, education and the awareness level of people and critical stakeholders. Other than these, techno-

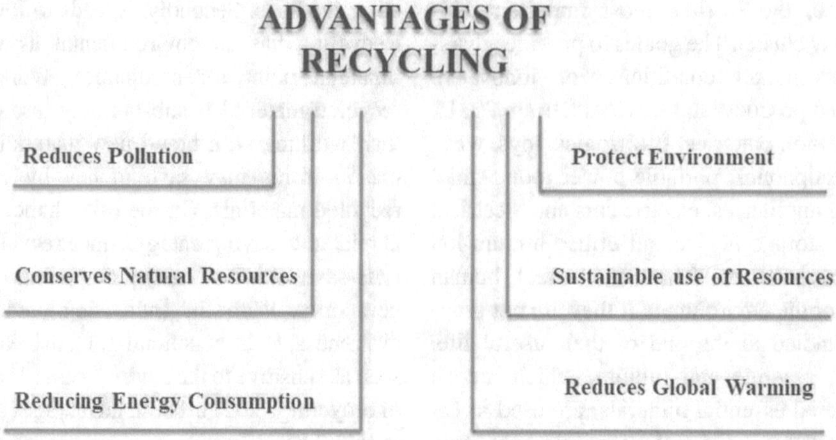

ADVANTAGES OF RECYCLING

Reduces Pollution

Conserves Natural Resources

Reducing Energy Consumption

Protect Environment

Sustainable use of Resources

Reduces Global Warming

Fig. 2.81   Advantages of recycling

logical issues significantly affect that action to minimise the environmental impact while using technologies in the recycling step (Srinivas 2015). Automation will have a considerable impact on the recycling industry. In the coming decades, substantial developments may be expected to provide identification and separation processes more carefully than humans. Indeed, 62.6% of jobs in the waste management sector will become automated by 2030, according to a study conducted by PwC (2017). The recycling process will become much more effective and safer with advances in robot and automation technologies because fewer items are directed towards the landfill, and fewer people are exposed to hazardous waste items ("The Future of Recycling Services 2021). In terms of economy, the circular economy concept has grown in popularity, and it only takes a little effort to put it into practice. As manufacturers and recycling efforts collaborate and achieve critical mass, the future of recycling will see a rise in circular products. Once they are accomplished, they may become standard, with 100% closed-loop recycling systems assuring that recovered components are utilised in products of equivalent value to the original. Additionally, the future of the recycling sector depends on reducing the amount of the different substances recycled together and improving the quality of the raw materials obtained by

the recycling process. One aspect of this future vision is standardising materials across products, which effectively reduces and redefines what we define as waste ("The Future of Recycling – Looking to 2020 and Beyond" 2020).

2.30   Robotic Process Automation

The most important debate in business and information systems engineering is which jobs will be done by humans and which jobs will be done by automation. This argument has become more critical as data science, machine learning and artificial intelligence have become widely used (van der Aalst et al. 2018). These technologies have shaped the structure of robotic process automation (RPA) solutions (Lamberton et al. 2017). The level of automation has risen by 75% in the factories since 1980, whereas in the office, automation has only grown by 3% ("Office 4.0 | RPA – the industrial revolution in the office" 2019). Because classical automation systems do not offer an effective solution for office work. In the classic system, the "inside-out" solution was used, but RPA provides an opposite solution, the "outside-in" approach (van der Aalst et al. 2018). RPA differs from other business automation systems in the following aspects (Willcocks and Lacity 2016):

- RPA does not affect the main system programming code because RPA operates over the top of the systems and accesses them via the presentation layer.
- RPA is about linking, dragging or dropping icons. Thus, RPA does not require software programming expertise.
- Data model or database is not important for RPA.

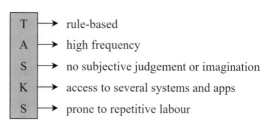

Fig. 2.82 Characteristics of RPA-appropriate tasks. (Fung 2014)

Even though the title "robot" refers to electromechanical devices, RPA is a software-based approach (Lacity and Willcocks 2016). For a better understanding, assume that there is a physical robot standing by the worker while the worker is doing a regular job performed as part of a process-related application, observing and learning about the job the worker is doing on the computer. The only difference from the robot is that it can perform this routine work with software without using computer hardware like a mouse or keyboard ("Robotic Process Automation (RPA)" 2021).

RPA systems use these key components which are (Tripathi 2018):

1. Recorder: It captures mouse and keyboard actions on the UI. The record keeping may then be replayed to repeat the same steps.
2. Extension and plugins: Several plugins and extensions are available on most platforms to make it easier to create and execute applications. RPA providers have created plugins and extensions.
3. Development studio: The robot configuration is created in the studio, and the robots are then trained there. In addition, robots are programmed with a set of instructions and decision-making logic.
4. Bot runner: This is what makes other components work and is called a "robot" and other components make it run.
5. Control centre: The management capabilities of the robot are provided at the centre by monitoring and controlling the operation of the robot.

Integrating automation into a process is costly. Figure 2.82 presents some of the RPA features that are needed to affect the business process positively.

If a process consists of tasks with these characteristics shown in Fig. 2.82, positive results are obtained when RPA automates this process. Also, RPA solutions differ for each business model as they should be designed for each company or industry according to their requirements (Madakam et al. 2019). So, a systematic approach is necessary to analyse the business model for RPA, and this approach consists of at least three main steps: proof of concept, pilot and leveraging. Firstly, the goal of an RPA is installation and identifying potential use cases inside the organisation. End-to-end processes and details are examined to find use cases. Depending on the process, it is seen for which parts of the process RPA can be a solution. At this stage, RPA use cases are established during the pilot phase. Procedures and technical requirements are completed. To ensure data flow, all necessary data must be in electronic form and missing data must be entered to ensure that the data is available. Also, the standardisation of data is essential. Lastly, the RPA system is tested at the leveraging stage. RPA is adjusted according to the tasks to be done. Once the procedures for usage and RPA have been determined, the RPA is then expected to be ready for use (Alberth and Mattern 2017). RPA office automation will perform effectively if these stages are followed carefully, and applicable areas are explained below:

- Business process outsourcing: RPA can replace outsourced workforce in business processes (Tripathi 2018).

- Insurance: Complex tasks in the insurance industry such as policy management can be accomplished with RPA (Tripathi 2018).
- Finance: RPA plays an important role in automating financial processes. Implementation of RPA provides efficiency and credibility (Tripathi 2018).
- Healthcare: Performing tasks such as data entry, patient planning, billing and cash flow by RPA increases the quality of healthcare (Rutaganda et al. 2017).
- Telecom: Tasks such as SIM swaps and order issuance are automated with RPA, providing time-saving, flexibility and scalability (Rutaganda et al. 2017).
- Manufacturing: RPA replaces FTEs (full-time employees), offering a more efficient and faster production process (Rutaganda et al. 2017).

RPA and other automation systems can automate various business processes of enterprises using structured data and specific business rules. With these implementations, the business hierarchy has changed and diamonds have replaced the triangle organisational model. This is shown in Fig. 2.83.

As explained in Fig. 2.83, the majority of the changes were in the medium portion of the market. The pyramid strategy is useful in terms of information storage, but it is also costly. The pyramid model tends to increase staff to fill skill gaps or expand resources. Robots are more flexible as they can more easily adapt to increases or decreases in service volumes. In the diamond model, SMEs and software robots work together to manage better processes that both require humans and are suitable for automation. The diamond-shaped enterprise needs more subject matter experts, quality assurance and management (quality/governance) to coordinate services with internal business units and RPA and business process outsourcing (BPO) providers (Willcocks et al. 2017).

RPA has positive effects on the business process if it is suitable for a business and the implementation steps are followed carefully. Several benefits of RPA are compiled in Fig. 2.84.

In Fig. 2.3, the positive effects of RPA are divided into three – on the company, the customer and the worker. RPA substantially boosts productivity while saving operational costs in terms of the company. Unlike workers, RPA can operate without any performance loss all day and is 15 times quicker than human beings (Engels et al. 2018). According to a Deloitte survey, after adopting RPA, the firm's profitability has increased by 86%. Moreover, the quality of the work done has increased by up to 90% and the consistency of the work by up to 92% (*The*

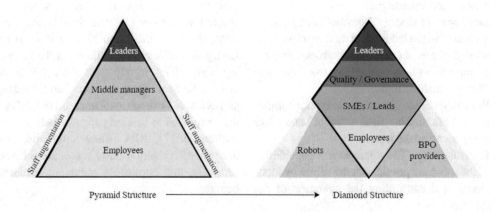

Fig. 2.83 Changing business hierarchy from the pyramid to diamond. (Willcocks et al. 2017)

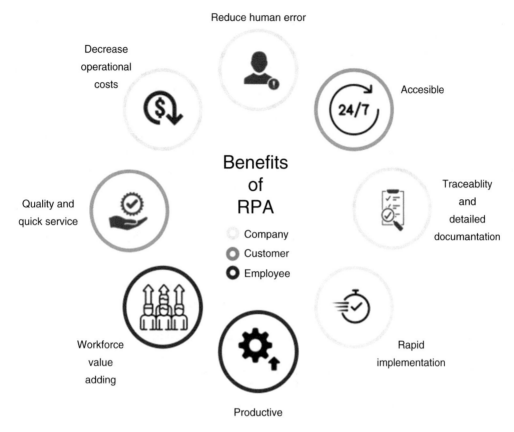

Fig. 2.84   Benefits of RPA to companies, customers and employees

Robots are Ready. Are You? 2017). RPA prevents human error in an ordinary process, avoiding missing steps and greater data input accuracy. Also, customers can get service whenever they want, since RPA is available 24 hours a day (Alberth and Mattern 2017). Employees get rid of repeated work owing to RPA. Thus, they can devote their time to self-development, and they can take part in more important tasks in their firms (Axmann and Harmoko 2020). RPA has an important place in office automation.

RPA has some drawbacks, just like other technologies. Firstly, RPA implementation in the business model is costly. Also, many people believe that to benefit from robotic process automation, the end-user needs a lot of technical knowledge. This misunderstanding frequently hinders individuals from using the numerous advantages accessible to them. Moreover, fears that robots will replace humans and the distributive effects of RPA make it harder for people to adopt RPA. These

disadvantages are due to a poor understanding of RPA (Sadaf et al. 2021). The main limitations of RPA ("The benefits (and limitations) of RPA implementation" 2017) include:

- RPA cannot read a non-electronic data source with an unstructured input. Use OCR (optical character reader) or demand for digital data as a solution to this problem.
- RPA will be problematic in terms of making RPA designs if there are variations in the format of the fields.
- Improvements (changes) in the process flow require a new design of RPA.
- RPA will not solve a faulty and inefficient process. Before using RPA, businesses should address the fundamental causes of their process or technological failures.

These present limits will not last indefinitely. RPA service providers will continue to attempt to eliminate these constraints to offer the leading product and participate in the RPA market (Axmann and Harmoko 2020). On the technical side, instead of being incorporated into an organisation's IT platform, RPA is located on top of IT (Aguirre and Rodriguez 2017). With the excessive accumulation of data in the business world and the evolution of new business processes, RPA will also be needed to automate processes that are not structured or not yet rule-bound. Businesses will be able to enhance quality, operational scalability dramatically and staff productivity with the integration of artificial intelligence and machine learning, due to big data and the cloud (Devarajan 2018).

2.31 Robotics

An autonomous mechanism capable of detecting its surroundings, doing calculations to make judgements and acting in the real world is called a robot. Typically, robots accomplish three things: detecting, calculating and acting. A sensor is used to detect its surroundings, a device that records, measures or indicates the physical properties and converts them into electrical signals. To calculate, robots can contain everything from a tiny analogue or digital circuit to high-performance multi-core processor units. To act, some robots are built to accomplish certain functions, while others are more versatile and can perform a variety of applications (Guizzo 2018). Thanks to the major advancements in silicon-based chips and AI, some robots can even execute basic decisions. Ongoing robotics research is aimed at building self-sufficient robots that can navigate and make judgments in an unstructured environment. The study of robots is called robotics. Robotics is an interdisciplinary topic as it is an integrated mechanism that typically includes sensors, actuators and computational platforms on a single physical chassis. An advanced robotic system is composed of elements on several levels (Mckee 2006):

1. The fundamental physical core of the system is defined by the materials and mechanical systems, which include motors and gears.

2. The control and measuring systems that allow the robot system to function in stable conditions.
3. Electronic systems that connect sensors, actuators and controls with higher-level computing systems and incorporate lower-level intelligence.
4. Computational systems, which are generally based on a real-time operating system, provide a platform for high-level programming, multithreaded and concurrent operations, sensor fusion and sensor integration.

In Fig. 2.85, these four-level explanations are grouped as robotics' subsystems.

Developments in these subsystems help the use of robotics in various fields. These fields are shown in Fig. 2.86.

Healthcare and Medical Advances in robotics have the potential to revolutionise a wide variety of healthcare processes, including surgery, rehabilitation, therapy, patient companionship and everyday tasks. Robotic devices in healthcare are not intended to take over the tasks of healthcare professionals; rather, they are intended to make these tasks easier for them ("Top 5 Industries Utilizing Robotics" 2021). Medical robotics is one of the fastest-growing segments of the medical device industry, with applications ranging from minimally invasive surgery, targeted treatment and hospital optimisation to disaster response, prosthetics and home support (Yang et al. 2017). Rehabilitation robotics, which includes assistive robots, prosthetics, orthotics and therapeutic robots, has made the most extensive use of robotic technology in medical applications. People with disabilities gain more freedom through assistive robots by assisting them with daily activities (Hillman 2004).

Agriculture According to Bechar and Vigneault (2017), agricultural robots are sensitive programmable devices that perform a range of agricultural tasks, such as cultivation, transplanting, spraying and selective harvesting (Santos Valle and Kienzle 2020). Agricultural robotics aims to achieve more

Fig. 2.85 Robotic subsystems

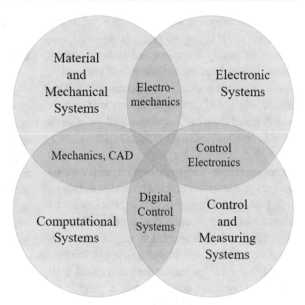

Fig. 2.86 Robotic use cases

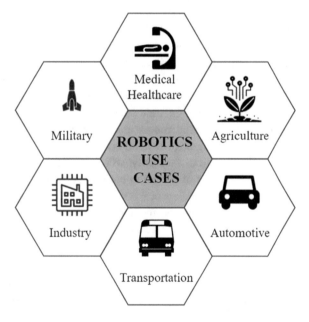

than only the application of robotics technologies to farming. Most of the agricultural vehicles that are utilised for weed detection, pesticide dissemination, terrain levelling, irrigation and other tasks are currently staffed (Cheein and Carelli 2013). Agriculture evolved from a labour-intensive business to mechanisation and power-intensive production systems throughout the last century, and the agricultural industry has begun to digitise in the last 15 years. A constant labour outflow from agri-culture occurred as a result of this transition, primarily from routine activities within the production process. Robots and artificial intelligence can now do non-standard jobs previously more suitable for human labour (e.g. fruit picking, selective weeding, crop sensing) at cost-effective levels (Marinoudi et al. 2019).

Automotive The automotive sector has been greatly influenced by advances in robotic machine vision technologies. Vehicle manufacturers and component suppliers are increasingly relying on 2D and 3D imaging to perform an expanding variety of quality assurance and assembly operations with increasing accuracy and complexity. As 3D systems become more commonly utilised and sophisticated, new applications will emerge, and demand for robotic systems will grow as the market becomes more competitive (Bogue 2013).

Transportation Transportation research committees are progressively coming around to the idea of using robots. A unique robotics tutorial session has been set aside for an upcoming conference on new technologies in transportation engineering. The synthesis of all robotics-related data to establish a complete knowledge base is the most critical short-term goal, which is mostly numeric, graphic and image data (Najafi and Naik 1989).

Industry Industry 4.0, also known as the fourth industrial revolution, is a new period in which industry will engage with technology such as robotics, automation and artificial intelligence (AI), among others. Robotics is becoming increasingly popular in industries around the world. More industrial robots are being developed using cutting-edge technology to aid the industrial revolution. Not only will intelligent robots take the role of people in basic organised processes in restricted spaces, but they will also take the place of humans in more complicated workflows (Bahrin et al. 2016). Precision and the ability to be reprogrammed for jobs of varying sizes and complexity are more important to these machines than speed. Robotic manufacturing technology is also becoming safer to employ. Robots can recognise and avoid people in the workplace thanks to cameras, sensors and automated shut-off capabilities ("Top 5 Industries Utilizing Robotics" 2021, p. 5).

Military Surveillance, reconnaissance, the detection and demolition of mines and IEDs as well as offensive and assault are all areas where military robots might be useful. Weapons are mounted on that last type of vehicle, which remote human controllers currently fire. Although there are many ethical concerns, unmanned

ground vehicles and robotics in many areas are being developed rapidly in the military sector (US Department of Defense 2007).

However, many people think that their jobs may be replaced with robots, but the situation is not like that ("7 Advantages of Robots in the Workplace" 2018). Robots have generated new occupations for previously employed folks on manufacturing lines as programmers. They've shifted staff away from repetitive, tedious tasks and placed them in more rewarding, demanding positions. In the workplace, robots provide more benefits than drawbacks. They enhance the lives of current workers who are still required to keep operations operating efficiently while also increasing a business's chances of success. Robots have also grown increasingly prevalent across various sectors, from manufacturing to healthcare, as a result of robotics advancements ("Benefits of Robots" 2021). The advantages of robotics in these fields are given in Fig. 2.87.

The advantages of using robots in various fields are explained in three main titles: productivity, safety and savings.

Productivity Robots perform work that is more precise and of higher quality. They are more accurate than human employees and produce fewer errors. A robotic pharmacist at the University of California, San Francisco, fills and dispenses prescriptions better than people. There was not a single mistake in over 350,000 dosages ("New UCSF Robotic Pharmacy Aims to Improve Patient Safety" 2011). As robots do not require breaks, days off or holiday time, they can produce more in less time.

Safety Robots remove the need for workers to do hazardous jobs. They may work in unsafe situations, such as low lighting, dangerous chemicals or tight areas. Robots, for example, are assisting with the clean-up of Fukushima, a nuclear power plant in Japan that was destroyed by an earthquake and tsunami in 2011. The Sunfish robot-assisted in finding missing fuel within a nuclear containment tank (Beiser 2018). Furthermore, robots can carry heavy loads without harming themselves or becoming tired.

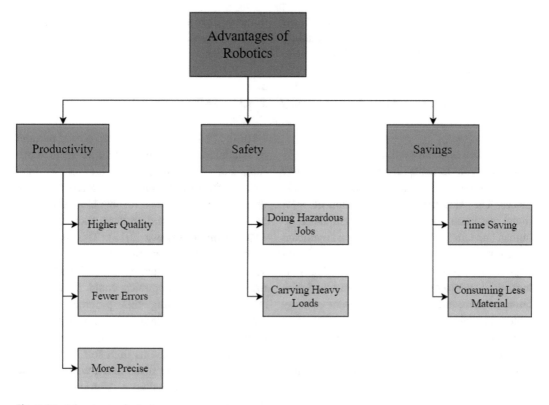

Fig. 2.87 Advantages of robotics

Savings Robots save time by producing more products in a shorter amount of time. Because of their precision, they also decrease the amount of wasted material. Robots save firms money in the long term by providing quick ROIs (returns on investment), reducing or eliminating worker's compensation and consuming fewer materials.

Robots have many more advantages apart from these. They also help businesses to remain competitive while maintaining jobs in the economy. As more industries use robots, the benefits of robotics continue to increase ("Benefits of Robots" 2021).

The number of robots has been exploited to such a degree that both the scientific and industrial sectors will be unable to absorb them, and we are about to approach a phase in which selection will be required (Pagliarini and Lund 2017). According to a study conducted by Oxford University, in the next 20 years, 47% of jobs in the USA may be automated (Frey and Osborne 2013). Additionally, according to 2021 robotics industry data, the number of robots used in the automotive industry is about 1290 per 10,000 employees ("US Robot Density in Car Industry Ranks 7th Worldwide" 2021). We can understand with this statistic, whether we like it or not, robotics will be a much more essential part of daily life. Sensors, motion control and machine learning advancements have increased the versatility of robotics and cognitive systems to unprecedented levels (Matthews 2019). Given the rapid advancement of robotics, now is an ideal time to consider what the future may hold.

1. *Artificial Intelligence Will Face Regulation*

Elon Musk has fuelled the fire by declaring artificial intelligence (AI) to be our "most existential concern" and comparing AI research to summoning demons (Palmer 2019). Although it can benefit humanity, AI will remain to be examined by those worried that robots will become more intelligent than their developers. Expect the fight over AI regulations to continue in the future and governmental or industry rules to arise (Worth 2016).

2. *Designers Will Move Away from Humanoid Robots*

Humanoid aesthetics will play an increasingly significant role in accepting new robots as they become more widespread in homes and workplaces. The "uncanny valley", a graph of emotional responses to a robot's resemblance to human appearance and behaviour, is a significant impediment to a seamless transition to a society populated by robots (Brenton et al. 2005). If humanoids are to become more intertwined with humans, more must be done to prevent uncanny valley. To make people more at ease with robots and make it simpler to engage with them, robotics developers will enhance semi-humanoid advancements that maintain essential human characteristics while also retaining conventional machine forms (Worth 2016).

3. *Aerial Robots Will Reach Widespread Adoption*

Safety concerns, efficiency and legislation have recently sparked much discussion about UAVs. Despite this, people and companies continue to experiment with unmanned drones in various ways, like lifeguards have flown them to keep swimmers safe throughout the summer months. However, there are concerns about the growing usage of that technology. A man from the USA attached a handgun to a consumer-grade drone and fired the gun remotely. While local authorities claim this type of drone usage is legal, the FAA (Federal Aviation Administration) claims it has laws prohibiting installing weapons on civil aircraft (Worth 2016).

The future of robotics will be reshaped with technological advancements and how society will react to these developments. Advances in robotics come with ethical issues. The ethical question will be critical and thoroughly explored in the future. On the other hand, experts have critical

duties such as initiating discussions amongst specialists from diverse fields and developing rules for this area.

2.32 Soilless Farming

Until the year 2050, the estimated total number of inhabitants of the world will be up to 9 billion, and meanwhile, nearly half of the land suitable for farming might be useless. To fulfil the need for sustenance for this blooming number of people, the food products should be up to 110% more (Okemwa 2015). This increasing number of the population needs an efficient solution to manage their needs. Soilless farming methods like hydroponics could be a proper suggestion regarding this problem. Soilless farming can be explained as a way of breeding plants by using water to nourish the roots while using different mediums than soil. The plant's needed nutrients are fused into a water solution in proper density, called a "nutrient solution" (Oyeniyi 2018). The goal is to create an environment where the plant can grow properly without soil as a growing medium. When using farming methods such as hydroponics, the process is, simply put, making the plant meet the nutrients it needs by carrying them to the roots as a water solution containing the needed oxygen and nutrients ("Soilless Agriculture" 2021). Many sorts of yields can be grown hydroponically. Herbs such as rosemary, sage, oregano, basil, chive, celery, mint and lavender; vegetables such as cabbage, cucumber, potatoes, cauliflower, cabbage eggplant, lettuce, peas and asparagus; fruits such as tomatoes, watermelon, blueberries, strawberries, blackberries and grapes (Mohammed 2018). Different techniques of hydroponic configurations and systems according to the way of giving nutrients are shown in Fig. 2.88.

Soilless farming techniques in Fig. 2.88 are explained below:

The Deep Water Culture The deep water culture makes the roots of the plant float on the nutrient water constantly circulating (Elkazzaz

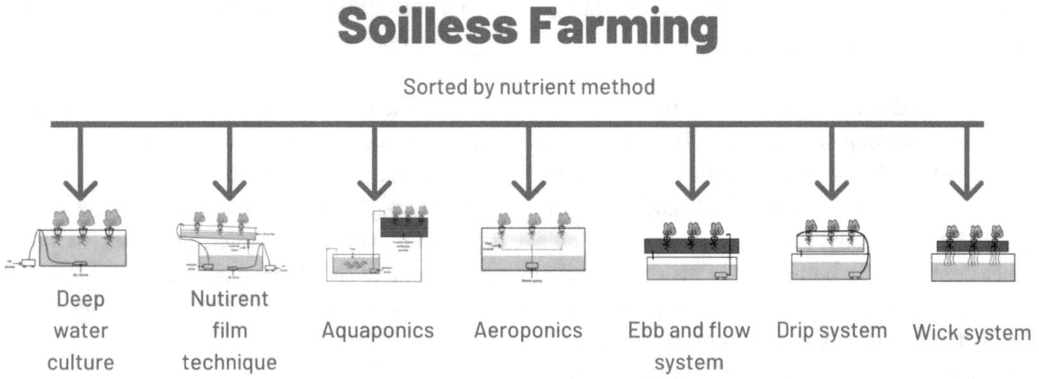

Soilless Farming

Sorted by nutrient method

Deep water culture | Nutirent film technique | Aquaponics | Aeroponics | Ebb and flow system | Drip system | Wick system

Fig. 2.88 Soilless farming methods according to the way of providing nutrients

2017). It is a method of growing plants on a floating or hanging support, such as rafts, panels or boards, in a container holding a 10–20 cm nutritional solution (van Os et al. 2021). The roots are dangled 6–18 inches into the nutrient water, which contains dense oxygen levels, to the time of harvesting. The deep water culture contains a large amount of water to balance inconsistencies in the saturation of the solution. In any failure in the pump mechanism, one would have plenty of hours fixing the pump without facing essential issues such as the roots not reaching the nutrients and water they need (Elkazzaz 2017).

Nutrient Film Technique (NFT) NFT is a hydroponic system based on sending nutrients to plants' roots by circulating water. The nutrient film technique system provides plants with an environment in which the plants are placed in a platform that is placed above the tank containing the nutrient solution in a slightly tilted manner. The nutrient film technique is focused on circulating the nutrient solution plants need to their roots through some routes such as pipes. The solution is contained in a tank, and through a pump, it is sent to pipes or watering equipment, and it turns back to the tank, constantly circulating (Elkazzaz 2017). For this technique, a level slope must be prepared, immune from depressions that could allow deep pools of the non-rotating solution to collect within a channel and multiple methods are utilised to achieve this (Spensley et al. 1978).

Aquaponics Aquaponics is a mutual system of plants and fish and their environments. Aquaponics consists of both aquaculture and hydroponics as symbiotic cooperation since the water used for the plants in the hydroponics system is also used for the animals living in the water tank of the aquaculture system. For aquaculture, the waste of the water animals should be expelled from the environment since the excrement is harmful to the animals due to the ammonia it contains. The aquaponics create a configuration where the water is purified by the organisms living in the roots of the plants, which are in the hydroponics, nitrifying the waste coming from the aquaponics systems and turning the waste into nutrients the plant can benefit from, therefore cleaning the water and circulating it back (Elkazzaz 2017).

Aeroponics Aeroponics systems work by sending nutrients to the plant by spraying a mist of nutrient water to their roots. Aeroponics is the method where the plants' hanging roots are exposed to a mist from the nutrient solution periodically; the main asset to this method is aeration (Modu et al. 2020). In aeroponics systems, the plants' roots are left dangling inside a cover where a powerful pump mists the nutrient solution and diffuses the solution to the entire root. Some aspects of aeroponics make it highly dependent on energy since the spraying mechanism is the most crucial part of the configuration. Since the roots constantly contact the air, they tend to be dehydrated very quickly if the pumping circle is disrupted (Elkazzaz 2017).

The Ebb and Flow The ebb and flow is a configuration that depends on initially surrounding the grow tray then discharging the nutrient solution to the nutrient solution tank and repeating this cycle. Generally, this cycle is regulated by a timer controlling a pump that is inside the nutrient solution tank. In regards to the type of plants, the heat and humidity of the environment and the kind of medium used for the plant to develop in, the daily periods of the pump are regulated (Dunn 2013).

Drip Systems Drip systems are most likely to be the most preferred kind of hydroponics worldwide. By a pump controlled by a timer, the nutrient solution drips to plants one by one thanks to a dripping pipe (Dunn 2013).

Wick System Wick system is the easiest type of hydroponic system. It doesn't have any parts that are moving and hence does not require any electrical types of equipment, including pumps. On the other hand, the wick is the link between the plant in the pot and the solution to nourish it in the existing reservoir. It is also beneficial in regions where electricity isn't available or is unstable because it doesn't require electricity to function. Plants are grown on a substrate in this method (Elkazzaz 2017).

Also, these systems can be designed by using two different methods according to the circulation of water as shown in Fig. 2.89.

Closed Soilless Systems Closed systems use the same nutrient water and circulate it relatively long. In these systems, the nutrients in the solution are cycled to reuse again, and the solution's concentration of nutrients are checked and regulated likewise. Balancing and regulating the nutrient solution is a challenge in a closed soilless system. The examination and analysis of the nutrients within the solution is a requirement per week (Elkazzaz 2017). This nutrition solution is going to be recycled regularly. Water and nutrients are constantly monitored. The disadvantage

of this technology is that it is electricity-dependent (Lee and Lee 2015).

Open Soilless Systems Open systems use a new nutrient solution for every irrigation period. In these systems, a recent nutrient solution is prepared via adjusting the ingredients in the watering system for every watering cycle. In these systems, there should be a goal specified to regulate the solution that is reaching the roots (Elkazzaz 2017).

On the other hand, the main benefits of this system are that there will be no risk of infection in the plant system as a result of the regular changes in the environment (Jones 2016). Vertical and smart farming can be used with soilless farming techniques. Firstly, vertical farming is one of the suitable configuration options when working with hydroponic systems. Vertical farming is the method of producing crops via a configuration that is multiple layered and controllable with being inside of the buildings where every aspect affecting the improvement of the plant-like saturation of carbon dioxide, amount of illumination, levels of heat and nutrition are under close control to get yields with great qualities and quantities for the entire year without being related on the sun for illumination and different outside obstacles ("Vertical Farming" 2020). Vertical farming could be a considerable solution for indoors and small spaces. Secondly, smart farming is the application of stand-by technology to agricultural production practices to reduce waste and increase productivity. Smart farms, as a result, draw on technical resources to assist in several steps in the production path with planting and soil management, irrigation, controlling the pesticides, monitoring the delivery and so on (Bhagat 2019). On the other hand, smart farming is based on a certain and resource-efficient technique that aims to increase production of agricultural goods efficiency while also improving quality on a long-term basis (Balafoutis 2017). Moreover, using IoT technologies provide solutions to increase productivity on yields. With the wireless sensors, data can be collected from sensing devices and delivered to the main servers

Fig. 2.89 Open and closed methods of soilless farming

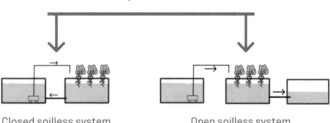

Soilless Farming

Sorted by water circulation

Closed soilless system   Open soilless system

using networks. By analysing these data, the most suitable growing conditions can be selected (Ojha et al. 2015).

Fertile lands are rapidly disappearing due to climate change and intensive farming. On the other hand, the world needs to feed more people. Hydroponics, also known as soilless farming, has the potential to convert agriculture by offering a more sustainable and effective alternative to conventional agriculture (van Os et al. 2021). The hydroponic systems have positive attributes, making the method more efficient than traditional farming. The positive attributes of farming without using soil are saving and using more qualified soil for growing the core crops, recovering at least 90% overrun water, working with a stabilised expense of converted water, and the achieved harvest is the most effective when compared to farming with soil (Elkazzaz 2017). There are several advantages of soilless farming. It can be sorted as (Tajudeen and Oyeniyi 2018):

- *Increased productivity*: It has been observed that the products grown with soilless agriculture are more efficient because they spend energy on leaf and fruit development instead of root system development.
- *Lower labour costs and no need for expensive machinery:* Less labour is needed in the soilless farming method because there is no need for ploughing or weeding. On the other hand, there is no need for expensive vehicles such as bulldozers.
- *Is not restricted by the seasons*: With this method, plants are constantly fed with the

nutrients and water they need during their growth process. Therefore, they are not affected by the season in this process.

- *Low-cost management*: The cost of running the systems is low, especially for the NFT system, because it is virtually automated.
- *No weed competing:* There is no weed problem in soilless agriculture as no soil is used, and all microbes are carefully selected.
- *There are no pests or diseases in the soil*: Soilless farming has been observed that soilless agriculture has very few pests and disease problems than conventional agriculture.
- *Sensibility for nutrient supply*: Nutrients are either released according to plant requirements or, in most cases, recycled. Therefore, no excessive fertiliser is used.
- *Pollution and preservation of water and land*: Less water and land is used up in soilless farming. Because there is no indiscriminate use of water for irrigation, there is no demand for traditional spacing regulations.
- *Sustentation life in space*: This method has been tested for use on other planets where there is no soil to grow on.
- *More suitable for research purposes*: With this method, it can be used to determine the amount of fertiliser, water or light a plant needs to grow and develop.
- *Greenhouse and vertical farming adaptability:* Soilless farming is commonly practised in greenhouses and is often grown vertically.

Soilless farming is a relatively new technique that has evolved significantly in the 70 years since its birth. Its multiple technologies can be used in both emerging and high-tech space stations. Progression in related technologies such as artificial lighting and agricultural plastics will increase crop yields and decrease production unit costs, so new cultivars with improved pest and disease resistance. On the other hand, governments are determined that there are politically desirable effects of hydroponics. Another desirable effect is job creation. This type of employment generates tax revenue as well as personal income and improves the quality of life ("Future" 2021). There are several things for governments to do for the success of the soilless farming process. The government should consider it as a vital part of the nation's food production chain. The government must provide credits to encourage entrepreneurs and young generations to invest (Tajudeen and Oyeniyi 2018). However, there is a dearth of technical knowledge of this new technique among farmers and horticulturists in many nations, and well-trained employees are required. Significant progress has been achieved recently in the development of economically proper soilless systems, and there is now a broad range of business applications in countries that have implemented farming technologies (Elkazzaz 2017). Accordingly, this deficiency can be eliminated by providing innovations in the field of education. More people can be employed in the future. Although methods that increase productivity, such as vertical farming and smart farming, continue to increase, some plants will depend on the soil to grow. With all technological advances, soilless farming is free of all of the current issues that soil has, making it a proper alternative to soil farming to reach a world without hunger by the year 2030 (Tajudeen and Oyeniyi 2018).

2.33 Spatial Computing

In 2003, Simon Greenwold, who was a researcher in MIT Media Lab, defined spatial computing as follows: human interaction with a machine in which the machine keeps and manipulates referents to real objects and environments (Greenwold 2003). We can explore the concept in detail in

three main parts, which are virtual reality (VR), augmented reality (AR) and extended reality (XR).

A. *VR (virtual reality)* is a cutting-edge human-computer interface that creates a lifelike world. It replicates a person's physical presence in both the real and virtual worlds (Zheng et al. 1998). It is a fully immersive, engrossing, interactive alternate reality experience in which the participant is completely engaged in sophisticated human-computer interface technology (Halarnkar et al. 2012). The virtual world allows the participants to move freely (Zheng et al. 1998). They may look at it from a variety of angles, reach into it, grab it and change it. They can engage with the virtual world via traditional input devices like a keyboard and mouse or multimodal equipment like a wired glove, omnidirectional treadmill and/or a Polhemus boom arm (Halarnkar et al. 2012). There is no little screen with symbols to manipulate nor are there any commands to input to make the computer do something (Zheng et al. 1998).

Based on the definitions above, it is reasonable to conclude that certain factors define the quality of any VR system. The three (3) characteristics referred to as the 3 "I" or triangle will be the subject of this review. Immersion, interaction and imagination, or presence, refer to the feeling of being involved in or a part of a computer-generated environment. This is due to the system's stimulation of the human senses (haptic, aural, visual, smell etc.). Interaction is a way of talking to the system, although unlike traditional human-computer interaction (HCI), which employs 1–2 dimensional (1D, 2D) means such as a keyboard, mouse or the keypad, VR interaction is commonly done using three dimensional (3D) means such as a head-mounted device (HMD) or a space ball. VR interaction systems have some characteristics, including effectiveness, real-time responsiveness and human engagement. The system designer's imagination might be thought of as a strategy for achieving a certain goal. VR systems are used broadly for solving complicated problems in a variety of sectors. Their effectiveness as a more efficient way of communicating concepts than traditional 2D drawings or text explanations cannot be denied (Bamodu and Ye 2013). Virtual reality is divided into three levels:

(i) *Non-immersive*

This level is typically encountered on a desktop computer, where the virtual environment is created without the need for any certain hardware or processes. It has the potential to be used for training reasons. Almost any event may be reproduced if the necessary technology is present, which eliminates any potential threats. Pilots can use flight simulators to experience and prepare for circumstances that are either impossible or too dangerous and expensive to replicate in real-world training. The illusion of immersion is created by the user's ability to interact with responsive computer-generated characters and behaviours (Halarnkar et al. 2012).

(ii) *Sensory-immersive (semi-immersive)*

In this method, real-world modelling is crucial in a variety of virtual reality applications, including robot navigation, building modelling and flight simulation. The user can navigate a visual depiction of himself within the virtual world. As an example, the CAVE is a 10'x10'x9' cube in which the user is surrounded by projected pictures, giving the impression of being completely immersed in the virtual environment (Halarnkar et al. 2012).

(iii) *Neural-direct (fully immersive)*

The most significant concept in virtual reality, as well as the ultimate aim, is neural-direct. Immersion into a world in which the human brain is directly linked to a database as well as the viewer's present position and orientation is the goal of this sort of virtual reality. It fully ignores the equipment and physical senses in favour of immediately transferring a sensory input (Halarnkar et al. 2012).

B. *Augmented reality (AR)* is defined as a real-time, indirect or direct view of a physical, real-world environment that has been enhanced by the addition of virtual computer-generated information (Carmigniani et al. 2011). Augmented reality (AR) is a broad term that encompasses a variety of technologies that display computer-generated content such as text, pictures and video onto users' perceptions of the actual environment. Academics first described AR in terms of particular enabling equipment, such as head-mounted displays (HMDs). However, claiming that such definitions were too basic for a changing and increasing industry, three criteria may be used to define AR experience: (a) a combination of real-world and virtual components, (b) that are interactive in real-time and (c) that are registered in three dimensions (3D) (Yuen et al. 2011). There are many different types of augmented reality, but they all share a few things in common: screens, graphical interfaces, trackers and processors. There has to be a way for users to comprehend both real situations and the digital format provided information, a pointing instrument (e.g. a mobile device), a way to guarantee that the digital information is properly coordinated with what the consumer is currently seeing in real-time, and software program to handle the display. Creating applications determines how these elements are assembled and then used (Berryman 2012).

The possibilities for spatial computing technology are endless: simulation of surgical procedures, archaeology with site reconstructions, virtual museum tours, architecture, phobia treatment, training with simulators and many sorts of learning. The importance of virtual reality in education and learning derives from its potential to improve and support learning, boost memory capacity and make better judgments while working in a fun and engaging setting. When we read textual material (such as a printed document), our brain participates in an interpretation process, which increases our cognitive efforts. Understanding gets clearer while utilising virtual reality since there are fewer symbols to comprehend. For example, picturing the process rather than reading a verbal description makes it simpler to understand how a machine performs. With further VR technology, it becomes much easier to visualise. Virtual reality-based learning has been shown to improve both test performance and boost learners' level of attention by 100% (Elmqaddem 2019). Virtual reality (VR) is a cutting-edge teaching method for medical professionals. It can be utilised to provide adequate medical communication for a variety of conditions. It is often used to identify and investigate bone structure in orthopaedics. Medical students are now able to perform surgery virtually by step-by-step learning. Furthermore, VR is beneficial in the treatment of cancer patients. The patient's chemotherapy is carried out easily with high precision. By using a VR headset, one may view the patient's body and parts from various angles. On the patient's side, when they wear VR glasses, this technology boosts their confidence by reducing their apprehension since it gives more realistic information. This technique allows a cardiac surgeon to monitor a patient's heartbeat and changes in rhythm (Javaid and Haleem 2020). AR may be utilised to present a real-world battlefield environment and augment it with annotation data for military applications. Liteye, a firm, studied and developed certain HMDs for military use. A hybrid optical and inertial tracker with tiny MEMS (micro-electro-mechanical systems) sensors was created for cockpit helmet tracking. The use of AR for military training planning in urban terrain is effective. Arcane, a company, created an AR approach for displaying an animated terrain that may be utilised for military intervention planning. The National Research Council of Canada's Institute for Aerospace Research (NRC-IAR) developed the helicopter night vision system using AR technology to extend the effective range of helicopters and improve pilots' capability to control inclement weather. Practising in massive battle situations and replicating real-time enemy action, as in the battlefield augmented reality system (BARS), may provide additional benefits particular to military users. The BARS system also includes tools for authoring new 3D

Fig. 2.90 Application areas of spatial computing

information into the environment, which other system users may see (Mekni and Lemieux 2014). Application areas of spatial computing can be seen in Fig. 2.90.

Fixed exterior systems, mobile interior systems, mobile exterior systems and mobile interior and exterior systems are the five types of systems of augmented reality. A mobile system can be defined as a system that allows the user to move around without being restricted to a single room, and therefore allows users to move using a wireless connection. Static systems cannot be relocated, and users must utilise them anywhere they are established, with any ability to move unless they are moving the entire system configuration. The sort of system to be developed is the first decision that programmers must consider since it will guide them in selecting the tracking system, display option and potentially interface to employ. For example, installed systems will not employ GPS tracking, but exterior mobility systems will (Carmigniani et al. 2011). Extending reality, which might be called the future technology platform, has altered the way we work, learn, engage and amuse by integrating the physical and digital worlds. It also affects how companies train their staff, service their customers, generate things and manage their value chain.

C. *Extended reality (XR)* stands for augmented reality (AR), virtual reality (VR) and mixed reality (MR) (MR). AR merges physical and

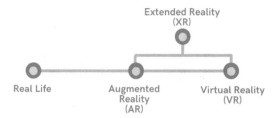

Extended Reality
(XR)

Real Life   Augmented   Virtual Reality
                  Reality   (VR)
                   (AR)

Fig. 2.91 The scope of XR

virtual components in a real-time projection, whereas VR enables users to control and steer their actions in a simulated actual or fictional world (Chuah 2018). Figure 2.91 shows the scope of XR.

The concept "mixed reality" was used to characterise a range of virtual and augmented reality (VR/AR) technologies that combine the real and digital worlds. MR is a concept that is occasionally defined variably, and based on its meaning, it might include or exclude specific VR or AR implementations. As a result, the concept of "extended reality" has lately gained popularity as a catch-all word for AR, VR and MR. The terms VR and AR are used to refer to gadgets that fully block the user's sight of the actual world and XR to relate to devices that allow users to see both the real and digital worlds. In Table 2.5, the characteristics of several XR device types are listed (Andrews et al. 2019).

VR, unlike AR, obscures the actual world and digitally places items (such as music, movies, images and messages) throughout the real world in a completely manufactured manner. Users may mentally immerse themselves in the simulated 3D environment and experience the sense of "being physically there", thanks to VR's capacity to mimic detailed real-life scenarios. As a result, virtual reality (VR) has been dubbed integrated virtual multimedia that produces a 3D, virtual imagined and interactive entertainment environment that a person perceives as being in the actual world (Chuah 2018). Whereas virtual reality innovation, or virtual environment as Milgram refers to it, totally immerses consumers in an arti-

ficial world without allowing them to see the reality. Augmented reality (AR) technology enhances the perception of reality by overlaying digital environments and cues onto the physical world in real-time (Carmigniani et al. 2011). The goal of AR is to make the person's life easier by introducing virtual knowledge not just to his surrounding environment but also to any indirect view of the physical environment, such as a video broadcast. The person's perspective of and connection with the physical environment is improved through augmented reality (Carmigniani et al. 2011). As a result, people engage with real-time virtual 3D items seamlessly and intuitively, seeing them as realistic and learning to comprehend, analyse and connect with the items (Halarnkar et al. 2012).

VR is a technology that draws inspiration from a wide range of areas. However, it is distinguished by a close relationship between human aspects and assisting technology as a discipline. Hardware, programming, human aspects and transmitting VR via the Internet are the most pressing issues (Zheng et al. 1998). Today, AR is everywhere in our daily life. The wide usage of AR has become feasible due to the low cost of smartphones and other technology designed to process and present data at high rates. In addition, experts anticipate that the advancement of portable devices that can deliver augmented reality content and experiences will keep accelerating. As the technologies that enable AR to continue to improve, research and development on how AR may be used in education will keep on improving simultaneously (Yuen et al. 2011). In the upcoming years, spatial computing offers a wide range of ground-breaking abilities; for instance, instead of focusing on the shortest path or the travel time, companies have come up with eco-routing, which involves determining paths that reduce greenhouse gas emissions. UPS saves over 3 million gallons of gasoline each year by using smart sequencing, which avoids left turns. When customers and owners offer eco-routing options, such savings can be increased substantially (Shekhar et al. 2015).

Table 2.5 Characteristics of several XR device types

Extended reality classification	Hardware examples	User interface	Technical strengths	Technical limitations
Virtual reality	• Oculus rift	• Handheld motion-tracked controllers	• Superior 3D graphics performance and highest resolution	• User has no direct view of physical environment
	• HTC vive			• Requires controller inputs
2D augmented reality (indirect)	• iPhone	• Touchscreen	• Widely available, inexpensive	• Phone or tablet must be held or mounted
	• iPad			• Requires touch input
	• Android devices			
2D augmented reality (direct)	• Google glass	• Side-mounted touchpad	• Lightweight head-mounted display	• 2D display
		• Voice		• UI does not interact with physical environment
3D augmented reality	• Microsoft HoloLens	• Voice	• Touch-free input	• Narrow field of view for 3D graphics
	• Magic leap	• Gaze	•3D display	
	• RealView Holoscope	• Gestures	• Full visibility of surroundings	

2.34 Wireless Power Transfer

Wireless power transfer (WPT) is currently one of the trendiest subjects in research, and it is getting a lot of attention. WPT has already made great progress in the field of charging mobile devices or charging electric vehicles, and in the future decades, the WPT sector is anticipated to develop steadily (Rim 2018). WPT is the transferal of electrical energy without the need for cables based on electric, magnetic or electromagnetic fields that change over time (Georgios and Evangelos 2019). The notion of WPT is based on Faraday's law of induction, which is well-known among engineers. According to this rule, alternating current (AC) generations are caused by the changing magnetic field (Würth Elektronik 2021). There have been a lot of fascinating initiatives in the history of WPT. Even though the inherent complexity of this technology prevented it from being commercialised, the improvements deserve consideration and are worth investigating. The infographic timeline in Fig. 2.92 summarises the history of WPT (IEEE 2021).

WPT is divided into two categories depending on the length of transmitter-to-antenna distance. The term "near-field WPT" refers to when the antenna is located close to the transmitter. On the other hand, if the transmitter and the antenna are further or far apart, this is referred to as "far-field WPT".

An inductive and a capacitive, also known as primary and secondary respectively, coils are used in close distance technology. These coils have been known as Tesla coils since Nikola Tesla invented them in the early 1890s. They can be used to transfer power between themselves by employing the transformer principle. That is electromagnetic induction, electric power through spark-excited ratio frequency resonant frequency. The alteration of the magnetic field and the passage of a current that oscillates through the primary coil conducts the secondary coil. The primary coil carries a magnetic flux around it, produced by the oscillating current it possesses. Because these coils are looped around the secondary coil, a voltage is induced in the secondary coil. The largest distance that may be covered by this way of delivering electricity via electromagnetic induction is 5 cm. This is since,

1864
A researcher and professor at University of Cambridge, James Clerk Maxwell suggested that electric and magnetic phenomena are not separate but are two aspects of the same entity. Maxwell also neatly described electromagnetic equations

1886-1888
After Maxwell, Heinrich Hertz experimentally verified Maxwell's electromagnetic equations. Hertz's experiment caught the attention of scientists and encouraged them to do research in this area.

1891
One of the scientists affected by Hertz's experiments was Nikola Tesla. He worked on electromagnetic radiation, especially the concept of resonance, and as a result of these studies, invented the device he called the Tesla coil.

1902
Tesla researched the conductivity of the Earth's atmosphere, and built the Magnifying Transmitter, which would become one of the largest Tesla coils.

1901-1906
Tesla built the Wardenclyffe Tower for research in wireless telegraphy. Then, he shifted the emphasis of the project after learning about Marconi's radio transmission in 1901, broadening it to meet new goals, including wireless power transfer. However, because of financial difficulties, the tower was closed in 1906.

1959
Until the 1950s, the technology for producing high-power microwave radiation had not progressed far enough to allow practical use. The Raytheon Company overcame this restriction by creating the crossed-fieldamplifier. As a result, the age of wireless powertransfer by microwave focusing began.

1964
The first microwave power transmission system was publicly demonstrated, and a helicopter soared on the vertical axis, tied by strings, thanks to microwave power transmission

1968
Peter Glaser proposed the solar space satellite (SPS) idea. This proposal involves harvesting solar energy in space using photovoltaic arrays and sending it to Earth via microwaves, where it is gathered by a huge rectenna

1987
The Canadian Communications Research Centre created the first totally untethered microwave powered aircraft

2001
A research project was suggested to wirelessly power a distant town on the island of La Reunion a WPT solution was proposed to give extra power supply.

2008
In a series of trials between two Hawaiian Islands, the longest WPT transmission occurred. 20 W were received at 2.45GHz at a distance of 148 kilometers.

Fig. 2.92 A brief story of wireless power transfer. (IEEE 2021)

as the distance between primary and secondary coils increase, the loss of magnetic flux exponentially increases, resulting in a significant power loss.

Long-distance technologies, sometimes known as radioactive technologies, are utilised to achieve long-range wireless uses with distances of several kilometres range. Microwaves and lasers are the most common technologies for long-range communication. The wavelength steers the transmitter in the receiver's direction, while diffraction limit is employed in radio frequency design, allowing this application to be free of electromagnetic radiation dangers and achieve nearly fully efficient transmission. Several methods of WPT technology are shown in Fig. 2.93 (Lokesh 2020).

There are several uses for WPT. To begin with, it has the potential to replace existing charging methods. Instead of using a power cord to charge a phone or laptop, wireless power may be applied in the house, allowing computers or smartphones to charge constantly and wirelessly. Another application of WPT is the charging of electric vehicles (EVs). Chargers for EVs are among the higher-level uses. As EVs grow increasingly common on the road, fixed and even mobile WPT devices can improve the viability of driving one (Mehrotra 2014). Implantable medical devices are also among the usage areas of WPT. These devices have a very low power demand and thus can be powered by WPT to substantially increase the operation time in vivo (a medical procedure that is done on a living organism), improve the

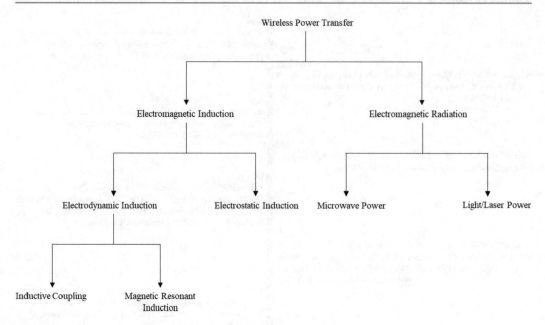

Fig. 2.93 Schematic of different methods of WPT. (Lokesh 2020)

accuracy of diagnosis and therapy, minimise the rate of misdiagnosis and achieve permanent operation in vivo, all while improving patient comfort. WPT can also be used in industrial environments where it is impossible to use wire for power transmission, such as underwater and chemical areas (Wang 2018).

Even though the notion of wireless power transmission has been around for over a century, it is regaining popularity in the twenty-first century as a result of the widespread use of smartphones and other mobile devices (Allied Components International 2019). The widespread use of WPT has the potential to bring many benefits. Such as the ability to decrease or eliminate the usage of wires and batteries. Therefore, reducing the use of copper- and aluminium-based electric cable (Sumi et al. 2018). In addition, thanks to WPT, the use of elements used in battery technology is reduced. Until recently, charging difficulties have made EVs less appealing to consumers, despite several government incentive programs. However, charging becomes the simplest process by wirelessly transmitting energy to the EV. Furthermore, projections indicate that the battery capacity of EVs equipped

with WPT could be reduced to 20% or less compared to those without WPTs (Li and Mi 2015).

WPT eliminates the need for power ports, which solves significant problems of conventional charging methods such as mating cycles, deterioration of connection points, thus greatly contributing to device longevity. Additionally, removing power docks and ports allows manufacturers to design completely sealed devices, advancing water-resistant to waterproof, thus increasing durability and longevity (Würth Elektronik 2021). WPT offers more efficient charging systems making it a must-have feature for IoT-enabled portable devices like cell phones, digital cameras and camcorders, laptops and tablets, wearable electronics and more (Patil et al. 2020). WPT technology has many advantages and disadvantages. Table 2.6 shows the detailed advantages and disadvantages of WPT (Bhardwaj and Ahlawat 2018).

Projections state that it will be possible to utilise electric equipment without the need for a wire in the future. The following section is a discussion of potential applications of wireless power transfer technology, summarised by Sumi et al. (2018).

Table 2.6 Advantages and disadvantages of WPT technology

Advantages	Disadvantages
Eliminates possible power transmission problems caused by short circuits in wired connections	Execution of WPT systems has significant capital costs
Reduces power losses attributable to wired connections	Less effective in comparison to customary charging
Enables and eases the global interconnectivity of power plants	The use of WPT might result in obstructions in correspondence frameworks

Bhardwaj and Ahlawat (2018) and Shanmuganantham et al. (2010)

(a) Solar Power Satellite

Satellites equipped with solar panels can be used to gather as much solar energy from the sun as possible in orbit. A microwave transmission device is used in satellites to convert electricity into microwaves for transmission to a microwave receiving antenna back on Earth. These microwaves transmitted from space can be converted into energy to power homes, workplaces and so on. Despite being in the early development stages, the use of WPT through satellites is a highly promising source of clean and renewable energy (Cuthbertson 2021).

(b) Home Appliances That Are Powered Wirelessly

The future developments of WPT are likely to produce a power transmission device within the home that will send electricity to the entirety of home appliances, including televisions, laptops, lamps, irons, sound boxes, fridges and mobile phones. The proposed transmitting device would send out electricity. Each appliance featuring WPT receivers could use it to power themselves.

(c) Wireless Charging for Electric Vehicles on Way

EVs are projected to be solely powered wirelessly in the future, which would make it possible to charge EVs while driving thus, eliminating the need to stop and charge. Power beam transmitters can be linked to roads and heavy traffic zones with power sources. These transmitters could transform electricity into power beams and project them onto EVs equipped with appropriate receivers to re-convert the power beam into electrical power, charging the car's battery in return.

(d) Wireless Charging Train

In the future, all trains may be powered wirelessly, which would eliminate the need to use wire to link the train. Sumi et al. (2018) propose a wirelessly powered future train model, according to this which a dual-mode power receiver and transmitter would be connected to train poles. All train stations would feature pole(s) equipped with a dual-mode transmitter and receiver. The power would come directly from the power station, to be collected and transmitted by dual-mode transmitters. When utilising a dual-mode transmitter, power reception and broadcast occur at the same time. Power would reach the train through reception made possible by receivers installed on their roofs. There would be no need to utilise wire in this model.

(e) Supplying Homes with Power from the Power Station Wirelessly

Renewable energy sources may be used to generate clean and green power in the future. The generated power from these power stations can be directly and wirelessly transmitted to residential areas. This would be possible through the use of antennas that convert electric power into microwaves and transmit them to another set of antennas that could receive and again transmit power. The final destination would be housing; however, to utilise the power transformed into microwaves, they would require an antenna that could re-transform microwave energy into electrical power.

(f) Controlling a Drone Wirelessly to Extinguish a Fire

Projections state that drones could prove highly useful in emergencies, especially fires. They can be used to carry water pipes and place them in precise locations via a remote-control system. It is thought that fire trucks would carry a transmitter to send power, to be received by an appropriately equipped drone. Drones are invaluable because they can fly where people can't go and take pictures and videos of the situation. Since it will not be possible to connect the drone with a wire in an emergency, it is foreseen that this process will be very useful.

(g) Medical Equipment Can Benefit from Wireless Power

The wireless power supply may be used for medical equipment in the future. A transmitter that is directly linked to the power station may be used to establish a viable power supply. This would transmit the power to a receiver installed in the hospital to create wireless electricity to power medical equipment.

Regarding economics, it is forecast that the WPT market will reach a size of 11.3 billion USD by 2022 and 35.2 billion USD by 2030, which is a substantial increase from 2.5 billion in 2016. This growth is expected to be mainly driven by advancements in far-field transmission, the need for effective charging systems that WPT can offer and increased needs for appliances powered by batteries (Markets and Markets 2017; Wankhede et al. 2021). An additional driving factor is presumed to be IoT-enabled devices, such as cell phones, digital cameras and camcorders, laptops and tablets, wearable electronics and home electronics, gaining more and more traction, and their WPT featuring design needs (Patil et al. 2020). If not addressed, the initial capital costs of installing WPT systems are expected to have detrimental effects on the growth of its market (Markets and Markets 2017).

References

7 Advantages of Robots in the Workplace [WWW Document] (2018). https://roboticstomorrow.com/story/2018/08/7-advantages-of-robots-in-the-workplace/12342/. Accessed 26 July 2021

A. Abdulkadir, T. Sarker, Q. He, Z. Guo, S. Wei, Mössbauer spectroscopy of polymer nanocomposites (2016), pp. 393–409. https://doi.org/10.1016/B978-0-323-40183-8.00013-6

A.H. Abedin, A critical review of thermochemical energy storage systems. Open Renew. Energy J. 4, 42–46 (2011). https://doi.org/10.2174/1876387101004010042

A. Abele, E. Elkind, B. Washom, J. Intrator, 2020 Strategic Analysis of Energy Storage in California (California Energy Commission, Sacramento, 2011)

H. Abie, An overview of firewall technologies 10 (2000)

M. Abualkibash, K. Elleithy, Cloud computing: The future of IT industry. Int. J. Distrib. Parallel Syst. 3, 1–12 (2012). https://doi.org/10.5121/ijdps.2012.3401

F. Adams, C. Barbante, Nanotechnology and analytical chemistry. Compr. Anal. Chem. 69, 125–157 (2015). https://doi.org/10.1016/B978-0-444-63439-9.00004-9

Advantages of recycling [WWW Document]. Business Waste (2021), https://www.businesswaste.co.uk/advantages-recycling/. Accessed 4 Aug 2021

S. Aguirre, A. Rodriguez, Automation of a business process using robotic process automation (RPA): A case study, in Applied computer sciences in engineering, communications in computer and information science, ed. by J. C. Figueroa-García, E. R. López-Santana, J. L. Villa-Ramírez, R. Ferro-Escobar, (Springer, Cham, 2017), pp. 65–71. https://doi.org/10.1007/978-3-319-66963-2_7

M.F. Ahern, Cybersecurity in power systems. IEEE Potentials 36, 8–12 (2017). https://doi.org/10.1109/MPOT.2017.2700239

M. Alawadhi, J. Almazrouie, M. Kamil, K.A. Khalil, Review and analysis of the importance of autonomous vehicles liability: A systematic literature review. Int. J. Syst. Assur. Eng. Manag. 11, 1227–1249 (2020). https://doi.org/10.1007/s13198-020-00978-9

M. Alberth, M. Mattern, Understanding robotic process automation (RPA). Capco Inst. J. Financial Transf., 54–61 (2017)

J.F. Allen, Natural language processing, in Encyclopedia of Computer Science, ed. by A. Ralston, E. D. Reilly, D. Hemmendinger, (Wiley, London, 2003), pp. 1218–1222

G. Allen, T. Chan, Artificial intelligence and national security. Natl. Secur. 132 (2017)

Allied Components International, Allied Components International. [Çevrimiçi] (2019), Available at: https://www.alliedcomponents.com/blog/the-growing-importance-of-wireless-energy-transmission. Erişildi 26 July 2021

D. Alves de Lima, A. Correa Victorino, A hybrid controller for vision-based navigation of autonomous vehicles in urban environments. IEEE Trans. Intell. Transp. Syst. 17, 2310–2323 (2016). https://doi.org/10.1109/TITS.2016.2519329

AMP Robotics [WWW Document]. AMP Robotics (2021), https://www.amprobotics.com. Accessed 4 Aug 2021

B. Anasori, M.R. Lukatskaya, Y. Gogotsi, 2D metal carbides and nitrides (MXenes) for energy storage. Nat. Rev. Mater. **2**, 1–17 (2017). https://doi.org/10.1038/natrevmats.2016.98

S.B. Andrew, Scoring dimension-level job performance from narrative comments: Validity and generalizability when using natural language processing. Organ. Res. Methods **3**(24), 572–594 (2021)

C. Andrews et al., Extended reality in medical practice. Curr. Treatment Options Cardiovasc. Med. **21**(4), 18 (2019). https://doi.org/10.1007/s11936-019-0722-7

A. Anupam, S. Kumar, N.M. Chavan, B.S. Murty, R.S. Kottada, First report on cold-sprayed AlCoCrFeNi high-entropy alloy and its isothermal oxidation. J. Mater. Res. **34**, 796–806 (2019). https://doi.org/10.1557/jmr.2019.38

APWG, Anti-Phishing Working Group Trends Report Q1 2021 (2021)

ArborMetrix, What is healthcare analytics and why is it important? [Online] (2020), Available at: https://www.arbormetrix.com/blog/intro-big-data-analytics-healthcare. Accessed 15 June 2021

Y. Arnautova, Digital twins technology, its benefits & challenges to information security. GlobalLogic, 11 June (2020), Available at: https://www.globallogic.com/insights/blogs/if-you-build-products-you-should-be-using-digital-twins/. Accessed 30 July 2021

S. Asha, C. Chellappan, Biometrics: An overview of the technology, issues. Int. J. Comput. Appl. **39**(10) (2012) Available at: https://www.dhs.gov/biometrics. Accessed 10 July 2021

B. Attallah, Z. Ilagure, Wearable technology: Facilitating or complexing education? Int. J. Inf. Educ. Technol. **8**(6), 433–436 (2018). https://doi.org/10.18178/IJIET.2018.8.6.1077

Axios, Say hello to the Internet of Behaviors: What is it & why you should care. [Online] (2020), Available at: https://axiosholding.com/say-hello-to-the-internet-of-behaviors-what-is-it-why-you-should-care/. Accessed 12 July 2021

B. Axmann, H. Harmoko, *Robotic process automation: An overview and comparison to other technology in industry 4.0*, in 2020 10th International Conference on Advanced Computer Information Technologies (ACIT). Presented at the 2020 10th International Conference on Advanced Computer Information Technologies (ACIT), IEEE (2020), pp. 559–562. https://doi.org/10.1109/ACIT49673.2020.9208907

T. Ayode Otitoju, P. Ugochukwu Okoye, G. Chen, Y. Li, M. Onyeka Okoye, S. Li, Advanced ceramic components: Materials, fabrication, and applications. J. Ind. Eng. Chem. **85**, 34–65 (2020). https://doi.org/10.1016/j.jiec.2020.02.002

S.A. Bagloee, M. Tavana, M. Asadi, T. Oliver, Autonomous vehicles: Challenges, opportunities, and future implications for transportation policies. J. Mod. Transp. **24**, 284–303 (2016). https://doi.org/10.1007/s40534-016-0117-3

M.A.K. Bahrin, M.F. Othman, N.H.N. Azli, M.F. Talib, Industry 4.0: A review on industrial automation and robotic. Jurnal teknologi, 78(6-13) (2016)

H.E. Bal, J.G. Steiner, A.S. Tanenbaum, Programming languages for distributed computing systems. ACM Comput. Surv. **21**, 261–322 (1989). https://doi.org/10.1145/72551.72552

A.T. Balafoutis, B. Beck, S. Fountas, Z. Tsiropoulos, J. Vangeyte, T.V.D. Wal, ... S.M. Pedersen, Smart farming technologies–description, taxonomy and economic impact. In Precision agriculture: Technology and economic perspectives (pp. 21–77). Springer, Cham (2017)

M. Balasingam, Drones in medicine-The rise of the machines. Int. Clin. Pract. **71**(9) (2017)

O. Bamodu, X. Ye, Virtual reality and virtual reality system components (2013), p. 4

D. Banerjee, Data science foundation. [Online] (2020), Available at: https://datascience.foundation/sciencewhitepaper/natural-language-processing-nlp-simplified-a-step-by-step-guide. Accessed 23 July 2021

R. Banerjee, I. Manna, *Ceramic Nanocomposites* (Elsevier, 2013)

E. Barbour, Investigation into the potential of energy storage to tackle intermittency in renewable energy generation (2013)

J. Bardeen, L.N. Cooper, J.R. Schrieffer, Theory of Superconductivity. Phys. Rev. **108**, 1175–1204 (1957). https://doi.org/10.1103/PhysRev.108.1175

M.R. Barrick, M.K. Mount, T.A. Judge, Personality and performance at the beginning of the new millennium: What do we know and where do we go next? Int. J. Sel. Assess. **9**, 9–30 (2001)

D.G. Bayyou, Artificially intelligent self-driving vehicle technologies. Benef. Challenges **26**, 10 (2019)

M. Beaudin, H. Zareipour, A. Schellenberglabe, W. Rosehart, Energy storage for mitigating the variability of renewable electricity sources: An updated review. Energy Sustain. Dev. **14**, 302–314 (2010). https://doi.org/10.1016/j.esd.2010.09.007

A. Bechar, C. Vigneault, Agricultural robots for field operations. Part 2: Operations and systems. Biosystems engineering, 153, 110–128 (2017)

N.H. Behling, Fuel cells: Current technology challenges and future research needs. Newnes (2012)

S. Beiker, R. Calo, Legal aspects of autonomous driving. SSRN Electron. J. (2010). https://doi.org/10.2139/ssrn.2767899

V. Beiser, No job for humans: The robot assault on Fukushima. Wired (2018)

Benefits of Robots [WWW Document]. RobotWorx (2021), https://www.robots.com/articles/benefits-of-robots. Accessed 26 July 2021

D.R. Berryman, Augmented reality: A review. Med. Ref. Serv. Q. **31**(2), 212–218 (2012). https://doi.org/10.1080/02763869.2012.670604

M. Bertoncello, D. Wee, Ten ways autonomous driving could redefine the automotive world | McKinsey [WWW Document] (2015), https://www.mckinsey.com/industries/automotive-and-assembly/our-insights/ten-ways-autonomous-driving-could-redefine-the-automotive-world. Accessed 21 July 2021

E. Bezirhan Arikan, H.D. Ozsoy, A review: Investigation of bioplastics. J. Civ. Eng. Archit **9** (2015). https://doi.org/10.17265/1934-7359/2015.02.007

M. Bhagat, D. Kumar, D. Kumar, Role of Internet of Things (IoT) in smart farming: A brief survey. 2019 Devices for Integrated Circuit (DevIC), 141–145 (2019)

M. Bhardwaj, A. Ahlawat, Wireless power transmission with short and long range using inductive coil. Wirel. Eng. Technol. **9**(1), 1–9 (2018)

Bioplastics – Are they truly better for the environment? [WWW Document]. Environment (2018), https://www.nationalgeographic.com/environment/article/are-bioplastics-made-from-plants-better-for-environment-ocean-plastic. Accessed 23 Oct 2021

BMW, Hydrogen cars, fuel cells, etc.: What you need to know | BMW.com [WWW Document] (2020), https://www.bmw.com/en/innovation/how-hydrogen-fuel-cell-cars-work.html

R. Bogue, Shape-memory materials: A review of technology and applications. Assem. Autom. **29**, 214–219 (2009). https://doi.org/10.1108/01445150910972895

R. Bogue, Robotic vision boosts automotive industry quality and productivity. Ind. Robot. **40** (2013). https://doi.org/10.1108/IR-04-2013-342

M. Borowski-Beszta, M. Polasik, Wearable devices: New quality in sports and finance Analizy metod płatności View project Growth through innovation or economies of scale? A survey of participants of the Polish payment system View project. J. Phys. Educ. Sport. (2020). https://doi.org/10.7752/jpes.2020.s2150

B. Boykin, Hydrogen: A clean, flexible energy carrier infographic [WWW Document]. Coroflot (2021), https://www.coroflot.com/BrandiBoykin/Hydrogen-A-Clean-Flexible-Energy-Carrier-Infographic

H. Brenton, M. Gillies, D. Ballin, D. Chatting, The uncanny valley: Does it exist (2005)

M. Brodin, M. Vallejos, M.T. Opedal, M.C. Area, G. Chinga-Carrasco, Lignocellulosics as sustainable resources for production of bioplastics – A review. J. Clean. Prod. **162**, 646–664 (2017). https://doi.org/10.1016/j.jclepro.2017.05.209

M. Brunn, Purdue innovation converts plastic waste into high-value fuels. RECYCLING magazine (2021), https://www.recycling-magazine.com/2021/06/16/purdue-innovation-converts-plastic-waste-into-high-value-fuels/. Accessed 4 Aug 2021

M. Budt, D. Wolf, R. Span, J. Yan, A review on compressed air energy storage: Basic principles, past milestones and recent developments. Appl. Energy **170**, 250–268 (2016). https://doi.org/10.1016/J.APENERGY.2016.02.108

R. Buyya, M.A.S. Netto, A.N. Toosi, M.A. Rodriguez, I.M. Llorente, S.D.C. Di Vimercati, P. Samarati, D. Milojicic, C. Varela, R. Bahsoon, M.D. De Assuncao, S.N. Srirama, O. Rana, W. Zhou, H. Jin, W. Gentzsch, A.Y. Zomaya, H. Shen, G. Casale, R. Calheiros, Y. Simmhan, B. Varghese, E. Gelenbe, B. Javadi, L.M. Vaquero, A manifesto for future generation cloud computing. ACM Comput. Surv. **51**(5), 1–38 (2018)

S. Byford, Sony's wearable air conditioner is pretty cool. The Verge (2020), Available at: https://www.theverge.com/2020/7/22/21333837/sony-reon-pocket-hands-on-wearable-air-conditioner-japan. Accessed 25 July 2021

C. Cagatay, How long should an electric car's battery last? [WWW Document] (2021), https://www.myev.com/research/ev-101/how-long-should-an-electric-cars-battery-last. Accessed 13 Nov 2021

S. Campana, Drones in archaeology. State-of-the-art and future perspectives. Archaeol. Prospect. **24**(4), 275–296 (2017)

T. Campbell, C. Williams, O. Ivanova, B. Garrett, *Could 3D Printing Change the World?* (Atlantic Council, s.l, 2011)

K. Campbell, J. Diffley, B. Flanagan, B. Morelli, B. O'Neil, F. Sideco, *The 5G Economy: How 5G Technology Will Contribute to the Global Economy* (IHS Economics/IHS Technology, s.l, 2017)

J. Carmigniani et al., Augmented reality technologies, systems and applications. Multimed. Tools Appl. **51**(1), 341–377 (2011). https://doi.org/10.1007/s11042-010-0660-6

M.C. Campion, M.A. Campion, E.D. Campion, M.H. Reider, Initial investigation into computer scoring of candidate essays for personnel selection. Journal of Applied Psychology, **101**(7), 958 (2016)

CBINSIGHTS, Blockchain startups absorbed 5X more capital via ICOs than equity financings in 2017. [Online] (2018), Available at: https://www.cbinsights.com/research/blockchain-vc-ico-funding/. Accessed 26 July 2021

CBINSIGHTS, The edge computing ecosystem: From sensors to the centralized cloud [WWW Document] (2019), https://www.cbinsights.com/research/edge-computing-ecosystem/

Cellular Agriculture Society, What is cellular agriculture? (2021), Available at: https://www.cellag.org/?p=m3. Accessed 25 July 2021

L. Ceschini, A. Dahle, M. Gupta, A.E.W. Jarfors, S. Jayalakshmi, A. Morri, F. Rotundo, S. Toschi, R.A. Singh, *Aluminum and Magnesium Metal Matrix Nanocomposites, Engineering Materials* (Springer, Singapore, 2017). https://doi.org/10.1007/978-981-10-2681-2

S. Chakravarthy, C. Nagaraj, Towards data science. [Online] (2020), Available at: https://towardsdatascience.com/tokenization-for-natural-language-processing-a179a891bad4. Accessed 23 July 2021

B. Chakroun et al., *Artificial Intelligence for Sustainable Development: Synthesis* (ODU Digital Commons, Teaching & Learning Faculty Publications, 2019), p. 50

C.-Y. Chan, Advancements, prospects, and impacts of automated driving systems. Int. J. Transp. Sci. Technol. **6**, 208–216 (2017). https://doi.org/10.1016/j.ijtst.2017.07.008

S. Chand, Importance of marketing for the economic development of a country. [Online] (2021), Available at: https://www.yourarticlelibrary.com/marketing/importance-of-marketing-for-the-economic-development-of-a-country/22145. Accessed 11 June 2021

P. Chauhan, M. Sood, Big data: Present and future. Computer **54**(4), 59–65 (2021). https://doi.org/10.1109/MC.2021.3057442

F.A.A. Cheein, R. Carelli, Agricultural robotics: Unmanned robotic service units in agricultural tasks. IEEE Ind. Electron. Mag., 7 (2013). https://doi.org/10.1109/MIE.2013.2252957

A. Chehri, H.T. Mouftah, Autonomous vehicles in the sustainable cities, the beginning of a green adventure. Sustain. Cities Soc. **51**, 101751 (2019). https://doi.org/10.1016/j.scs.2019.101751

Y.J. Chen, Bioplastics and their role in achieving global sustainability. J. Chem. Pharm. Res. **6**(1), 226–231 (2014)

J. Chen, Investment crowdfunding definition. Investopedia (2021), Available at: https://www.investopedia.com/terms/i/investment-crowdfunding.asp. Accessed 25 July 2021

H. Chen, T.N. Cong, W. Yang, C. Tan, Y. Li, Y. Ding, Progress in electrical energy storage system: A critical review. Prog. Nat. Sci. **19**, 291–312 (2009). https://doi.org/10.1016/j.pnsc.2008.07.014

X. Chen et al., An Internet of Things based energy efficiency monitoring and management system for machining workshop. J. Clean. Prod. **199**, 957–968 (2018). https://doi.org/10.1016/j.jclepro.2018.07.211

K.-T. Cheng, T.-C. Huang, What is flexible electronics? ACM SIGDA Newsl. **39**(4), 1–1 (2009). https://doi.org/10.1145/1862894.1862895

D.S. Chiaburu, I.-S. Oh, C.M. Berry, N. Li, R.G. Gardner, The five-factor model of personality traits and organizational citizenship behaviors: A meta-analysis. J. Appl. Psychol. **96**, 1140–1166 (2011)

R. Cho, How drones are advancing scientific research. [online] State of the Planet (2021), Available at: https://news.climate.columbia.edu/2017/06/16/how--drones-are-advancing-scientific-research/. Accessed 12 July 2021

R. Chopra, Introduction to cloud computing, in: *Cloud Computing: An Introduction* (2017)

K. Chopra, K. Gupta, A. Lambora, *Future Internet: The Internet of Things-A Literature Review*, in 2019 International Conference on Machine Learning, Big Data, Cloud and Parallel Computing (COMITCon). 2019 International Conference on Machine Learning, Big Data, Cloud and Parallel Computing (COMITCon) (IEEE, Faridabad, 2019a), pp. 135–139. https://doi.org/10.1109/COMITCon.2019.8862269

K. Chopra, K. Gupta, A. Lambora, Proceedings of the International Conference on Machine Learning, Big Data, Cloud and Parallel Computing: Trends, Perspectives and Prospects: COMITCON-2019: 14th-16th February, 2019. Institute of Electrical and Electronics Engineers (2019b), Available at: https://ieeexplore.ieee.org/servlet/opac?punumber=8851231. Accessed 28 July 2021

S. Choudhuri, What is a data hub and why should you care? [Online] (2019), Available at: https://www.actian.com/company/blog/what-is-a-data-hub-and-why-should-you-care/. Accessed 26 July 2021

G. Chowdhury, Natural language processing. Annu. Rev. Inf. Sci. Technol. **37**(1), 51–89 (2003)

Christianlauer, What is a data hub? [Online] (2021), Available at: https://towardsdatascience.com/what-is-a-data-hub-41d2ac34c270. Accessed 19 July 2021

S.H.-W. Chuah, Why and who will adopt extended reality technology? Literature review, synthesis, and future research agenda. SSRN Electron. J. (2018). https://doi.org/10.2139/ssrn.3300469

L. Chunli, L. DongHui, Computer network security issues and countermeasure 4 (2012)

Civil Aviation Cybersecurity [WWW Document], Int. Civ. Aviat. Organ. (2021), https://www.icao.int/Cybersecurity/Pages/default.aspx. Accessed 11 July 2021

G.I. Clark, Imperial College London and Duncan, 2012. What's the "hydrogen economy"? [WWW Document]. The Guardian (2012), http://www.theguardian.com/environment/2012/oct/11/hydrogen-economy-climate-change. Accessed 28 July 2021

P. Cohn, A. Green, M. Langstaff, M. Roller, *Commercial Drones Are Here: The Future of Unmanned Aerial Systems. Capital Projects & Infrastructure* (McKinsey & Company, 2017)

Committee on National Security Systems (CNSS), Committee on National Security Systems Glossary. Comm. Natl. Secur. Syst. Gloss (2015)

N. Connor, What is sensible heat storage – SHS – Definition [WWW Document] (2019), https://www.thermal-engineering.org/what-is-sensible-heat-storage-shs-definition/. Accessed 22 May 2019

J.W. Cortada, D. Gordon, B. Lenihan, *The value of analytics in healthcare* (IBM Institute for Business Value, New York, 2012)

С. Стельмах, Edge vs. Cloud: в чем разница? [WWW Document] (2020), https://www.itweek.ru/its/article/detail.php?ID=212193. Accessed 20 July 2021

A.E. Creamer, B. Gao, Carbon dioxide capture: an effective way to combat global warming (p. 62). Cham, Switzerland: Springer (2015)

A. Cuthbertson, Solar power satellites that could beam energy to anywhere on Earth successfully tested by Pentagon scientists. [Çevrimiçi] (2021), Available at: https://www.independent.co.uk/climate-change/news/space-laser-satellite-solar-power-b1806680.html. Erişildi 26 Oct 2021

K. Dalamagkidis, K.P. Valavanis, L.A. Piegl, Aviation history and unmanned flight, in *On Integrating Unmanned Aircraft Systems into the National Airspace System*, (Springer, Dordrecht, 2012), pp. 11–42

R.P. Dameri, Comparing smart and digital city: initiatives and strategies in Amsterdam and Genoa. Are they digital and/or smart? in *Smart city, progress in IS*, ed. by R. P. Dameri, C. Rosenthal-Sabroux,

(Springer, Cham, 2014), pp. 45–88. https://doi.org/10.1007/978-3-319-06160-3_3

R. Dashurov, Crowdfunding infographic 10 steps concept startup Vector Image (2021), Available at: https://www.vectorstock.com/royalty-free-vector/crowdfunding-infographic-10-steps-concept-startup-vector-27893258. Accessed 25 July 2021

P. Davidson, A. Spinoulas, Autonomous vehicles – What could this mean for the future of transport? 15 (2015)

Dawex, Data hub. [Online] (2021), Available at: https://www.dawex.com/en/data-exchange-platform/data--hub/. Accessed 10 June 2021

J.B. de Vasconcelos, Á. Rocha, Business analytics and big data. Int. J. Inf. Manag. **46**, 320–321 (2019). https://doi.org/10.1016/j.ijinfomgt.2018.10.019

P. Delaney, C.J. Pettit, *Urban data hubs supporting smart cities* (University of Melbourne, Canberra, 2014)

L. Delicado, J. Sanchez-Ferreres, J. Carmona, L. Padro, *NLP4BPM – Natural Language Processing Tools for Business Process Management* (Universitat Politecnica de Catalunya, Barcelona, 2017)

Deloitte, *The Robots Are Ready. Are You?* (Deloitte Development LLC, London, 2017)

Delta Impact, Stretchable electronics: The future of wearables – delta impact (2021), Available at: https://deltaimpact.com/blog/blog-stretchable-electronics/. Accessed 25 July 2021

D. Demner-Fushman, M.S. Simpson, Biomedical text mining: A survey of recent progress, in *Mining Text Data*, ed. by C. Z. Charu, C. Aggarwal, (Springer, Maryland, 2012), pp. 465–517

Dena, *Blockchain in The Integrated Energy Transition* (Deutsche Energie-Agentur GmbH (dena), (2019)

Y. Devarajan, A study of robotic process automation use cases today for tomorrow's business. Int. J. Comput. Techniq. **5**, 12–18 (2018)

A. Dhillon, D. Kumar, 2 – Recent advances and perspectives in polymer-based nanomaterials for Cr(VI) removal, in *New Polymer Nanocomposites for Environmental Remediation*, ed. by C. M. Hussain, A. K. Mishra, (Elsevier, 2018), pp. 29–46. https://doi.org/10.1016/B978-0-12-811033-1.00002-0

DHS, Department of Homeland Security. [Online] (2021)

A. Di Bartolo, G. Infurna, N.T. Dintcheva, A review of bioplastics and their adoption in the circular economy. Polymers **13**, 1229 (2021). https://doi.org/10.3390/polym13081229

T. Dillon, C. Wu, E. Chang, *Cloud Computing: Issues and Challenges*, in: Proceedings – International Conference on Advanced Information Networking and Applications, AINA (IEEE, 2010), pp. 27–33. https://doi.org/10.1109/AINA.2010.187

B. Dorn, B. MacWhirter, Shifting the focus from end-of-life recycling to continuous product lifecycles 9 (2016)

B. Dunn, Hydroponics (2013)

N. Duque Ciceri, M. Garetti, S. Spérandio, From product End-of-Life sustainable considerations to design management. Presented at the IFIP advances in information and communication technology (2009), pp. 152–159. https://doi.org/10.1007/978-3-642-16358-6_20

EEA Glossary – European Environment Agency [WWW Document] (2021), https://www.eea.europa.eu/help/glossary/eea-glossary. Accessed 4 Aug 2021

J. Eisert, M.M. Wolf, Quantum computing, in *Computational Complexity: Theory, Techniques, and Applications*, (Springer, Berlin/Heidelberg, 2013), pp. 2388–2405. https://doi.org/10.1007/978-1-4614-1800-9_148

W. Ejaz et al., Efficient energy management for the Internet of Things in smart cities. IEEE Commun. Mag. **55**(1), 84–91 (2017). https://doi.org/10.1109/MCOM.2017.1600218CM

E. Ekudden, Edge computing [WWW Document]. Ericsson (2021), https://www.ericsson.com/en/blog/2021/2/5g-and-cloud

Electronic Textiles (Elsevier Science & Technology, Cambridge, 2015)

A. Elkazzaz, Soilless agriculture a new and advanced method for agriculture development: An introduction. Agric. Res. Technol. Open Access J. **3** (2017). https://doi.org/10.19080/ARTOAJ.2017.03.555610

N. Elmqaddem, Augmented reality and virtual reality in education. Myth or reality? Int. J. Emerg. Technol. Learn. (iJET) **14**(3), 234 (2019). https://doi.org/10.3991/ijet.v14i03.9289

H. El-Sayed, S. Sankar, M. Prasad, D. Puthal, A. Gupta, M. Mohanty, C.T. Lin, Edge of things: The big picture on the integration of edge, IoT and the cloud in a distributed computing environment. IEEE Access **6**, 1706–1717 (2017). https://doi.org/10.1109/ACCESS.2017.2780087

Energy Storage Association, Benefits of Energy Storage | Energy Storage Association. Energy Storage Assoc (2021), https://energystorage.org/why-energy-storage/benefits/. Accessed 28 July 2021

F. Engels, W. Hach, J. Distler, C. Kiefer, RPA – Tomorrow's must-have technology. Roland Berger Focus 8 (2018)

Equinor, Northern Lights CCS – CO2 transport and storage (2020), Available at: https://www.equinor.com/en/what-we-do/northern-lights.html. Accessed 25 July 2021

European Bioplastics (2020), Available at: https://www.european-bioplastics.org/bioplastics/. Accessed 11 June 2022

European Commission, Rewards-based crowdfunding (2021), Available at: https://ec.europa.eu/growth/tools-databases/crowdfunding-guide/types/rewards_en. Accessed 25 July 2021

European Parliament, What is Carbon Neutral and how can it be achieved by 2050 (2019), https://www.europarl.europa.eu/news/en/headlines/society/20190926STO62270/what-is-carbon--neutrality-and-how-can-it-be-achieved-by-2050

M. Fairhurst, *Biometrics: A Very Short Introduction*, 2nd edn. (Oxford University Press, Oxford, 2019)

FCHEA, Unlocking the potential of hydrogen energy storage [WWW Document]. Fuel Cell Hydrog. Energy Assoc. (2021), https://www.fchea.org/in-transition/2019/7/22/unlocking-the-potential-of-hydrogen-energy-storage

C. Fernández Gil-Delgado, The use of military drones: The impact on land forces and legal implications. [online] finabel.org (2021), Available at: https://finabel.org/the-use-of-military-drones-the-impact-on-land-forces-and-legal-implications/. Accessed 12 July 2021

J. Ferrari, *Electric utility resource planning: Past, present and future* (Elsevier, 2020)

R.P. Feynman, Simulating physics with computers. Int. J. Theor. Phys. **21**(6–7), 467–488 (1982). https://doi.org/10.1007/BF02650179

D. Floreano, R. Wood, Science, technology and the future of small autonomous drones. Nature **521**(7553), 460–466 (2015)

R. Francis, *Recycling of polymers: Methods, characterization and applications* (Wiley, Weinheim, 2016). https://doi.org/10.1002/9783527689002

C.B. Frey, M.A. Osborne, *The Future of Employment: How Susceptible Are Jobs to Computerisation* (University of Oxford, Oxford, 2013)

A. Fridman, The future of crowdfunding | Inc.com (2016), Available at: https://www.inc.com/adam-fridman/the-future-of-crowdfunding.html. Accessed 25 July 2021

A.V. Frolov, Can a quantum computer be applied for numerical weather prediction? Russ. Meteorol. Hydrol. **42**(9), 545–553 (2017). https://doi.org/10.3103/S1068373917090011

H.P. Fung, Criteria, use cases and effects of Information Technology Process Automation (ITPA). Adv. Robot. Autom. **3**, 1–10 (2014). https://doi.org/10.4172/2168-9695.1000124

Future, [WWW Document] (2021), https://cals.arizona.edu/hydroponictomatoes/future.htm. Accessed 20 July 2021

A.B. Gallo, J.R. Simões-Moreira, H.K.M. Costa, M.M. Santos, E. Moutinho dos Santos, Energy storage in the energy transition context: A technology review. Renew. Sust. Energ. Rev. **65**, 800–822 (2016). https://doi.org/10.1016/j.rser.2016.07.028

V.P. Gao, H.W. Kaas, D. Mohr, D. Wee, Automotive revolution – Perspective towards 2030 | McKinsey [WWW Document] mckinsey.com (2016), https://www.mckinsey.com/industries/automotive-and-assembly/our-insights/disruptive-trends-that-will-transform-the-auto-industry/de-de. Accessed 21 July 2021

K. Gasteratos, 90 reasons to consider cellular agriculture (2019), Available at: https://dash.harvard.edu/handle/1/38573490. Accessed 25 July 2021

M. Gavrilescu, Y. Chisti, Biotechnology – A sustainable alternative for. Biotechnol. Adv. **23**(7-8), 471–499 (2005)

A.T. Gebremariam, F. Di Maio, A. Vahidi, P. Rem, Innovative technologies for recycling End-of-Life concrete waste in the built environment. Resour. Conserv. Recycl. **163**, 104911 (2020). https://doi.org/10.1016/j.resconrec.2020.104911

S. Geetha, S. Gouthami, Internet of things enabled real time water quality monitoring system. Smart Water **2**(1), 1 (2016). https://doi.org/10.1186/s40713-017-0005-y

B. Georgios, P. Evangelos, Upper-limb prosthetic devices, in *Handbook of biomechatronics*, ed. by J. Segil, (Academic, London, 2019), p. 208

R. Gibb, What is a distributed system? stackpath (2019)

M. Gill, Bioplastic: A better alternative to plastics. IMPACT Int. J. Res. Appl. Nat. Soc. Sci. IJRANSS **2**, 115–120 (2014)

S.S. Gill, et al., Quantum computing: A taxonomy, systematic review and future directions (2020), pp. 1–37. Available at: http://arxiv.org/abs/2010.15559

A.S. Gillis, Biometrics (2020a), Available at: https://searchsecurity.techtarget.com/definition/biometrics

A.S. Gillis, What is IoT (Internet of Things) and how does it work? [WWW Document]. IoT Agenda (2020b), https://internetofthingsagenda.techtarget.com/definition/Internet-of-Things-IoT. Accessed 11 July 2021

F. Giones, A. Brem, From toys to tools: The co-evolution of technological and entrepreneurial developments in the drone industry. Business Horizons **60**(6), 875–884 (2017)

Global Electronic Services, The future of flexible electronics (2021), Available at: https://gesrepair.com/future-of-flexible-electronic/. Accessed 25 July 2021

B. Gludovatz, E.P. George, R.O. Ritchie, Processing, microstructure and mechanical properties of the CrMnFeCoNi high-entropy alloy. JOM **67**, 2262–2270 (2015). https://doi.org/10.1007/s11837-015-1589-z

M.M. Gobble, Big data: The next big thing in innovation. Res. Technol. Manag. **56**(1), 64–67 (2013). https://doi.org/10.5437/08956308X5601005

E.M. Goel, M. Sudhakar, R.V. Shahi, *Carbon Capture, Storage & Utilization: A Possible Climate Change Solution for Energy Industry* (Teri Press, New Delhi, 2015)

M. Gowda, N. Sheety, P. Kumar, Working of Autonomous Vehicles. Int. Res. J. Eng. Technol. **6**, 13–18 (2019)

Grand View Research, Flexible electronics market by components, by application and segment forecast to 2024 (2016), Available at: https://www.grandviewresearch.com/industry-analysis/flexible-electronics-market/methodology. Accessed 25 July 2021

S. Greengard, *The Internet of Things* (The MIT Press, London, 2015)

S. Greenwold, Spatial computing (2003), Available at: https://acg.media.mit.edu/people/simong/thesis/SpatialComputing.pdf

M. Grieves, J. Vickers, Digital twin: Mitigating unpredictable, undesirable emergent behavior in complex systems, in *Transdisciplinary Perspectives on Complex Systems*, ed. by F.-J. Kahlen, S. Flumerfelt, A. Alves, (Springer, Cham, 2017), pp. 85–113. https://doi.org/10.1007/978-3-319-38756-7_4

E. Guizzo, What is a Robot? – ROBOTS: Your guide to the world of robotics [WWW Document] (2018), https://robots.ieee.org/learn/what-is-a-robot/. Accessed 29 July 2021

T. Guo, Alan Turing: Artificial intelligence as human self-knowledge. Anthropol. Today **31**, 3–7 (2015). https://doi.org/10.1111/1467-8322.12209

R. Gupta, The future of driving: What to expect in 2021. Automot. World. (2021), http://www.automotive-world.com/articles/the-future-of-driving-what-to-expect-in-2021/. Accessed 21 July 2021

A. Gupta, R.K. Jha, A survey of 5G network: Architecture and emerging technologies. IEEE Access **3**, 1206–1232 (2015)

S. Guran, R.L. Mersky, S.K. Ghosh, Status of plastics waste in circular economy in the USA, in *Circular Economy: Global Perspective*, ed. by S. K. Ghosh, (Springer, Singapore, 2020), pp. 413–421. https://doi.org/10.1007/978-981-15-1052-6_21

C.J. Haboucha, R. Ishaq, Y. Shiftan, User preferences regarding autonomous vehicles. Transp. Res. Part C Emerg. Technol. **78**, 37–49 (2017). https://doi.org/10.1016/j.trc.2017.01.010

I. Hadjipaschalis, A. Poullikkas, V. Efthimiou, Overview of current and future energy storage technologies for electric power applications. Renew. Sust. Energ. Rev. **13**, 1513–1522 (2009). https://doi.org/10.1016/j.rser.2008.09.028

C. Hadnagy, *Social Engineering: The Science of Human Hacking*, 2nd edn. (Wiley, Indianapolis, 2018)

P. Halarnkar, et al., A review on virtual reality (2012), p. 1

P. Hamet, J. Tremblay, Artificial intelligence in medicine. Metabolism **69**, S36–S40 (2017). https://doi.org/10.1016/j.metabol.2017.01.011

P. Hartmann, C.L. Bender, M. Vračar, A.K. Dürr, A. Garsuch, J. Janek, P. Adelhelm, A rechargeable room-temperature sodium superoxide (NaO_2) battery. Nat. Mater. **12**, 228–232 (2013). https://doi.org/10.1038/nmat3486

Q.F. Hassan, A.R. Khan, S.A. Madani, Internet of things challenges, advances, and applications (2018), Available at: https://www.taylorfrancis.com/books/e/9781498778534. Accessed: 28 July 2021

A. Hauer, *Thermal energy storage* (IEA-ETSAP IRENA, 2012)

J.Y. He, H. Wang, H.L. Huang, X.D. Xu, M.W. Chen, Y. Wu, X.J. Liu, T.G. Nieh, K. An, Z.P. Lu, A precipitation-hardened high-entropy alloy with outstanding tensile properties. Acta Mater. **102**, 187–196 (2016). https://doi.org/10.1016/j.actamat.2015.08.076

Healthcare Analytics Market, Healthcare analytics market by type (Descriptive, prescriptive, cognitive), application (Financial, operational, RCM, fraud, clinical), component (Services, hardware), Deployment (On-premise, cloud), End-user (Providers, payer) – Global forecast to 2026. [Online] (2021), Available at: https://www.researchandmarkets.com/reports/5439282/healthcare-analytics-market--by-type-descriptive?utm_source=BW&utm_medium=PressRelease&utm_code=mgmtw6&utm_campaign=1592920+-+The+Worldwide+Healthcare+Analytics+Industry+is+Expected+to+Reach+%2475.1+Billio. Accessed 7 July 2021

E. Henderson, Tubeless, wearable insulin pump can improve blood sugar control for people with type 1 diabetes. The Endocrine Society (2021), Available at: https://www.news-medical.net/news/20210321/Tubeless-wearable-insulin-pump-can-improve-blood--sugar-control-for-people-with-type-1-diabetes.aspx. Accessed 25 July 2021

L. Hickman, et al., *Automated Video Interview Personality Assessments: Reliability, Validity, and Generalizability Investigations* (PsyArXiv, s.l., 2021)

M. Hillman, 2 Rehabilitation robotics from past to present – A historical perspective, in *Advances in Rehabilitation Robotics: Human-Friendly Technologies on Movement Assistance and Restoration for People with Disabilities*, Lecture notes in control and information science, ed. by Z. Z. Bien, D. Stefanov, (Springer, Berlin/Heidelberg, 2004), pp. 25–44. https://doi.org/10.1007/10946978_2

A. Ho, J. McClean, S.P. Ong, The promise and challenges of quantum computing for energy storage. Joule **2**(5), 810–813 (2018). https://doi.org/10.1016/j.joule.2018.04.021

M. Hossain, G. Oparaocha, Crowdfunding: Motives, definitions, typology and ethical challenges. Entrep. Res. J. **7**(2), 20150045 (2017). https://doi.org/10.1515/erj-2015-0045

E. Hossain, H. Faruque, M. Sunny, N. Mohammad, N. Nawar, A comprehensive review on energy storage systems: Types, comparison, current scenario, applications, barriers, and potential solutions, policies, and future prospects. Energies **13**, 3651 (2020). https://doi.org/10.3390/en13143651

W.M. Huang, Z. Ding, C.C. Wang, J. Wei, Y. Zhao, H. Purnawali, Shape memory materials. Materials today, 13(7-8), 54–61 (2010)

X. Huang, X. Qi, F. Boey, H. Zhang, Graphene-based composites. Chem. Soc. Rev. **41**, 666–686 (2012). https://doi.org/10.1039/C1CS15078B

C. Hughes, et al., What is a qubit? **2**(1), 7–16 (2021) https://doi.org/10.1007/978-3-030-61601-4_2

J. Huiskens, What is the future of data and analytics in healthcare after COVID-19? [Online] (2020), Available at: https://www.healthitoutcomes.com/doc/what-is-the-future-of-data-and-analytics-in--healthcare-after-covid-0001. Accessed 5 July 2021

J. Hurwitz, R. Bloor, M. Kaufman, F. Halper, Cloud Computing (2012)

A. Huth, J. Cebula, The basics of cloud computing (2011). https://doi.org/10.1201/9781351049221-1

IAE, *20 Years of Carbon Capture and Storage: Accelerating Future Deployment, 20 Years of Carbon Capture and Storage* (OECD, Paris, 2016). https://doi.org/10.1787/9789264267800-EN

IBM, What is distributed computing [WWW Document] (2017), https://www.ibm.com/docs/en/txseries/8.1.0?topic=overview-what-is-distributed-computing. Accessed 16 July 2021

IBM, Natural Language Processing (NLP). [Online] (2020), Available at: https://www.ibm.com/cloud/learn/natural-language-processing. Accessed 15 July 2021

IEA, *Going Carbon Negative: What Are the Technology Options?* (IEA, Paris, 2020) https://www.iea.org/

commentaries/going-carbon-negative-what-are-the-technology-options

IEA, *Global Energy Review 2021* (IEA, Paris, 2021). https://www.iea.org/reports/global-energy-review-2021

IEEE, IEEWPT. [Çevrimiçi] (2021), Available at: https://wpt.ieee.org/wpt-history/. Erişildi 8 July 2021

V. Ilková, A. Ilka, Legal aspects of autonomous vehicles – An overview. Presented at the 21st International Conference on Process Control (PC), Štrbské Pleso, Slovakia (2017)

International Banker, What is the Internet of Behaviour? [Online] (2021), Available at: https://internationalbanker.com/technology/what-is-the-internet-of-behaviour/. Accessed 12 June 2021

Invitrogen & Gibco, CELL CULTURE BASICS Handbook Cell Culture Basics Cell Culture Basics Cell Culture Basics | i (2021), Available at: https://www.vanderbilt.edu/viibre/CellCultureBasicsEU.pdf. Accessed 25 July 2021

IoT Desing Pro, IoT desing pro. [Online] (2021), Available at: https://iotdesignpro.com/articles/what-is-internet-of-behavior-iob. Accessed 9 June 2021

S. Islam et al., A systematic review on healthcare analytics: Application and theoretical perspective of data mining. MDPI **6**(2), 1–43 (2018)

N. Ivanov, Forbes. [Online] (2020), Available at: https://www.forbes.com/sites/forbestechcouncil/2020/08/21/digital-transformation-and-the-role-of-data-hubs-data-lakes-and-data-warehouses/?sh=3761b7447cd9. Accessed 22 July 2021

J3016B: Taxonomy and definitions for terms related to driving automation systems for on-road motor vehicles – SAE International [WWW Document] (2018), https://www.sae.org/standards/content/j3016_201806/. Accessed 29 July 2021

F. Jackson, The future of carbon capture is in the air. Forbes (2020), Available at: https://www.forbes.com/sites/feliciajackson/2020/12/02/is-the-future-of-carbon-capture-in-the-air/?sh=604a6e6b2c4a. Accessed 25 July 2021

N.A. Jackson, K.E. Kester, D. Casimiro, S. Gurunathan, F. DeRosa, The promise of mRNA vaccines: a biotech and industrial perspective. npj Vaccines, **5**(1), 1–6 (2020).

A. Jahejo, Distributed computing | Advantages and disadvantages [WWW Document] (2020), https://computernetworktopology.com/distributed-computing/. Accessed 16 July 2021

A.K. Jain, A. Kumar, Biometrics of next generation: An overview. Second Gen. Biometr. **12**(1) (2010)

M. Javaid, A. Haleem, Virtual reality applications toward medical field. Clin. Epidemiol. Glob. Health **8**(2), 600–605 (2020). https://doi.org/10.1016/j.cegh.2019.12.010

G. Jevtic, Edge computing vs Cloud computing: Key differences (2019), https://phoenixnap.com/blog/edge-computing-vs-cloud-computing

R. Jiang, R. Kleer, F.T. Piller, Predicting the future of additive manufacturing: A Delphi study on economic and societal implications of 3D printing for 2030. Technol. Forecast. Soc. Chang. **117**, 84–97 (2017). https://doi.org/10.1016/j.techfore.2017.01.006

J. Jiménez López, M. Mulero-Pázmány, Drones for conservation in protected areas: Present and future. Drones **3**(1), 10 (2019)

M. Jiménez, L. Romero, I.A. Domínguez, M.M. Espinosa, M. Domínguez, Additive manufacturing technologies: An overview about 3D printing methods and future prospects. Complexity **2019** (2019). https://doi.org/10.1155/2019/9656938

J. Johnson, Artificial intelligence & future warfare: implications for international security. Def. Secur. Anal. **35**, 147–169 (2019). https://doi.org/10.1080/14751798.2019.1600800

W.L. Jolly, Hydrogen – Isotopes of hydrogen [WWW Document]. Encycl. Br. (2020), https://www.britannica.com/science/hydrogen

J.B. Jones, *Hydroponics: A Practical Guide for the Soilless Grower* (CRC Press, Boca Raton, 2016)

T. Jones, R. Bishop, The future of autonomous vehicles. Open Foresight (2020)

T.A. Judge et al., Hierarchical representations of the five-factor model of personality in predicting job performance: integrating three organizing frameworks with two theoretical perspectives. J. Appl. Psychol. **98**, 875–925 (2013)

Juniper Networks, What is 5G? (2021), Available at: https://www.juniper.net/us/en/research-topics/what-is-5g.html

S. Jusoh, A study on NLP applications and ambiguity. J. Theor. Appl. Inf. Technol. **96**(6), 1486–1499 (2018)

G. Kamiya, S. Hassid, P. Gonzalez, Energy storage – Analysis [WWW Document] (IEA, 2021), https://www.iea.org/reports/energy-storage. Accessed 13 Nov 2021

M. Kamran, S. Abhishek , A comprehensive study on 3D printing technology. MIT Int J Mech Eng 6.2, 63–69 (2016)

Y. Kanamori, S.-M. Yoo, Quantum computing: Principles and applications. J. Int. Technol. Inf. Manage. **29**(2), 2020 (2020)

H. Karan, C. Funk, M. Grabert, M. Oey, B. Hankamer, Green bioplastics as part of a circular bioeconomy. Trends Plant Sci. **24**, 237–249 (2019). https://doi.org/10.1016/j.tplants.2018.11.010

S. Kavanagh, What is enhanced Mobile Broadband (eMBB). [online] 5g.co.uk (2021), Available at: https://5g.co.uk/guides/what-is-enhanced-mobile-broadband-embb/. Accessed 6 July 2021

S. Kaza, L.C. Yao, P. Bhada-Tata, F. Van Woerden, *What a Waste 2.0: A Global Snapshot of Solid Waste Management to 2050* (World Bank, Washington, DC, 2018). https://doi.org/10.1596/978-1-4648-1329-0

É. Kelly, UK government to fund four carbon capture and storage clusters. Science|Business (2020), Available at: https://sciencebusiness.net/news/uk-government-fund-four-carbon-capture-and-storage-clusters. Accessed 25 July 2021

A. Kennedy, J. Brame, T. Rycroft, M. Wood, V. Zemba, C. Weiss, M. Hull, C. Hill, C. Geraci, I. Linkov, A definition and categorization system for advanced materials: The foundation for risk-informed environmental health and safety testing. Risk Anal. **39**, 1783–1795 (2019). https://doi.org/10.1111/risa.13304

Kent State University, The importance of data analytics in healthcare. [Online] (2021), Available at: https://onlinedegrees.kent.edu/ischool/health-informatics/community/why-is-data-analytics-important-in-healthcare. Accessed 27 July 2021

J. Kerry, P. Butler, Smart packaging technologies for fast moving consumer goods smart packaging technologies for fast moving consumer goods (2008)

F.A. Khan, *Introduction to Biotechnology. Biotechnology Fundamentals* (CRC Press, s.l, 2011), pp. 1–16

A. Khan, Cellular agriculture: The future of food. Medium (2017), Available at: https://medium.com/cellagri/cellular-agriculture-the-future-of-food-ab4710eced9b. Accessed 25 July 2021

W.Z. Khan, E. Ahmed, S. Hakak, I. Yaqoob, A. Ahmed, Edge computing: A survey. Futur. Gener. Comput. Syst. **97**, 219–235 (2019). https://doi.org/10.1016/j.future.2019.02.050

K.C.A. Khanzode, R.D. Sarode, Advantages and disadvantages of artificial intelligence and machine learning: A literature review 8 (2020)

Z.X. Khoo, J.E.M. Teoh, Y. Liu, C.K. Chua, S. Yang, J. An, K.F. Leong, W.Y. Yeong, 3D printing of smart materials: A review on recent progresses in 4D printing. Virtual Phys. Prototyping **10**(3), 103–122 (2015). https://doi.org/10.1080/17452759.2015.1097054

C. Kidd, What is the Internet of Behaviors? IoB explained. [Online] (2019), Available at: https://www.bmc.com/blogs/iob-internet-of-behavior/. Accessed 11 June 2021

S.J. Kim, Y. Jeong, S. Park, K. Ryu, G. Oh, A survey of drone use for entertainment and AVR (augmented and virtual reality), in *Augmented reality and virtual reality*, (Springer, Cham, 2018), pp. 339–352

H. Kitonsa, S.V. Kruglikov, Significance of drone technology for achievement of the United Nations sustainable development goals. R-Economy **4**(3), 115–120 (2018)

М.С. Косяков, Введение В Распределенные Вычисления 155 (2014)

Kolomiiets, K., 2021. What is the Internet of Behavior And Why is it Important for Business?. [Online]. Available at: https://gbksoft.com/blog/internet-of-behaviors/. [Accessed 11 June 2021].

A.D. Kshemkalyani, M. Singhal, Distributed Computing: Principles, Algorithms, and Systems 748 (2007)

W. Kuckshinrichs, J.F. Hake, *Carbon Capture, Storage and Use: Technical, Economic, Environmental and Societal Perspectives* (Springer, New York, 2015). https://doi.org/10.1007/978-3-319-11943-4

S. Küfeoğlu, *The Home of the Future*, 1st edn. (Springer, Cambridge, 2020)

S. Küfeoğlu, *The Home of the Future: Digitalization and Resource Management*, 1st edn. (Springer, London, 2021)

S. Küfeoglu, Ş. Üçler, Designing the business model of an energy datahub. Electr. J. **II**(34), 1–9 (2021)

S. Küfeoğlu, Ş. Üçler, *Designing the business model of an energy datahub* (University of Cambridge, Cambridge, 2021)

S. Küfeoğlu, G. Liu, K. Anaya, M.G. Pollitt, *Digitalisation and New Business Models in Energy Sector* (Cambridge Working Papers In Economics, Cambridge, 2019)

M.C. Lacity, L.P. Willcocks, A new approach to automating services. MIT Sloan Manag. Rev. **58**, 41–49 (2016)

J.M. Lagaron, A. Lopez-Rubio, Nanotechnology for bioplastics: Opportunities, challenges and strategies. Trends Food Sci. Technol. **22**, 611–617 (2011). https://doi.org/10.1016/j.tifs.2011.01.007

C. Lamberton, D. Brigo, D. Hoy, *Impact of Robotics, RPA and AI on the Insurance Industry: Challenges and Opportunities* (SSRN Scholarly Paper No. ID 3079495) (Social Science Research Network, Rochester, 2017)

L. Lamport, N. Lynch, Distributed computing: Models and methods, in *Formal Models and Semantics*, (Elsevier, 1990), pp. 1157–1199. https://doi.org/10.1016/B978-0-444-88074-1.50023-8

Laser-based sensor technology for recycling metals [WWW Document]. RECYCLING magazine (2021), https://www.recycling-magazine.com/2021/01/25/laser-based-sensor-technology-for-recycling-metals/. Accessed 1 Aug 2021

S. Latre, et al., *City of Things: An Integrated and Multi-Technology Testbed for IoT Smart City Experiments*, in 2016 IEEE International Smart Cities Conference (ISC2). 2016 IEEE International Smart Cities Conference (ISC2) (IEEE, Trento, 2016), pp. 1–8. https://doi.org/10.1109/ISC2.2016.7580875

C. Lauer, Towards data science. [Online] (2021), Available at: https://towardsdatascience.com/what-is-a-data-hub-41d2ac34c270. Accessed 16 July 2021

A. Lee, Towards Data Science Inc.. [Online] (2019), Available at: https://towardsdatascience.com/why--nlp-is-important-and-itll-be-the-future-our-future--59d7b1600dda. Accessed 14 July 2021

S. Lee, J. Lee, Beneficial bacteria and fungi in hydroponic systems: Types and characteristics of hydroponic food production methods. Sci. Hortic. **195**, 206–215 (2015). https://doi.org/10.1016/j.scienta.2015.09.011

D. Lee, T. Miyoshi, Y. Takaya, T. Ha, 3D microfabrication of photosensitive resin reinforced with ceramic nanoparticles using LCD microstereolithography. J. Laser Micro Nanoeng. **1**(2), 142–148 (2006). https://doi.org/10.2961/jlmn.2006.02.0011

J. Lee, D. Kim, H.-Y. Ryoo, B.-S. Shin, Sustainable wearables: Wearable technology for enhancing the quality of human life. Sustainability **8**(466) (2016). https://doi.org/10.3390/su8050466

L.Y. Leong et al., Predicting the determinants of the NFC-enabled mobile credit card acceptance: A neural networks approach. Expert Syst. Appl. **40**(14),

5604–5620 (2013). https://doi.org/10.1016/J.ESWA.2013.04.018

D.Y.C. Leung, G. Caramanna, M.M. Maroto-Valer, An overview of current status of carbon dioxide capture and storage technologies. Renew. Sust. Energ. Rev. **39**, 426–443., ISSN 1364-0321 (2014). https://doi.org/10.1016/j.rser.2014.07.093

G. Lewis, Basics about cloud computing. Software engineering institute carniege mellon university, Pittsburgh (2010)

S. Li, C.C. Mi, Wireless power transfer for electric. IEEE J. Emerg. Select. Topics Power Electron. **1**, 4–17 (2015)

J.W.Y. Lim, P.K. Hoong, E.-T. Yeoh, I.K.T. Tan, *Performance Analysis of Parallel Computing in a Distributed Overlay Network*, in: TENCON 2011 – 2011 IEEE Region 10 Conference. Presented at the TENCON 2011 – 2011 IEEE Region 10 Conference (IEEE, Bali, 2011), pp. 1404–1408. https://doi.org/10.1109/TENCON.2011.6129040

Liskov, Primitives for distributed computing. Proc. 7th ACM Symp. Oper. Syst. Princ SOSP **1979**, 33–42 (1979). https://doi.org/10.1145/800215.806567

S. Liu, Statista. [Online] (2020), Available at: https://www.statista.com/statistics/607891/worldwide-natural-language-processing-market-revenues/. Accessed 3 Aug 2021

K. Liu, L. Dong, Research on cloud data storage technology and its architecture implementation. Proc. Eng. **29**, 133–137 (2012). https://doi.org/10.1016/j.proeng.2011.12.682

C. Liu, F. Li, L.-P. Ma, H.-M. Cheng, Advanced materials for energy storage. Adv. Mater. **22**, E28–E62 (2010). https://doi.org/10.1002/adma.200903328

L. Liu, J. Zhang, C. Ai, Nickel-based superalloys, in *Reference Module in Materials Science and Materials Engineering*, (2020). https://doi.org/10.1016/B978-0-12-803581-8.12093-4

M.V.D. Lokesh, The future of mobile charging: wireless infrared. Int. J. Eng. Dev. Res. **8**(3), 138–141 (2020)

M. Lopa, J. Vora, Evolution of mobile generation technology: 1G to 5G and review of upcoming wireless technology 5G. Int. J. Modern Trends Eng. Res. **2**(10), 281–290 (2015) Available at: http://www.danspela.com/pdf/p113.pdf

A. Luccioni, E. Baylor, N. Duchene, *Analyzing Sustainability Reports Using Natural Language Processing* (Tackling Climate Change with Machine Learning workshop at NeurIPS 2020, s.l., 2020)

P.D. Lund, J. Lindgren, J. Mikkola, J. Salpakari, Review of energy system flexibility measures to enable high levels of variable renewable electricity. Renew. Sust. Energ. Rev. **45**, 785–807 (2015). https://doi.org/10.1016/j.rser.2015.01.057

X. Luo, J. Wang, M. Doonet, J. Clarke, Overview of current development in electrical energy storage technologies and the application potential in power system operation. Appl. Energy, 511–536 (2015)

B. Lutkevich, Natural Language Processing (NLP). [Online] (2021), Available at: https://searchenter-priseai.techtarget.com/definition/natural-language-processing-NLP. Accessed 12 July 2021

P.R. Lutui, B. Cusack, G. Maeakafa, *Energy Efficiency for IoT Devices in Home Environments*, in 2018 IEEE International Conference on Environmental Engineering (EE). 2018 IEEE International Conference on Environmental Engineering (EE) (IEEE, Milan, 2018), pp. 1–6. https://doi.org/10.1109/EE1.2018.8385277

Z. Lv et al., Next-generation big data analytics: State of the art, challenges, and future research topics. IEEE Trans. Industrial Inf. **13**(4), 1891–1899 (2017). https://doi.org/10.1109/TII.2017.2650204

X. Lv, Y. Wu, M. Gong, J. Deng, Y. Gu, Y. Liu, J. Li, G. Du, R. Ledesma-Amaro, L. Liu, J. Chen, Synthetic biology for future food: Research progress and future directions. Future Foods **3**(March), 100025 (2021). https://doi.org/10.1016/j.fufo.2021.100025

N. Ma, S. Liu, W. Liu, L. Xie, D. Wei, L. Wang, L. Li, B. Zhao, Y. Wang, Research progress of titanium-based high entropy alloy: Methods, properties, and applications. Front. Bioeng. Biotechnol. (2020). https://doi.org/10.3389/fbioe.2020.603522

S. Madakam, R.M. Holmukhe, D.K. Jaiswal, The future digital work force: Robotic Process Automation (RPA). J. Inf. Syst. Technol. Manag. **16** (2019). https://doi.org/10.4301/S1807-1775201916001

G. Malandrino, Chemical vapour deposition. Precursors, processes and applications. Edited by Anthony C. Jones and Michael L. Hitchman. Angew. Chem. Int. Ed. **48**, 7478–7479 (2009). https://doi.org/10.1002/anie.200903570

S. Mallakpour, S. Rashidimoghadam, Carbon nanotubes for dyes removal (2019), pp. 211–243. https://doi.org/10.1016/B978-0-12-814132-8.00010-1

MAN, Power-to-Gas energy storage [WWW Document]. MAN Energy Solut. (2021), https://www.man-es.com/discover/decarbonization-glossary%2D%2D-man-energy-solutions/power-to-gas-energy-storage. Accessed 28 July 2021

K. Mannanuddin, M. Ranjith Kumar, S. Aluvala, Y. Nagender, S. Vishali, *Fundamental Perception of EDGE Computing*, in: IOP conference series: Materials science and engineering (2020), pp. 1–7. https://doi.org/10.1088/1757-899X/981/2/022035

V. Marinoudi, C.G. Sørensen, S. Pearson, D. Bochtis, Robotics and labour in agriculture. A context consideration. Biosyst. Eng. **184** (2019). https://doi.org/10.1016/j.biosystemseng.2019.06.013

Markets and Markets, Wireless power transmission market by technology (Induction, magnetic resonance), implementation, transmitter, and receiver application (Smartphones, electric vehicles, wearable electronics, and furniture) and geography – Global forecast to 2022. [Çevrimiçi] (2017), Available at: https://www.marketsandmarkets.com/Market-Reports/wireless--power-market-168050212.html. Erişildi 26 Oct 2021

Markets and Markets, Markets and Markets. [Online] (2021), Available at: https://www.marketsandmarkets.

com/Market-Reports/natural-language-processing--nlp-825.html. Accessed 3 Aug 2021

Marklogic, Marklogic. [Online] (2021), Available at: https://www.marklogic.com/product/comparisons/data-hub-vs-data-warehouse/. Accessed 26 July 2021

M. Martonosi, M. Roetteler, Next steps in quantum computing (2019), Available at: http://arxiv.org/abs/1903.10541

D. Maslov, Y. Nam, J. Kim, An outlook for quantum computing [Point of view]. Proc. IEEE **107**(1), 5–10 (2019). https://doi.org/10.1109/JPROC.2018.2884353

F. Mattern, C. Floerkemeier, From the internet of computers to the Internet of Things, in *From Active Data Management to Event-Based Systems and More*, Lecture notes in computer science, ed. by K. Sachs, I. Petrov, P. Guerrero, (Springer, Berlin/Heidelberg, 2010), pp. 242–259. https://doi.org/10.1007/978-3-642-17226-7_15

K. Matthews, 6 industries where demand for robotics developers will grow by 2025 [WWW Document]. The Robot Report (2019), https://www.therobotreport.com/6-industries-demand-robotics-developers--grow-2025/. Accessed 26 July 2021

J. McCarthy, Artificial intelligence, logic and formalizing common sense, in *Philosophical Logic and Artificial Intelligence*, (Springer, Dordrecht, 1989), pp. 161–190

G.T. Mckee, The maturing discipline of robotics. Int. J. Eng. Educ. **22** (2006)

McKinsey Global Institute, Big data: The next frontier for innovation, competition, and productivity (2011), Available at: https://www.mckinsey.com/~/media/mckinsey/business%20functions/mckinsey%20digital/our%20insights/big%20data%20the%20next%20frontier%20for%20innovation/mgi_big_data_full_report.pdf

R. Mehrotra, Cut the cord: Wireless power transfer, its applications, and its limits. [Çevrimiçi] (2014), Available at: https://www.cse.wustl.edu/~jain/cse574-14/ftp/power/#kesler13. Erişildi 26 July 2021

M. Mekni, A. Lemieux, Augmented reality: Applications, challenges and future trends (2014), p. 10

T. Mekonnen, P. Mussone, H. Khalil, D. Bressler, Progress in bio-based plastics and plasticizing modifications. J. Mater. Chem. A **1**, 13379–13398 (2013). https://doi.org/10.1039/C3TA12555F

M.A. Memon et al., Big data analytics and its applications. Proc. Comput. Sci. **1**(1), 10 (2017)

MindSea 2021. How to Organize a Successful Crowdfunding Campaign in 6 Steps. [WWW Document] https://mindsea.com/crowdfunding/

Y. Mintz, R. Brodie, Introduction to artificial intelligence in medicine. Minim. Invasive Ther. Allied Technol. **28**, 73–81 (2019). https://doi.org/10.1080/13645706.2019.1575882

G. Miragliotta, A. Perego, A. Tumino, Internet of Things: Smart present or smart future? (2012), pp. 1–6

S.P. Mirashe, N.V. Kalyankar, Cloud computing can simplify HIT infrastructure management. J. Comput. **2**, 78–82 (2010)

F. Modu, A. Adam, F. Aliyu, A. Mabu, M. Musa, A survey of smart hydroponic systems (2020), https://astesj.com/v05/i01/p30/. Accessed 19 July 2021

A. Mohamed, Chapter eight – Synthesis, characterization, and applications carbon nanofibers, in *Carbon-Based Nanofillers and Their Rubber Nanocomposites*, ed. by S. Yaragalla, R. Mishra, S. Thomas, N. Kalarikkal, H. J. Maria, (Elsevier, 2019), pp. 243–257. https://doi.org/10.1016/B978-0-12-813248-7.00008-0

S. Mohammed, Introduction to nutrient film technique, in *Tomorrow's Agriculture: "NFT Hydroponics"-Grow Within Your Budget*, SpringerBriefs in plant science, ed. by S. Mohammed, (Springer, Cham, 2018), pp. 7–11. https://doi.org/10.1007/978-3-319-99202-0_2

M. Mohammed, M.B. Khan, E.B.M. Bashier, *Machine Learning: Algorithms and Applications*, 1st edn. (CRC Press, Boca Raton, 2016)

MonkeyLearn, Major Challenges of Natural Language Processing (NLP). [Online] (2021), Available at: https://monkeylearn.com/blog/natural-language-processing-challenges/. Accessed 20 July 2021

G.E. Moore, Cramming more components onto integrated circuits. Proc. IEEE **86**(1), 82–85 (1998). https://doi.org/10.1109/JPROC.1998.658762

Morgan Stanley 2021. [WWW Document] https://www.morganstanley.com/about-us/sustainability-at-morgan-stanley

O. Morrison, 'We want to make sure that when people live on Mars, we'll be there too': Aleph Farms plans to accelerate cell-based meat production in outer space. FoodNavigator (2020), Available at: https://www.foodnavigator.com/Article/2020/10/21/We-want-to--make-sure-that-when-people-live-on-Mars-we-ll--be-there-too-Aleph-Farms-plans-to-accelerate-cell-based-meat-production-in-outer-space. Accessed 25 July 2021

T. Moser et al., Hydrophilic organic electrodes on flexible hydrogels. ACS Appl. Mater. Interfaces **8**(1), 974–982 (2016). https://doi.org/10.1021/ACSAMI.5B10831

S.L. Moskowitz, *The Advanced Materials Revolution: Technology and Economic Growth in the Age of Globalization* (Wiley, New York, 2014)

T.P. Mpofu, C. Mawere, M. Mukosera, The impact and application of 3D printing technology. Int. J. Sci. Res. (IJSR) **3**(6), 2148–2152 (2014) https://www.researchgate.net/publication/291975129

M. Mukherjee, Towards data science. [Online] (2020), Available at: https://towardsdatascience.com/nlp-meets-sustainable-investing-d0542b3c264b. Accessed 3 Aug 2021

A. Mukherjee, S. Phalora, H. Jadav, The case for a smart enterprise data hub: Why IT and business need to collaborate. [Online] (2021), Available at: https://www.wipro.com/consulting/the-case-for-a-smart--enterprise-data-hub-why-it-and-business-need-to-collaborate/. Accessed 20 July 2021

K. Mylvaganam, Y. Chen, W. Liu, M. Liu, L. Zhang, Hard thin films: Applications and challenges. Anti-Abras.

Nanocoatings Curr. Future Appl., 543–567 (2015). https://doi.org/10.1016/B978-0-85709-211-3.00021-2

P.M. Nadkarni, L. Ohno-Machado, W. Chapman, Natural language processing: An introduction. J. Am. Med. Inform. Assoc. **18**, 544–551 (2011)

F.T. Najafi, S.M. Naik, Potential applications of robotics in transportation engineering. Transportation Research Record (1989)

S. Namani, B. Gonen. *Smart Agriculture Based on IoT and Cloud Computing*, in 2020 3rd International Conference on Information and Computer Technologies (ICICT). 2020 3rd International Conference on Information and Computer Technologies (ICICT) (IEEE, San Jose, 2020), pp. 553–556. https://doi.org/10.1109/ICICT50521.2020.00094

National Biometric Security Project, *Biometric Technology Application Manual*, 1st edn. (National Biometric Security Project, s.l., 2008)

R.C. Neville, Semiconductors. Sol. Energy Convers., 71–118 (1995). https://doi.org/10.1016/B978-044489818-0/50003-X

New UCSF Robotic Pharmacy Aims to Improve Patient Safety [WWW Document]. New UCSF Robotic Pharmacy Aims to Improve Patient Safety | UC San Francisco (2011), https://www.ucsf.edu/news/2011/03/98776/new-ucsf-robotic-pharmacy-aims-improve-patient-safety. Accessed 26 July 2021

Nibusinessinfo.co.uk, Advantages and disadvantages of crowdfunding | nibusinessinfo.co.uk. [online] (2021), Available at: https://www.nibusinessinfo.co.uk/content/advantages-and-disadvantages-crowdfunding. Accessed 31 July 2021

M. Noohani, K. Magsi, A review of 5G technology: Architecture, security and wide applications. Int. Res. J. Eng. Technol. **7**(5) (2020)

A. O'Brien, L. Liao, J. Campagna, *Responsible Investing: Delivering Competitive Performance* (Nuveen TIAA Investments, New York, 2018)

O. O'Callaghan, P. Donnellan, Liquid air energy storage systems: A review. Renew. Sust. Energ. Rev. **146**, 111113–111113 (2021). https://doi.org/10.1016/J.RSER.2021.111113

Office 4.0 | RPA – the industrial revolution in the office, Servicetrace (2019), https://www.servicetrace.com/blog/office-4-0-rpa-the-industrial-revolution-in-the-office/. Accessed 23 July 2021

T. Ojha, S. Misra, N.S. Raghuwanshi, Wireless sensor networks for agriculture: The state-of-the-art in practice and future challenges. Comput. Electron. Agric. **118**, 66–84 (2015). https://doi.org/10.1016/j.compag.2015.08.011

E. Okemwa, Effectiveness of aquaponic and hydroponic gardening to traditional gardening [WWW Document] (2015), https://www.ijsrit.com/uploaded_all_files/3563230518_m3.pdf. Accessed 19 July 2021

O. Okhrimenko, 8 crowdfunding trends you need to know in 2020, JUSTCODED (2020), Available at: https://justcoded.com/blog/8-crowdfunding-trends-you-need-to-know/. Accessed 25 July 2021

C. Oliveira et al., Benchmarking business analytics techniques in big data. Proc. Comput. Sci. **160**, 690–695 (2019). https://doi.org/10.1016/j.procs.2019.11.026

J. Ondruš, E. Kolla, P. Vertaľ, Ž. Šarić, How do autonomous cars work? Transp. Res. Procedia **44**, 226–233 (2020). https://doi.org/10.1016/j.trpro.2020.02.049

S. Onen Cinar, Z.K. Chong, M.A. Kucuker, N. Wieczorek, U. Cengiz, K. Kuchta, Bioplastic production from microalgae: A review. Int. J. Environ. Res. Public Health **17**, 3842 (2020). https://doi.org/10.3390/ijerph17113842

Орлов, С., Периферийные вычисления перемещаются в центр внимания [WWW Document]. Cnews (2019), https://www.cnews.ru/articles/2019-10-18_ot_shaht_do_pokemonov_periferijnye. Accessed 20 July 2021.

S. Oyeniyi, Soilless farming – A key player in the realisation of Zero Hunger of the Sustainable Development Goals in Nigeria (2018)

Г.И. Радченко, Распределенные Вычислительные Системы. (Фотохудожник, Челябинск, 2012)

L. Pagliarini, H. Lund, The future of robotics technology. J. Robot. Netw. Artif. Life **3**, 270 (2017). https://doi.org/10.2991/jrnal.2017.3.4.12

S. Painuly, P. Kohli, P. Matta, S. Sharma, *Advance Applications and Future Challenges of 5G IoT*, in Proceedings of the 3rd International Conference on Intelligent Sustainable Systems, ICISS 2020 (Institute of Electrical and Electronics Engineers Inc., 2020), pp. 1381–1384. https://doi.org/10.1109/ICISS49785.2020.9316004

N. Pal, Internet of Behavior- Consumer behavior analysis in the age of Internet. [Online] (2021), Available at: https://www.cronj.com/blog/internet-of-behavior-consumer-behavior-analysis-in-the-age-of-internet/. Accessed 1 June 2021

A. Palmer, Most people think AI poses a threat to the human race, study claims [WWW Document]. Mail Online (2019), https://www.dailymail.co.uk/sciencetech/article-6960317/Most-people-think-artificial-intelligence-poses-threat-human-race-study-claims.html. Accessed 26 July 2021

D. Paret, P. Crégo, Wearables, smart textiles and smart apparel: Smart textiles and smart apparel (2018), pp. 1–372. https://doi.org/10.1016/C2017-0-01438-6

Y. Parihar, E. Prasad, Global recycled textiles market by type, and end-user industry: Global opportunity analysis and industry forecast, 2020–2027 (2021)

O. Parlak, A. Salleo, A. Turner, *Wearable Bioelectronics, Wearable Bioelectronics* (Elsevier, 2020). https://doi.org/10.1016/C2017-0-00863-7

A. Parrott, L. Warshaw, Industry 4.0 and the digital twin. Deloitte Insights (2017), Available at: https://www2.deloitte.com/us/en/insights/focus/industry-4-0/digital-twin-technology-smart-factory.html. Accessed 30 July 2021

K.K. Patel, S.M. Patel, P.G. Scholar, Internet of Things-IOT: Definition, characteristics, architecture, enabling technologies, application & future challenges. Int.

J. Eng. Sci. Comput. **6**(5), 1–10 (2016). https://doi.org/10.4010/2016.1482

A. Patil, R. Humbare, V. Kumar, Allied Market Research. [Çevrimiçi] (2020), Available at: https://www.allied-marketresearch.com/wireless-charging-market. Erişildi 27 July 2021

A. Perdomo-Ortiz et al., Opportunities and challenges for quantum-assisted machine learning in near-term quantum computers. Quantum Sci. Technol. **3**(3), 1–13 (2018). https://doi.org/10.1088/2058-9565/aab859

Petrofac, The difference between green hydrogen and blue hydrogen [WWW Document]. Petrofac (2021), https://www.petrofac.com/en-gb/media/our-stories/the-difference-between-green-hydrogen-and-blue--hydrogen/

S. Phadke, The importance of a biometric authentication system. SIJ Trans. Comput. Sci. Eng. Appl. (CSEA) **1**(04), 18–22 (2013). https://doi.org/10.9756/SIJCSEA/V1I4/0104550402

P.V. Pham, Medical biotechnology: Techniques and applications. Omics Technol. Bio-Eng., 449–469 (2018). https://doi.org/10.1016/b978-0-12-804659-3.00019-1

Phishing attacks: defending your organisation [WWW Document]. Natl. Cyber Secur. Cent. (2018), https://www.ncsc.gov.uk/guidance/phishing. Accessed 24 July 2021

J. Pisarov, G. Mester, The impact of 5G technology on life in the 21st century. IPSI BgD Trans. Adv. Res. (TAR) **16**(2), 11–14 (2020)

T. Pobkrut, T. Eamsa-ard, T.Kerdcharoen, Sensor drone for aerial odor mapping for agriculture and security services. [online] (2016), Available at: https://ieeexplore.ieee.org/abstract/document/7561340?casa_token=iQtMWbVs95IAAAAA:s9Wsffg2h7iHX1_a06d6gDLBqupmLM58CSyguQKQcaA90_-_iETL04FFsYcx2edONlZMPSZPxQ. Accessed 13 July 2021

H. Porwal, R. Saggar, Ceramic matrix nanocomposites, in *Reference Module in Materials Science and Materials Engineering*, (2017). https://doi.org/10.1016/B978-0-12-803581-8.10029-3

Precisely Editor, Precisely. [Online] (2021), Available at: https://www.precisely.com/blog/integrate/the-enterprise-data-hub. Accessed 25 July 2021

Projected CO2 emissions worldwide 2050, Statista (2019), Available at: https://www.statista.com/statistics/263980/forecast-of-global-carbon-dioxide--emissions/. Accessed 23 July 2021

Pulp and paper mill waste becomes fish feed, energy and more [WWW Document]. RECYCLING magazine (2019), https://www.recycling-magazine.com/2019/04/25/pulp-and-paper-mill-waste--becomes-fish-feed-energy-and-more/. Accessed 4 Aug 2021

PURESTORAGE, What is a data hub? [Online] (2021). Available at: https://www.purestorage.com/knowledge/what-is-a-data-hub.html. Accessed 16 July 2021

V. Puri, A. Nayyar, L. Raja, Agriculture drones: A modern breakthrough in precision agriculture. J. Stat. Manage. Syst. **20**(4), 507–518 (2017)

PwC, Up to 30% of existing UK jobs could be impacted by automation by early 2030s, but this should be offset by job gains elsewhere in economy [WWW Document] (2017), https://pwc.blogs.com/press_room/2017/03/up-to-30-of-existing-uk-jobs-could-be-impacted-by--automation-by-early-2030s-but-this-should-be-offse.html. Accessed 4 Aug 21

C. Pylianidis, S. Osinga, I.N. Athanasiadis, Introducing digital twins to agriculture. Comput. Electron. Agric. **184**, 105942 (2021). https://doi.org/10.1016/j.compag.2020.105942

Y. Raban, A. Hauptman, Foresight of Cybersecurity threat drivers and affecting technologies. Foresight **20**, 353–363 (2018). https://doi.org/10.1108/FS-02-2018-0020

S.A. Rackley, *Carbon Capture and Storage*, 2nd ed. (2017), pp. 23–36. Available at: https://www.sciencedirect.com/science/article/pii/B9780128120415000027. Accessed 24 July 2021

P. Raj, A.C. Raman, *The Internet of Things: Enabling Technologies, Platforms, and Use Cases* (CRC Press/Taylor & Francis Group, Boca Raton, 2017)

R. Ramakala, et al., *Impact of ICT and IOT Strategies for Water Sustainability: A Case study in Rajapalayam-India*, in 2017 IEEE International Conference on Computational Intelligence and Computing Research (ICCIC). 2017 IEEE International Conference on Computational Intelligence and Computing Research (ICCIC) (IEEE, Coimbatore, 2017), pp. 1–4. https://doi.org/10.1109/ICCIC.2017.8524399

J.J. Ramsden, Chapter 9 – Carbon-based nanomaterials and devices, in *Nanotechnology*, Micro and nano technologies, ed. by J. J. Ramsden, 2nd edn., (William Andrew Publishing, Oxford, 2016), pp. 231–244. https://doi.org/10.1016/B978-0-323-39311-9.00015-7

T. Randall Curlee, S. Das, Advanced materials: Information and analysis needs. Res. Policy **17**, 316–331 (1991). https://doi.org/10.1016/0301-4207(91)90016-O

C.N.R. Rao, A.K. Sood, K.S. Subrahmanyam, A. Govindaraj, Graphene: The new two-dimensional nanomaterial. Angew. Chem. Int. Ed. **48**, 7752–7777 (2009). https://doi.org/10.1002/anie.200901678

G.N. Reddy, G.J.U. Reddy, A study of cybersecurity challenges and its emerging trends on latest technologies 7 (2014)

S. Rehman, L.M. Al-Hadhrami, M.M. Alam, Pumped hydro energy storage system: A technological review. Renew. Sust. Energ. Rev. **44**, 586–598 (2015). https://doi.org/10.1016/J.RSER.2014.12.040

ReinhardtHaverans 2021. From 1G to 5G: A Brief History of the Evolution of Mobile Standards [WWW Document] https://www.brainbridge.be/en/blog/1g-5g-brief-history-evolution-mobile-standards

Renault Group, The future of energy storage: are batteries the answer? Renault Group [WWW Document] (2021), https://www.renaultgroup.com/en/news-on-air/news/the-future-of-energy-storage-are-batteries--the-answer/. Accessed 13 Nov 2021

Report on Industrial Biotechnology and Climate Change: Opportunities and Challenges – OECD (2011),

Available at: https://www.oecd.org/sti/emerging-tech/reportonindustrialbiotechnologyandclimatechangeopportunitiesandchallenges.htm. Accessed 1 Aug 2021

A. Reshamwala, D. Mishra, P. Pawar, Review on natural language processing. IRACST Eng. Sci. Technol. (ESTIJ) **III**(1), 113–116 (2013)

C.J. Rhodes, Current commentary; carbon capture and storage. Sci. Prog. **95**(4), 473–483 (2012). https://doi.org/10.3184/003685012X13505722145181

K. Rieck, T. Holz, C. Willems, P. Düssel, P. Laskov, Learning and classification of malware behavior, in *Detection of Intrusions and Malware, and Vulnerability Assessment*, Lecture notes in computer science, ed. by D. Zamboni, (Springer, Berlin/Heidelberg, 2008), pp. 108–125. https://doi.org/10.1007/978-3-540-70542-0_6

C.T. Rim, Science Direct [Çevrimiçi] (2018), Available at: https://www.sciencedirect.com/science/article/pii/B9780128114070000386. Erişildi 9 June 2021

Robotic Process Automation (RPA): definition, benefits and usage [WWW Document] i-SCOOP (2021), https://www.i-scoop.eu/robotic-process-automation--rpa/. Accessed 18 July 2021

E. Röös et al., Greedy or needy? Land use and climate impacts of food in 2050 under different livestock futures. Glob. Environ. Chang. **47**, 1–12 (2017). https://doi.org/10.1016/J.GLOENVCHA.2017.09.001

R. Rosen et al., About the importance of autonomy and digital twins for the future of manufacturing. IFAC-PapersOnLine **48**(3), 567–572 (2015). https://doi.org/10.1016/j.ifacol.2015.06.141

P. Rueckert, Pegasus: The new global weapon for silencing journalists | Forbidden Stories [WWW Document]. Forbid. Stories (2021), https://forbiddenstories.org/pegasus-the-new-global-weapon-for-silencing--journalists/. Accessed 23 July 2021

J. Ruer, E. Sibaud, T. Desrues, P. Muguerra, Pumped heat energy storage (2010)

P. Russom, Exploring the benefits of a modern data hub. [Online] (2019), Available at: https://tdwi.org/articles/2019/09/09/arch-all-benefits-of-a-modern-data--hub.aspx. Accessed 21 July 2021

L. Rutaganda, R. Bergstrom, A. Jayashekhar, D. Jayasinghe, J. Ahmed, Avoiding pitfalls and unlocking real business value with RPA. J. Financ. Transf. **46**, 104–115 (2017)

S. Sadaf, A. Rana, A. Pathak, Robotic process automation (2021)

A. Saeed, A. Younes, C. Cai, G. Cai, A survey of hybrid Unmanned Aerial Vehicles. Prog. Aerosp. Sci. **98**, 91–105 (2018)

S. Sagiroglu, D. Sinanc, *Big Data: A Review*, in 2013 International Conference on Collaboration Technologies and Systems (CTS). 2013 International Conference on Collaboration Technologies and Systems (CTS) (IEEE, San Diego, 2013), pp. 42–47. https://doi.org/10.1109/CTS.2013.6567202

A. Saha, A. Subramanya, H. Pirsiavash, Hidden trigger backdoor attacks. Proc. AAAI Conf. Artif. Intell. **34**, 11957–11965 (2020). https://doi.org/10.1609/aaai.v34i07.6871

S.A. Sajjadi, H.R. Ezatpour, H. Beygi, Microstructure and mechanical properties of Al–Al$_2$O$_3$ micro and nano composites fabricated by stir casting. Mater. Sci. Eng. A **528**, 8765–8771 (2011). https://doi.org/10.1016/j.msea.2011.08.052

J.F. Salgado, The big five personality dimensions and counterproductive behaviors. Int. J. Sel. Assess. **10**, 117–125 (2002)

Sannacode | Web & Mobile App Development, Why is the Internet of Behaviors so important? [Online] (2021), Available at: https://sannacode.medium.com/why-is-the-internet-of-behaviors-so-important-13807b39a97c. Accessed 26 July 2021

S. Santos Valle, J. Kienzle, Agriculture 4.0 Agricultural robotics and automated equipment for sustainable crop production. Integr. Crop Manag. **24** (2020)

J. Saqlain, IoT and 5G: History evolution and its architecture their compatibility and future (2018)

Sara Castellanos, Quantum computing holds promise for banks, executives say, Wall Street J. (2019). Available at: https://www.wsj.com/articles/quantum--computing-holds-promise-for-banks-executives--say-11573230983

B. Sarkan, O. Stopka, J. Gnap, J. Caban, Investigation of exhaust emissions of vehicles with the spark ignition engine within emission control. Proc. Eng. **187**, 775–782 (2017). https://doi.org/10.1016/j.proeng.2017.04.437

A. Saxena, The role of big data analytics and AI in the future of healthcare. [Online] (2019), Available at: https://www.dataversity.net/the-role-of-big-data--analytics-and-ai-in-the-future-of-healthcare/#. Accessed 6 July 2021

A. Saxena, (Ed.). Biotechnology Business-Concept to Delivery. Cham, Switzerland: Springer (2020)

I. Sayol, Ignasi Sayol. [Online] (2021), Available at: https://ignasisayol.com/en/behavioral-internet-iob-the-evolution-of-personalization/. Accessed 9 June 2021

N. Sazali, Emerging technologies by hydrogen: A review. Int. J. Hydrog. Energy **45**, 18753–18771 (2020). https://doi.org/10.1016/j.ijhydene.2020.05.021

B. Schleich et al., Shaping the digital twin for design and production engineering. CIRP Ann. **66**(1), 141–144 (2017). https://doi.org/10.1016/j.cirp.2017.04.040

Schlumberger Energy Institute (SBC), *Electricity Storage Factbook. Leading the Energy Transition* (Schlumberger Energy Institute (SBC), 2013)

P. Schmidt et al. Power-to-liquids: Potentials and perspectives for the future supply of renewable aviation fuel. German Environment Agency, 2016.

A. Schnettler, At the dawn of the hydrogen economy. POWER Mag. (2020), https://www.powermag.com/at-the-dawn-of-the-hydrogen-economy/

P. Schürch, L. Philippe, Composite metamaterials: Types and synthesis. Encycl. Mater. Compos., 390–401 (2021). https://doi.org/10.1016/B978-0-12-803581-8.11750-3

S.M. Schwartz et al., Digital twins and the emerging science of self: Implications for digital health experience design and "small" data. Front. Comput. Sci. **2**, 31 (2020). https://doi.org/10.3389/fcomp.2020.00031

Securities and Exchange Commission, Crowdfunding (2015), Available at: https://www.sec.gov/rules/final/2015/33-9974.pdf. Accessed 25 July 2021

V. Sedlakova, M. Ruel, E.J. Suuronen, Therapeutic use of bioengineered materials for myocardial infarction. Nanoeng. Mater. Biomed. Uses, 161–193 (2019). https://doi.org/10.1007/978-3-030-31261-9_9

Semarchy, Why the data hub is the future of data management. [Online] (2021), Available at: https://blog.semarchy.com/why-the-data-hub-is-the-future-of-data-management. Accessed 10 June 2021

C. Semeraro et al., Digital twin paradigm: A systematic literature review. Comput. Ind. **130**, 103469 (2021). https://doi.org/10.1016/j.compind.2021.103469

S. Seo, J. Won, E. Bertino, Y. Kang, D. Choi, *A Security Framework for a Drone Delivery Service*, in: Proceedings of the 2nd Workshop on Micro Aerial Vehicle Networks, Systems, and Applications for Civilian Use (2016)

A. Sestino et al., Internet of Things and Big Data as enablers for business digitalization strategies. Technovation **98**, 102173 (2020). https://doi.org/10.1016/j.technovation.2020.102173

P. Seuwou, E. Banissi, G. Ubakanma, The future of mobility with connected and autonomous vehicles in smart cities, in *Digital twin technologies and smart cities, Internet of Things*, ed. by M. Farsi, A. Daneshkhah, A. Hosseinian-Far, H. Jahankhani, (Springer, Cham, 2020), pp. 37–52. https://doi.org/10.1007/978-3-030-18732-3_3

J. Shabbir, T. Anwer, Artificial intelligence and its role in near future. ArXiv180401396 Cs 14 (2018)

N. Shahrubudin, T.C. Lee, R. Ramlan, An overview on 3D printing technology: Technological, materials, and applications. Proc. Manuf. **35**, 1286–1296 (2019). https://doi.org/10.1016/j.promfg.2019.06.089

T. Shanmuganantham, K. Ramasamy, S.S. Mohammmed, Wireless power transmission – A next generation power transmission system. Int. J. Comput. Appl. **1**(13), 102–105 (2010)

R. Sharma, *Artificial Intelligence in Agriculture: A Review*, in: 2021 5th International Conference on Intelligent Computing and Control Systems (ICICCS). Presented at the 2021 5th International Conference on Intelligent Computing and Control Systems (ICICCS) (IEEE, Madurai, 2021), pp. 937–942. https://doi.org/10.1109/ICICCS51141.2021.9432187

C.J. Shearer, A. Cherevan, D. Eder, Application and future challenges of functional nanocarbon hybrids. Adv. Mater. **26**, 2295–2318 (2014). https://doi.org/10.1002/adma.201305254

S. Shekhar, S.K. Feiner, W.G. Aref, Spatial computing. Commun. ACM **59**(1), 72–81 (2015). https://doi.org/10.1145/2756547

D. Shi, Z. Guo, N. Bedford, 8 – Superconducting nanomaterials, in *Nanomaterials and Devices, Micro and Nano Technologies*, ed. by D. Shi, Z. Guo, N. Bedford, (William Andrew Publishing, Oxford, 2015), pp. 191–213. https://doi.org/10.1016/B978-1-4557-7754-9.00008-1

W. Shi, J. Cao, Q. Zhang, Y. Li, L. Xu, Edge computing: Vision and challenges. IEEE Internet Things J. **3**, 637–646 (2016). https://doi.org/10.1109/JIOT.2016.2579198

W. Shi, G. Pallis, Z. Xu, Edge computing. Proc. IEEE **107**, 1474–1481 (2019). https://doi.org/10.1109/MC.2017.3641639

F. Shrouf, G. Miragliotta, Energy management based on Internet of Things: Practices and framework for adoption in production management. J. Clean. Prod. **100**, 235–246 (2015). https://doi.org/10.1016/j.jclepro.2015.03.055

I.S. Sidek, S.F.S. Draman, S.R.S. Abdullah, N. Anuar, Current development on bioplastics and its future prospects: An introductory review. INWASCON Technol. Mag.-TECH MAG **1**, 3–8 (2019)

P. Simon, Y. Gogotsi, B. Dunn, Where do batteries end and supercapacitors begin? Science **343**, 1210–1211 (2014). https://doi.org/10.1126/science.1249625

U. Singh, Carbon capture and storage: An effective way to mitigate global warming on JSTOR. JSTOR (2013), Available at: https://www.jstor.org/stable/24098511. Accessed 23 July 2021

K. Singh, Business standart. [Online] (2021), Available at: https://www.business-standard.com/content/specials/internet-of-behaviors-is-critical-to-remodeling-customer-experience-and-business--innovation-121032400443_1.html. Accessed 26 July 2021

S. Singh, G. Singh, C. Prakash, S. Ramakrishna, Current status and future directions of fused filament fabrication. J. Manuf. Process. **55**, 288–306 (2020). https://doi.org/10.1016/j.jmapro.2020.04.049

Sinu, Sinu Your IT Department. [Online] (2020), Available at: https://www.sinu.com/blog/2020/12/7/internet--of-behavior-predicted-to-grow-in-2021. Accessed 26 July 2021

Sisense, What is healthcare analytics? [Online] (2021), Available at: https://www.sisense.com/glossary/healthcare-analytics-basics/. Accessed 10 July 2021

D. Skilskyj, 3 advantages of flexible electronics that consumers love (2018), Available at: https://info.promerus.com/blog/3-advantages-of-flexible-electronics-that-consumers-love. Accessed 25 July 2021

D. Skilskyj, How are flexible electronics manufactured? (2019), Available at: https://info.promerus.com/blog/how-are-flexible-electronics-manufactured. Accessed: 25 July 2021

Small Business Guide: Cybersecurity [WWW Document]. Natl. Cyber Secur. Cent. (2018), https://www.ncsc.gov.uk/collection/small-business-guide. Accessed 24 July 2021

J. Smith, Hydrogen and fuels cells for transport [WWW Document]. Mobil. Transp. Eur. Comm. (2016), https://ec.europa.eu/transport/themes/urban/vehicles/road/hydrogen_en

K. Smith, How scientists are growing meat and saving endangered animals at the same time. LiveKindly (2021), Available at: https://www.livekindly.co/how-scientists-growing-meat-saving-endangered-animals/. Accessed 25 July 2021

Soilless Agriculture: An In-depth Overview [WWW Document]. AGRITECTURE (2021), https://www.agritecture.com/blog/2019/3/7/soilless-agriculture-an-in-depth-overview. Accessed 19 July 2021

SourceFuse, Healthcare analytics matters: A unique data analytics approach. [Online] (2021), Available at: https://www.sourcefuse.com/blog/healthcare-analytics-matters-a-unique-data-analytics-approach/. Accessed 27 July 2021

K. Spensley, G.W. Winsor, A.J. Cooper, Nutrient film technique – Crop culture in flowing nutrient solution. Outlook Agric. **9**, 299–305 (1978). https://doi.org/10.1177/003072707800900608

M. Spilka, A. Kania, R. Nowosielski, Integrated recycling technology. J. Achievement. Mater. Manuf. Eng. **31**, 6 (2008)

H. Srinivas, The 3R concept and waste minimization [WWW Document] (2015), https://www.gdrc.org/uem/waste/3r-minimization.html. Accessed 4 Aug 21

S.D. Stoller, M. Carbin, S. Adve, K. Agrawal, G. Blelloch, S. Dan, K. Yelick, M. Zaharia, Future directions for parallel and distributed computing: SPX 2019 Workshop Report 32 (2019)

F.H. Sumi, L. Dutta, D. Sarker, F., Future with wireless power transfer technology. J. Electrical Electron. Syst. **7**(4), 1–7 (2018)

C.-J. Sun (Simon), Cambridge analytica: A property-based solution (2020), https://doi.org/10.5281/ZENODO.3764717

S. Suoniemi et al., Big data and firm performance: The roles of market-directed capabilities and business strategy. Inf. Manag. **57**(7), 103365 (2020). https://doi.org/10.1016/j.im.2020.103365

R.C. Tabata, The future challenges of big data in healthcare. [Online] (2021), Available at: https://www.forbes.com/sites/forbestechcouncil/2021/06/18/the-future-challenges-of-big-data-in-healthcare/?sh=24ada90a46b2. Accessed 6 July 2021

Tableu, 8 natural language processing (NLP) examples. [Online] (2021), Available at: https://www.tableau.com/learn/articles/natural-language-processing-examples. Accessed 14 July 2021

Tadviser, Периферийные вычисления Граничные вычисления [WWW Document] (2019), https://www.tadviser.ru/index.php/Статья:Периферийные_вычисления_(Edge_computing). Accessed 20 July 2021

A. Tajudeen, S. Oyeniyi, Soilless farming-a key player in the realisation of "Zero Hunger" of the Sustainable Development Goals in Nigeria. Int. J. Ecol. Sci. Environ. Eng. **5**, 1–7 (2018)

Talview, Here's why you need to leverage NLP during recruiting. [Online] (2018), Available at: https://blog.talview.com/heres-why-you-need-to-leverage-nlp-during-recruiting. Accessed 8 June 2021

Tech The Day, Understanding the concept of the Internet of Behaviors. [Online] (2021), Available at: https://techtheday.com/understanding-the-concept-of-the-internet-of-behaviors/. Accessed 1 June 2021

Techvice Company, The Internet of Behavior. [Online] (2021), Available at: https://techvice.org/blog/popular/internet-of-behavior/. Accessed 1 June 2021

D. Tezza, M. Andujar, The State-of-the-Art of Human-Drone Interaction: A Survey. IEEE Access (2019)

The benefits (and limitations) of RPA implementation, [WWW Document] Financial Services Blog (2017), https://financialservicesblog.accenture.com/the-benefits-and-limitations-of-rpa-implementation. Accessed 29 July 2021

The Future of Recycling – Looking to 2020 and Beyond [WWW Document]. Recycle Track Systems (2020), https://www.rts.com/blog/the-future-of-recycling-looking-to-2020-and-beyond/. Accessed 4 Aug 2021

The History of Biometrics: From the 17th Century to Nowadays, RecFaces (2020), Available at: https://recfaces.com/articles/history-of-biometrics. Accessed 26 July 2021

M. Tirrell, E. Kokkoli, M. Biesalski, The role of surface science in bioengineered materials. Surf. Sci. **500**, 61–83 (2002). https://doi.org/10.1016/S0039-6028(01)01548-5

D. Todaro, How agile will drive the Internet of Behaviors (IoB) in 2021 and beyond. [Online] (2021), Available at: https://www.forbes.com/sites/forbestechcouncil/2021/02/11/how-agile-will-drive-the-internet-of-behaviors-iob-in-2021-and-beyond/?sh=705e44a36636. Accessed 1 June 2021

Top 5 Industries Utilizing Robotics, Ohio University (2021), https://onlinemasters.ohio.edu/blog/5-industries-utilizing-robotics/. Accessed 29 July 2021

A.M. Tripathi, *Learning Robotic Process Automation: Create Software robots and automate business processes with the leading RPA tool – UiPath* (Packt Publishing Ltd., Birmingham, 2018)

P. Tripicchio, M. Satler, G. Dabisias, E. Ruffaldi C.A. Avizzano, *Towards Smart Farming and Sustainable Agriculture with Drones*, in: 2015 International Conference on Intelligent Environments (IEEE, 2015, July), pp. 140–143

K.-Y. Tsai, M.-H. Tsai, J.-W. Yeh, Sluggish diffusion in Co–Cr–Fe–Mn–Ni high-entropy alloys. Acta Mater. **61**, 4887–4897 (2013). https://doi.org/10.1016/j.actamat.2013.04.058

A.M. Turing, Computing machinery and intelligence, in *Parsing the Turing Test*, (Springer, Dordrecht, 2009), pp. 23–65

K. Uchino, *Advanced Piezoelectric Materials: Science and Technology* (Woodhead Publishing, Cambridge/Philadelphia, 2010)

G. Udeanu, A. Dobrescu, M. Oltean, *Unmanned Aerial Vehicle in Military Operations*, in: The 18th International Conference "Scientific Research and Education in the Air Force–AFASES", Brasov, Romania (2016, May), pp. 199–205

UN 2017. Department of Economic and Social Affairs. [WWW Document] https://www.un.org/development/desa/en/news/population/world-population-prospects-2019.html

UN News, FEATURE: Does drone technology hold promise for the UN? [online] (2021), Available at: https://news.un.org/en/story/2017/09/564452-feature-does-drone-technology-hold-promise-un. Accessed 12 July 2021

United Nations, Universal Declaration of Human Rights 6 (1948)

University of Pittsburgh, The role of data analytics in health care. [Online] (2021). Available at: https://online.shrs.pitt.edu/blog/data-analytics-in-health-care/. Accessed 15 June 2021

US Department of Defense, Unmanned systems roadmap 2007–2032 (2007)

US EPA, O., Used Lithium-Ion Batteries [WWW Document] (2019), https://www.epa.gov/recycle/used-lithium-ion-batteries. Accessed 1 Aug 2021

US Robot Density in Car Industry Ranks 7th Worldwide [WWW Document] IFR International Federation of Robotics (2021), https://ifr.org/ifr-press-releases/news/us-robot-density-in-car-industry-ranks-7th-worldwide. Accessed 26 July 2021

A. Vahdat, D. Clark, J. Rexford, A purpose-built global network: Google's move to SDN: A discussion with Amin Vahdat, David Clark, and Jennifer Rexford. Queue **13**, 100–125 (2015)

K. Valavanis, G. Vachtsevanos (eds.), *Handbook of unmanned aerial vehicles* (Springer, Dordrecht, 2015), p. 44

H. van der Aa et al., *Challenges and Opportunities of Applying Natural Language Processing in Business Process Management* (Association for Computational Linguistics, New Mexico, 2018)

W.M.P. van der Aalst, M. Bichler, A. Heinzl, Robotic process automation. Bus. Inf. Syst. Eng. **60**, 269–272 (2018). https://doi.org/10.1007/s12599-018-0542-4

J. Van Hoof, J. Peterseim, O. Hatop, Green hydrogen economy – Predicted development of tomorrow [WWW Document]. PwC (2021), https://www.pwc.com/gx/en/industries/energy-utilities-resources/future-energy/green-hydrogen-cost.html

D.S. Van Os, et al., Soilless farming: A new sustainable agriculture method – Mega [WWW Document]. Megatrends (2021), https://mega.online/en/articles/soil-less-farming-the-future-of-agriculture. Accessed 19 July 2021

K. Vassakis, E. Petrakis, I. Kopanakis, Big data analytics: Applications, prospects and challenges, in *Mobile Big Data*, Lecture notes on data engineering and communications technologies, ed. by G. Skourletopoulos et al., (Springer, Cham, 2018), pp. 3–20. https://doi.org/10.1007/978-3-319-67925-9_1

Vector ITC, Vector ITC Group. [Online] (2021), Available at: https://www.vectoritcgroup.com/en/tech--magazine-en/user-experience-en/what-is-the-internet-of-behaviour-iob-and-why-is-it-the-future/. Accessed 26 July 2021

B. Vergouw, H. Nagel, G. Bondt, B. Custers, Drone technology: Types, payloads, applications, frequency spectrum issues and future developments, in *The future of drone use*, (TMC Asser Press, The Hague, 2016), pp. 21–45

O. Vermesan (ed.), *Internet of Things: Converging Technologies for Smart Environments and Integrated Ecosystems* (River Publishers (River Publishers series in information science and technology), Aalborg, 2013)

Vertical Farming: Moving from Genetic to Environmental Modification | Elsevier Enhanced Reader [WWW Document] (2020) https://doi.org/10.1016/j.tplants.2020.05.012

P.B.M. Vindis, C. Rozman, J.F.C. Marjan, Biogas production with the use of mini digester. J. Achievement. Mater. Manuf. Eng. **28** (2008)

Y.V. Wang, *A View of Research on Wireless Power Transmission* (IOP Publishing, Bristol, 2018)

Y. Wang, B.T. Hazen, Consumer product knowledge and intention to purchase remanufactured products. Int. J. Prod. Econ. Recent Dev. Sustain. Consump. Prod. Emerg. Mark. **181**, 460–469 (2016). https://doi.org/10.1016/j.ijpe.2015.08.031

S. Wankhede, S. Sachan, V. Kumar, Wireless power transmission market by technology, (Near-field technology and far-field technology), type (Devices with battery and devices without battery), and application (Receiver and transmitter): Global opportunity analysis and industry forecast, 202. [Çevrimiçi] (2021), Available at: https://www.alliedmarketresearch.com/wireless-power-transmission-market. Erişildi 26 Oct 2021

J. Warren, Les utilisations industrielles des cultures: les bioplastiques [WWW Document]. GreenMaterials (2011), https://www.greenmaterials.fr/les-utilisations-industrielles-des-cultures-les-bioplastiques/. Accessed 23 Oct 2021

S. Watkins, et al. The mind-bending future of flexible electronics. (Australian Academy of Science, 2021), Available at: https://www.science.org.au/curious/technology-future/flexible-electronics. Accessed 25 July 2021

What is Biotechnology? Types and Applications – Iberdrola (2021), Available at: https://www.iberdrola.com/innovation/what-is-biotechnology. Accessed 10 July 2021

T. Whelan, U. Atz, T. Van Holt, C. Clark, *Esg and Financial Performance: Uncovering the Relationship by Aggregating Evidence from 1,000 Plus Studies Published Between 2015–2020* (NYU Stern Center for Sustainable Business, New York, 2020)

L. Willcocks, M. Lacity, Service automation robots and the future of work (2016)

L. Willcocks, M. Lacity, A. Craig, Robotizing global financial shared services at Royal DSM. J. Financ. Transf. **46**, 62–75 (2017)

S. Winterfeld, J. Andress, *The Basics of Cyber Warfare: Understanding the Fundamentals of Cyber Warfare in Theory and Practice*, Syngress basics series (Syngress/Elsevier, Amsterdam/Boston, 2013)

R. Wolff, 26. MonkeyLearn. [Online] Available at: https://monkeylearn.com/blog/what-is-natural-language-processing/. Accessed 14 July 2021

World Economic Forum, Global agenda council on the future of software & society. [Online] (2015), Available at: http://www3.weforum.org/docs/WEF_GAC15_Technological_Tipping_Points_report_2015.pdf. Accessed 26 July 2021

J. Worth, Top 3 predictions for the future of robotics [WWW Document] (2016), https://www.rodon-group.com/blog/top-3-predictions-for-the-future-of-robotics. Accessed 26 July 2021

R. Wright, L. Keith, Wearable technology: If the tech fits, wear it. Electron. Resour. Med. Libr. 11(4), 204–216 (2014). https://doi.org/10.1080/15424065.2014.969051

Würth Elektronik, Würth Elektronik. [Çevrimiçi] (2021), Available at: https://www.we-online.com/web/en/electronic_components/news_pbs/blog_pbcm/blog_detail-worldofelectronics_100414.php. Erişildi 9 July 2021

Q. Xiao, Biometrics-technology, application, challenge, and computational intelligence solutions. IEEE Comput. Intell. Mag. 2(2), 5–10 (2007). https://doi.org/10.1109/MCI.2007.353415

M. Xylia, J. Svyrydonova, S. Eriksson, A. Korytowski, Beyond the tipping point: Future energy storage [WWW Document]. Urban Insight (2021), https://www.swecourbaninsight.com/urban-energy/beyond-the-tipping-point-future-energy-storage/. Accessed 13 Nov 2021

J. Yaacoub, H. Noura, O. Salman, A. Chehab, Security analysis of drones systems: Attacks, limitations, and recommendations. Internet of Things 11, 100218 (2020)

M. Yamada, S. Morimitsu, E. Hosono, T. Yamada, Preparation of bioplastic using soy protein. Int. J. Biol. Macromol. 149, 1077–1083 (2020). https://doi.org/10.1016/j.ijbiomac.2020.02.025

G.Z. Yang, J. Cambias, K. Cleary, E. Daimler, J. Drake, P.E. Dupont, N. Hata, P. Kazanzides, S. Martel, R.V. Patel, Medical robotics – Regulatory, ethical, and legal considerations for increasing levels of autonomy. Sci. Robot. 2 (2017). https://doi.org/10.1126/scirobotics.aam8638

I. Yaqoob et al., Big data: From beginning to future. Int. J. Inf. Manag. 36(6), 1231–1247 (2016). https://doi.org/10.1016/j.ijinfomgt.2016.07.009

H. Yoo, S. Chankov, Drone-delivery using autonomous mobility: An innovative approach to future last-mile delivery problems. [online] (2021), Available at: https://ieeexplore.ieee.org/abstract/document/8607829/authors#authors. Accessed 12 July 2021

J. Yu, L.X.L. Chen, The greenhouse gas emissions and fossil energy requirement of bioplastics from cradle to gate of a biomass refinery. Environ. Sci. Technol. 42, 6961–6966 (2008)

H. Yu, H. Lee, H. Jeon, What is 5G? Emerging 5G mobile services and network requirements. Sustainability (Switzerland) 9(10), 1–22 (2017). https://doi.org/10.3390/su9101848

S.C.-Y. Yuen, G. Yaoyuneyong, E. Johnson, Augmented reality: An overview and five directions for ar in education. J. Educ. Technol. Dev. Exchange 4(1) (2011). https://doi.org/10.18785/jetde.0401.10

H.-S. Yun, T.-H. Kim, T.-H. Park, Speed-bump detection for autonomous vehicles by Lidar and Camera. J. Electr. Eng. Technol. 14, 2155–2162 (2019). https://doi.org/10.1007/s42835-019-00225-7

S.H. Zaferani, 1 – Introduction of polymer-based nanocomposites, in Polymer-Based Nanocomposites for Energy and Environmental Applications, Woodhead publishing series in composites science and engineering, ed. by M. Jawaid, M. M. Khan, (Woodhead Publishing, 2018), pp. 1–25. https://doi.org/10.1016/B978-0-08-102262-7.00001-5

M. Zellner, D. Massey, E. Minor, M. Gonzalez-Meler, Exploring the effects of green infrastructure placement on neighborhood-level flooding via spatially explicit simulations. Computers, Environment and Urban Systems, 59, 116–128 (2016)

Y. Zeng, R. Zhang, T. Lim, Wireless communications with unmanned aerial vehicles: Opportunities and challenges. IEEE Commun. Mag. 54(5), 36 (2016)

Y.H.P. Zhang, J. Sun, Y. Ma, Biomanufacturing: History and perspective. J. Ind. Microbiol. Biotechnol. 44(4–5), 773–784 (2017). https://doi.org/10.1007/s10295-016-1863-2

J.M. Zheng, K.W. Chan, I. Gibson, Virtual reality: A real-world review on a somewhat touchy subject. IEEE. doi: 0278-6648/98 (1998)

P. Zhu, L. Wen, X. Bian, H. Ling, Q. Hu, Vision meets drones: A challenge. [online] (2018), Available at: https://arxiv.org/abs/1804.07437. Accessed 12 July 2021

Zillner et al., Big data-driven innovation in industrial sectors, in New Horizons for a Data-Driven Economy, ed. by M. J. Cavanillas, E. Curry, W. Wahlster, (Springer, Cham, 2016). https://doi.org/10.1007/978-3-319-21569-3

S. Zisk, Beyond the marketing stack: Why a data hub is in your future. [Online] (2016), Available at: https://www.redpointglobal.com/blog/beyond-marketing-stack-data-hub-future/. Accessed 10 June 2021

I. Zohdy, R.K. Kamalanathsharma, H. Rakha, Intersection management for autonomous vehicles using cooperative adaptive cruise control systems. Presented at the Transportation Research Board 92nd Annual Meeting Transportation Research Board (2013)

Abstract

Poverty, which has taken shape in different dimensions under the influence of the conditions from the past to the present, can be defined as the deficiency experienced by people in fulfilling their life functions or the living standards being below the average level. This chapter presents the business models of 39 companies and use cases that employ emerging technologies and create value in SDG-1, No Poverty. We should highlight that one use case can be related to more than one SDG and it can make use of multiple emerging technologies.

Keywords

Sustainable Development Goals · Business models · No poverty · Sustainability

Poverty, which has taken shape in different dimensions under the influence of the conditions from the past to the present, can be defined as the deficiency experienced by people in fulfilling their life functions or the living standards being below the average level. Poverty results from a lack of critical capabilities, such as insufficient income or education, bad health, insecurity, low self-confidence, a sense of powerlessness or the lack of rights such as freedom of expression (Haughton and Khandker 2009). In its various versions, poverty has long been one of humanity's most significant conflicts. Especially with the increased bad conditions, poverty has gained a higher level called extreme poverty. Extreme poverty is a more multidimensional concept than poverty. According to Sachs (2015), it should be defined more broadly as the inability to satisfy fundamental human requirements such as food, water, sanitation, safe energy, education and a means of subsistence. With the spread of technology and the rise in living standards, there was an expectation that the problem of poverty would be overcome throughout the world. However, the gap between rich and poor people is being further opened day by day with unequal economic distributions in most countries, leading numerous people to live below the poverty line despite the improvements in science and technology. Many hypotheses have been proposed to explain why inequality is bad for growth. Still, the OECD study focuses on one in particular: as the wealth gap widens, low-income households would spend less on education and skills (Keeley 2015). Last but not least, according to the World Bank (2021), a large part of the world's population is still living below the international poverty line of $1.90 a day.

The author would like to acknowledge the help and contributions of Berkay İspir, Ekrem Gümüş, Zeynel Salman, Begüm Nur Okur, Canberk Özemek, Elif Berra Aktaş and Beyza Özerdem in completing this chapter. They also contributed to Chapter 2's Autonomous Vehicles, Cybersecurity and Drones sections.

In the early 2000s, the United Nations announced the Millennium Development Goals (MDGs) in the face of major problems affecting humanity. Until 2015, these goals were expected to solve these problems or at least reduce them. The MDGs aim to boost global awareness, governmental responsibility, improved measurements, social feedback and public demands by putting these objectives into a simply comprehensible set of eight goals and creating quantifiable and measurable targets (Sachs 2012). Research state that with the MDGs there has been an impressive poverty reduction, such as, in a quarter of a century time frame between 1990 and 2015, the extreme poverty rate fell by 33% in developing countries (United Nations 2015). According to Sachs (2012) research, in a world where severe global warming and other significant environmental disruptions are now a reality, there is indeed a broad recognition that sustainability goals must be prioritised with poverty reduction goals. In this chapter, the definition and different versions of poverty, the multidimensional concept of extreme poverty, effects of poverty on human life with the inclusion of technological developments and historical activities made on poverty through MDGs are stated.

As a result, in addition to the improvements made in 15 years by MDGs, new requirements bring new updates about methods of taking action for the UN, which led to the emergence of sustainable development goals (SDGs). As good news, the percentage of the world's population living in extreme poverty has decreased, from 15.7% in 2010 to 10.0% in 2015. The rate of global poverty reduction, on the other hand, has slowed. It is claimed that the worldwide poverty rate will be around 7.4% in 2021; however, with the emergence of the COVID-19 pandemic, forecasts are affected negatively with an increase to 8.7% (United Nations Department for Economic and Social Affairs 2020). In January 2021, it was estimated that the pandemic would push between 119 and 124 million people into extreme poverty around the globe in 2020 (World Bank 2021). Due to the new requirements of the current world order and pandemic conditions to humanity, significant

updates are planned in SDGs from the beginning of the COVID-19 pandemic.

As the United Nations have mentioned in their official website and publications, the first sustainable development goal is to eliminate poverty in all its forms everywhere (2015). This goal originally took place under the millennium development goals, which are predecessors of sustainable development goals. As MDGs were practised for 15 years (2000–2015) and fulfilled their schedule, SDGs were presented as an improved and more detailed version of MDGs. Sharma et al. state that poverty is a social problem that affects people's standard of living, food consumption and other aspects of their lives. The disadvantaged community cannot produce more and support their families since it lacks the resources to access quality and nutritious food and health services (2016). According to Sachs (2012), eliminating poverty includes providing safe and sustainable water and sanitation, sufficient nutrition, basic health services and basic infrastructures such as electricity, transportation and access to the global information network. These sub-goals are planned to be achieved by 2030 at the latest. This target might seem utopic, but some popular theories claim that it is well within reach with the help of technological advances and economic growth. One of the notable facts about poverty nowadays is that over half of the one billion people with a low income are living in middle-income countries. This means those people are living in societies with the financial and technological means to address their remaining poverty (as Brazil and China have effectively and notably done in recent years). Although hundreds of millions of impoverished people still live in the least developed countries, they are a dwindling proportion of the world's poorest people, such that small financial and technological transfers from high-income and middle-income countries could alleviate their plight. The UN's perspective of targets for SDG-1 can be found in Fig. 3.1.

Significant steps were taken to achieve this goal, as the number of people living in extreme poverty has decreased from 1.9 billion to 736 million between 1990 and 2015. Still, numerous people strive to provide for their basic human

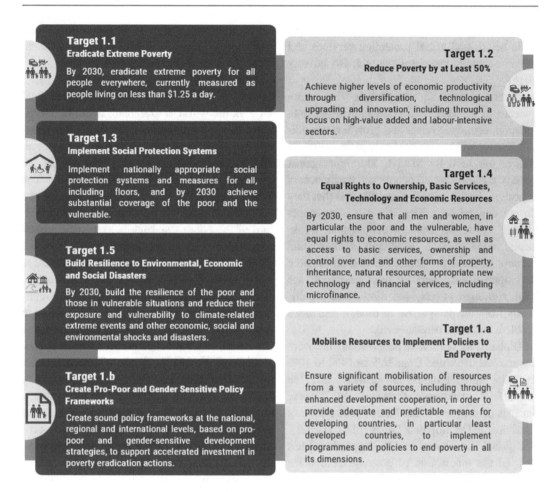

Fig. 3.1   Targets of SDG-1. ("Goal 1: No Poverty" 2016)

needs (United Nations 2015). Poverty reduction has been the centre of worldwide discussions since the 1970s. The sustainable development goals (SDGs) were established by 193 UN member states in 2015, with Goal 1 being to "end poverty in all its forms and everywhere" (UN 2016). According to Alkire et al. (2014), rural areas account for 85% of all poor people in 105 countries.

As shown in Fig. 3.1, there are several subgoals and indicators under Goal 1, "No Poverty". These include eliminating extreme poverty for all people everywhere by 2030, which is currently measured as the number of people living on less than $1.25 a day and halving the number of people living in poverty. Another sub-objective of this aim is to minimise the sensitivity of the poor

and the vulnerable to climate-related, economic and social events by 2030. Implementing poverty-eradication programs and policies and mobilising resources for developing and underdeveloped nations are other sub-objectives.

The main reason for aiming to end all types of poverty is that it affects individual welfare and living conditions, such as accessing basic necessities like health, nourishment, clothing and accommodation. With the COVID-19 pandemic, it is associated with increased extreme poverty. Fighting against poverty ranks as the first sustainable development goal in the United Nations remarks. Poor people are more vulnerable to the impacts of natural disasters and all examples of extreme poverty such as hunger and malnutrition, fuel poverty, limited access to education, social isolation and discrimination and exclusion from basic health and social protection services and decision-making processes (Goal 1: No Poverty 2016).

Poverty-causing factors such as the consequences of globalisation, population expansion, inflation, economic crises and growth rates indicate that we will confront poverty in the future, as well as our need for SDG-1. The fight against poverty ensures that all people can easily access education, health, nutrition and life and that they do not depend on others. Poverty assessments contain epistemic and methodological weaknesses according to academic studies, and because of these problems, the first sustainable development goal and its implementation are vital. Although poverty has diminished significantly from the perspective of time, it remains a problem that must be addressed.

Globally, the number of people living in severe poverty has decreased from 36% in 1990 to 10% in 2015. Still, progress is slowing, and the COVID-19 pandemic threatens to reverse decades of progress in the battle against poverty (Sustainable Development Goals (SDG 1) 2021). The global recession caused by the pandemic has delayed the eradication of poverty. Companies and other workplaces began to close in April 2020. Eighty-one percent of workers and 66% of freelancers were affected (SDG 1: No poverty – Iberdrola n.d.). Only 87 nations had unemployment insurance schemes in place as part of their national legislation in February 2020, and only 34 of those countries protected freelancers (SDG 1: No poverty – Iberdrola n.d.).

SDG-1, unlike the others, is directed at those who are unable to satisfy their fundamental requirements, who have a lower quality of life than the average and who are underprivileged in most ways. Poverty is important for poor people and everyone, privileged or disadvantaged. The elimination of extreme poverty will result in economic progress and a greater level of education among the population. Society must take global poverty seriously and take steps to achieve SDG-1 before 2030. Extreme poverty is a univer-

sal conflict, and every person should meet their basic needs. Therefore, Sustainable Development Goal 1 is an essential goal for all humanity.

Economic growth in the world or even in a country is not always predictable. If the total GDP in the world is examined, it can be recognised that the actual total GDP in 2009 is about two-thirds of the predicted value (Jolliffe et al. 2015). This kind of deviation may have a massive effect on the process of fulfilling SDGs. The World Bank advocates SDG-1 because of the adamant relationship between growth and welfare of the poor. A setback in economic growth could also cause a significant setback on the global goal for tackling poverty. This problem is not necessarily global, as some countries such as Bangladesh, China, India, etc. have a huge number of people living in poverty. A slowdown in the economic growth in one of these countries can have a major effect on the global poverty goal (Jolliffe et al. 2015). Tourism is considered a potential solution for this issue (by preventing economic setbacks and helping underdeveloped countries build up their economies). Scheyvens and Hughes define tourism as a promising economic sector to help create strategies for decreasing poverty, and statistics back up their opinion. It is known to help improve rural areas' economy and well-being in developing countries. It is important to recognise that certain conditions need to be provided for tourism to contribute to SDG-1. Powerful actors such as local elites, company directors and government leaders must do their best to ensure factors like corruption and dictatorship do not undermine the sector of tourism in the relevant countries (Scheyvens and Hughes 2019).

Opposing the opinions claiming that technology and economic growth can make SDG-1 possible, current data predicts that eradicating poverty by 2030 seems unrealistic. Under the most optimistic possibilities predicted before the COVID-19 pandemic, 6.1% of the world's population will most likely remain in extreme poverty by 2030 (Castaneda et al. 2020). The results of comparing poverty scenarios in different countries, as a result of the analysis to monitor progress towards the achievement of the "No Poverty" sustainable development goal, point to the difficulty of achieving this goal unless extra development policy efforts are implemented (Crespo Cuaresma et al. 2018). One of the recommended actions to help this situation in the future is to involve children in the process. According to UNICEF, with the right information and tools, children and youth can play a critical role in implementing the SDGs by driving action in their communities. As millions of children and young people realise the goals, more and more people around the world will start taking action ("How can we achieve the Sustainable Development Goals for and with children?" 2021). When it comes to achieving this target over the long term (between today and 2100), climate change will have devastating consequences for urban and rural regions in Sub-Saharan Africa and Southeast Asia, resulting in new poverty in both developing and developed countries and threatening long-term development (Olsson et al. 2014). Decent work has been carried out during COVID-19, whereby many companies have lent a hand in supporting the healthcare system's response. Pharmaceutical businesses collaborate with governments to improve testing capabilities, while mask and ventilator makers are willing to transfer or build new production lines. To combat the outbreak, tech businesses give critical digital tools to eliminate social isolation, increase social cohesion and raise awareness about health and safety rules. Private sector innovation can make a substantial contribution to the pandemic response, both in the short and long term, as well as long-term resilience. Big data and artificial intelligence, in particular, must be used to develop digital public goods such as actionable real-time and predictive insights (UN 2020).

Sumner et al. (2020) underline that according to UN estimates, COVID-19 poses a formidable challenge to the UN's goal of ending poverty by 2030, as global poverty could rise for the first time since 1990. Such an increase could represent a reversal of approximately a decade in global poverty reduction progress depending on the poverty line. The negative consequences could lead to poverty levels equivalent to those seen 30 years ago in some areas. In the most extreme case of a 20% reduction in income or consumption, the number of people living in poverty might rise by 420–580 million, compared to the most recent official data for 2018.

3.1 Companies and Use Cases

Table 3.1 presents the business models of 39 companies and use cases that employ emerging technologies and create value in SDG-1. We should highlight that one use case can be related to more than one SDG and it can make use of multiple emerging technologies. In the left column, we present the company name, the origin country, related SDGs and emerging technologies that are included. The companies and use cases are listed alphabetically.[1]

---

[1]For reference, you may click on the hyperlinks on the company names or follow the websites here (Accessed Online – 2.1.2022):

http://www.lifebankcares.com/; http://www.oneconcern.com/; https://about.wefarm.com/; https://avy.eu/; https://banqu.co/; https://beam.org/; https://childgrowthmonitor.org/; https://climate.com/; https://efarms.com.ng/en; https://hellotractor.com; https://hellotractor.com; https://kamworks.com/; https://leafglobalfintech.com/; https://monkee.rocks/; https://nextfood.co/; https://perfectday.com/; https://pesitho.com/; https://solarcreed.com/; https://swoop.aero/; https://tala.co/; https://wagestream.com/; https://waka-waka.com/; https://whr.loans/; https://wingcopter.com/; https://www.acrecx.com/; https://www.aerofarms.com/; https://www.beyondmeat.com/; https://www.borlaug.ws/; https://www.crop2cash.com.ng/; https://www.donationmatch.com/; https://www.edenagri.co.th/; https://www.pikadiapers.com/; https://www.redefinemeat.com/; https://www.truvito.io/; https://www.unhoused.org/; https://www.veritas.com/360; https://www.vodafone.com/; https://www.wildtypefoods.com/; https://zolaelectric.com/; https://www.shambarecords.com/

Table 3.1 Companies and use cases in SDG-1

No	Company info	Value proposal (what?)	Value creation (how?)	Value capture
1	AcreCX The Netherlands 1, 8, 11 Blockchain	The company offers a blockchain-based marketplace for agricultural sales for both the spot and future.	Using blockchain, the company offers a digital marketplace supported by smart contracts that allow farmers and buyers to purchase and sell commodities at future dates while avoiding transaction risks.	The company connects global markets and enables agricultural enterprises to gather various contracts in one place, securely negotiate payment terms and save time with blockchain. Environmental value is attained by creating a community that supports sustainable agribusiness and empowers smallholder farmers.
2	AeroFarms USA 1, 2, 6 AI, big data, Internet of things, soilless farming	The indoor farming company grows food using an aeroponic growing method that enables water to flow through plant roots without using any soil. They utilise big data and IoT for efficiency and correction while growing.	The company monitors multiple data points every harvest to review and improve the growing system and use that data to enhance the process of fresh produce farmed locally and worldwide through machine learning and IoT integration.	Value is captured by producing a wide range of high-quality food products grown with advanced technology.
3	Avy The Netherlands 1, 3, 8, 13 Drones	The company uses a drone network for creating solutions in healthcare logistics, emergency services, last-mile medical delivery and urgent logistics.	The company provides temperature-controlled drones that collect samples for analysis from any place and make emergency deliveries while avoiding traffic and geographical barriers. Furthermore, firefighters on the ground may watch the crisis unfold live on the company's heat camera, allowing them to react quickly and precisely.	In numerous areas, value is produced by delivering faster and better transportation with drones. Reduced emissions and the use of a totally electric aerial fleet capture environmental value.
4	BanQu USA 1 Blockchain	The company offers a blockchain-based supply chain management platform that enables poverty-stricken individuals to gain financial inclusion and economic possibilities.	The company creates a blockchain solution to construct an economic profile based on training, work experience and referral or loyalty systems. Blockchain transforms supply chain management into a more efficient and transparent system that turns cyclical poverty into a circular economy.	The firm helps customers achieve their goals and establish a social presence by creating a blockchain-based loyalty chain. It also generates monetary value by charging a fee for this service.

No	Company info	Value proposition (what?)	Value creation (how?)	Value capture
5	Beam UK 1, 4 Crowdfunding	The company provides education to homeless people according to their career plans.	The company first examines the jobseekers physically and mentally to figure out if they are prepared for a job. Then, a volunteer helps them with a tailor-made career plan. Required education for the job they want is covered by donations.	Value is captured by delivering people education and facilities they could not access otherwise.
6	Beyond Meat USA 1, 2, 14 Biotech and biomanufacturing	It is a biotechnology company focused on mimicking the texture of meat that offers its customers animal-free and plant-based meat-like products.	The company provides animal-free and plant-based meat products with biotechnology and biomanufacturing processes. A meat-like texture that resembles muscle fibres is developed in heat and pressure processes via food extrusion machine plant proteins to turn into fibrous.	The company captures value by using biotechnology to create meat-like products. It also creates environmental value by reducing green gas emissions compared to the actual meat process. The company creates social value with healthier products for a healthy society.
7	Borlaug Web Services The Netherlands 1, 8, 15 Big data, blockchain	It is a blockchain software service company that offers transparency and traceability as a solution to the problems of those operating in the agricultural sector, such as food waste and food fraud.	For sustainability and investor trust, the company helps businesses set up, analyse and monitor certified environmental, social and governance data. On the blockchain, decentralised identities generate irrefutable verification evidence for financial inclusion and trustworthy identity records that can be shared among stakeholders.	Creating third-party applications opportunities is offered to customers by providing an open-source structure of the company. With the transparency through blockchain, economic gain possibilities are increased for customers while contributing to the long-term sustainability of food systems. Monetary value creation is based on its availability to both large enterprises and transactions of individuals.
8	Climate FieldView USA 1, 12, 15 Big data, drones	The company offers a digital platform for farmers that comes into service by Climate Corporation. The platform provides accurate agriculture information and advice relevant to smallholders, helping them make more grounded decisions to reduce costs and increase yield.	The platform creates a map of the user's field to make it easier to track crops. Drones are used to track the field and provide detailed real-time insights. Moreover, data processing is used for providing accurate predictions and instructions for the users.	The value is captured through using drone technology to obtain detailed insights about the fields of the users and collecting and processing data for helping the users make better decisions.

(continued)

Table 3.1 (continued)

No	Company info	Value proposal (what?)	Value creation (how?)	Value capture
9	Crop2Cash Nigeria 1, 2, 8, 13, 15 AI, cloud computing, cybersecurity	The company provides access to financing and market for smallholders. They also digitalise the value chain of food crops by building a platform that eases the management of farmers' supply chains by using an encrypted platform based on AI.	The company offers two solutions. The first one introduces a digital finance platform to small farm owners. This platform allows users to check their balance and send/receive money using USSD (unstructured supplementary service data). The second solution authorises farmers to control their supply chains digitally and get smart suggestions prepared with the help of an AI system. The user data is encrypted and stored safely in the cloud.	The value is captured by providing farmers with AI-powered financial advice. They also add social value by recommending solutions to farmers using their digital technology.
10	Donation Match USA 1, 2 AI, big data, cloud computing, crowdfunding	The company offers a platform that brings companies and nonprofit organisations together to efficiently distribute donated products and services for fundraising and community impact, with crowdfunding support.	Data collection, digital gift certificate generation, e-delivery, recordkeeping, mass changes, messaging and reminders are all solutions available for donor companies with the use of AI-based automated models and storing all data in their cloud service.	Value is captured by saving time and money through AI-based solutions that ensure security through managing big data in the cloud service. They also provide improved communication between fundraisers and volunteers.
11	Eden Agritech Thailand 1, 2, 12 Biotech and biomanufacturing	The company offers natural ingredients that enable the shelf-life extension of fresh produce.	The company is working on an anti-fouling coating that can triple the shelf life of fresh fruit. The coating protects the fruits against oxidation, browning and microbial formation.	Value is captured by keeping fresh produce in good conditions for a long time, which decreases waste and spoilage in the agricultural supply chain.
12	eFarms Nigeria 1 Crowdfunding	The company offers an agriculture platform that aggregates funding by crowdfunding and uses these funds to smallholders with expertise in different value chains. They also share the profit with the investors and farmers.	The company digitises African agriculture to ensure food security through empowering smallholders in distant pocket locations. The company presents a preselected list of farms consisting of different crops and animals, chosen after careful consideration to demand, yield, return of investment (ROI) and tenure of investment. These are divided into sub-units corresponding to minimum investments for each farm.	Monetary value is captured by using digital agriculture to create a dynamic food market and by providing funds to farmers with the help of crowdfunding in Sub-Saharan Africa. Social value is captured by empowering smallholders in rural pockets of Africa to maintain food security and ensure the nutrition of people in those countries.

No	Company info	Value proposal (what?)	Value creation (how?)	Value capture
13	Hello Tractor Nigeria 1, 8, 12, 13 AI, big data, Internet of things	The company offers its customers a marketplace that includes renting or sales services on agriculture equipment, mainly tractors. Also, its services increase farmer productivity with IoT and incorporate the sharing economy into agriculture.	They use AI and big data to guide farmers in providing fuel, driver and fleet management services. IoT technology also enables equipment owners to keep track of their customers and improve efficiency.	Value is capture by generating solutions that increase user productivity through AI and IoT and serving as a marketplace for farmers, dealers, banks and farm equipment owners.
14	Kamworks Cambodia 1, 7 AI, cloud computing, energy storage	It is a solar technology firm that offers large-scale grid systems to PAYGO (a budget guideline)-enabled solar home systems. The company has designed and installed solar energy projects using AI, cloud computing and energy storage technologies. They are also specialised in feasibility studies, maintenance and energy monitoring.	The company leverages technological innovations in off-grid solar systems by developing pay-as-you-go solutions, a data monitoring cloud platform, machine learning and predictive maintenance and an ERP (enterprise resource planning) software.	The company sells SHS (solar home systems) that use AI, cloud computing and energy storage, with PAYGO capabilities in Cambodia, and offers it on credit to consumers who cannot afford the upfront investment.
15	Leaf USA 1, 11 Blockchain	The company provides virtual banking services with blockchain to vulnerable refugees and migrants crossing borders.	The Fintech company offers secure accounts encrypted with blockchain, a virtual bank service that allows people, especially refugees and migrants across borders, to send and receive money via USSD (unstructured supplementary service data) without a need for a smartphone. With the company's wallet system, users can safely store, send and receive money in supported currencies without paying transaction fees.	Value is captured by delivering cheap and secure banking services to vulnerable and unbanked people through blockchain-based platforms.

(continued)

Table 3.1 (continued)

No	Company info	Value proposal (what?)	Value creation (how?)	Value capture
16	LifeBank Nigeria 1, 3, 8, 9 AI, blockchain	The company offers multi-modal AI-powered medical products (blood, test kits, etc.) and a distribution network in Africa. Additionally, they aim to make healthcare professionals and patients securely access the required information.	Using deep techs like AI and blockchain, the company aims to create tools to organise distribution systems. SmartBag is a blockchain-based platform that assists patients and healthcare practitioners in discovering blood and blood product safety information. The processes involved in blood supply are stored on the blockchain, which ensures the integrity of the data.	Value is created through using an AI-powered distribution system for medical products. In addition, social value is captured with its blockchain-based information system that enables hospitals and facilities to provide more reliable, more secure and cheaper services. Lastly, donations of critical medical supplies to the low-income African people contribute to the social value.
17	Monkee Austria 1 AI	The company aims to help people improve their financial health with a mobile application. The app combines behavioural design, gamification and machine learning to help people balance their finances and save more for their future.	The application allows users to transfer their savings to their e-money accounts to save greater amounts of money. The financial health coach in the mobile application provides the user insights on their progress and constantly adapts itself according to the user. It uses ML to obtain and analyse the data of targets and progress of users.	Value is created through increased insights and the feature of transferring savings into e-money. Also, social value is created by making the application affordable for all since it is completely free to use.
18	Nextfood Denmark 1, 2, 8, 10 Soilless farming	It is an agritech firm that uses soilless aeroponic farming techniques (the plants' roots are suspended in mid-air and misted by nutritious water) to produce vegetables locally and sustainably.	The company specialises in soilless farming using an aeroponic approach that allows for same-day fresh, local, tasty, healthy and pesticide-free production. Precision control of nutrients, light, water and other climatic aspects ensures that the crop is grown under ideal conditions for optimal taste and nutritional content, as well as removing production uncertainties and ensuring predictable on-time delivery.	Value is created by cost-effective and local production (in-store). The aeroponic system will have lucrative prospects in the food space while lowering carbon emissions from production. Also, the environmental value is created by using 95% of the water and eliminating 40% of the food waste compared to the traditional farming techniques.

No	Company info	Value proposal (what?)	Value creation (how?)	Value capture
19	One Concern USA 1, 3, 13, 15 AI, big data, digital twins	The company provides a platform that uses disaster science and ML for better decision-making. The company uses data analytics, digital twins and AI to predict potential disasters and their effects. They aim to make taking precautions possible and make the disasters less devastating.	The company provides an algorithm that aims to examine data obtained from disasters that occurred before. This AI/ML-integrated algorithm is capable of predicting potential disasters. Digital twins of cities are used for generating simulations and estimating the risks of these potential disasters.	Making AI-based disaster forecasts and damage assessments based on digital simulations captures value.
20	Perfect Day USA 1, 2 Biotech and biomanufacturing, cellular agriculture	The company provides animal-free milk protein, which can be used in any dairy product, for its customers by using biotechnology.	The company obtains animal protein from an organism called flora. This organism is fostered with plant sugar, fermented and processed efficiently with plant-based sugar to produce milk protein. The outcome is an animal-free milk protein identical to the protein found in cow's milk that can be used to make any dairy product.	Environmental value is captured by producing animal-free and sustainable milk protein by enhanced cellular agriculture and biomanufacturing processes.
21	Pesitho Denmark 1, 7, 13 Energy storage	The company developed a furnace to provide sustainable and clean energy to families in need. The furnace is a compact, self-contained, multipurpose home cooking unit, run by solar energy.	The company enables people not to be required to collect or pay for biomass to fuel open fires since the solar-powered cookstove runs on electricity generated from the corresponding solar panel. The solar cooker helps refugees save money with the benefits it provides to their families, enabling them to devote more time and money to other works and activities.	The furnace power generated by the solar panel and stored in the battery is sufficient to meet the usual cooking demands of a home. The first environmental benefit is that it does not emit harmful fumes or $CO2$ that can be dangerous in closed areas. The second one is the sustainability and healthiness of the option because it works entirely with solar energy.

(continued)

Table 3.1 (continued)

No	Company info	Value proposal (what?)	Value creation (how?)	Value capture
22	Pika Israel 1, 6, 7, 9, 11, 12 Biotech and biomanufacturing, recycling	It is a sterilising capsule provider to clean reusable diapers to assist the recycling of single-use materials using biotechnology.	The company provides a machine that cleans reusable diapers in one step with biotechnology. Developed products have sensors that provide controllable nutrition insights straight to mobile phones. After the cleaning process, the machine transmits nutrition analysis results to your mobile phone by using wireless connection. Through sterilising capsules, diapers can be recycled.	Value is created via energy efficiency, nutrition analysis and easy control. Recycling and less energy consumption can contribute to environmental value creation.
23	Redefine Meat Israel 1, 2, 14 3D printing, biotech and biomanufacturing, cellular agriculture	It is an animal-free meat company that creates and serves an alternative named alt-meat.	The company provides a product named alt-meat, an alternative created through cellular agriculture and biomanufacturing technologies. Alt-meat is acquired with two different procedures: cell-cultivated and plant-based. Furthermore, the company produces alt-meat via industrial-size 3D printers.	The company creates value from cell-cultivated or plant-based 3D-printed alt-meat. They create environmental value through sustainable meat production by reducing water use in production processes, which leads to 95% more efficient meat production.
24	Shamba Records Kenya 1, 8, 12 AI, big data, blockchain, cloud computing	The company offers a distributed ledger that runs on blockchain and uses big data and AI to collect farmer's data and process payments to farmers.	The company offers a distributed ledger that runs on blockchain and uses big data and AI to collect farmer's data and process payments to farmers.	Value is captured by providing data storage services with cloud technologies. They manage and process the data with AI. They also process farmer's transactions of payments and use escrow services to secure financials.
25	SolarCreed The Netherlands 1, 7, 8 Energy storage, flexible electronics and wearables	The firm works in African countries to give solar energy instruments to off-grid farmers. These instruments assist off-grid farmers in growing, preserving, packaging and distributing food in a sustainable manner.	The company has a scale of solar-powered products including power batteries, solar panels, pumps, fans, cameras, watches, etc.	They capture value by helping farmers in African countries to use clean and sustainable energy with the technology they produce.

No	Company info	Value proposal (what?)	Value creation (how?)	Value capture
26	Swoop Aero Australia 1 AI, autonomous vehicles, cloud computing, drones, digital twins	The company provides essential supplies and medicines to needy people in rural and isolated areas by transporting them with a drone fleet.	The company's drone fleet offers different solutions to air logistics in autonomously controlled drones via ground operators. Drones equipped with multiple sensors can process real-time data with AI and transfer the data captured in a flight to its digital twin via cloud computing. Withal, unmanned aerial vehicles that can simultaneously monitor a pilot from various parts of the world.	The company captures value with autonomous drones carrying emergency aid. Those drones land/take off autonomously after delivering/receiving carry loads that minimise the human interaction in flights. In addition, the company captures social value through ensuring the delivery of humanitarian aids, disaster relief and urgent medications in a fast turnaround transport.
27	Tala USA 1 AI, big data	The company provides loans to people in need in Kenya, Mexico, the Philippines and India. Capital is transferred directly to the borrower's mobile phone.	The company provides a proprietary algorithm to decide whether people are creditworthy or not. Credits are transferred directly to the borrowers' smartphones, making the process easier and faster.	Value is captured by automatising the process of deciding the creditworthiness of applicants.
28	TruVito The Netherlands 1, 2, 5, 6, 12, 15, 16 Blockchain	The company creates blockchain-based solutions for farmers and agriculture stakeholders.	They serve three different products. The first one serves as a blockchain-based platform for auditions that optimises safety, transparency, cost and efficiency. The second product is an input-management platform for farmers which includes identity registry, production records, crop and harvest planning and a map of the relevant field. The third product enables producers to trace their products until they meet the end consumer, based on batch numbers.	The value is captured through serving farmers blockchain-powered tools to optimise their businesses, both in and out of the field.

(continued)

Table 3.1 (continued)

No	Company info	Value proposal (what?)	Value creation (how?)	Value capture
29	Unhoused UK 1 Crowdfunding, flexible electronics and wearables	The company is a clothing brand that donates a product when the same type is bought.	Their products are supported with nanotech, including self-cleaning fabric. This helps people in poverty, as they cannot easily wash or renew their clothes. They also use their profit to provide clothes to people in need, which means their customers indirectly donate for poor people's clothes.	Value is captured by making convenient clothes for the poor and funding their donations with earnings from all around the world.
30	Veritas USA 1, 8 Cloud computing, cybersecurity	The company works on solving data protection problems by using cybersecurity such as ransomware (protect the business from malware that encrypts, or locks, valuable digital files and demands a ransom to release them), modern workloads, software-defined storage, multi-cloud (data protection and availability across leading cloud platforms) for governments, hospitals and schools.	The company's data services platform automates data preservation and recovery, claims 24/7 availability of business-critical applications and gives businesses the knowledge they need to stay compliant with changing data rules. They support many data sources, operating systems, storage targets and cloud platforms. They also have a deployment approach to meet any demand.	Monetary value is captured by delivering storage and infrastructure software products and services that use cloud computing and cybersecurity for governments, schools, entrepreneurs and healthcare providers. The company provides low-cost and scalable tools without hardware agenda requirements.
31	Vodafone for Business UK 1, 2, 8 5G, AI, drones, Internet of things	The company hosts a project that aims to contribute to sustainability with sub-projects such as connecting all members of the society to the digital world including IoT devices with 5G, providing solutions with AI algorithms in agriculture and supporting farmers with the use of drones.	Customers of network operators may now save money on 5G network planning by using 360-degree HD representations of the planet and high-precision components and surroundings. They can also leverage this infrastructure to facilitate the usage of drones in agriculture. Mobile network infrastructure can be used to put cellular-connected drones to good use in a variety of ways.	Low-cost and fast connection features of IoT devices are aimed with 5G to solve economic problems. For digital agriculture, it is aimed to detect and interpret values such as humidity and temperature through sensors with AI and to offer more autonomous agriculture by reducing physical intervention with the use of drones.

No	Company info	Value proposal (what?)	Value creation (how?)	Value capture
32	Wagestream UK 1, 8 AI, big data, Internet of things	The company provides a platform that enables its employees to track their wages and access them quickly when they need, save directly on salaries and receive financial training using big data, IoT and AI.	The company enables each colleague access to, up to a certain portion of, their monthly payments, which they can use in the case of an unanticipated expenditure (funded by Wagestream and provided via the app in a matter of seconds). They can take out an advance loan alternatively. AWS KMS (AWS Key Management Service) encrypts all data to control all devices. The script algorithm is used to salt and hash all user passwords, and all data centres are easily compatible. The technologies outlined are used for all transfers.	Monetary value is captured by allowing employees to rapidly get to gathered profit and progress financial well-being through other product offerings while the changes are secure and conceivable with AI and big data. The environmental benefit is the technology-based business with limited supply chain conditions and a small environmental footprint.
33	WakaWaka The Netherlands 1, 2, 7, 13 Energy storage	The company produces renewable clean solar energy source products using energy storage technology in regions with limited electricity access.	Ultraviolet light charges WakaWaka. The solar panel utilised is a mono-crystalline solar panel with a 22% efficiency, which implies that 22% of the energy absorbed by the solar panel is turned into electricity and stored in the lithium polymer battery.	Value is created via storing solar energy, which is greener and cheaper. The company offers products such as lighting devices, power banks and solar panels that are open to personal purchase with competitive prices for people with low income. It also creates social and environmental value with projects like solar libraries for children with investments in solar energy storage.

(continued)

Table 3.1 (continued)

No	Company info	Value proposal (what?)	Value creation (how?)	Value capture
34	Wefarm UK 1, 2, 8, 13, 15 AI, big data, natural language processing	The company offers a P2P (peer-to-peer) knowledge-sharing platform for small-scale farmers using machine learning, big data and NLP.	The company provides a platform where independent farms thrive. By connecting farms, the community open-sources its collective expertise and resources, unlocking better outcomes for themselves and their farms. The community can communicate online or via SMS and access a trusted marketplace of physical retailers. Farmers can access Wefarm anytime, for free, even without access to the Internet.	The company generates value through the unique information and data generated by the millions of organic interactions on the system.
35	WeltHungerHilfe – Child Growth Monitor Germany 1, 3, 10 AI, big data, spatial computing	The company offers an application that creates ways to monitor the health status of children with the help of AI, big data and augmented reality and provides the necessary support in societies where families and states cannot prevent children from being malnourished.	The application focuses on measurement and data processing for impoverished infants under the age of 5. It's a smartphone app that mixes augmented reality with artificial intelligence. The software can detect malnutrition by predicting a child's weight and height from a 3D image.	Value is created via a mobile application, which requires only a smartphone with AR capabilities. Sensors in smartphones and AI contribute to high-accuracy results in detecting health problems in children. The organisation aims to create transparent and democratic solution alternatives by providing fast actions against poverty.
36	Whrrl India 1 Blockchain, Internet of things	The company provides loans to producers such as farmers and commodity traders in need. These loans can be in terms of grains, spices, seeds, metals, etc.	The company provides cameras connected to an IoT system to track the status of the goods stored in warehouses. Also, blockchain is used to generate smart contracts and easily track the arrangements between the company and producers that received loans. Thus, the loans are recorded and stored in an irrevocable system.	The monetary value is captured through using IoT and blockchain in the loaning field to create security and traceability. Value is also generated by using goods instead of capital as loans.

No	Company info	Value proposal (what?)	Value creation (how?)	Value capture
37	Wildtype USA 1, 2, 12, 14 Biotech and biomanufacturing, cellular agriculture	It is a cellular agriculture company specifically focused on salmon that offers its customers animal-free seafood.	The company provides animal-free salmon to its customers through cellular agriculture. In the company's facility, cell cultivators create conditions being the same as that exist in nature. These procedures make producing cell-cultivated salmon possible.	The company captures value from cell-cultivated salmon via cellular agriculture. In addition, it contributes to the formation of a healthy society as the salmon they produce does not contain metal toxins, antibiotics and microplastics that exist in the seas. Furthermore, reducing the consumption of animal food creates an environmental value by lowering the carbon footprint.
38	Wingcopter Germany 1, 3, 8, 13 Drones	The company provides medicine and vital supplies in robust areas to needy people by using drone fleets. Drones reduce delivery time and transportation costs in rural areas.	The company creates a procurement service with drones for the people in rural areas. They assist in cutting down the time it takes to get essential supplies. They work as a drone technology maker and a service provider for several drone deployments. These services include medical, parcel and food deliveries. They also provide spare part distribution, mapping, surveying and inspection/monitoring.	Value is captured by producing efficient and fast solutions with drones in areas such as e-commerce, food service and logistics. Additionally, by collaborating with UNICEF in a pilot project, they create social value through delivering vaccines in such a challenging environment.
39	ZOLA Electric The Netherlands 1, 7, 9, 11, 13 Energy storage	The company offers affordable and renewable energy to supply 7/24 electricity to remote and off-grid homes and facilities in Africa.	The company produces and provides reliable primary power sources to replace unstable power grids. They also offer products such as plug-and-play energy sources. The products are based on solar panels, alternating current and several enlarged power sources.	The company's different products offer its customers reliable and renewable energy and storage at affordable prices. Social value is captured by energising remote locations where a robust power grid lacks.

References

S. Alkire, M. Chatterje, A. Conconi, S. Seth, A. Vaz, (2014) Poverty in rural and urban areas: direct comparisons using the global MPI 2014

R.A. Castaneda, A. Whitby, E. Wang, The near future of global poverty [WWW Document]. Atlas Sustain. Dev. Goals 2020 (2020). https://datatopics.worldbank.org/sdgatlas/goal-1-no-poverty/. Accessed 8 June 2021

J. Crespo Cuaresma, W. Fengler, H. Kharas, K. Bekhtiar, M. Brottrager, M. Hofer, Will the Sustainable Development Goals be fulfilled? Assessing present and future global poverty. Palgrave Commun. **4**, 1–8 (2018). https://doi.org/10.1057/s41599-018-0083-y

Goal 1: No Poverty [WWW Document]. . . Glob. Goals (2016). https://www.globalgoals.org/1-no-poverty. Accessed 13 Aug 2021

J.H. Haughton, S.R. Khandker, *Handbook on Poverty and Inequality* (World Bank, Washington, DC, 2009)

How can we achieve the Sustainable Development Goals for and with children? [WWW Document] (2021). https://www.unicef.org/sdgs/how-achieve-sdgs-for-with-children. Accessed 8 Sept 2021

D. Jolliffe, P. Lanjouw, World Bank, World Bank Group (eds.), *A Measured Approach to Ending Poverty and Boosting Shared Prosperity: Concepts, Data, and the Twin Goals, Policy Research Report* (World Bank Group, Washington, DC, 2015)

B. Keeley, *Income Inequality: The Gap between Rich and Poor, OECD Insights* (OECD, 2015). https://doi.org/10.1787/9789264246010-en

L. Olsson, M. Opondo, P. Tschakert, A. Agrawal, S.E. Eriksen, Livelihoods and poverty (2014)

J.D. Sachs, *The Age of Sustainable Development* (Columbia University Press, New York, 2015)

J.D. Sachs, From millennium development goals to sustainable development goals. Lancet **379**, 2206–2211 (2012). https://doi.org/10.1016/S0140-6736(12)60685-0

R. Scheyvens, E. Hughes, Can tourism help to "end poverty in all its forms everywhere"? The challenge of tourism addressing SDG1. J. Sustain. Tour. **27**, 1061–1079 (2019). https://doi.org/10.1080/09669582.2018.1551404

SDG 1: No poverty – Iberdrola [WWW Document] (n.d.). https://www.iberdrola.com/sustainability/committed-sustainable-development-goals/sdg-1-no-poverty. Accessed 14 Aug 2021

A. Sumner, C. Hoy, E. Ortiz-Juarez, (2020) Estimates of the impact of COVID-19 on global poverty. WIDER Working Paper 2020/43. Helsinki: UNU-WIDER.

Sustainable Development Goals (SDG 1) [WWW Document] (2021). . U. N. West. Eur. https://unric.org/en/sdg-1/. Accessed 14 Aug 2021

United Nations, *The Millennium Development Goals Report 2015* (2015)

United Nations Department for Economic and Social Affairs, *SUSTAINABLE DEVELOPMENT GOALS REPORT 2020* (United Nations, S.l., 2020)

UN 2016 [WWW Document] https://www.un.org/sustainabledevelopment/blog/2015/09/historic-new-sustainable-development-agenda-unanimously-adopted-by-193-un-members/

UN 2020. Report of theSecretary-GeneralRoadmapfor Digital Cooperation. https://www.un.org/en/content/digital-cooperation-roadmap/assets/pdf/Roadmap_for_Digital_Cooperation_EN.pdf

World Bank, *Global Economic Prospects, June 2021 198* (2021). https://doi.org/10.1596/978-1-4648-1665-9

SDG-2 Zero Hunger

4

Abstract

People's lives, communities and civilisations have all been defined by constant danger. Hunger is the menace, a plague that causes weakness, despair and death in the worst-case scenarios. One of the primary common threads has been hunger throughout history, which has resulted in large-scale migration, wars, conflicts and great sacrifices. This chapter presents the business models of 40 companies and use cases that employ emerging technologies and create value in SDG-2, Zero Hunger. We should highlight that one use case can be related to more than one SDG and it can make use of multiple emerging technologies.

Keywords

Sustainable development goals · Business models · Zero Hunger · Sustainability

Today, people's lives, communities and civilisations have all been defined by constant danger.

The author would like to acknowledge the help and contributions of Sedef Güraydın, Esra Çalık, Gülen Mine Demiralp, Hasan Serhat Bayar, İbrahim Alperen Karataylı, İbrahim Yusuf Yıldırım and Tuana Özten in completing this chapter. They also contributed to Chapter 2's Bioplastics, Recycling, Robotic Process Automation, Robotics and Soilless Farming sections.

Hunger is the menace, a plague that causes weakness, despair and death in the worst-case scenarios. One of the primary common threads has been hunger throughout history, which has resulted in large-scale migration, wars, conflicts and great sacrifices (FAO 2019). Since the early colonial era, keeping populations fed has been a key administrative concern (Nally 2011). It helped communities strengthen their bonds of friendship and solidarity. For this important issue, the first UN conference on food and agriculture was called by US President Franklin D. Roosevelt in 1943. The conference specifically stated that countries must develop a food and nutrition policy to set their own intermediate goals gradually. Afterwards, the escalating Cold War and Malthusian worries that food shortages would fuel communism became the reason for the 1960s to be dubbed as the "fighting hunger decade" by the United Nations Food and Agriculture Organization (FAO) in its Freedom from Hunger Campaign (Byerlee and Fanzo 2019). Two significant advances on the way to SDG-2 occurred in the 1980s and 1990s. First, the conversation switched from food supply to food availability. The FAO modified its definition of food security in 1982 to guarantee that all people have physical and economic access to the food they need at all times (Sen 1982; Shaw 2007). After these new developments, global goals have become important for a new millennium. The WHO member nations approved six global targets for promoting

S. Küfeoğlu, *Emerging Technologies*, Sustainable Development Goals Series,
https://doi.org/10.1007/978-3-031-07127-0_4

maternal, baby and early child nutrition in 2012 and committed to tracking progress towards those goals (WHO 2014). Zero Hunger, developed by economist and agronomist José Graziano da Silva, has been considered as one of the most significant achievements in the fight against hunger and poverty on a global scale (FAO 2019).

Many people worldwide think that hunger can be eradicated in the coming decades, and they are working together to achieve this objective (UN 2021). In many countries, undernutrition has decreased nearly half due to the increasing agricultural productivity and rapid growth of the economy. Yet, some people have encountered starvation and malnutrition (Paramashanti 2020). World leaders affirmed the right of everyone to have access to safe and nutritious food at the 2012 Conference on Sustainable Development (Rio+20), which is compatible with the right to enough food and the basic right of everyone to be invulnerable to hunger. "End hunger, achieve food security and enhanced nutrition and promote sustainable agriculture" are the objectives of SDG-2. Because it is inextricably tied to society, the economy and the environment, SDG-2 is critical to the entire SDG agenda's success. Even though undeveloped countries rely more heavily on agricultural operations, food production and consumption are important to every economy and permeate all cultures (Gil et al. 2019).

More coordinated decision-making mechanisms at the national and local levels are necessary to collaborate and effectively address trade-offs between climate change, water, agriculture, land and energy. To avoid large-scale future shortages and to ensure food security and excellent nutrition for everybody, local food systems must be strengthened (UN 2021). The United Nations Committee on World Food Security defines food security as all people having social, physical and financial access to adequate, clean and nutritional food that meets their dietary requirements at all times to live a healthy life (UNDESA 2021). The goal of SDG-2, which is to ensure global food security and agricultural sustainability, necessitates prompt and coordinated action from both developing and developed countries. This, in turn, is contingent on clear,

broadly applicable objectives and indicators, which are now in short supply. The SDGs' new and sophisticated character complicates its implementation on the ground, especially in light of interlinkages across SDG targets and scales (Gil et al. 2019).

The Zero Hunger objective, in particular, highlights a long-overdue realisation that industrial agriculture threatens fundamental ecological processes on which food supply depends by including sustainable agriculture targets into the larger endeavour to end hunger (Blesh et al. 2019). As shown in Fig. 4.1, there are eight targets within the context of SDG-2.

While the increasing population growth was 7.339 billion in 2015, this number increased to 7.753 billion in 2020. This rise carries dozens of new issues, such as decreasing per capita income and increasing consumption of natural resources ("Population, total | Data" 2021). According to UN estimates, there are over 690 million hungry people globally, accounting for 8.9% of the global population, an increase of ten million in a year and over 60 million in 5 years (Goal 2 2021). When it comes to calculating needed calorie-based consumption, the rise in food demand has overtaken population growth. To consume food, one must also produce it, which necessitates agricultural and animal resources (Fukase and Martin 2020). The world's arable land rose from 1523 million hectares to 1562 million hectares between 1992 and 2012. As a result, arable land per capita has decreased, as has the effect of population expansion and rising food consumption. In other words, the food will be produced with more difficulty, and this will also cause the reduction of forests (Fukase and Martin 2020). Agriculture for food production and the food consumed have harmed the environment, for example, GHG emissions and land conversion. These challenges can help solve many problems through agricultural research, resource management and infrastructure improvement, but they are insufficient. Focusing on the agriculture sector and doing complementary research outside of it can help solve both the environmental and hunger problems. This is a global issue as well as one that affects regional economies. Reduced hunger

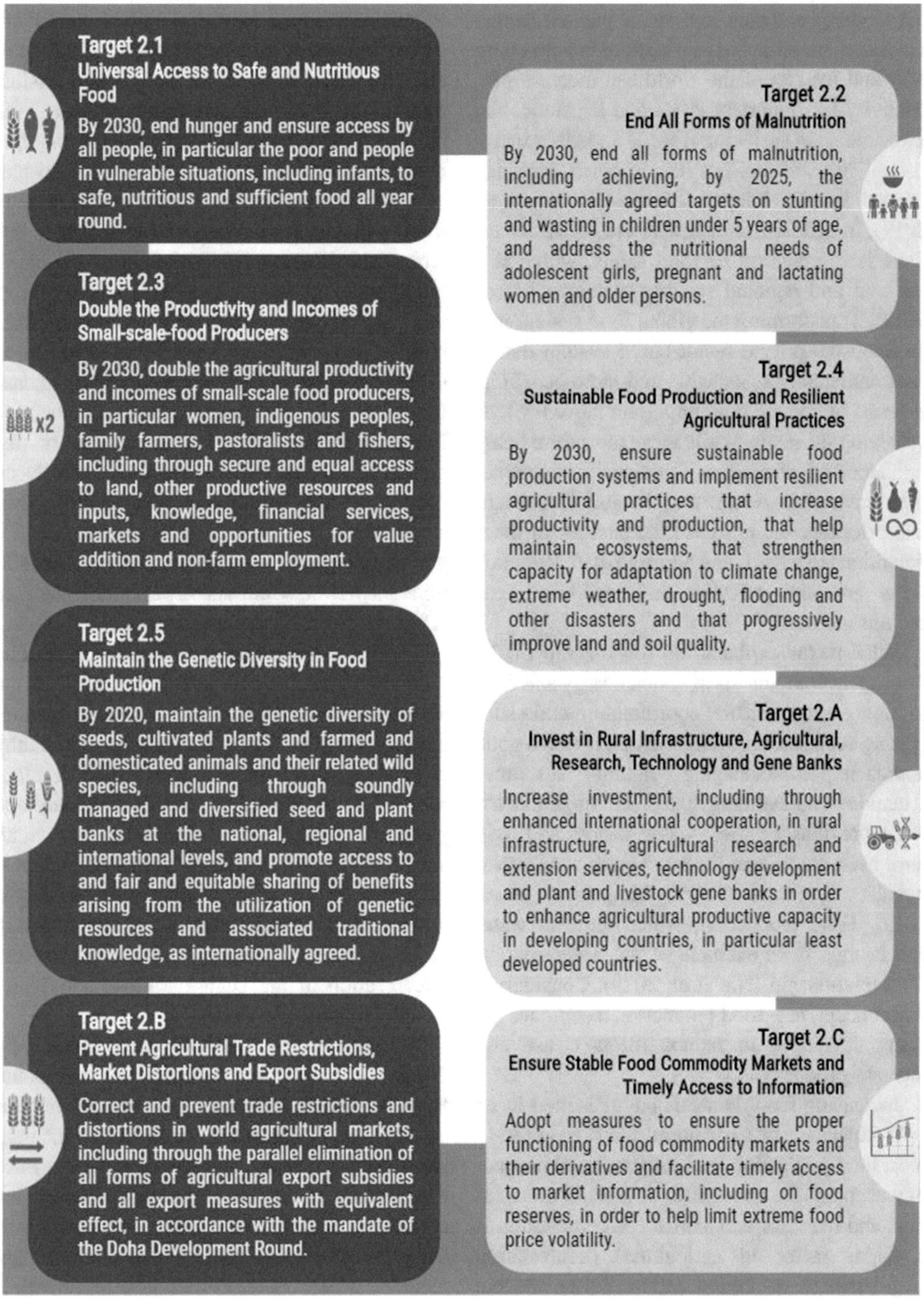

Target 2.1
Universal Access to Safe and Nutritious Food

By 2030, end hunger and ensure access by all people, in particular the poor and people in vulnerable situations, including infants, to safe, nutritious and sufficient food all year round.

Target 2.2
End All Forms of Malnutrition

By 2030, end all forms of malnutrition, including achieving, by 2025, the internationally agreed targets on stunting and wasting in children under 5 years of age, and address the nutritional needs of adolescent girls, pregnant and lactating women and older persons.

Target 2.3
Double the Productivity and Incomes of Small-scale-food Producers

By 2030, double the agricultural productivity and incomes of small-scale food producers, in particular women, indigenous peoples, family farmers, pastoralists and fishers, including through secure and equal access to land, other productive resources and inputs, knowledge, financial services, markets and opportunities for value addition and non-farm employment.

Target 2.4
Sustainable Food Production and Resilient Agricultural Practices

By 2030, ensure sustainable food production systems and implement resilient agricultural practices that increase productivity and production, that help maintain ecosystems, that strengthen capacity for adaptation to climate change, extreme weather, drought, flooding and other disasters and that progressively improve land and soil quality.

Target 2.5
Maintain the Genetic Diversity in Food Production

By 2020, maintain the genetic diversity of seeds, cultivated plants and farmed and domesticated animals and their related wild species, including through soundly managed and diversified seed and plant banks at the national, regional and international levels, and promote access to and fair and equitable sharing of benefits arising from the utilization of genetic resources and associated traditional knowledge, as internationally agreed.

Target 2.A
Invest in Rural Infrastructure, Agricultural, Research, Technology and Gene Banks

Increase investment, including through enhanced international cooperation, in rural infrastructure, agricultural research and extension services, technology development and plant and livestock gene banks in order to enhance agricultural productive capacity in developing countries, in particular least developed countries.

Target 2.B
Prevent Agricultural Trade Restrictions, Market Distortions and Export Subsidies

Correct and prevent trade restrictions and distortions in world agricultural markets, including through the parallel elimination of all forms of agricultural export subsidies and all export measures with equivalent effect, in accordance with the mandate of the Doha Development Round.

Target 2.C
Ensure Stable Food Commodity Markets and Timely Access to Information

Adopt measures to ensure the proper functioning of food commodity markets and their derivatives and facilitate timely access to market information, including on food reserves, in order to help limit extreme food price volatility.

Fig. 4.1 SDG-2 targets. (Goal 2 2021, p. 2)

in Africa, for example, might be shown as a goal. Accepting some conditions is important to go confidently towards this objective. These situations can be handled as follows: climate change has hindered and may continue to prevent hunger reduction, investment in agriculture in poor countries and the rest of the world can increase productivity for important crops and livestock, and investments in agricultural R&D and other incremental investments, not just in agriculture, are needed to end hunger (Mason-D'Croz et al. 2019). In addition, to deal with the global issues, there is the concept of food safety, which is a national and regional security concept (Tansey 2013). Transforming the global food system into an inclusive private sector-based system that is environmentally sustainable and more beneficial in terms of climate is an important move towards achieving the goals. While these moves are being implemented, changes or situations may complement each other. As a result, the goal is to establish priorities and optimise the success of these definitions (Rickards and Shortis 2019). Research areas are also effective in providing these optimisations.

WEF nexus explains the relationship of the three main components of water, energy and food to improve intersectoral coordination while supporting sustainable development. It is also a good approach to managing natural resources (Hamidov and Helming 2020). Although this triangle previously varied, water, energy and food have been considered the most basic triad due to unbalanced access (Sharifi Moghadam et al. 2019). Food production, which requires water and energy, is an example of the water-energy-food relationship (Nie et al. 2019). Considering this concept, new food production techniques are being developed to reduce resource use and increase product yield.

Increasing food demands have resulted in an over-expansion of agricultural lands required to meet food production goals. Agricultural production accounts for about 80% of global deforestation, and livestock and animal feed production is a major factor in agricultural deforestation (Agribusiness & Deforestation 2021). People settlement and agriculture have changed the majority of the natural ecosystems (Ellis and

Ramankutty 2008). Studies should be carried out to prevent these adverse effects for the most effective use of agricultural lands. Food production is sensitive to climate change. With climate change, temperatures have increased, ecosystem boundaries have changed, and invasive species have emerged. As a result, both livestock productivity and the nutritional quality of grains and crop yield decrease (Climate-Smart Agriculture 2021). Food production accounts for between 1/5 and 1/3 of greenhouse gas emissions from humans (Agriculture and Food Production Contribute Up to 29 Percent of Global Greenhouse Gas Emissions According to Comprehensive Research Papers 2012). CSA focuses on the effects of climate change and food security on parts of the food supply such as agriculture, livestock and fisheries. Food supply aims to increase productivity, increase resilience to harsh conditions and minimise emissions per calorie obtained (Climate-Smart Agriculture 2021).

Dietary patterns that support all aspects of an individual's health and well-being while being environmentally friendly are known as sustainable healthy diets. The goal of sustainable healthy diets is to ensure optimal growth and development of all people; to support functionality and physical, mental and social well-being at all stages of life; to prevent all forms of malnutrition; to reduce the risk of diet-related diseases; and to maintain biodiversity and planetary health while providing nutrients (Food and Agriculture Organization of the United Nations and World Health Organization 2019).

The aim to reach SDG-2, Zero Hunger by 2030, will not be accomplished (Grebmer and Bernstein 2020). Still, hopeful future predictions could be made regarding the positive processes that have been made already. Even in the most dangerously vulnerable countries to hunger, the conditions have gotten significantly better over the years. Our problematic global food arrangement has a share in the current position, which is the limits of the planet's ecology and social connections in the sense of being no longer suitable for the population to be safe and develop equally (Grebmer and Bernstein 2020). Decreasing hunger is a crucial means to extend the growth further globally, but ending hunger holds an underlying position of bringing everyone the right to fair living conditions, including nutritional needs they deserve, as stated by the Universal Declaration of Human Rights (Cohen 2019). Producing food creates an imminent compromise in protecting nature. However, diminishing hunger can be considered the core element of sustainable development. By definition, sustainable development is creating growth that will satisfy the demands of the present generations while preserving its potential to satisfy the needs of the later generations. Therefore supplying enough food is a primary demand towards sustainable development. Attentively planned distribution of cropland in prospective would affect compromises to function better between producing the food and protecting the biological diversity (Zhang et al. 2021). A comprehensive solution for agriculture and food arrangement internationally is demanded to feed the current 690 million food-deprived people with the predicted addition to the global population of two billion people by 2050. The dangers of hunger could be relieved by more productive agriculture systems and more sustainable management of food supplies (Goal 2 2021). Closing the yield gap would both create a great saving of soil and decrease the species that are going extinct (Zhang et al. 2021). Being careless about food security comes at a cost; hunger creates great expenses in respect of patients' well-being, diminishes the capacity of human force and decreases sustainable growth (Cohen 2019). An important outcome that could be seen through the projects aiming to eradicate hunger that did not get funded properly and fulfil its purpose throughout the last 20 years is that the electorate's support is crucial for powerful policies (Cohen 2019). It is crucial to create proper policies for the smaller-scale farmers and women in underprivileged regions of the world, which are experiencing the worst of global hunger, for them

to present themselves politically and become actors in the actions that they get affected by (Cohen 2019). FAO predicts the need for food internationally will grow by 70% until 2050. This higher need for food will be caused by Asia, Eastern Europe and Latin America, which are growing areas in terms of the predicted increase in the residents' incomes (Linehan et al. 2012).

According to former UN Secretary-General Ban Ki-moon, business is a critical partner in accomplishing sustainable development goals. People want organisations to assess their impact, set ambitious targets and communicate honestly about the results through companies' key activities. The SDGs attempt to reroute global public and private investment flows towards the challenges they represent. As a result, they define expanding markets for businesses that provide creative solutions and dramatic change (SDG Compass 2021).

The most crucial two aspects of the "Zero Hunger" goal is agricultural production and food supply, and they mostly depend on the activities of the private sector. That means, to achieve SDG-2, major involvement in the private sector is needed. There are many different types and sizes of businesses in the agriculture and food sector. Some businesses use the most conventional methods, while others prefer the most modern methods. The sizes of these businesses may range from smallholder farmers to global multi-billion-dollar companies. Considering these factors, investors, customers and end consumers have a wide range of needs and expectations. If nations, continents, sectors and professions join forces and act on evidence, the world can attain Zero Hunger. Agricultural, fishing and forestry employ 80% of the world's poor. As a result, achieving "Zero Hunger" requires a rural economic revolution. Governments must provide possibilities for more private sector investment in agriculture, as well as strengthen social protection programs for the poor and connect food farmers with metropolitan areas (A #ZeroHunger world by 2030 is possible 2021). Since 2015, the worldwide food and beverage industry has grown at a compound annual growth rate (CAGR) of 5.7%, reaching almost $5943.6 billion in 2019. From 2019 to 2023, the market is expected to grow at a CAGR of 6.1%, reaching $7525.7 billion. In 2025, the market is expected to reach $8638.2 billion, and in 2030, $11,979.9 billion (Food and Beverages Global Market Opportunities and Strategies to 2030: COVID-19 Impact and Recovery 2020). There are many opportunities in the food market for businesses that attach importance to the targets of SDG-2, agile to adapt and use the emerging technologies, social responsibility values and sustainability, thanks to its massive size.

4.1 Companies and Use Cases

Table 4.1 presents the business models of 40 companies and use cases that employ emerging technologies and create value in SDG-2. We should highlight that one use case can be related to more than one SDG and it can make use of multiple emerging technologies. In the left column, we present the company name, the origin country, related SDGs and emerging technologies that are included. The companies and use cases are listed alphabetically.[1]

[1]For reference, you may click on the hyperlinks on the company names or follow the websites here (Accessed Online – 2.1.2022):

http://betahatch.com/; http://notco.com/; http://www.scadafarm.com/; https://agrograph.com/; https://asirobots.com/; https://bensonhill.com/; https://biomemakers.com/; https://brouav.com; https://future-meat.com/; https://get-nourished.com/; https://gussag.com/; https://impossiblefoods.com/; https://indigodrones.com/; https://orbisk.com/en/; https://orbital.farm/; https://plantix.net/en/; https://sunbirds.aero/; https://www.agbotic.com/; https://www.apeel.com/; https://www.beehex.com/; https://www.biomilq.com/; https://www.gamaya.com/; https://www.ibm.com/blockchain/solutions/food-trust; https://www.nokia.com/networks/services/wing/; https://www.ifarm360.com/; https://www.infyulabs.com/; https://www.intelligentgrowthsolutions.com/; https://www.novolyze.com/; https://www.nrgene.com/; https://www.odd.bot/; https://www.phytech.com/; https://www.plantiblefoods.com/; https://www.rapidpricer.com/; https://www.refucoat.eu/about/; https://www.smallrobotcompany.com/; https://xfarm.ag/?lang=en; https://algamafoods.com/; https://www.indigoag.com/; http://skymaps.cz/main.php?content=agrimatics

Table 4.1 Companies and use cases in SDG-2

No	Company info	Value proposal (what?)	Value creation (how?)	Value capture
1	Agbotic USA 2, 12 AI, robotics	It is a B2B company building intelligent and automatic farms with its platform SmartFarm which combines robotics, ML and AI to make quality organic products	The heme that gives meat its meat flavour is derived from soy leghemoglobin via fermentation of genetically engineered yeast. Discovering heme, an iron-containing molecule that occurs naturally in the cells of animals and plants creates the unique flavour of the meat. It adds a plant gene to yeast cells, using fermentation to produce a heme protein naturally found in plants, called leghemoglobin, without harming animals	Value is captured through continuous improvements in plant quality, fewer resources, higher yields and faster crop cycles through the use of robotics, ML and AI
2	Agrograph USA 2 AI	It is a platform combining AI with satellite imagery to generate millions of acres worth of global field-level predictions	The software takes a simple satellite image and turns it into high-value data that helps people make better decisions. ML enables the technology to grow and increase its library of evidence-based insights that predict trends, manage risk proactively and help firms make decisions that offer them a competitive advantage	Value is captured by helping risk management of agricultural banks, crop insurers, land investment trusts, grain distributors, biofuels companies and ag service providers via ML-based satellite imagery
3	Algama France 2, 3, 12, 13, 14 Biotech and biomanufacturing	It is a sustainable food-tech company producing algae-based foods and drinks to alternate animal-based ones	The company selects the most promising algae and transforms them into actionable ingredients. These ingredients are then used to shape sustainable alternatives to animal-based products, like eggs or meat, or as emulsifiers	Value is captured through harvesting the potential of algae and then producing vegan products by using them which provides sustainable food production

(continued)

Table 4.1 (continued)

No	Company info	Value proposal (what?)	Value creation (how?)	Value capture
4	Apeel USA 2 AI, biotech and biomanufacturing	The firm coats fruits and vegetables with a biotechnology-derived covering substance that is tasteless, odourless and edible and retains moisture while keeping oxygen out. The company bought a start-up that employs spectroscopy, picture recognition and predictive learning to deliver information on the quality and attributes of various foods	The coating mimics cutin, a waxy barrier made of fatty acids – lipids and glycerolipids – that link together to form a seal around the plant that helps keep moisture in. Hyperspectral imaging combines digital imaging with a chemical technique called spectroscopy which takes a picture of a food item and understands the nutritional content and freshness levels	Monetary value is generated through the produced plants lasting longer than untreated produce, lengthening the shelf life of produce, reducing food waste. It is attained also through making suppliers know the ripening window for each fruit and vegetable, enabling them to sort and ship accordingly, giving insights of internal quality, phytonutrient content of the produces
5	Asirobots USA 2 AI, autonomous vehicles	It is an automation solution provider combining robotic hardware components, AI-based unmanned command and control systems, sensor-based obstacle detection and avoidance system. The company offers the solution for autonomous farming vehicles to increase agricultural production	Unmanned farm vehicle hardware and software integrate with current by-wire systems, using control systems to connect vehicles to a central command centre, allowing the operator to handle several vehicles at once. Obstacle detection and avoidance systems are installed on farming vehicles to protect them from unanticipated threats. The area coverage pattern for a field is calculated using AI-based path-generating algorithms that take into account the kind of task, vehicle, size of implements and number of vehicles	Value is captured through AI-based autonomous farming vehicles that enable increased productivity, safety and non-stop agricultural production
6	BeeHex USA 2, 3 3D printing, AI, robotics	The company designs and builds autonomous bakery equipment for personalised nutrition making by using 3D printing, robotics and AI. Their 3D food printers can be reconfigured for a wide range of use cases	Their process takes customised web orders and converts them into production-ready machine commands with 3D printing, robotics, AI and ML	Value is captured by using autonomous 3D food printing systems to produce nutrition bar production that uses robotics, AI and ML technologies to increase throughput and reduce costs. This system is also built to operate efficiently and meet the industry safety standards

No	Company info	Value proposal (what?)	Value creation (how?)	Value capture
7	Benson Hill USA 2, 15 AI, big data	The company is the developer of the CropOS technology platform, which combines data science, plant science and food science to develop and commercialise food and ingredients using big data and AI	CropOS, a cognitive engine using cloud biology to empower a new era of plant genomics innovation, enables proprietary phenotyping, predictive breeding and environmental modelling algorithms. CropOS uses data analytics and biological knowledge to identify the most promising plant genetics	Value is captured by enabling greater control and precision to breeding practices, developing and commercialising food and ingredients that are more nutritious, better tasting, making the process sustainable and affordable by leveraging the natural genetic diversity of plants and influencing characteristics such as taste and texture while reducing processing steps and the need for additives. Products are bred to have fewer trade-offs and reach the market faster
8	Beta Hatch USA 2, 13 Biotech and biomanufacturing	It is a mealworm farming company that manufactures and distributes insect-based protein products for animal feed and agricultural fertiliser	Their genetic selection method produces stronger bugs. They can modify the nutritional profiles of their products to satisfy specific industry and animal feed demands, thanks to selectively bred insects and innovative diet mixtures	Value is captured through insect-rearing technology that converts organic waste directly into high-value proteins, oils and nutrients for poultry and aquaculture
9	Biome Makers USA 2, 15 AI, biotech and biomanufacturing	The company leads precision medicine for plant health, using biotechnology and AI, by modelling soil functionality to enhance the productivity of arable soils	The company profiles a soil sample of the network of microbes to an all-in-one report using AI and data science to provide soil insights. It uses ecological computing utilising the soil microbiome and develops ecological algorithms decoding microbial networks	Enhancing the productivity of arable soils and recovering soil health, measuring the biological quality of the soil and delivering agronomic insights to optimise farm operations and output generate monetary and environmental value by arable soil management
10	Biomilq USA 2, 3 Biotech and biomanufacturing, cellular agriculture	It is a company that targets infant nutrition by reproducing mothers' breast milk in a lab with biomanufacturing	Breast milk is produced from culturing mammary epithelial cells. The product boasts macronutrient profiles that can match the proportions and types of proteins, carbohydrates, fatty acids and bioactive lipids that can be found in real human breast milk	Value is captured by the new alternative method to nourish infants more healthily and close the nutritional gap between infant feeding options

(continued)

Table 4.1 (continued)

No	Company info	Value proposal (what?)	Value creation (how?)	Value capture
11	Brouav China 2, 7, 13, 15 AI, big data, drones	It is a company producing AI-based drones for agricultural spraying	The aircraft adopts intelligent control via AI, and the operator controls it through the ground remote control and GPS positioning. The downward airflow generated by the rotor helps to increase the penetration of the fog flow to the crop	Environmental value is captured through saving pesticide usage and water consumption. The method of spraying with drones reduces the risk of contact caused by the gathering of people
12	Future Meat Israel 2 Cellular agriculture	The company is the world's first industrial line for cultured meat produced from animal cells, without the need to raise or harvest animals by cellular agriculture	Cultured meat is made by putting an animal's stem cells into a culture medium that feeds the cells and allows them to proliferate. The medium is then placed in a bioreactor to help the cells develop	With cellular agricultural technology, monetary value is created by removing waste products and allowing animal cells to proliferate, lowering capital investment and production costs
13	Gamaya Switzerland 2 AI, drones	It is a digital agronomy company for large farming businesses, enabled by remote sensing and AI	Data collected from fields based on drone, aircraft and satellite-based remote sensing images and historical climate and weather records are analysed by crop and variety and region-specific using comprehensive crop models and AI. Analyses create maps characterising agronomic issues for optimum land and crop management	Facilitating optimal decision-making, including the use of pesticides and fertilisers, and attaining bigger and better yields while limiting environmental effects are all ways to capture value
14	GUSS USA 2 Autonomous vehicles, Internet of Things	The company provides an unmanned autonomous orchard spray system combined with vehicle sensors and LiDar technologies as a solution for the shortage of labour in agriculture	They begin by drawing a map of the field. The process is then completed with a visual examination on the farm to "ground truth" the map. After that, GUSS sprayers must be set for the application they will be used for. They choose the spray nozzles to utilise before configuring the GUSS program. Finally, each sprayer is given a path that specifies which rows it will spray. In addition to GPS, GUSS has vehicle sensors and LiDar for increased precision	Value is captured by autonomous spraying that uses fewer resources and reduces costs by eliminating operator error and downtime. Also, GUSS eliminates the chance for operators to be exposed to the used materials and reduces the need for personal protective equipment

No	Company info	Value proposal (what?)	Value creation (how?)	Value capture
15	IBM Food Trust USA 2 Blockchain	IBM Food Trust is the cloud-based blockchain solution that allows members to share food data, draw value from other people's contributions, build breakthrough features and decide where and how. It provides an open, flexible and trustworthy approach to exchanging food data	Authorised users get quick access to actionable food supply chain data, from farm to store and finally to the customer, thanks to the solution. Once uploaded to the blockchain, every individual food item's whole history and present location, as well as associated information such as certificates, test results and temperature data, are instantly available	Built on a blockchain, the IBM Food Trust benefits members in the network with a safer, more intelligent and sustainable food ecosystem. Transaction digitalisation and data provide an effective means for producers, processors, suppliers, dealers, regulators and customers to cooperate along the supply chain
16	ifarm360 Kenya 2 Crowdfunding	The company provides a crowdfunding platform that connects African smallholders with major purchasers through tech-enabled hubs	The company aims to de-risk investments in agriculture and enable farming as a business through alternative financing, market linkages, integrated advisory services, solar irrigation and remote sensing technology for agro-analytics	Environment and social value are captured by contributing to local food production, providing, financing and thus protecting food security in the region
17	Impossible Foods USA 2, 12 Biotech and biomanufacturing, cellular agriculture	They recreate meat, fish and dairy foods, from plants using biotechnology, with a much lower carbon footprint than their animal counterparts	The heme that gives meat its meat flavour is derived from soy leghemoglobin via fermentation of genetically engineered yeast. Discovering heme, an iron-containing molecule that occurs naturally in the cells of animals and plants creates the unique flavour of the meat. It adds a plant gene to yeast cells, using fermentation to produce a heme protein naturally found in plants, called leghemoglobin, without harming animals	Value is captured by creating an alternative to meat made using a small fraction of the land, water and energy required to produce meat from a cow and captivating the defining taste of the traditional meat

(continued)

Table 4.1 (continued)

No	Company info	Value proposal (what?)	Value creation (how?)	Value capture
18	Indigo Agriculture USA 1, 2, 3, 13, 15 AI, big data	The company provides a way in that farmers can buy seeds that have been treated with microorganisms to promote plant health and productivity	The company creates a database of which microbes operate best to encourage increased agricultural yields using machine learning and data analytics. By boosting the plant's ability to respond to drought stress, heat stress, nitrogen use efficiency and natural plant immunity, the microbial seed inoculant boosts corn production potential	Value is created through supporting farmers in their transition to more beneficial practices, delivering technological solutions and sponsoring investments to accelerate soil enrichment
19	INDIGO Drones Costa Rica 2 Drones, Internet of Things	It is an agricultural intelligence company using drones, and IoT sensor devices, to create "smart maps" that allow farmers to make data-driven decisions in the field	The company goes to the field on demand, gathers data with drones utilising remote sensing and creates reports that help farmers better manage irrigation, fertigation and crop health	Value is captured through increased plant quality by drone technology that leads to better decision-making and product management
20	InfyU Labs India 2 AI, Internet of Things	It is an agritech company with IoT-based products for quality assurance of fruits and vegetables to get accurate results as per customers' business requirements with deep learning algorithms	The company's product InfyZer provides QA tests based on IoT. The test results include the chemical composition of tested fruit or vegetable. With the test report, the chemical composition of fruits and vegetables can be analysed using deep learning algorithms to make critical decisions at the time of accepting the consignment	Value is created through helping stakeholders in the food safety and supply chain to carry out tests at scale and that too in a cost-effective manner with a combination of IoT and deep learning
21	Intelligent Growth Solutions UK 2 AI, Internet of Things, robotics	The company provides a platform for indoor farming that uses IoT, AI and ML for controlling environmental elements, including lighting, watering, CO2 levels and nutrient delivery, and for some operations. The platform can use a robotic arm when needed	Many environmental, soil and air-related parameters are metered via IoT nodes, and this data is computed via AI and ML algorithms for productive indoor farming	Environmental value is created through enhanced productivity and efficiency. The platform also interacts with external energy sources to deliver optimised power usage day and night, with advanced power management to ensure energy costs are as low as possible

No	Company info	Value proposal (what?)	Value creation (how?)	Value capture
22	IronOx USA 2, 15 AI, robotics, soilless farming	It is an autonomous hydroponic farming service provider aiming to create sustainable food production systems	Each facility uses robotics, ML and AI to ensure each plant receives the optimal levels of sunshine, water and nutrients. With data models which enabled the estimation of crop yields, the schedule of harvesting is regulated, using the data to reduce the number of time plants needed to spend in their initial growth phase (called "propagation")	Environmental value is created by using less water and less energy and emitting less CO_2, increasing waste at stages of the process. Value is created using the data being collected from plants which is used to improve the process, improving plant quality and quantity, while reducing environmental impacts
23	Nokia Finland 2 Internet of Things	The company offers a smart agriculture-as-a-service solution, which utilises Nokia's Worldwide IoT Network Grid (WING) solution, ensuring that precise and practical data is sent to farmers enabling them to enhance productivity	WING solution provides smart agriculture with an IoT network. The solution includes soil probes, weather stations, insect traps and crop cameras to help drive the productivity of soy and cotton crops	Value is captured by IoT networks providing affordable subscription-based access to regional climate and posting data for farmers to mitigate risks
24	NotCo USA 2, 15 AI, biotech and biomanufacturing	It is a food-tech company producing plant-based alternatives to animal-based food products	With the help of AI and biomanufacturing, the company uses pineapple, coconut, cabbage, peas, bamboo, beets, chickpeas and seeds in their products to replicate animal-based products' taste, texture, colour and aroma, enabling manufacturers to discover a new source of fibre, calcium and proteins from the vegetables	Value is captured by faster, more sustainable and resourceful food production that has a smaller impact on the environment

(continued)

Table 4.1 (continued)

No	Company info	Value proposal (what?)	Value creation (how?)	Value capture
25	Nourished UK 2, 3 3D printing	It is a personalised vitamin gummy provider utilising patented 3D printing technology and a vegan encapsulation formula	It has 3D printing technology which allows combining seven different, high-impact active ingredients into gummy nutrition stacks. This combination, which changes based on consumers' lifestyle questionnaire on their website contains 28 active ingredients for consumers to choose from, all sourced from ethical and vegan food sources in the UK and Europe	Value is captured through less waste, hassle and cost than purchasing all the active ingredients separately in addition to helping reduce malnutrition and food insecurity
26	Novolyze France 2, 12 AI, Internet of Things	It is a food safety enterprise utilising IoT and machine learning. The company develops dry and ready-to-use surrogate microorganisms that aid in the safe storage of food	For food safety and quality, the firm depends on cutting-edge microbiological solutions paired with technological advancements in digital, IoT and AI	Value is captured through enabling food companies to produce clean food products by increasing productivity, improving quality and reducing environmental impact
27	NRGene Israel 2 AI, big data, cloud computing	It is a company developing and commercialising AI-based genomic tools, cloud-based software solutions to analyse big data generated by sequencing technologies	The company offers AI software solutions with cloud-based big data analytical technology for managing the full genomic diversity of species to enhance plants' and animals' natural breeding processes without genetic engineering. It provides an analytic service that delivers the mapping and gene space comparison between several assembled genomes of the same organism	Value is captured by maximising the agricultural yield, optimising and accelerating breeding programs in less time for lower costs
28	Odd.Bot The Netherlands 2, 15 AI, robotics	It is a provider of AI-based robots developed to remove weeds at an early stage without any chemical pesticides	With AI-based software, the weed whacker robots determine the route and the unwanted weeds. Weeds are removed by the robots at an early stage without using any chemicals	Environmental value is created by preventing soil and water pollution caused by the use of chemical fertilisers. Also, the robots provide a higher yield with less manual labour in organic food production

No	Company info	Value proposal (what?)	Value creation (how?)	Value capture
29	Orbisk The Netherlands 2, 12 AI	The company provides the foodservice industry with detailed insights into their food waste by using AI	The food waste monitor has a camera with AI that automatically recognises which products are thrown away. Based on data, customers optimise the supply, the purchasing process or the menu	It is an online reporting service that gives the customer an idea about how food waste creates monetary value through waste monitoring, subscription, repair service with AI via B2B
30	Orbital Farm Canada 2, 3 Biotech and biomanufacturing, soilless farming	It is a closed-loop farming company providing management of food, water and energy for long-term space missions using biotechnology, energy storage and soilless agriculture	CO_2 is extracted from waste streams generated by its industry partners, such as food waste, wastewater, methane and even waste heat, as well as for biotechnology and energy; resulting in the creation of a mini-ecosystem. The fish are then raised in massive recirculating tanks, with the water being pumped into large greenhouses. Without the need for pesticides or fossil fuel-based fertilisers, fish give nutrients for plants to live and grow. In the system, water is recycled, and fish, fruit and vegetables are gathered	Large commercial farms developed using biotechnology and soilless farming create value in tackling climate change, fighting world hunger and enhancing humans' ability to become multi-planetary species

(continued)

Table 4.1 (continued)

No	Company info	Value proposal (what?)	Value creation (how?)	Value capture
31	Phytech Israel 2 AI, big data, Internet of Things	It is an agriculture IoT company that develops plant-based farming applications. Their alert-driven mobile platform combines predictive algorithms and data analysis tools with ML to integrate continuous crop health and supportive environmental data	The application includes real-time plant status push alerts and plant AI-based irrigation requests, as well as direct continuous monitoring of plant growth parameters and applied irrigation, microclimate variables, forecasts and imaging analysis. They use algorithms to convert raw data into plant status for specific crops. Micro-variations of stem diameter, which are scientifically verified stress indicators, are continuously monitored by sensors on selected plants. Irrigation schedule recommendations are provided by machine learning algorithms to keep plant status in the ideal zone with the least amount of resources	Value is captured by optimising irrigation based on accurate real-time data from plants and letting them tell farmers where, when and how much water they need using AI. Also, using pressure sensors placed on the irrigation systems provides farmers with the actual pressures and sends irrigation alerts when the pressure drops. So, farmers can save water, energy, time, money and other resources
32	Plantible Foods USA 2, 3, 6, 9, 12, 13, 15 Biotech and biomanufacturing	It is a food technology company aiming to develop functional and applicable plant-based protein by harnessing *Lemna*: one of the most sustainable and nutrient-dense plants in the world with biomanufacturing	Scalable processes and environmentally friendly indoor aqua farms to grow *Lemna* are used. An organic and complete protein that will be a sustainably produced plant-based protein is created with biomanufacturing technology. Any chemical substances and valuable farmland are not necessary for production	Value is captured through keeping the CO2 footprint and impact on the environment low such as no pesticides, hexane or other toxic chemicals and not requiring valuable farmland or irrigation to grow our crop by using biomanufacturing
33	Plantix Germany 2 AI, big data	The company offers a digital agro-ecosystem platform around smallholder farming based on AI and big data that collects data from farmers	It provides crop management solutions. Farmers upload photos of their crops and diagnose problems in their crops by applying its AI to those photos	Value is captured through disease detection, pest control and yield growth services. They aim to be more efficient in crop production

No	Company info	Value proposal (what?)	Value creation (how?)	Value capture
34	RapidPricer The Netherlands 2 AI, Internet of Things	The company offers a mathematical platform to process data generated via IoT devices. It helps automate pricing and promotions utilising DL and AI algorithms for retailers	IoT provides real-time information to generate insights. Data generated in retail stores is leveraged and used for automated decisions. With the weights and algorithms learned, DL has the retail knowledge to make decisions for each product-store location. Algorithms continuously price fresh products based on life stages and inventory to reduce unsold fresh produce	Real-time data from retail stores with IoT and AI generates value by making the retail world a more efficient place with less food wasted
35	RefuCoat Spain 2, 12, 15 Bioplastics	The company aims to develop fully recyclable food packaging with enhanced gas barrier properties and high-performance coatings	Alternative to current metalised and modified atmosphere packages (MAP), the company develops hybrid bio-based high oxygen/water barriers and active coatings to be used in monolayer bio-based food packages (films and trays)	Improved food packaging and product preservation, reduced landfill waste and the avoidance of non-renewable resources in multilayer constructions are all ways to extract value
36	SCADAfarm New Zealand 2, 7, 9 5G, AI, Internet of Things	The company controls irrigation and wastewater systems by using 5G, AI and IoT remotely	Through industrial-scale automation, controlling and monitoring capabilities, cloud-based operation, IoT and mobile applications, farmers may operate, monitor and record their irrigation and wastewater systems from anywhere in the globe	Saving time spent with equipment, improving irrigation water utilisation and lowering energy consumption for pumping water are all ways to extract value
37	SkyMaps Agrimatics Czechia 2, 15 AI	The company is a precision farming solutions provider that uses remote sensing and advanced image analysis	Farmers can use application maps for variable rate pesticide spraying with the company's web-portal Cultiwise. Data from drone and satellite imagery is also supported by the software. They use machine learning models and computer vision techniques to reduce the use of herbicides in farming	Environmental value is captured by reducing the negative impact of herbicides and pesticides in farming

(continued)

Table 4.1 (continued)

No	Company info	Value proposal (what?)	Value creation (how?)	Value capture
38	Small Robot Company UK 2 AI, robotics	It is a company developing robots utilising AI for the farming industry	Their AI advice engine turns field data into an intelligence system, enabling decisions that take into account agronomy, soil science and market conditions. Smart robots which plant, monitor and treat arable crops autonomously make food production more accurate, sustainable and profitable	Environmental value is generated by ecologically harmonious, efficient, profitable and sustainable food production. Monetary value is captured through using robots in place of heavy-duty farming equipment to decrease the costs, reducing reliance on fossil fuels and soil erosion caused by tractors
39	Sunbirds France 2, 7, 13, 15 AI, big data, drones	The company offers solutions in livestock farming, invasive weed detection and precision forestry where surveillance on very wide areas is needed	Instead of planes or helicopters, they are using ultra long-distance electric drones, designed by the company, for mapping and surveillance tasks. Also, images are processed through AI and utilised as needed	By the use of solar-powered drones for inspection, value is captured by saving money, time, gas and water. The company provides efficient production options for farmers and contributes to the right decision in the food production chain
40	xFarm Italy 2, 9, 12 Big data, Internet of Things	The company offers a digital platform for farmers, based on free management software in the cloud, IoT field sensors and value-added services	It is a SaaS platform that collects data from farms by using satellite and IoT sensors. Simplifying data collection and analysis provides visualisation of farms on digital platforms	The value is generated by reducing paperwork, improving efficiency and sustainability and allowing the traceability of agricultural products by using IoT and big data

References

A #ZeroHunger world by 2030 is possible, [WWW Document] (2021), https://sdg2advocacyhub.org/actions/zerohunger-world-2030-possible. Accessed 11 Aug 2021

Agribusiness & Deforestation, *Greenpeace USA* (2021), https://www.greenpeace.org/usa/forests/issues/agribusiness/. Accessed 24 Aug 2021

Agriculture and Food Production Contribute Up to 29 Percent of Global Greenhouse Gas Emissions According to Comprehensive Research Papers, *Climate Change, Agriculture and Food Security* (2012), https://ccafs.cgiar.org/media/press-release/agriculture-and-food-production-contribute-29-percent-global-greenhouse. Accessed 11 Aug 2021

Blesh, J., Hoey, L., Jones, A.D., Friedmann, H., Perfecto, I., 2019. Development pathways toward "zero hunger." World Dev. 118, 1–14. doi:https://doi.org/10.1016/j.worlddev.2019.02.004

D. Byerlee, J. Fanzo, The SDG of zero hunger 75 years on: Turning full circle on agriculture and nutrition. Glob. Food Sec. **21**, 52–59 (2019). https://doi.org/10.1016/j.gfs.2019.06.002

Climate-Smart Agriculture [WWW Document]. World Bank (2021), https://www.worldbank.org/en/topic/climate-smart-agriculture. Accessed 11 Aug 2021

M.J. Cohen, Let them eat promises: Global policy incoherence, unmet pledges, and misplaced priorities undercut Progress on SDG 2. Food Ethics **4**, 175–187 (2019). https://doi.org/10.1007/s41055-019-00048-2

E.C. Ellis, N. Ramankutty, Putting people in the map: Anthropogenic biomes of the world. Front. Ecol. Environ. **6**, 439–447 (2008). https://doi.org/10.1890/070062

FAO, *From Fome Zero to Zero Hunger : A Global Perspective* (FAO, Rome, 2019). https://doi.org/10.4060/CA5524EN

Food and Agriculture Organization of the United Nations, World Health Organization, *Sustainable Healthy Diets: Guiding Principles* (Food and Agriculture Organization of the United Nations: World Health Organization, 2019)

Food and Beverages Global Market Opportunities and Strategies to 2030: COVID-19 Impact and Recovery, (The Business Research Company, Global, 2020)

E. Fukase, W. Martin, Economic growth, convergence, and world food demand and supply. World Dev. **132**, 104954 (2020). https://doi.org/10.1016/j.worlddev.2020.104954

J.D.B. Gil, P. Reidsma, K. Giller, L. Todman, Sustainable development goal 2: Improved targets and indicators for agriculture and food security. Ambio **48**, 685–698 (2019)

Goal 2: Zero Hunger, United Nations Sustainable Development (2021), https://www.un.org/sustainabledevelopment/hunger/. Accessed 11 Aug 2021

K. Grebmer, J. Bernstein, *Global Hunger Index 2020 One Decade to Zero Hunger – Linking Health and Sustainable Food Systems* (Deutsche Welthungerhilfe e.V. & Concern Worldwide, 2020)

A. Hamidov, K. Helming, Sustainability considerations in water–energy–food nexus research in irrigated agriculture. Sustainability **12**, 6274 (2020). https://doi.org/10.3390/su12156274

V. Linehan, S. Thorpe, N. Andrews, Y. Kim, F. Beaini, *Food Demand to 2050: Opportunities for Australian Agriculture–Algebraic Description of Agrifood Model* (Australian Bureau of Agricultural and Resource Economics and Sciences, Canberra, 2012), p. 21

D. Mason-D'Croz, T.B. Sulser, K. Wiebe, M.W. Rosegrant, S.K. Lowder, A. Nin-Pratt, D. Willenbockel, s. Robinson, T. Zhu, N. Cenacchi, S. Dunston, R.D. Robertson, (2019) Agricultural investments and hunger in Africa modeling potential contributions to SDG2 – Zero Hunger. World Development 11638-53 10.1016/j.worlddev.2018.12.006

D.P. Nally, *Human Encumbrances: Political Violence and the Great Irish Famine* (University of Notre Dame Press, Notre Dame, 2011)

Y. Nie, S. Avraamidou, X. Xiao, E.N. Pistikopoulos, J. Li, Y. Zeng, F. Song, J. Yu, M. Zhu, A food-energy-water nexus approach for land use optimization. Sci. Total Environ. **659**, 7–19 (2019). https://doi.org/10.1016/j.scitotenv.2018.12.242

B.A. Paramashanti, Challenges for Indonesia zero hunger agenda in the context of COVID-19 pandemic. Kesmas Natl. Pub. Health J. **15**, 24–27 (2020) https://doi.org/10.21109/kesmas.v15i2.3934

Population, total | Data, [WWW Document]. World Bank (2021), https://data.worldbank.org/indicator/SP.POP.TOTL?end=2020&start=1960. Accessed 11 Aug 2021

L. Rickards, E. Shortis, *SDG 2: End Hunger, Achieve Food Security and Improved Nutrition and Promote Sustainable Agriculture* (undefined, 2019)

SDG Compass, *A Guide for Business Action to Advance the Sustainable Development Goals* (2021), https://sdgcompass.org/. Accessed 11 Aug 2021

A. Sen, The food problem: Theory and policy. Third World Q. **4**, 447–459 (1982). https://doi.org/10.1080/01436598208419641

E. Sharifi Moghadam, S.H. Sadeghi, M. Zarghami, M. Delavar, Water-energy-food nexus as a new approach for watershed resources management: A review. Environ. Resour. Res. **7**, 129–136 (2019)

D.J. Shaw, World food crisis, in *World Food Security: A History since 1945*, ed. by D. J. Shaw, (Palgrave Macmillan, London, 2007), pp. 115–120. https://doi.org/10.1057/9780230589780_10

G. Tansey, Food and thriving people: Paradigm shifts for fair and sustainable food systems. Food Energy Sec. **2**, 1–11 (2013). https://doi.org/10.1002/fes3.22

UN, *Food Security and Nutrition and Sustainable Agriculture* [WWW Document] (2021), https://sdgs.un.org/topics/food-security-and-nutrition-and-sustainable-agriculture. Accessed 11 Aug 2021

UNDESA, *Water and Food Security* [WWW Document] (2021), https://www.un.org/waterforlifedecade/food_security.shtml. Accessed 11 Aug 2021

WHO, *Comprehensive Implementation Plan on Maternal, Infant and Young Child Nutrition* [WWW Document] (2014), http://apps.who.int/iris/bitstream/handle/10665/113048/WHO_NMH_NHD_14.1_eng.pdf?ua=1. Accessed 11 Aug 21

Y. Zhang, R.K. Runting, E.L. Webb, D.P. Edwards, L.R. Carrasco, Coordinated intensification to reconcile the 'zero hunger' and 'life on land' sustainable development goals. J. Environ. Manag. **284**, 112032 (2021). https://doi.org/10.1016/j.jenvman.2021.112032

Abstract

There is a consensus that health is a fundamental human right. The extent of the countries seeking to improve the health conditions of their people is one of the indications of sustainable development. Poor health systems jeopardise a country's citizens' rights, hinder their involvement in educational programs, limit their ability to participate in economic activities and engage in meaningful work fully and ultimately raise poverty regardless of gender. This chapter presents the business models of 55 companies and use cases that employ emerging technologies and create value in SDG-3, Good Health and Well-Being. We should highlight that one use case can be related to more than one SDG and it can make use of multiple emerging technologies.

Keywords

Sustainable development goals · Business models · Good Health and Well-Being · Sustainability

The author would like to acknowledge the help and contributions of İlayda Zeynep Mert, Ömer Sami Temel, Abdullah Aykut Kılıç, Abdullah Enes Ögel, Veysel Ömer Yıldız and Enes Ürkmez in completing this chapter. They also contributed to Chapter 2's Data Hubs, Healthcare Analytics, Internet of Behaviours, Natural Language Processing and Wireless Power Transfer sections.

There is a consensus that health is a fundamental human right. The extent of the countries seeking to improve the health conditions of their people is one of the indications of sustainable development. Poor health systems jeopardise a country's citizens' rights, hinder their involvement in educational programs, limit their ability to participate in economic activities and engage in meaningful work fully and ultimately raise poverty regardless of gender. The current state of health is concerning; women all over the world continue to face barriers to general and reproductive healthcare, billions of people lack access to important medicines, hundreds of millions of adults and children lack access to safe drinking water, and many suffer from malnutrition (Filho et al. 2019). Furthermore 2020 demonstrated to the world how infectious diseases might spread from a small group of people to a health issue of international significance in a matter of days. Infectious diseases' global effect has steadily declined since 2000, yet they were still responsible for more than 10.2 million fatalities in 2019, accounting for 18% of all deaths. Investments in the diagnosis, treatment and control of major infectious diseases such as HIV/AIDS, malaria and tuberculosis (TB), as well as child and maternal problems, have had positive effects over the past 20 years, with global declines in their prevalence, incidence and rates. However, in 2019, these diseases remain among the top 10 causes of mortality in low-income countries (LICs) (World Health Organization 2021a, b).

One of the primary aims of the United Nations' Sustainable Development Goals (SDG) principles is to create healthy living conditions and ensure well-being at every stage of human life. Despite the significant development of technology in health, the impact of which will be felt globally in all areas of life, various health problems may cause permanent or temporary damage to people, which dramatically affects global functioning. To overcome these problems, the main principles to be followed are to focus on more efficient financing of health systems, improving sanitation and hygiene and providing greater access to doctors (United Nations 2021a, b, c, d).

SDG-3 seeks to "make sure healthy lifestyles and promote well–being for every generation". Other than the millennium development goals (MDGs), however, SDG-3 looks at health and well-being in a broader sense by looking beyond a narrow range of disorders (Seidman 2017). According to SDG-3 (Good Health and Well-Being), ensuring health and well-being across all generations is crucial for sustainable development, and only rigorous and continuing healthcare monitoring will be able to do this (Papa et al. 2018). SDG-3 also asks for greater research and innovation, increased healthcare costs and stronger ability in all nations to reduce and manage health risks (UN Office for Outer Space Affairs 2021). The SDG-3's main objective is to prevent 40% of premature deaths in each nation (i.e. at 2010 mortality rates, deaths before the age of 70 years would be witnessed in the 2030 population) and to enhance healthcare for all ages. To strengthen this main objective, four sub-objectives have been targeted, such as avoiding two-thirds of child and maternal deaths; preventing two-thirds of deaths caused by tuberculosis, HIV and malaria; refraining one-third of premature deaths caused by non-infectious diseases; and preventing one-third of deaths caused by other causes (other infectious diseases, malnutrition and injuries) (Alleyne, et al. 2015). Figure 5.1 demonstrates the targets set by the UN in the field of health under the name SDG-3.

Health is a fundamental human right and an important measure of long-term growth. Poor health jeopardises children's rights to education, limits economic prospects for men and women and fosters poverty around the world (SDG Compass 2021). In addition to disrupting the well-being of the individual, diseases harm the resources of the family and society and reduce the potential of people. The health of each individual and society plays a key role in long-term development. Avoiding diseases is important for survival, promoting wealth and economic growth (United Nations 2021a, b, c, d). Building thriving communities requires ensuring healthy lives and fostering well-being (United Nations 2021a, b, c, d). The link between health and well-being, which is one of the drivers of human capital, and financial growth was studied from 1991 to 2014 in the following high-income nations (Luxemburg, Israel, Australia, Switzerland, Spain, Denmark, Hungary, Sweden, Portugal and Poland). According to the findings of the research, health factors have a long-term impact on financial sub-variables. On the other hand, income has no direct impact on the health variable. However, through financial sub-variables, the income factor indirectly impacts the health variable (Kuloglu and Ecevit 2017).

SDG-3 is viewed in conjunction with SDG-1 No Poverty, SDG-14 Life Below Water and SDG-15 Life on Land as a direct outcome of progress towards the other objectives in terms of social and environmental benefits. They are also important since regressing these goals restricts and limits the human and natural resources needed to maintain a stable global system (Cernev and Fenner 2019). According to WHO (World Health Organization 2021a, b), several factors influence a person's health, including lifestyle, financial position, social status, available healthcare services and facilities, degree of education, nutritional access, communal life and genetic composition. Understanding health via these variables can aid in predicting medical status using factors that are measured, evaluated and compared across groups. Individual and community health can be improved through specific treatments. This is inextricably connected to SDG 3.4, which specifies the goal of reducing early death due to non-communicable illnesses

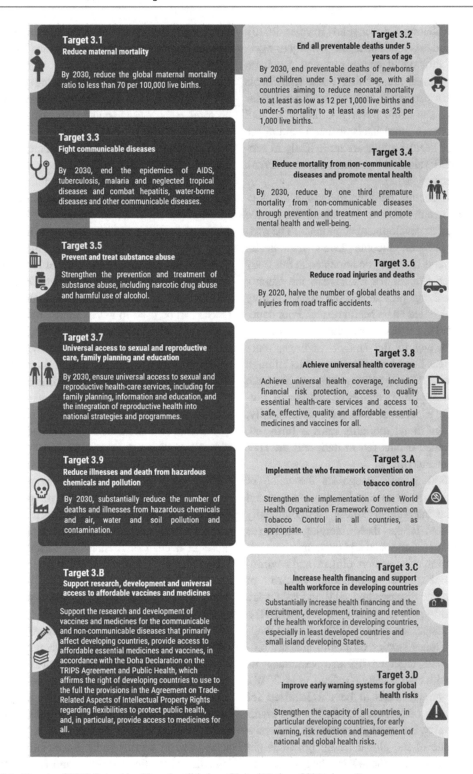

Fig. 5.1 Targets of SDG-3 good health and well-being. (United Nations 2021a, b, c, d)

(NCDs) (Sharma-Brymer and Brymer 2020). Illnesses are likely to hinder one's ability to work, which in turn could create financial disadvantages or worsen pre-existing ones. Individuals who create economic income by working for themselves can avoid poverty. Children without any disease are more likely to learn, and at the same time, healthy adults are more likely to generate added value and generate income (Frenk and de Ferranti 2012).

The connection between economic development and health outcomes is a significant avenue of research in development economics. The relationship between income levels and health improvements is widely documented in this body of research (Vu 2020). Each year, the County Health Rankings provide a score to each county in the USA based on its health results. Poor counties are said to perform worse than wealthy counties. According to the study's findings, more affluent areas fared better rankings (McCullough and Leider 2017). Significant progress has been made against primary causes of illness and mortality. Life expectancy has risen substantially; maternal and infant mortality rates at birth fell, malaria-related deaths fell in half, and the HIV epidemic changed.

The 2030 agenda recognises the value of health in the path of sustainable development and reflects the interconnectedness and complexity between them. It takes these emerging challenges into account; rising social and economic inequality, increasing urbanisation, climate and environmental risks, the continuing burden of communicable diseases and HIV and NCDs are considered to pose new concerns (UNDP 2021). The implementation of SDG-3 "Health and Well-Being" includes studies on health and the biomedical field together. It also examines its relations with civil society (Guégan, et al. 2018). Health coverage needs to be universalised for SDG-3's goal of reducing inequality and eradicating poverty. For example, antimicrobial resistance is one of the global health priorities not directly mentioned in the SDGs. The gap between and even within countries is wide in advances. There is a 31-year difference in life expectancy between countries with the shortest life expec-

tancy and the countries with the longest life expectancy. While there are countries that have made significant progress, national averages hide that many countries lag. Addressing inequities and promoting good health for all requires multisectoral, rights-based and gender-sensitive methods (UNDP 2021).

Numerous aspects require significant consideration regarding the future of good health and well-being. Initially, maternal mortality, one of the targets identified by the UN, is considered to be an important criterion. Comprehensive cross-sectional research reports that there have been great reductions in global maternal mortality and a substantial increase in the ratio of births under healthcare services such as hospitals (Souza et al. 2013). Furthermore, the maternal severity index (MSI) can be used to assess whether the performance of health facilities is sufficient with regard to their ability to care for pregnant women and deal with possible complications. However, it is also noted to reduce deaths during childbirth, global coverage of vital treatments must be accompanied by complete emergency care.

Another critical factor that determines an individual's well-being is mental health. Certain complications come across when seeking and administering mental healthcare. The accessibility to mental healthcare is among the greatest of these challenges. Financial resources are vital for most, if not all, healthcare facilities to provide for those in need. Comprehensive research by the World Health Organization (WHO) reports that, on a global scale, mental health makes up for less than 2% of the total government health expenditure.

Furthermore, a great gap between various regions still exists, showing that there is a great lack of funding for those on the lower end. The findings above indicate a positive relationship between income and accessibility to mental health services. This means that individuals in lower-income regions and countries have less access due to certain variables, including funding, mental health workforce and costs (World Health Organization 2018). Another identified factor that affects the accessibility of mental healthcare is the stigma around mental illnesses and treatments. Evidence demonstrates that around one-third of the global population suffers from various mental illnesses. Many of these individuals do not receive any treatment, which holds even for countries that have adequate resources (Thornicroft 2011). Additionally, men have been reported to be less likely to seek treatment or help regarding mental health (Terlizzi and Zablotsky 2020), which can be partially explained by the worldwide masculine norms (Chatmon 2020), indicating that stigma regarding mental health might have more severe results for males. There are various governmental programs utilised to decrease the stigma that surrounds mental health. For instance, the UK's Time to Change campaign has been reported to be successful as it was related to increased help-seeking and comfort in disclosing mental health issues (Henderson et al. 2018). However, there have been suggestions to change into alternative frameworks regarding these policies, campaigns or programs to increase their comprehensiveness and efficiency (Stangl et al. 2019).

Tracking and preventing the spread of infectious diseases is another great concern that needs to be addressed to achieve SDG-3. That is to make sure that all individuals of the whole generation have healthy lives and well-being is promoted. Although it is not exhaustive, a list of diseases of major concern has been made. These include HIV, malaria, tuberculosis and more. Additionally, the recent COVID-19 pandemic, which has had devastating effects on numerous aspects of society, is a great example that further underlines the importance of precautions and measures regarding infectious diseases. For instance, studies indicate that it has reduced life expectancy, both directly by fatal infections and indirectly, through unemployment that is due to measures taken to stop the spreading of the virus, most notably lockdowns (Bianchi, et al. 2020). Furthermore, it has been reported that it has disrupted essential health services globally (World Health Organization 2021a, b). Thus, it's a major disruption to the progress made regarding SDG-3, if not a great setback. This issue can be addressed by using personal, mobile and wearable technological devices that can track and report vital health information. These data can be

delivered to the corresponding healthcare facilities (Savelyeva et al. 2019). Additionally, the processing and gathering of this data could be facilitated by using emerging technologies such as AI, big data and cloud computing. These technologies are suggested as highly efficient solutions to the points mentioned earlier.

5.1 Companies and Use Cases

Table 5.1 presents the business models of 55 companies and use cases that employ emerging technologies and create value in SDG-3. We should highlight that one use case can be related to more than one SDG and it can make use of multiple emerging technologies. In the left column, we present the company name, the origin country, related SDGs and emerging technologies that are included. The companies and use cases are listed alphabetically.[1]

---

[1]For reference, you may click on the hyperlinks on the company names or follow the websites here (Accessed Online – 2.1.2022):

http://cipherskin.com/; http://genomes.io; http://graft3d.com/patient-specific-implant/; https://albert.health/; https://augmedics.com/; https://autonomoushealthcare.com/; https://biontech.de/; https://captionhealth.com/; https://carbonhealth.com/; https://cloudmedxhealth.com/; https://engage.3m.com/3M-Respiratory-Tracker; https://healx.io/; https://in-medprognostics.com/; https://iryo.network/#network; https://jetbrain.ai/#/robots/amro; https://khealth.com; https://kinsahealth.com/; https://kumovis.com/; https://optellum.com/; https://prognoshealth.com/; https://proximie.com/; https://raslabs.com/; https://rsresearch.net/about-us/; https://siramedical.com/; https://solve.care/; https://www.philips.nl; https://www.aleph-farms.com/; https://www.atomwise.com/; https://www.babylonhealth.com/; https://www.bagmo.in/; https://www.basebit.ai/en/; https://www.battelle.org/; https://www.behavr.com/about-us/; https://www.benevolent.com/; https://www.biomodex.com/; https://www.dualgoodhealth.com/healthcare/; https://www.emedgene.com/; https://www.epigenelabs.com/; https://www.fathomhealth.com/; https://www.freenome.com/; https://www.implicity.com/; https://www.insight-rx.com/; https://www.intenseye.com/; https://www.medimsight.com/; https://www.micromek.net/; https://www.nextgen.com/products-and-services/health-data-hub; https://www.operoo.com; https://www.paige.ai/; https://www.prellisbio.com/; https://www.prokarium.com/; https://www.tempus.com/; https://www.univfy.com/; https://www.unlearn.ai/; https://www.vergegenomics.com/; https://www.xr.health/

Table 5.1 Companies and use cases in SDG-3

No	Company info	Value proposal (what?)	Value creation (how?)	Value capture
1	3M USA 3 Crowdfunding, flexible electronics and wearables	The company utilises crowdfunding to produce a gadget that captures, monitors and trends respiratory quality in various conditions and locales	3M Respiratory Tracker attaches to garments. It tracks and trends the users' breathing, activity, weather, air quality and pollen. It also provides personalised trigger forecasts as well as specific weather, allergies and air quality through its smartphone application	Value is captured by monitoring respiratory activity in various settings and scenarios. It encourages good breathing patterns and assists in the identification of trends by analysing air and air quality data to determine how the environment influences breathing. Thus, it helps maintain overall respiratory health
2	Albert Health Turkey 3 AI, natural language processing	The company offers an audio-based healthcare assistant that helps patients take their medicine at the right time and dose	Thanks to AI-powered NLP algorithms, the healthcare assistant detects voice commands, records the entered data and creates a medical diary. Through the comprehension of the questions asked, it can give appropriate answers. Also, it provides a healthy life, nutrition and exercise recommendations according to the disease	Value is captured by facilitating the registration and follow-up of treatment for chronic patients while allowing relatives to follow patients through the app. It can benefit public health greatly through preventive features such as water drinking reminders, step tracking, medication reminder
3	Aleph Farms Israel 2, 3, 12 3D printing, cellular agriculture	It is a company producing beef steaks from non-genetically modified cells extracted from a living cow. It reduces animal slaughter and environmental effects	The 3D technology relies on creating a complex tissue composed of the four meat cell types found in conventional cuts of meat, including vascular and connective tissues. These cells, extracted from living cows, are grown on an intricate proprietary three-dimensional platform using a similar process to beef processing in large tanks, called bio farms, thanks to cellular agriculture	Environmental value is captured by growing beef on bio farms that reduces the environmental impact and resource consumption of modern meat production and facilitates beef production, making it easier for disadvantaged people in many parts of the world to access beef. In addition, the beef production process becomes more controllable, efficient and cheaper. Also, less use of animals for meat production leads to a decrease or even disappearance of diseases transmitted through animals

#	Company	Description	Value
4	Atomwise USA 3 AI	The company offers an AI-based drug discovery platform with an emphasis on preclinical development	AtomNet platform leverages AI, making it possible to study the properties of existing drugs to match their formulas with one another. These studies can provide insightful recommendations regarding the development of more effective medication as well as novel medication that can be used to treat formerly unaddressed diseases	The value is created by enabling massive scales and unprecedented speeds needed to create a deep and broad pipeline of drugs to improve human health. By using AI, the platform drastically reduces the time required for preclinical drug development
5	Augmedics Israel 3 Spatial computing	The company provides a system that projects a 3D copy of a computerised tomography (CT) scan to the patient during spine surgery	The AR-powered Xvision Spine System locates surgical instruments in real time and overlays them on the patient's CT data. The navigation data is then projected onto the surgeon's retina via the headset, allowing the surgeon to look at the patient and view the navigation data without looking away from the patient	Value is captured through AR devices that display real-time CT scans of patients during surgery, giving physicians a 3D augmented perspective of the patient's anatomy and reducing tissue damage. This novel approach greatly increases the efficiency and accuracy of surgical processes
6	Autonomous Healthcare USA 3 AI, autonomous vehicles	Based on AI and mathematical modelling, the company develops automated medical technologies for critical care, non-invasive patient monitoring and controlled-substance tracking	AstroSight™ provides video analytics and non-invasive vital sign monitoring for astronauts during space exploration missions. TraCS™ leverages recent advances in ML to provide seamless medication tracking and analytics for tracking controlled substances. E-Fusion™ is a smart fluid and vasoactive drug management system, with the capability to provide semi-automated (clinical decision support) and fully automated (closed-loop) fluid and vasoactive drug management. Syncron-E is an AI-based software technology that analyses pressure and flow waveforms in real time to detect various types of patient-ventilator asynchrony	Value is captured by preventing patients from staying in the hospital more due to ventilator incompatibility. It prevents excessive drug use and provides as much use as necessary. It allows astronauts to operate more healthily

(continued)

Table 5.1 (continued)

No	Company info	Value proposal (what?)	Value creation (how?)	Value capture
7	Babylon UK 3 AI	The company provides an AI-powered platform that connects patients with doctors and healthcare providers combined with virtual clinical operations	Using anonymised, aggregated and validated medical datasets, patient health records and consultation notes made by clinicians, the AI-powered platform helps doctors make decisions about triage, causes of symptoms and future health predictions	Value is captured through an app that brings together healthcare providers that are available 7/24 and patients
8	Bagmo India 3, 12 Cloud computing, Internet of Things	The company offers an IoT-based blood bank management system that is designed to help ensure sufficient conditions during transportation and storage of blood bags through detailed monitoring	A tracking system is installed on the blood bags. The current location and temperature of the bag are constantly monitored. It produces daily, weekly and monthly reports. Thanks to IoT technology, dangerous conditions are reported to authorities immediately. Since it is cloud-based, it can be followed from anywhere	Value is captured by assisting blood centres to reduce costs and wastage and prevent the collection and storage of excess inventory. Continuous monitoring and tracking ensure the safety of blood samples and minimise wastage during both storage and transportation processes. Therefore improving the low availability of blood in rural areas eliminates avoidable maternal mortality and morbidity
9	BaseBit.ai China 3 AI, big data, cloud computing	The company offers a medical imaging platform addressing practical challenges through the use of AI, cloud computing and big data	Perceptor helps doctors by providing an image review tool that is designed for numerous organs and lesions, powered by DL models and cloud computing	Value is captured by the automatic report feature of the platform that rapidly generates clinically acceptable diagnostic results. It helps doctors save time while simultaneously improving the quality of their work
10	Battelle USA 3 Cybersecurity	The company offers a cybersecurity platform designed to evaluate new medical systems, identify vulnerabilities in existing systems and implement security and anti-tampering measures into medical product designs	The cybersecurity suite DeviceSecure enables to characterise, assess, model, predict and measure a broad spectrum of threats and vulnerabilities posed by a medical device through a thorough analysis of hardware and software	Value is captured by reducing risks and security compliance costs. It automates the monitoring of newly detected and reported vulnerabilities in medical device hardware and related software

11	BehaVR USA 3 Spatial computing	The company offers a VR-powered platform for patients who are trying to overcome their mental health issues such as stress, worry and fear	Various treatment programs prepared by specialists are applied to patients through VR technology. With this method, patients are guided to solve their problems by being alone with themselves, independent of any therapist	Value is captured by giving people agency over their health and through utilising VR for generating value, changing and better clinical outcomes for all stakeholders in the healthcare system
12	BenevolentAI UK 3 AI	The company provides an exploratory and predictive AI-based platform that facilitates the research and development processes of new medications	The platform is trained using ML techniques, with data from publicly available sources and internal experiments, serving as a bridge between them. Using AI, the platform allows users to predict, track and validate optimal hypotheses for drug discovery	Value is captured by enabling scientists to explore data and disease networks, ask questions and find new insights and hypotheses. Thus, it provides ways for faster and more efficient drug development
13	Biomodex France 3 3D printing	The company provides 3D-printed surgical simulators for medical education and patient-specific preoperative preparation	3D modelling and printers are used to create precise copies of organs for use in medical education and preoperative simulations	It generates value streams from various solutions it creates through 3D-printed models for neurovascular and structural heart applications
14	BioNTech Germany 3 Biotech and biomanufacturing	It is a therapeutics company that harnesses the power of the immune system to develop novel therapies against cancer and infectious diseases	The performance of mRNA is influenced by its structural features. This comprises the molecule's immunogenicity, translation efficiency and stability. They use their significant experience to develop, synthesise, manufacture and formulate therapeutic mRNA, tailoring its composition to the application	The company's pipeline comprises mRNA-based immune activators, antigen-targeting T cells and antibodies and specified immunomodulators of diverse immune cell processes, all of which generate value
15	Caption Health USA 3 AI	It is a company that empowers healthcare providers with new capabilities to acquire and interpret ultrasound exams, thanks to AI	Caption Health's AI empowers the broader healthcare professional to perform ultrasound exams and unlocks a critical tool for early disease detection	The platform incorporates AI and health sciences to provide healthcare providers with new capabilities to receive and interpret ultrasound exams. The platform allows for earlier diagnosis of certain diseases, consequently more effective interventions made in good time

(continued)

Table 5.1 (continued)

No	Company info	Value proposal (what?)	Value creation (how?)	Value capture
16	Carbon Health USA 3, 10 AI, healthcare analytics	It is a healthcare company providing tech-enabled solutions to primary care, urgent care and virtual care	Carbon Health blends virtual and in-person services in clinics and provides personalised healthcare services, thanks to AI. Via the company's mobile app, patients may access their health information, schedule appointments, make payments or schedule a video visit for ailments. Patients may message Carbon doctors on the app to obtain treatment for illnesses. They can also access in-person treatments such as primary care, urgent care, mental health, women's health and LGBTQ+ health	Value is captured by making the health system, which has become cumbersome due to bureaucracy, more accessible, efficient, effective and solution-oriented for everyone. The application increases the harmony between insurance companies and health institutions and reduces wastes arising from incompatibility. Widespread use of the platform can address significant healthcare system problems of countries that have dysfunctional or inefficient systems. No subscription or membership is required to use the platform, and fees are paid per treatment by health insurance or patients
17	Cipher Skin USA 3 Biometrics, flexible electronics and wearables	It is a biometrics company that offers a network of sensors that can be wrapped around human limbs to capture motion and biometrics data in real time	The "BioSleeve" (network of sensors printed on textile) captures movement and retrieves internal metrics, and the software "Digital Mirror" translates this data into actionable, visualised insights, delivering 3D visualisations and diagnostics. Through continuous gathering and analysis of motion and biometric data, the company allows physical therapists to access the most accurate patient recovery data	Value is created by enabling the building and monitoring of efficient rehab programs for both remote and in-clinic sessions. Under the provision of medical specialists, the programs can be used to enhance recovery and prevent sending athletes back to training before rehab is complete and predict injury

18	CloudMedx	The company offers Aligned Intelligence, an AI-based DL platform that offers RPA solutions to enhance existing workflows and generate automated clinical insights, improving operations, case management and patient engagement for healthcare organisations	Massive sums of structured and unstructured data gathered directly from electronic medical records (EMRs), labs, corporations and customers' proprietary sources are ingested into the clinical AI assistant. Through DL methods, the platform then curates and converts this data to identify key patterns and actionable insights that result in superior patient outcomes and operational excellence	Value is created by automising the coding process which makes it much faster and accurate. The predictive AI analytics platform provides clinicians, nurses and frontline staff relevant insights to improve patient outcomes
	USA			
	3			
	AI, robotic process automation			
19	Dual Good Health	The company offers a VR-powered interactive healthcare training platform targeted to the general staff, medical specialists and students	By using VR, the platform delivers interactive scenario training with a user-centred approach. Some training scenarios also incorporate "hands-on" approaches by using external objects to deliver tactile feedback	Value is created by experiencing a real-life scenario by VR simulation that improves learning outcomes, allows healthcare staff to train more often and aims to improve knowledge retention
	UK			
	3			
	Spatial computing			
20	Emedgene	The company provides an AI-powered genomic analysis platform that delivers solutions for genomic diagnostics and discovery at scale	The explainable AI platform eliminates noise and focuses analysis on the data points most likely to answer the problem, giving scientists more time to focus on areas where their clinical judgement is critical. Running on an AI knowledge graph updated with information from books and databases, accurate and validated ML models locate causative variations with proof	Value is generated by reducing the time needed to process data and increasing the accuracy of analyses. Their solutions effectively increase the efficiency of clinical labs, translational and research labs and children's hospitals
	USA			
	3			
	AI			

(continued)

Table 5.1 (continued)

No	Company info	Value proposal (what?)	Value creation (how?)	Value capture
21	Epigene Labs France 3 AI	The company offers an AI-based platform with a multi-source approach for analysing data to design drugs for cancer patients	The company combines information from publicly available data with its partners' proprietary data and expertise. Data is translated into actionable insights to design novel medication with precision oncology approaches using an AI-based platform	Value capture is made possible through AI-enabled aggregation, analysis and visualisation of massive amounts of genomic data. Their solution tackles major challenges of precision oncology, which allows more people to benefit from it. The company operates in partnership with world-leading cancer centres and high-profile biotechs and more to help an increasing number of cancer patients
22	Fathom USA 3 AI, big data, natural language processing	The company offers a coding platform designed to accelerate and automate the process of medical coding	Using NLP models, the ML-trained AI, DL and big data methods, the platform transforms information contained in electronic health records into medical coding while automatically evaluating incoming charts, allowing healthcare professionals to automate their coding	Value is captured by the AI system that can translate unstructured texts into medical coding reducing coding costs by up to 70% while boosting speed, accuracy and security
23	Freenome USA 3 AI	The company offers a multi-omics platform designed to ensure the early diagnosis of cancer through AI-based analyses of blood tests	Data from healthy and cancer-positive samples are used to train the algorithm with ML techniques. Following a multidimensional approach with computational biology techniques, they can distinguish between tumour-induced and non-tumour-induced manifestations. Thus, broader and clearer information that can be used in the early diagnosis of cancer is presented to clinicians with products called the multi-omics platform	Early diagnosis of cancer plays a key role in preventing the disease before it spreads in the body. The success of the treatment is increased, and the risk of death of patients is reduced, thanks to the AI-based diagnosis method

24	Genomes.io UK 3 Blockchain	The company offers a data bank that provides secure storage and interpretation of human DNA for individuals and anonymised datasets to researchers	DNA sequence data of customers are uploaded and encrypted to be stored and analysed in a virtual vault. Analyses provide individuals with personalised reports on health risks, ancestry and carrier status. This vault is owned and controlled by the customer via the ethereum blockchain. Thus, the data is only accessible through explicit consent by the owner. Researchers can use these anonymised genetic datasets in various studies upon being granted permission	With a blueprint of human DNA uploaded to the data bank, researchers and pharmaceutical companies can easily and anonymously access the data needed for DNA research. In addition, the DNA data that is uploaded to the data bank can be used in various researches as long as the owners allow, and DNA owners can both earn money and have information about their DNA in return
25	Graft3D India 3, 9 3D printing	The company manufactures patient-specific implants (PSIs) using metal additive manufacturing technology with 3D printing	They convert images from CT or MRI scans into software files compatible with 3D modelling and printing. These models are fed into a metal printer for creating the required implants. The direct metal laser sintering (DMLS) technique allows for a high degree of customisation	PSIs that improve the form, function and aesthetics of implants create value by tailoring them to the needs of the patient. Through 3D modelling and printing, the processes of personalised modelling, shaping and placing of implants are automatised
26	Healx UK 3 AI, big data, healthcare analytics, natural language processing	The company provides an AI-powered platform that focuses on patient-inspired treatments and drug discoveries for rare diseases	Healnet uses NLP and big data methods to analyse all publicly available biomedical databases to identify important knowledge gaps in uncommon diseases. Using AI as an analysis instrument, the platform is able to cross-validate between all available disease- and drug-related data findings and deliver pre-formed, data-driven theories to pinpoint new treatments for rare conditions	Value is generated through the AI platform that makes drug discovery for rare diseases faster, smarter and safer, while also supporting their team and partners in discovering and translating new treatments towards the clinic

(continued)

Table 5.1 (continued)

No	Company info	Value proposal (what?)	Value creation (how?)	Value capture
27	Implicity France 3 AI	The company offers an AI-based remote monitoring platform for data collected through pacemakers and defibrillators connected to patients suffering from cardiovascular diseases	Real-time data, collected from cardiac implantable electronic devices, is transformed and aggregated into simplified and screened reports to be used in diagnosis, rehabilitation and surveillance of cardiovascular patients. Heart failure module uses AI to analyse the data, predicting and generating automated alerts for heart failures	Value is generated through facilitated monitoring of cardiovascular information in real time and from all devices, consequently increasing the time efficiency and productivity of medical staff. The predictive algorithm allows for preventive management. By doing so, it challenges healthcare costs with reduced hospitalisation
28	InMed Prognostics India 3 AI	The company offers a cloud workflow platform that integrates data from several modalities powered by AI to deliver smart insights and endpoints for precision medicine	The Neuroshield software works with medical imaging data to support clinicians in diagnosing and proactively managing patients. The platform creates a biomarker system for various neurological illnesses, starting with paediatric epilepsy and dementia, using data analytics and DL techniques	The Neuroshield platform helps physicians reach more reliable and precise diagnoses. Thus, it allows them to provide patients with better treatment options while increasing the efficiency of time and resource usage of hospitals
29	InsightRX USA 3 AI, big data	The company offers an AI and big data-based clinical decision support tool that allows physicians to tailor treatment at the point of care	To comprehend each patient's unique pharmacological profile, InsightRX uses patient-specific data, PK/PD models and Bayesian estimates and keeps these models up-to-date with big data methods. ML methods are used to train its AI that analyses continuously updated data, thus providing the most optimal dose schedule	It provides clinicians with a comprehensive understanding of a patient's pharmacology to individualise treatment decision-making. It improves the quality of patient care by helping them individualise dosing for complex drugs
30	Intenseye USA 3 AI	The company provides an AI-powered HSE (Health and Safety Executive) platform	ML and computer vision are particularly capable of processing real-time streams of video data to quickly identify anomalous events and producing insights for EHS (environment, health and safety) teams to improve. High-performance software and IP-networked cameras also provide real-time incident notifications and 7/24 automated inspections	Value is captured by helping save lives in the workplace through AI-powered monitoring and detection of the most common unsafe act types across industries. Their solutions enable businesses to grow employee health and safety throughout their facility while saving time and resources through automation

31	Iryo.Network	The company offers a blockchain-based zero-knowledge health record storage platform with an anonymous query interface	Monetary value is generated through digitisation of all paper-based documents and saving information directly at the patient's profile. The platform facilitates payment of medical processes with certain features, such as overdue invoice tracking and income tracking per doctor or patient
	Slovenia	The platform uses blockchain to give patients complete ownership over their health information while also providing healthcare providers with real-time medical history to enhance treatment quality	
	3		
	Blockchain		
32	Jetbrain Robotics	The company offers an autonomous robot delivering hospital services safely and rapidly to different locations when they are needed	Value is captured by assisting in the management of nursing and internal logistics, along with advanced patient care. Transports drugs, samples and other materials throughout the hospital safely and securely. Its built-in power capacity allows for self-charging and continuous service provision
	India	AMRO is an autonomous robot with which a user can communicate regarding authorisation, predetermined destinations and material receipt. It does not require human intervention as it has location awareness and location detection capabilities. It includes sensors on the front, sides and rear that provide 360-degree awareness and can precisely detect obstacles and plan a route around them	
	3		
	Robotics		
33	K Health	The company offers an AI-based data-driven digital primary care system that delivers personalised primary care	The value is captured by enabling affordable medical tips and insights to users at any given time. It also facilitates the connection between specialists and individuals by serving as a bridge
	USA	Utilising AI, vast amounts of medical information data collected from users are analysed to provide the relevant medical insights and tips for deciding if there is a need for medical attention from an appropriate specialist	
	3		
	AI		

(continued)

Table 5.1 (continued)

No	Company info	Value proposal (what?)	Value creation (how?)	Value capture
34	Kinsa USA 3 Internet of Behaviours, Internet of Things	The company offers an illness tracking system that uses smart thermometers to calculate the risk of epidemics and provide various measures to be taken	The thermometer continuously tracks body temperature and makes recommendations according to the symptoms of students or employees in an institution. This data simultaneously appears in the school/institution's account that collectively allows mass monitoring and tracking. Through the use of IoT and IoB, stratified reports of epidemic relevant data are produced by the system, consequently allowing preventive action to be taken by authorities	Value is captured through the prevention of epidemics before they spread further and the protection of public and individual health by making recommendations to patients. The illness support system consists of a network of millions of smart thermometers that aggregate real-time population health insights across the USA to track the spread of illness and forecast long-term outbreaks. This assists people, businesses and societies in predicting, preparing for and preventing infectious illnesses
35	Kumovis Germany 3, 9 3D printing	The company offers a 3D printing system that enables the manufacturing of customised medical products and implants	For a range of thermoplastic materials, the printing technology employs fused-layer production. To allow its clients to just give a model file and receive the result, they use unique slicing software and printing technologies	The 3D printing system enables users to monitor and optimise their processes using industrial PLC technology. Featuring an integrated clean room, the printer is ideal for healthcare applications that capture monetary value through the effective production of personal implants and medical tools of high quality
36	Medimsight UK 3, 9 AI, cloud computing, natural language processing	The company offers an AI-based medical imaging cloud marketplace, operating through Google Cloud infrastructure	By using cloud computing, the platform allows unlimited data storage and analysis. NLP methods are used to integrate manual measurements with the platform, alongside normative and longitudinal data. The preliminary AI analysis performs a quality check and determines the best image analysis among the offers on the cloud marketplace. All of the options on the marketplace utilise AI to identify biomarkers and analyse imaging data	The marketplace allows medical experts and researchers to discover, store and analyse all of their data analysis needs in one place, with a single account. The company enables unlimited storage and free (for research only) PACS service running over the Google Cloud infrastructure for all of their data. The system is built on a pay-per-use model that allows quick analysis without large expenditures

37	MicroMek Malawi 1, 3, 11 Drones	It is a drone startup producing unmanned aerial vehicles (UAVs) to deliver healthcare services to communities in Malawi and across Africa	EcoSoar, a low-cost fabrication drone, is used in delivery services for remote medicine, dried blood samples, vaccines and other critical medicines from clinics in the UNICEF Kasungu drone corridor in Malawi to the Kasungu hospital as well as for environmental sensing, to serve the healthcare and environmental monitoring communities in Africa	The utilisation of drones is used to capture value. To assist hospitals and aid organisations in addressing healthcare difficulties impacting children in Malawi's hard-to-reach remote areas, the company decreases the time for delivery of diagnostics, vaccines and medicines by providing a lower-cost delivery option and enhancing delivery reliability. Furthermore, their drone can be utilised for environmental monitoring, such as on-demand evaluation of flooded regions, agricultural area assessment and hydrogeological mapping, urban mapping to analyse people for health clinic location planning and disaster response with pre-planning. Learned and shared with others is a clear appraisal and analysis of the data obtained
38	NextGen USA 3 Datahubs	The company offers a datahub platform that exchanges, normalises and aggregates data at large scales to provide a longitudinal picture of patient records	NextGen® Health Data Hub works by providing e-prescription, links to major labs, room status dashboard consolidated patient data, billing and reporting. The platform meets the needs of health information exchanges (HIEs), hospitals and big outpatient clinics by providing full and continuous access to aggregated patient health data	All clinical data types used by inpatient and ambulatory health systems are supported by HDH. Its extensive data normalisation features ensure that all connected systems can successfully interoperate and create value

(continued)

Table 5.1 (continued)

No	Company info	Value proposal (what?)	Value creation (how?)	Value capture
39	Operoo Australia 3, 4 Big data	The company offers a big-data-based platform that automatically keeps medical and emergency contact information up-to-date for schools and institutions	With Operoo, parents complete a comprehensive digital health form only once, which can be quickly accessed and updated whenever needed. It's the one form continuously verified and updated throughout the child's tenure at the school. Parents use the camera on their devices to easily upload action plans for the school. Anaphylaxis, asthma, epilepsy, diabetes or any other action plan can be attached to students' medical data and viewed by staff in an emergency	It supports thousands of schools and various establishments around the world in eliminating time-consuming, costly and repetitive paper-based processes. It creates value by enabling schools to save time and resources by automating operational processes using digital workflows, as well as increasing staff efficiency, parental involvement and student engagement
40	Optellum UK 3 AI	The company offers AI-based early lung cancer diagnostics and decision support software	Relevant patient data are combined in the platform and analysed by AI by taking into account both present and prior CT scans. A prediction score is obtained by this analysis, helping clinicians make informed decisions and diagnose cancer	Social value is captured by diagnosing cancer diseases faster through the platform. It ensures the efficient use of invaluable time and health resources available
41	Paige USA 3 AI, cloud computing	The company offers an AI-based digital pathology diagnostics software service delivered via an imaging platform	It utilises disease-specific modules by using visual and clinical data with DL algorithms, recurrent neural networks, generative models, cloud computing and web services. They also utilise ML and DL algorithms on petabyte-scale datasets to discover new insights in pathology images and transform the pathologists' workflow	Their AI-based solutions help pathologists identify, quantify and characterise pathological diseases and make precise diagnoses more efficiently
42	Philips The Netherlands 3 Healthcare analytics	The company provides predictive maintenance and aims for zero unplanned equipment downtime with predictive analytics	Machine and service data is continually evaluated using data analytics algorithms to find patterns and trends. These signals are predictive, allowing a service action to be planned ahead of time without disrupting routine clinical workflow	Reduced downtime of healthcare machinery results in uninterrupted service to patients, which adds value

43	Prellis USA 3 3D printing	The company offers a high-resolution laser projection technology to print ultra-thin tissue scaffolds from biocompatible materials. The 3D printing technology is used to make it possible to recreate a physiologically compatible organ	In the development of therapeutics, living tissues made in 3D that replicate biological function can be exploited. Prellis has overcome the problem of related speed and resolution by printing up to 2.5 million voxels (points of light) per second by utilising holographic image projection to create ultrafine biomimetic tissue scaffolds	It offers solutions to organ needs by developing synthetic tissue products for R&D, therapeutic production and organ transplantation, using a holographic printing system that can match and accurately replicate human organ and tissue structures
44	Prognos Health USA 3 AI, healthcare analytics, natural language processing	The company offers an analytics platform that enables purchasing and analysis of vast amounts of health records from anonymised patients	Making use of ML techniques and NLP methods, the AI-powered platform prognosFACTOR is used to aggregate, normalise and provide clinical interpretations of multi-sourced clinical diagnostics data derived from the integrated database, the Prognost Registry	Value is captured by categorising and analysing patient data derived from a multi-sourced database. The availability of processed healthcare data facilitates research and increases efficiency
45	Prokarium UK 3 Biotech and biomanufacturing	Biotech startup focusing on researching and developing genetically engineered microbial treatments for cancer patients and vaccines	Through proprietary genetic engineering, biotech and interventions, the natural ability of our bacterial strains to seek out and colonise solid tumours is enhanced. It allows the development of attenuated bacterial strains that are capable of targeting tumours without causing pathology in normal tissues. It also delivers specific immunostimulatory cargo aimed at activating the patients' immune systems to destroy tumours	Value is captured through the development of microbial immunotherapy for solid tumours and vaccines against infectious diseases, with some in collaboration with the UK government

(continued)

Table 5.1 (continued)

No	Company info	Value proposal (what?)	Value creation (how?)	Value capture
46	Proximie UK 3 5G, AI, spatial computing	The company offers a technology platform that uses a combination of AI, 5G and AR to allow for virtual and global collaboration of clinicians	Proximie uses AI, AR and 5G technology to allow multiple people in different locations to interact virtually, simulating what they would experience if they were in the same operating room. They can show one another where to make an incision in real time, or they can explain a method using bodily gestures	Through a multi-technology-based approach, the platform is designed to address connectivity problems in healthcare. They improve healthcare and make it more efficient, effective and most importantly accessible, thus creating value
47	Ras Labs USA 3 Advanced materials, AI, biotech and biomanufacturing, robotics	Human-precision electroactive polymers (EAPs) that mimic biological properties and behaviour, equipped with AI and ML-based nanotechnological sensor systems	FingerTip™ senses pressure from small touches to high-pressure touches. It generates immediate feedback at the first point of contact. The contact point is gentle, yet firm, as the pads in the contact area provide a soft, compliant interface. The sensors here also make it possible to detect and prevent slipping by sensing the change in pressure position on its surface, adjusting grip strength without stuttering due to both feedback and pad compatibility	This project can be used in products capable of lifelike tactile feedback and biomimetic action and withstand harsh environments such as extreme temperatures, pressures, radiation and incurable infectious areas. Value is created by potentially optimising prosthetic hands
48	RS Research Turkey 3 Biotech and biomanufacturing	It is a clinical-stage biotechnology enterprise that develops new nanomedicine to be used in the treatment of cancer patients	The Sagitta is a platform that targets tumours with a high cytotoxic burden using polymer-drug conjugates. The nano-drug ingested into the cell releases the active ingredient in such a way that it exhibits its full impact only in the tumour, thanks to the targeting module that identifies receptors on the tumour surface	Value is captured by helping maintain the life quality of cancer patients going under treatment, by ensuring that the active substances of medicines are more effective

49	Sira Medical USA 3, 4 Spatial computing	The company offers an AR-based software system that provides clinicians with patient-specific high-fidelity 3D holograms to assist in pre-surgical planning	Thanks to the AR software app "RadHA" (Radiology with Holographic Augmentation), doctors can study 3D radiological scans by superimposing them onto a real-world context	By generating anatomically correct patient-specific holograms, the company solves the barrier of unexpected anatomy, which accounts for 25% of all surgical errors, allowing physicians to reduce patient complications and increase surgical efficiency. Using this software, surgeons can better explain operations to patients while also highlighting key locations to avoid during surgery. The company also offers its solutions to educational institutions, to be used in the training of future surgeons, medical students and radiologic technicians
50	Solve.Care Ukraine 3 Blockchain, healthcare analytics	The company offers a patient-centric and decentralised blockchain platform that coordinates benefits, care and payments between all parties in the chain of healthcare	Based on blockchain, it provides a digital currency called SOLVE that is used in transactions on the "Solve Care Platform" for healthcare. The token supply is fixed and the price is variable, determined by market supply and demand. SOLVE runs natively on the ethereum blockchain and is designed to follow pre-set standards. Also, the platform provides another service called "Care Wallet" that enables patients and physician to communicate	Through the use of the platform and cryptocurrency, health transactions and spending in the field of health become more transparent, traceable and secure for every stakeholder of the industry. Thanks to the platform, the examination becomes more flexible for doctors and patients, and patients can meet with expert doctors around the world

(continued)

Table 5.1 (continued)

No	Company info	Value proposal (what?)	Value creation (how?)	Value capture
51	Tempus USA 3 AI, healthcare analytics	The company offers an AI-powered healthcare analytics platform designed to generate, collect, structure and analyse vast amounts of data to advance precision medicine processes	The explainable AI platform eliminates noise and focuses analysis on the data points most likely to answer the problem, giving scientists more time to focus on areas where their clinical judgement is critical. Running on an AI knowledge graph that is updated with information from books and databases, accurate and validated ML models locate causative variations with proof	With collected and analysed oncology datasets, the company accelerates cancer research. Extra value is captured by evaluating clinical and molecular data points, connecting clinicians with up-to-date treatment choices and providing useful information for patients based on their molecular profiles and advanced analytics by using AI algorithms
52	Univfy USA 3 AI	The company offers an AI-powered platform that provides patients with scientifically proven, personalised information about their chances of having a baby through in vitro fertilisation (IVF)	Using AI algorithms, specific prognostics are provided for each patient, showing their probability of having a baby with three IVF stages. The holistic health profile of the woman or couple is used to calculate the probability, including their age, BMI, reproductive history, ovarian reserve test results and the partner's semen analysis. Clinical IVF outcome data from the patient's treatment centre are used to validate the predictions	By using AI, it makes fertility expenditures and outcomes more reasonable and predictable for women and couples who are considering having a baby, which enables families who want to have a baby to make more rational decisions
53	Unlearn USA 3 Digital twins	The company provides a digital twin platform that generates virtual synthetic control arms made up of participants from digital twins	PROCOVA (prognostic covariate adjustment) is an adaptor that incorporates digital twins into statistical analysis plans to offer a more precise assessment of the treatment effect. Each patient in the trial has a digital twin, an AI-generated placebo result prediction. Digital twins maintain randomisation and blinding while boosting certainty without introducing bias	Value is captured through reduced sample sizes, increased power and confidence for clinical trials, made possible by using digital twins. Additionally, the hybrid trial design of the platform requires fewer human participants, resulting in accelerated, smaller and more efficient trials

54	Verge Genomics USA 3 AI	The company offers an AI-based drug discovery platform specialised in helping patients suffering from neurodegenerative diseases	The platform uses AI to identify patient genomes, gene expression and epigenomics, which enables determining novel therapeutic goals and making accurate predictions about the effectiveness of the medication	Value is generated by reducing the time and resources required to develop new drugs and their approval processes, made possible through utilising AI
55	XRHealth USA 3 Spatial computing	The company offers a virtual clinic using telehealth to deliver medical and therapeutic treatments	Services about the sessions are provided, carried out through the VR telehealth kit that consists of a VR headset and two hand controllers, created by the company and sent to its customers by meeting online with one of the therapists within the company. Therapies are directed and provided via FDA-registered medical VR apps	Monetary value is streamed by the therapies provided to the patients with XR applications related to many health problems such as physical therapy; occupational therapy; behavioural health management; pain management; Parkinson's, ADHD, memory and cognitive education; post-COVID-19 rehabilitation; and support groups. It also contributes to the accessibility of healthcare by providing access to therapies online

References

G. Alleyne, R. Beaglehole, R. Bonita, Quantifying targets for the SDG health goal. Lancet **385**(9964), 208–209 (2015)

F. Bianchi, G. Bianchi, D. Song, *The Long-Term Impact of the COVID-19 Unemployment Shock on Life Expectancy and Mortality Rates* (National Bureau of Economic Research, Cambridge, MA, 2020)

T. Cernev, R. Fenner, The importance of achieving foundational sustainable development goals in reducing global risk. Futures **115**, 11 (2019)

B.N. Chatmon, Males and mental health stigma. Am. J. Mens Health **9**(2), 14 (2020)

W.L. Filho et al., *Good Health and Well Being*, 1st edn. (Springer, Zug, 2019)

J. Frenk, D. de Ferranti, Universal health coverage: Good health, good economics. Lancet **380**(9845), 862–864 (2012)

J.-F. Guégan et al., Sustainable development goal #3, "Health and well-being", and the need for more integrative thinking. Veterinaria México OA **5**(2) (2018). https://doi.org/10.21753/vmoa.5.2.443

C. Henderson, E. Robinson, S. Evans-Lacko, G. Thornicroft, Relationships between anti-stigma Programme awareness, disclosure comfort and intended help-seeking regarding a mental health problem. Br. J. Psychiatry **211**(5), 316–322 (2018)

A. Kuloglu, E. Ecevit, The Relationship between health development index and Financial development index: Evidence from High income Countries. J. Res. Econ. Polit. Finance **2**(2), 83–95 (2017)

J.M. McCullough, J.P. Leider, Associations between county wealth, health and social services spending, and health outcomes. Am. J. Prev. Med. **53**(5), 592–598 (2017)

A. Papa, M. Mital, P. Pisano, M. Del Giudice, E-health and Well-being monitoring using smart healthcare devices: An empirical investigation. Technol. Forecast. Soc. Chang. **153**, 1–10 (2018)

T. Savelyeva, S.W. Lee, H. Banack, *SDG3 – Good Health and Well-being: Re-calibrating the SDG Agenda*, 1st edn. (Emerald Publishing Limited, Bingley, 2019)

SDG Compass, *Ensure Healthy Lives and Promote Well-Being for All at All Ages* [Online] (2021), Available at: https://sdgcompass.org/sdgs/sdg-3/. Accessed 11 Aug 2021

G. Seidman, Does SDG 3 have an adequate theory of change for improving health systems performance. J. Glob. Health **7**(1), 010302 (2017)

V. Sharma-Brymer, E. Brymer, Flourishing and eudaimonic well-being, in *Good Health and Well-Being*, ed. by W. Leal Filho et al., (Springer International Publishing, Cham, 2020), pp. 205–214

J.P. Souza, A.M. Gülmezoglu, J. Vogel, G. Carroli, P. Lumbiganon, Z. Qureshi, M.J. Costa, B. Fawole, Y. Mugerwa, I. Nafiou, I. Neves, J.J. Wolomby-Molondo, H.T. Bang, K. Cheang, K. Chuyun, K. Jayaratne, C.A. Jayathilaka, S.B. Mazhar, R. Mori, M.L. Mustafa, L.R. Pathak, D. Perera, T. Rathavy, Z. Recidoro, M. Roy, P. Ruyan, N. Shrestha, S. Taneepanichsku, N.V. Tien, T. Ganchimeg, M. Wehbe, B. Yadamsuren, W. Yan, K. Yunis, V. Bataglia, J.G. Cecatti, B. Hernandez-Prado, J.M. Nardin, A. Narváez, E. Ortiz-Panozo, R. Pérez-Cuevas, E. Valladares, N. Zavaleta, A. Armson, C. Crowther, C. Hogue, G. Lindmark, S. Mittal, R. Pattinson, M.E. Stanton, L. Campodonico, C. Cuesta, D. Giordano, N. Intarut, M. Laopaiboon, R. Bahl, J. Martines, M. Mathai, M. Merialdi, L. Say, Moving beyond essential interventions for reduction of maternal mortality (The who multicountry survey on maternal and newborn health): A cross-sectional study. Lancet **381**, 1747–1755 (2013)

A.L. Stangl et al., The health stigma and discrimination framework: A global, crosscutting framework to inform research, intervention development, and policy on health-related stigmas. BMC Med. **17**(1) (2019)

E.P. Terlizzi, B. Zablotsky, *Mental Health Treatment Among Adults: United States, 2019* (National Center for Health Statistics, Hyattsville, 2020)

G. Thornicroft, Stigma and discrimination limit access to mental health care. Epidemiol. Psychiatr. Sci. **17**(1), 14–19 (2011)

UN Office for Outer Space Affairs, *Sustainable Development Goal 3: Good Health and Well Being* [Online] (2021), Available at: https://www.unoosa.org/oosa/en/ourwork/space4sdgs/sdg3.html. Accessed 12 Aug 2021

UNDP, *The SDGs In Action* [Online] (2021), Available at: https://www.undp.org/sustainable-development-goals#good-health. Accessed 11 Aug 2021

United Nations, *Goal 3: Ensure Healthy Lives and Promote Well-Being for All at All Ages* [Online] (2021a), Available at: https://unric.org/en/sdg-3/. Accessed 1 Aug 2021

United Nations, *The Global Goals For Sustainable Development* [Online] (2021b), Available at: https://www.globalgoals.org/3-good-health-and-well-being. Accessed 17 Aug 2021

United Nations, *United Nations SDG 3: Good Health and Well-Being* [Online] (2021c), Available at: https://in.one.un.org/page/sustainable-development-goals/sdg-3-2/. Accessed 2 Aug 2021

United Nations, *United Nations Sustainable Development Goals* [Online] (2021d), Available at: https://www.un.org/sustainabledevelopment/wp-content/uploads/2017/03/3_Why-It-Matters-2020.pdf. Accessed 31 July 2021

T.V. Vu, Economic complexity and health outcomes: A global perspective. Soc. Sci. Med. **265**, 1–32 (2020)

World Health Organization, *Mental Health ATLAS 2017* (WHO, Geneva, 2018)

World Health Organization, *Health Determinants* [Online] (2021a), Available at: https://www.euro.who.int/en/health-topics/health-determinants/pages/health-determinants. Accessed 7 Aug 2021

World Health Organization, *World Health Statistics 2021* (World Health Organization, Geneva, 2021b)

SDG-4 Quality Education

6

Abstract

Education is a component of sustainable development with its strong effects at global, regional and local levels. The biggest challenge the world faces in this context is the preservation and continuous improvement of the effort put forward to provide sustainable education in studies on education. The lack of chances for learning stymies social, economic and sustainable development and long-term stability and peace. This chapter presents the business models of 49 companies and use cases that employ emerging technologies and create value in SDG-4, Quality Education. We should highlight that one use case can be related to more than one SDG and it can make use of multiple emerging technologies.

Keywords

Sustainable development goals · Business models · Quality education · Sustainability

Education is a component of sustainable development with its strong effects at global, regional and local levels. The biggest challenge the world faces in this context is the preservation and con-

tinuous improvement of the effort put forward to provide sustainable education in studies on education (Franco et al. 2020). The lack of chances for learning (also education) stymies social, economic and sustainable development and long-term stability and peace. Learning is especially essential for individuals who have been banned from formal schooling or who have not achieved basic skills and education. Learning is required to accomplish the 2030 Agenda for Sustainable Development, titled Transforming Our World (UN 2015), including 17 sustainable development goals (SDGs) and 169 related targets. The goal of providing opportunities for lifelong learning for everyone emphasises the global education agenda's comprehensive character and its importance for achieving all SDGs by 2030. To provide comprehensive learning opportunities and systems, this integrated approach supports the concept that bridges must be built among and amongst actors, institutions, processes, learning places and times (Hanemann 2019).

The main purpose of the fourth SDG, under the title of "Quality Education", put forward by the United Nations, is to encourage the principles and practices of sustainable development to create societies with exceptional opportunities in all fields of education (Franco et al. 2020). Today, almost 262 million children and adolescents are out of school. Sixty percent of school goers do not acquire basic numeracy and literacy skills in their first few school years. Seven hundred fifty million adults in the world are illiterate, which adversely affects the welfare of societies and reveals marginalisation (UNESCO 2021).

The author would like to acknowledge the help and contributions of İlayda Zeynep Mert, Ömer Sami Temel, Abdullah Aykut Kılıç, Abdullah Enes Ögel, Veysel Ömer Yıldız and Enes Ürkmez in completing this chapter.

S. Küfeoğlu, *Emerging Technologies*, Sustainable Development Goals Series,
https://doi.org/10.1007/978-3-031-07127-0_6

Despite this situation, the enrolment rate in regions that continue to develop rose to 91% in 2015 because of active work carried out since 2000. As a result of these efforts, the number of children who are out of school has nearly halved. In addition, the significant increase in girls' school enrolment and literacy rates are also among the remarkable achievements (United Nations Development Programme 2021). At the micro-level of society, the impact of epidemics of acute infectious disease on people, families and communities may be enormous. Children may lose their chance of going to school due to consequences or demands at home in the event of a large epidemic, at least until they are older (Kekić and Miladinovic 2013). These adverse outcomes of epidemics can easily be caused by pandemics as well. For example, the COVID-19 outbreak caused widespread school cancellations in 188 countries, affecting almost 1.5 billion children and adolescents. Only 30% of low-income countries have built a national distance learning platform. Nevertheless, over 65% of countries have done so. Almost 33% of young people in the world were already digitally excluded before the crisis. Also, girls have less access to digital technology than boys, restricting their online learning opportunities. It is especially challenging to reach children with mental or physical disabilities through online education programs. Distance education quality and accessibility will vary considerably within and between nations. Only 15 countries worldwide offer distance education in several languages (United Nations 2020). As a result, education plays a massive role in bringing societies to a certain level of resilience. All kinds of education are essential in generating sustainable development and in environmental problems, employment problems and industrial operation (UN Environment Programme 2021).

The objectives of SDG-4 concerning the problems as mentioned above are to present equality of opportunity based on literacy, numeracy and broader learning competencies, which are the most basic learning levels of education in general, from kindergarten, or nursery, to vocational schools and university (Unterhalter 2019). Expanding possibilities throughout all levels of education (preschool, primary, secondary, vocational, higher and adult education) is one of SDG-4's goals. The goals expand the definition of education as a worldwide enterprise to include objectives in reading, numeracy and other areas such as global citizenship, sustainability and gender equality (Unterhalter 2019). Figure 6.1 demonstrates the targets set by the UN in the field of quality education under the name SDG-4 (United Nations 2021a, b, c).

"Quality education", one of 17 different development goals, emphasises an egalitarian, inclusive, quality and lifelong education content. Achieving the goals set in the scope of SDG-4 is also of great importance in terms of achieving other sustainable development goals. Along with literacy and access to primary education, higher educational institutions are considered to be highly influential in achieving sustainable development, with a social responsibility to bring forth a setting that cultivates sustainable development amidst their students and communities (Ferguson and Roofe 2020). In addition to SDG targets, trade activities in countries are directly related to education. The lack of educational opportunities in a particular region, that is, the lack of professional and personal skills of the people living in that region, has a significant impact on the creation of new business areas in the region and the disruption of various entrepreneurial and investment activities. Investing in people is of great importance for faster economic developments (Cervelló-Royo et al., 2020).

Although primary school attendance in developing nations has reached 91%, 57 million children are excluded from school. Many of the other SDGs can only be achieved through a good education. If people can get a good education, they can break the cycle of poverty (United Nations 2021a, b, c; Patel20 2019). Due to high poverty levels, armed conflict and other emergencies, progress has also been hampered in developing regions. The number of youngsters out of school has risen due to the continuous violent situations in West Asia and North Africa. Although Sub-Saharan Africa has accomplished the most improvement of any developing region regarding primary school enrolment, substantial inequities

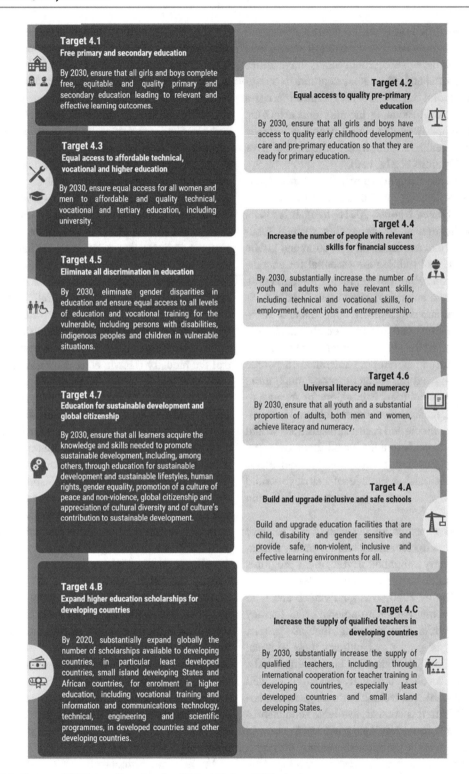

Fig. 6.1 Targets of SDG-4 quality education (United Nations 2021a, b, c)

still exist. Children from the poorest homes are four times more likely to drop out of school than those from the wealthiest households. Inequalities between rural and urban areas continue to be significant (Joint Sdg Fund 2021). Education has a critical role in reducing inequity and achieving gender equality. It also allows people worldwide to lead healthier and more sustainable lives (United Nations 2021a, b, c).

Education also plays a role in fostering intercultural tolerance, promoting a more peaceful society (United Nations 2021a, b, c). It is also a potent instrument for enhancing societal resilience. Formal and informal education and public awareness and training are essential for encouraging sustainable development, strengthening people's and countries' capacity to handle environmental and development concerns and establishing green and decent employment and industries (UNEP 2021). Education plays a significant part in developing tolerance in people interactions and the development of much more friendly communities. Fair access for females to education, medical care, decent jobs and involvement in economic and political institutions would improve humanity and the world economy's sustainability. Funding in educational initiatives for females and raising the age of marriage would provide a fivefold return on investment (Koßmann 2019).

SDG-4 aims for all boys and girls to have equal access to elementary and secondary education and early childhood development programs and accessible university education for both men and women by 2030. This goal's main aim is to increase young people's numeracy and literacy abilities while also ensuring that all people, regardless of gender or handicap, have an equal chance (Joint SDG Fund 2021). Simultaneously, increased access to university education, as well as vocational and technical training, is emphasised. Within this context, available scholarships for students from developing nations to enrol in higher education, vocational training programs and other science programs in developed or developing countries are gradually increasing (Patel20 2019).

SDG-4's main goal, which is to ensure that everyone, regardless of their race, gender, age or other characteristics, has access to inclusive and equal quality education, is ambitious and challenging to achieve. The way knowledge is passed down is presumed to change dramatically due to technological advancements, with a big move towards online platforms.

As an alternative to conventional methods of education, online education can be used to address specific challenges of SDG-4. Projections are ambiguous regarding the mix of online materials available to students in the future. Existing patterns indicate that a lot more online educational information is accessible, but it appears that considerably less of it would be used successfully by students. The ratio of rationales and ideologies between public and private content will continue to shift, but it appears that a few international content producers will start to control the industry (Unwin et al. 2017). A growing body of research aims to understand and explain the aspect of gender in online learning (Latchem 2014). Some suggest that online education methods are non-sexist and more gender-inclusive (Margolis and Fisher 2002). In contrast, others report that it does not solve pre-existing problems of traditional methods (Anderson 2004). Nevertheless, there is consensus that online platforms may offer more accessible knowledge, free exchange of information, networks and learning communities without regard to gender (Latchem 2014). Despite offering promising solutions, online education systems are not perfect. Literature suggests that, in developing regions and countries, women face the same challenges regardless of the educational platform, e.g. online vs traditional (Glen and Cédric 2003). It has been suggested that providing women with training and support in creating content that is appropriate to their needs and addresses their particular viewpoints, experiences and concerns would greatly help prevent their absence in online educational platforms (Latchem 2014).

Furthermore, virtual reality (VR) and different methods may also be used in the classroom. This would permit students to learn how to negotiate difficulties and communicate ideas online using new platforms. Forecasts regarding the future of education and the use of VR suggest that as gaming technologies are being created for the classrooms, augmented reality (AR) and VR are likewise expected to become much more common (Unwin et al. 2017). Campuses, as we know them today, may cease to exist. This would free learning from the confines of a physical school. A new campus would likely consist of mobile classrooms and a real-world setting. On the other hand, city libraries and laboratories would coexist to assist students in completing their assignments. Games that teach youngsters how to code, toys that teach robotics and various apps that help teachers quickly deliver knowledge to children are highly likely to become commonplace. The use of technology in education is expected to increase exponentially, aiding teaching and learning processes, thus evolving learning into being more creative and practical as time goes on. Conventional methods of performance and learning evaluations, such as tests, will likely be replaced by evaluations of students' critical thinking and problem-solving abilities through their performance in creative projects (Nerdy Mates 2021).

Forecasts indicate that by 2025, the use of information and communication technologies (ICTs) in schools will be substantially more diversified. This makes predicting how it will be utilised in any given situation exceedingly challenging. Similarly, there will be some imaginative and unique situations in exceedingly disadvantaged contexts, where well-trained, incredibly inspiring educators will use ICTs to encourage kids to critically discover a wealth of information and thoughts, allowing them to build the abilities and understanding required to change the world in which they live. Furthermore, many governments' educational systems will probably change. Many of these systems will expressly urge wider use of ICT in schools, driven in part by the interests of big multinational businesses and by a growing understanding of the impact advantages such technologies may provide. Educators will continue to play a critical role in education systems that schools still control. In

the finest systems, even so, their function will have shifted from that of knowledge suppliers to that of mentors, assisting youngsters in learning to navigate the universe of digital data. This is especially important when working with disadvantaged children who may lack the parental and community support needed to organise and socialise education (Unwin et al. 2017).

6.1 Companies and Use Cases

Table 6.1 presents the business models of 49 companies and use cases that employ emerging technologies and create value in SDG-4. We should highlight that one use case can be related to more than one SDG and it can make use of multiple emerging technologies. In the left column, we present the company name, the origin country, related SDGs and emerging technologies that are included. The companies and use cases are listed alphabetically.[1]

---

[1]For reference, you may click on the hyperlinks on the company names or follow the websites here (Accessed Online – 2.1.2022):

http://www.solarpak.net/; https://alchemyimmersive.com/; https://bluecanoelearning.com/; https://bridge-u.com/; https://campuslogic.com/; https://coachhub.io/en/; https://codecombat.com/; https://cubomania.io/; https://delphia.com/; https://edutekno.com.tr/; https://elevateu.ai/; https://elsaspeak.com/en/; https://en.duolingo.com/; https://eonreality.com/; https://gethownow.com/; https://inurture.co.in/; https://learnwithhomer.com; https://locorobo.co/index.html; https://odem.cloud/; https://photomath.com/en/; https://riiid.com/en/main; https://roybi-robot.com/; https://scanmarker.com/; https://shop.robolink.com/; https://tab.gladly.io/; https://tinalp.com/; https://wondertree.co/; https://www.12twenty.com/; https://www.applyboard.com/; https://www.arduino.cc/; https://www.aurum3d.com/; https://www.avidbots.com/; https://www.betterup.com/; https://www.brainscape.com/; https://www.brightbytes.net; https://www.century.tech/; https://www.civitaslearning.com/; https://www.coursera.org/; https://www.disciplina.io/; https://www.grammarly.com/; https://www.immerse.online/; https://www.mereka.my/; https://www.odilo.us/; https://www.ossovr.com/; https://www.packback.co/; https://www.talespin.com/; https://www.transfrvr.com/; https://www.verizon.com/; https://ziotag.com/

Table 6.1 Companies and use cases in SDG-4

No	Company info	Value proposal (what?)	Value creation (how?)	Value capture
1	12twenty USA 4 Big data	The company creates a big-data-focused platform for university career centres	By increasing student survey response rates, the platform improves the data available to colleges. Career counsellors can use this information to produce industry or bespoke reports, analyse student employment and avoid the time-consuming task of creating spreadsheets. 12twenty also gives students access to wage data and an interview question database, so they can leverage big data technology after graduation to get the information they can use beyond graduation	The company is changing the way university career centres, students and employers approach career recruitment. The platform facilitates recruitment processes for everyone; students can use it to achieve a comprehensive understanding of recruitment. It captures value by providing more effective management of employment processes for career counsellors and facilitating the hiring process for employers by helping them find candidates
2	Alchemy Immersive UK 4 Spatial computing	The company offers a VR-powered experimental educational platform	By using the user's smartphone camera to project animations into living rooms using VR technology, the Museum Alive app allows users to place extinct species and natural environments in the actual world. Other learning tools, such as voice-overs, soundscapes, maps, textual information and pre-recorded videos, are mixed with this	Integrating VR experiences with lesson plans has immense benefits for learners who might not be able to learn solely from a textbook. Through the immersive content created, Alchemy VR generates value by making the learning process more enjoyable and captivating
3	ApplyBoard Canada 4 AI, natural language processing	The company offers a portal that connects foreign students and recruiting partners with educational opportunities at universities around the world	ApplyBoard uses NLP to process and AI to analyse students' data to match them with the course, university, scholarship opportunities and job vacancies that best meet their background and interests	Students can apply to multiple schools and programs at the same time and save time by simply creating a profile using the portal. The platform generates value by facilitating the advisory and assisting processes regarding scholarships, visa procedures and more, by connecting students and experts. Students can apply to more than a thousand educational institutions worldwide using this portal

(continued)

Table 6.1 (continued)

No	Company info	Value proposal (what?)	Value creation (how?)	Value capture
4	Arduino Italy 4 3D printing, flexible electronics and wearables, Internet of Things	The company offers hardware- and software-based open-source electronics platforms where users can program or design robots and systems like IoT applications, wearable electronic equipment and 3D printing systems that interact with their environment	Arduino boards can take inputs – such as light from a sensor, a finger on a button or a Twitter message – and convert them to outputs, such as turning on an LED, triggering a motor or publishing anything online. By providing a set of instructions to the board's microcontroller, users may tell it what to do. The Arduino programming language (based on wiring) and the Arduino software (IDE) (based on processing) are used for accomplishing this	It generates value by facilitating global access to STEM education regardless of age and making education easier. It makes performing computational work, a costly and inefficient process, more efficient and cheaper for everyone using electronics boards
5	Aurum3D India 4 3D printing	It is a company that offers tailored 3D printing solutions for any level of education	The company provides personalised 3D printing solutions based on the specific demands of the model, prototype or product by allowing students/educators to choose from a variety of 3D printing services based on their specific requirements	Compared to the old-fashioned cardboard-based models, 3D-printed physical models allow different students to understand engineering and complex concepts details with ease. They also allow teachers and students to study and discuss their different viewpoints, aspects and various educational concepts. Moreover, they offer high-quality real-life visuals and give a better perspective for different age groups. Value is captured by helping the stakeholders in the educational system
6	Avidbots Canada 4, 10 AI, autonomous vehicles, Internet of Things, robotics	The company provides industrial-grade autonomous floor scrubbing robots used in various locations such as schools	Neo, the self-driving floor scrubbing robot, scans the designated area through its equipped sensors. This data is analysed through the Avidbots AI platform to create the most efficient cleaning plan, which is then carried out by Neo	The product provides measurable cleaning reports and boosts cleaning frequency. The Neo platform generates value by creating the most optimal plans for cleaning and disinfection, increasing efficacy and efficiency. The automatisation of these processes maximises the cleaning productivity while minimising human intervention

(continued)

Table 6.1 (continued)

No	Company info	Value proposal (what?)	Value creation (how?)	Value capture
7	BetterUp USA 4 AI, Internet of Behaviours	The company provides an AI-based platform that offers personalised coaching	After a personal analysis test is carried out, AI interprets the test result and recommends three coaches in line with the demands. After the coaches are determined, personal programs are prepared regarding the analyses	Value is created by maximising development through increasing individual resilience, adaptability and effectiveness in personal development. For business, it increases productivity by making employees more productive, thus maximising profits
8	Blue Canoe USA 4 AI	The company offers an ML-based English language learning platform	It creates entertaining smartphone games that effectively help non-native English speakers learn and practice to acquire clarity and confidence when speaking English, using a proven brain-science methodology and scaling it with speech recognition and machine learning	Through AI, the platform matches words that may evoke each other to facilitate memorising and learning processes. Combined with voice recognition, value is created by enabling users to practice conversational language skills. The convenient platform provided through an app increases access to quality language education
9	Brainscape USA 4 AI	The company offers a digital education platform that increases learning speed by using smart flashcards	It classifies online study cards according to the most efficient working time with the help of AI. It specifies which topic will be studied when and at what intervals	Value is captured by providing an online working environment, preventing the loss of study notes and ensuring the permanence of the learned information. Also, it facilitates planning and tracking of the learning process, thanks to flashcards created by the learner
10	BridgeU UK 4 AI	The company offers an AI-powered educational platform designed to provide career counselling and help students prepare and submit university applications	Prospective students must fill out the information about their interests, hobbies, grades, career and learning experiences. The BridgeU will then match their profiles with colleges or courses that best suit them using ML	Through the delivery of AI-enabled assistance regarding university and career choices, the platform captures value by helping young people fulfil their potential on a global scale
11	BrightBytes USA 4 AI, big data	The company offers a data analytics and infrastructure platform that aims to enhance student learning outcomes for all students	The ML-based platform creates predictive models and dashboards by analysing students' historical data to identify ideal strategies for students. Also, according to the results, coaches are assigned to the person for support purposes	Value is captured by generating relevant data and providing proper suggestions on what the data means and how to move on to more successful integration. Therefore, it facilitates developmental stages for people

(continued)

Table 6.1 (continued)

No	Company info	Value proposal (what?)	Value creation (how?)	Value capture
12	CampusLogic USA 4 AI, crowdfunding	It is an enterprise that creates AI tools and resources for college administrators to assist students in managing scholarships, loans and other financial aid	The platform digitises and simplifies financial transactions with tools such as a net price calculator, complete scholarship management (including crowdfunded scholarships), personalised digital communications, simplified financial aid verification, 24×7 personalised virtual advising, tuition and scholarship crowdfunding and integrated data visualisations	CampusLogic makes forms easier to fill out and allows them to be submitted securely from any device. It also provides the consolidation of aid applications or scholarships into a single system. It creates value by making the process easier and more efficient for students and colleges. In addition, it offers financial assistance to students
13	CENTURY Tech UK 4 AI	The company offers an education platform that uses learning science and AI to provide adaptive pathways for students and assessment data for teachers for various levels of education	Century's learning platform tracks the learning process of pupils and analyses their data through AI to provide recommendations on learning pathways depending on the pupils' strengths and areas of improvement	Value is generated by making the learning process easier and faster as it offers students a personalised and adaptable learning path
14	Civitas Learning USA 4 AI, big data	The company offers an intelligence platform that provides academic and career planning, student support, efficacy measurement and data-informed advisors	The company's connected infrastructure and tools use big data technologies to coordinate student accomplishment strategies in higher education, give proactive and collaborative care, provide holistic guidance and swiftly measure what works for whom	The platform generates value by assisting users in improving enrolment, retention and graduation rates, as well as empowering leaders, mentors, faculty and students
15	CoachHub Germany 4, 8 AI	The company offers a talent development platform that uses AI to match people to business coaches	Through AI-based matching algorithms, individuals and employees are automatically matched by finding the right coaches on the platform	Regular participation in AI-matched coaching sessions can result in many benefits for organisations including increased employee engagement, higher productivity, enhanced job performance and higher retention. Value is captured by the economic growth, achieved by allocating more productive employees to companies

(continued)

Table 6.1 (continued)

No	Company info	Value proposal (what?)	Value creation (how?)	Value capture
16	CodeCombat USA 4 AI	It is a programming game for learning how to code in a multiplayer setting	The CodeCombat AI League is a competitive AI battle simulator and game engine that blends project-based, standards-aligned curriculum, a compelling adventure-based coding game and an annual AI coding worldwide tournament into a structured academic competition. While playing, students learn coding and computer science, which they subsequently apply in arena combat as they practice and play on the same platform	Value is captured by bringing more users into the field of computer programming by making the logic and syntax more accessible and enjoyable to learn
17	Coursera USA 4, 8, 10 AI	The company provides an online education platform for both individuals and organisations	The company helps its users in gaining competencies through the training and programs it provides on various subjects and its AI-based recommendation system	Value is created via paid/free training offered by AI. The company offers a variety of financial aid options for users who want to access the content of the training as a social value with a limited budget
18	Cubomania Ukraine 4 AI, blockchain, Internet of Things, robotics	The company offers a blockchain-powered online platform for the creation and distribution of personalised educational content by using AI-based characters	CuboBuddy, a robotic educational toy that is connected with IoT, enables interaction between a user and an AI-based character with integrated educational programs. The programs include tasks with various levels of complexity, allowing children to play and develop as they progress. The CuboApp is a mobile application that connects the source of educational content on the blockchain and CuboBuddy	Through blockchain, the platform captures value by facilitating the creation and distribution of educational content for children, while providing teachers and pedagogues maximum protection of their intellectual property. The educational content is delivered via games, making it more enjoyable for children to learn and do their homework. The courses help develop motor abilities, creativity and social skills
19	Delphia Canada 4 AI	The company offers a platform with AI tools that assists users to make important academic and career choices	Delphia's AI, trained through a survey of recent college graduates, analyses high school students through another survey. The results provide advisory reports tailored to high school students. Furthermore, Delphia's AI is strengthened with feedback reports of students, in the case that they don't like their given advice	Realistic road maps are drawn for high school students who have difficulty in creating their career paths. Value is created by directing students to the fields that are suitable for them

(continued)

Table 6.1   (continued)

No	Company info	Value proposal (what?)	Value creation (how?)	Value capture
20	Disciplina Estonia 4 Blockchain	The company offers a platform that uses blockchain to create and maintain a unified register of academic achievement and qualifications for universities	DISCIPLINA, an open-source two-layered blockchain, creates an ecosystem, uniting students, educational institutions (including private tutors), employers and recruiters. Its decentralised algorithm automatically assigns a score to someone based on his or her achievements and qualifications. Universities can use those scores to determine individualised learning plans based on what the student has or hasn't learned and achieved	Through a universal blockchain, the company enables digital storage for personal achievements. Thus, it creates value by assuring their permanence and objective credibility while creating the opportunity for targeted education. The decentralised algorithm offers an effective method for candidate search by their fields of expertise and recruiters
21	Duolingo USA 4 AI	The company offers an ML-powered language learning platform that includes more than 30 language tutorials	The platform, which can be accessed through an app or website, leverages ML methods to deliver course content tailored to the individual user	The platform captures value by increasing accessibility to quality language education by offering its services to anyone who has an Internet connection. Despite the premium plans that require existing payments, the company offers their basic services for free. Content is delivered through games and quizzes, facilitating the learning process while also making it fun and easy to track users' progress
22	EduTekno Turkey 4 Big data	It is a big data-based company that reveals potential talents with genetic analysis and provides information about their characteristics	Students' fingerprints are tested using the genetic test. Students are categorised based on data that demonstrates a link between a fingerprint combination and a skill, which has been created through many years of data collection. These classes demonstrate which fingerprint combinations are effective or unsuccessful in certain sectors, as well as what traits they possess	Value is captured by setting the right goals for students in line with their abilities and offering the opportunity to develop their careers in the right direction
23	ElevateU USA 4 AI	The company offers an AI-based digital publishing platform that generates personalised textbooks for students	The platform performs a test to determine the learning style (visual, auditory, kinaesthetic) of students. The results are analysed using AI to generate dynamic and personalised textbooks by using resources provided by students' professors or universities	With personalised books, it becomes easier for students to understand lectures and for instructors to observe their students. Value is generated by reducing the costs of books and preventing paper waste. It also increases the number of money authors and universities make from copyrights

(continued)

Table 6.1 (continued)

No	Company info	Value proposal (what?)	Value creation (how?)	Value capture
24	ELSA USA 4, 10 AI	The company offers a mobile application that uses AI for English language learning	ELSA can listen to you speak English and identify specific faults on an individual sound level, as well as intonation, rhythm and pitch. It will then provide you with immediate feedback and suggestions on how to correct the issues. All of the input is automated by deep-learning algorithm-based speech recognition technology	The AI-powered assistant helps to improve pronunciation, reduce accent and enrich English skills. It generates value by making language learning more democratic by giving people who don't have a chance to talk to a native speaker
25	EON Reality USA 4 Cloud computing, spatial computing	The company offers an education platform where users can access, create, share and monitor training modules in AR and VR	Using spatial and cloud computing, EON-XR enables educators, trainers, employers and other users to create interactive and immersive AR&VR lessons needless of coding or advanced technological knowledge. The lessons can then be distributed to their audience of students, trainees, employees or the general public for consumption on common devices ranging from smartphones to laptops to publicly available headsets	Eon Reality states that knowledge is a human right, and through their mission and products, they create value by allowing anyone to create and share immersive lessons, which is an effective method of education improving learning and performance. Their tools can be used by corporations, governments, universities or any other educational institution
26	Grammarly USA 4 AI, natural language processing	The company offers a platform that uses AI and NLP to correct grammatical errors, checks tone consistency and suggests synonyms	Grammarly's algorithms identify potential flaws in the text and provide context-specific recommendations for grammar, spelling and punctuation, as well as plagiarism. The software explains why each suggestion is made, which allows users to make informed decisions about how to fix a problem	It generates value by enabling the emotions that are difficult to be perceived in texts to be transferred more comfortably and correctly. At the same time, it reveals grammatical or punctuation errors important in business and education life that people cannot notice while writing, thanks to AI. It also makes texts more readable and precise
27	HOMER USA 4 AI	The company offers a platform that provides a unique learning adventure for young people (2–8) that reassures and befriends them	With the Learn & Grow App, kids develop reading, maths, social and emotional learning, creativity and thinking skills through thousands of activities that kids love. Lessons go beyond memorisation to teach basic skills and how to apply them. It is powered by AI	It generates value by allowing youngsters to demonstrate their knowledge in a new or different situation and preparing them to use their talents in the actual world

(continued)

Table 6.1 (continued)

No	Company info	Value proposal (what?)	Value creation (how?)	Value capture
28	HowNow UK 4 AI	It is an EdTech company that connects experts and students by offering interactive video lessons and leverages ML technology	The course website can be accessed by course content creators and learners, on mobile phones, tablets, laptops or any other device. ML methods are used to deliver the content appropriate for the user	It increases the productivity and effectiveness of organisations by providing interactive online video lessons, which facilitates the acquisition of new skills. Value is captured by creating a platform that is accessible for everyone and contains all needed resources, which can cut time spent searching for information by up to 35%
29	iNurture India 4 AI	The company offers an AI-based learning management platform that enables universities to create and deliver job-oriented undergraduate or postgraduate courses	An AI-based student engagement platform called KRACKiN oversees Skill-X, an end-to-end employability solution that allows students to identify changing skill requirements for various industries and receive training in industry-relevant abilities, resulting in a talent pool that is industry-ready	On the platform, students are trained according to their interests and abilities, leading to creating a talent pool. Value is captured by the increases in education level and the increases in the workforce efficiency. As a result, businesses find more qualified candidates
30	Immerse USA 4 Spatial computing	The company offers a platform that provides VR-supported solutions to improve language skills through structured learning opportunities	Immerse allows students to use avatars to speak and practise their English through a variety of topics and themes while wearing a VR headset, enabling students to learn interactively	The company offers students an immersive English learning experience, on-campus or from a distance. Through VR, the platform creates value by building a suitable learning setting where students can socially interact to practice speaking, deepen emotional bonds and improve language proficiency

(continued)

Table 6.1 (continued)

No	Company info	Value proposal (what?)	Value creation (how?)	Value capture
31	LocoRobo USA 4 AI, drones, Internet of Things, robotics, flexible electronics and wearables	LocoDrone is created to allow educators to teach various skills such as Python or Java coding, sensor data analysis, data visualisation and multi-robotics using data obtained from the drone through its codable controller. The company sells a variety of educational devices that can be used to teach a variety of skills to students in grades K through 12	LocoWear, a wearable electronic that collects motion data, allows learning Python while programming the device, anatomical motion physics, statistics and data analysis with the generated data. LocoXtreme is a robot designed to deliver an interactive learning experience for robotics, programming and more at various levels through the six attached sensors. MyLoopy is also a robot equipped with sensors and can learn from its environment and human interactions through AI. It is designed to educate children on a range of topics such as early coding and STEM concepts. The LocoIoT allows students to learn to build an IoT system from scratch with a hands-on approach through the provided building blocks for an Internet-connected device ecosystem	LocoRobo's tools may differ in the content they are designed to teach, but they share the same goal: to bring an immersive and hands-on learning experience. Their educational programs, powered by numerous technologies, are offered to students at young ages and at the university level. By allowing them to interact and apply the skills taught through the provided tools, LocoRobo facilitates learning and reinforces knowledge. Learning goals and skills meet industry-level standards so that participation contributes significantly to finding employment, which is how LocoRobo generates value
32	Me.reka Malaysia 1, 4, 5, 8, 17 3D printing, big data, crowdfunding, Internet of Things, spatial computing	It is a collaborative learning hub for academics, industry players and entrepreneurial communities. Designing and developing digital platforms that use augmented reality, big data, IoT, 3D printing and delivering educational programs	Through an interesting, innovative and skills-based approach, Me. reka combines the ideas of sustainability with STEAM education (science, technology, engineering, arts and maths). All of Me.reka's programs adhere to the United Nations' sustainable development goals	To solve the problems of the future, training and skill development programs are organised using emerging technologies. Value is captured by making training accessible to everyone and putting this training into practice with projects. By doing so, the number of people who can deal with future problems can increase
33	ODEM Switzerland 4 Blockchain	The company offers a platform that allows students to access and own their academic records indefinitely	Thanks to blockchain, academic data of students and teachers is stored indefinitely in secure blocks. Based on smart contracts approved by both parties, the ODEM ledger locks students' course selection as well as the courses each professor teaches. The "skill badges" awarded by the ODEM serve as proof of students' and instructors' advanced abilities in certain areas. By searching for these badges, students can find professors and courses that are right for them, and instructors can track and review student progress by searching	Students and professors become able to have a permanent record of their success. So, value is captured by enhancing their reputation and preventing academic fraud

(continued)

Table 6.1 (continued)

No	Company info	Value proposal (what?)	Value creation (how?)	Value capture
34	Odilo Spain 4,10 AI	The company offers an intelligent content platform that enables any library, school, institution, company or municipality to provide its users with digital material such as ebooks, audiobooks, periodicals, films, podcasts, courses and other forms	Providers upload their content to the platform, which allows users to access them. The platform analyses user data with AI to determine and offer the most relevant content	Value is captured by democratising access to high-quality education on a global scale, improving literacy and critical thinking abilities and strengthening lifelong learning
35	Osso VR USA 3, 4 Spatial computing	It is a company that uses VR to allow medical device companies and healthcare professionals to share, test and acquire new skills and procedures	The surgical simulation training analytics from Osso VR help in tracking engagement and proficiency. This training data gives healthcare professionals feedback to help them perform at their best. It also allows medical device businesses to provide positive early case experiences, which foster trust and loyalty	Sales teams, surgeons, providers and hospital personnel may train together in the same virtual operating room – dozens of people at once – from anywhere in the world, thanks to Osso VR's digitisation of the HCP (Health Care Policy) learning journey. Value is captured by making health education easier and more accessible compared to traditional education
36	Packback USA 4 AI	The company offers an AI-powered online discussion platform	The AI-enabled Digital TA (Teaching Assistant) directs questions to students and adapts to their response, which delivers a scalable asynchronous discussion experience, thus automating the process of moderation and coaching	Thanks to AI, the Digital TA offers adaptive questions to elicit curiosity and interest in pupils. Bringing forth a very supportive learning environment adds value. It assists educators by automating the regulating and coaching procedures, allowing them to save time. Additionally, algorithmic evaluations allow students and instructors to track the growth of students' interest, dependability, communication and tradition

(continued)

Table 6.1   (continued)

No	Company info	Value proposal (what?)	Value creation (how?)	Value capture
37	Photomath USA 4 AI	The company provides an application that analyses, calculates and intuitively explains printed and handwritten calculus problems to users using step-by-step explanations	Photomath must first read any problem before it can solve it. The program takes advantage of Microblink's superior OCR (optical character recognition) technology to read and recognise both handwritten and printed characters in an issue. The detected characters, such as numbers, letters and maths symbols, are then passed through Photomath's proprietary algorithm, which analyses each character concerning the others and calculates the scanned problem's formula. The solution and solving stages are provided by applying a problem-solving algorithm to the formula	Thanks to the application, students can solve their calculus problems and understand the ways of solving the problem without the need for an instructor or anyone. Value is created by enhancing self-learning and making knowledge more accessible for everyone. In addition, the application contributes to making online education more efficient by reducing the dependency on the instructor
38	Riiid South Korea 4 AI	The company offers an AI-based tutoring platform, providing personalised language education	The Riiid AI Tutor analyses the user answers data to make predictions of their exam scores. Then, it provides courses tailored to address the weaknesses of users	The platform shortens the time needed to improve language exam scores by creating an education program designed specifically for each user. It also enables the users to view their learning history and thus use it for effective review and updates the courses according to the users' progress. The app can be downloaded by everyone and can be accessed at any given time, thus greatly increasing accessibility to quality language education
39	Robolink USA 4 AI, robotics	The company offers robotics building and development kits to students who want to learn how to code and build their own robots	Rokit Smart is a programmable robot kit designed to teach robotics, motors, infrared sensors, mechanics and tools. Zumi, a self-driving car kit that helps children learn about self-driving cars, robotics, ML, computer vision and mainly AI while playing with it. Codrones are designed to educate children about drones, sensors, LED lights and flight. All products can be programmed with various coding languages such as Blockly, Python or Arduino to display various features such as colour recognition, learning gestures and face recognition, while also teaching coding	Value is captured by facilitating learning coding through providing both fundamental and detailed knowledge regarding AI, ML and robotics by delivering education through interactive and programmable toys, thus rendering learning processes more enjoyable

(continued)

Table 6.1 (continued)

No	Company info	Value proposal (what?)	Value creation (how?)	Value capture
40	ROYBI Robot USA 4 AI, robotics	The company offers an AI-powered robotic toy designed to contribute to language learning and early childhood education	Roybi Robot uses AI to give customised entertainment to children based on their preferences and pace. Roybi Robot delivers a fun and interactive learning experience for youngsters with over 500 courses including basic STEM, stories, games and songs	Roybi Robot provides a personalised learning experience for each child by emphasising their unique talents and interests. Value is generated by creating a strong educational foundation for children in early childhood while also encouraging creativity and developing communication skills
41	ScanMarker USA 4, 10 Internet of Things	The company offers a reader pen that scans texts to computers or phones and translates them	Scanmarker Air is a portable reading tool that can scan texts from books, papers or other documents directly to the devices connected through Bluetooth. The text can be edited and translated into more than 110 languages through the desktop application or the website. The pen can also read aloud the text while scanning, providing audio through text to speech (TTS)	Scanmarker is 30 times faster than manual typing; thus it helps readers save time. By reading the selected words and developing automaticity and fluency skills, Scanmarker improves the decoding and phonological processing of readers. Value is generated by equipping readers with the ability to translate texts in different languages, which facilitates language learning and increases the accessibility of educational content. The conversion of TTS greatly increases the accessibility to quality education for individuals with visual or reading impairments
42	Solarpak Ivory Coast 4, 7 Flexible electronics and wearables	The company provides schoolbags equipped with solar panels powering attached light when needed	On their way to and from school, students carry solar-panelled backpacks. This offers enough energy to power the associated light for about 4 hours when needed	Value is created by providing a tool to combat inequality by allowing every student to study, even when they do not have electricity at home. It assists African children in their education by providing them access to a light source at night

(continued)

Table 6.1 (continued)

No	Company info	Value proposal (what?)	Value creation (how?)	Value capture
43	Tab for a Cause USA 1, 4, 16, 17 Cloud computing, crowdfunding	The company offers a browser extension that allows its users to raise money for charitable organisations working for sustainability in their respective ways	When the extension is installed, it changes users' default "new tab" page to the company website. It is a customisable page with widgets, pictures and more. Every time a new tab is opened, users earn "hearts", tokens that can be donated to a charity of choice. Among these, there are two charities concerned with access to quality and equal education. Ads shown on this page generate the revenue for the donations	The company allows everyone to generate revenue, simply by opening a new tab on your browser and donating it to charitable organisations of their choice. The platform generates value by acting as a bridge between organisations with a diverse array of missions
44	Talespin USA 4 AI, natural language processing, spatial computing	The company offers an educational VR platform that provides solutions for employee learning and engagement	The platform introduces users to a virtual human avatar designed to create simulations for employee training purposes and interact with them throughout the process. Powered by AI, speech recognition and NLP, the avatar tracks trainees' eye and body movements along with facial expressions. This enables the platform to provide them with guidance through branching narratives	Platform-generated VR simulations reduce on-the-job time, provide flexible remote learning for employees and increase learners' confidence. Value is created by enabling businesses to build a more collaborative, inclusive and productive business future
45	TINALP Italy 4 5G, spatial computing	The company offers an AR- and VR-based eLearning platform that uses 5G to provide services such as mobile cloud classrooms and virtual presence	The platform uses mixed reality and augmented intelligence to create remote virtual classrooms powered by 5G. These classrooms simulate real-life conditions with instructor-trainees and real-time interactions for a tailored learning experience	The AR and VR experiences are more realistic, thanks to the low latency and high bandwidth of 5G, allowing teachers to gain real-time information about each student to tailor education to their unique needs and skills. The virtual classrooms generate value by offering an immersive learning experience that can be delivered remotely, increasing access to quality education

(continued)

Table 6.1 (continued)

No	Company info	Value proposal (what?)	Value creation (how?)	Value capture
46	TransfrVR USA 4 AI, spatial computing	The company offers an AR&VR-based educational platform that delivers simulations along with an AI-driven digital coach	The company provides trainees with a virtual reality platform that allows them to engage in hands-on learning in a simulated environment. When a trainee makes a mistake, a digital coach powered by machine learning and artificial intelligence (AI) gives them feedback on how to improve their performance on the job. In addition, the intern is given a score that shows his or her job preparedness in a real-world setting. Hospitality, surgical and culinary fields are examples of where it is used	The platform offers a more comprehensive understanding of the material by promoting interactivity and participation in practical training. It allows users to learn lessons from their mistakes with guidance from its digital coach. The delivery of detailed evaluations of progress and feedback allows users to be aware of their knowledge and skills. It generates value by providing a hands-on learning experience, especially important for students with limited access to campus facilities or job sites
47	Verizon USA 4 5G, edge computing	The company offers a 5G wideband network that provides optimal bandwidth and refreshes rates for demanding educational uses of the Internet	Verizon provides 5G networks with sites that are excellent for storing educational institutions' edge computing resources. These are used to power and support educational tools like AR and VR that require a lot of bandwidth and low frequency	Utilising 5G, Verizon enables the use of demanding educational tools that allow immersive education and provide access to fast Internet to educational institutions
48	WonderTree Pakistan 3, 4, 10 AI, spatial computing	It is a software enterprise that creates AR-based interactive games for children who have special needs	Games may be personalised to a player's ability/needs, and they come with a reporting system that allows them to track their cognitive and/or motor skill improvement. Each game is based on psychological research and uses technology to track specific metrics such as attention span, hand-eye coordination, muscle mobility and so on. Thanks to AI, the measurements can be used to assess performance and determine future educational, therapeutic and/or rehabilitation strategies	Value is captured by measuring, tracking and interpreting a child's development through games that make the rehabilitative and instructional programs more enjoyable for children with special needs. Also, it may be used anywhere with a camera, Internet connection and computer, thus making rehabilitation more convenient and accessible

(continued)

Table 6.1   (continued)

No	Company info	Value proposal (what?)	Value creation (how?)	Value capture
49	ZIoTag USA 4 AI	The company offers an AI-powered video player producing categorised transcripts from videos	ZIoTag instantly generates transcripts once users copy and paste the URL of any Internet video (or audio). Then, its powerful AI creates an Actionable Table of Contents (AToCs) that makes every word and subject of the movie completely searchable	The AI-produced transcripts and AToCs render every word in the video searchable. Results appear as clickable timestamps to the exact moment the search phrase. Their entire video library is indexed and categorised to enable students to easily search any section of a video. Thus, it increases search time efficiency for educational content. The platform captures value by facilitating education, especially distance learning as it greatly depends on educational videos

References

B. Anderson, Writing power into online discussion. Comput. Compos. **23**(1), 108–124 (2004)

R. Cervelló-Royo, I. Moya-Clemente, M. Perelló-Marín, G. Ribes-Giner, Sustainable development, economic and financial factors, that influence the opportunity-driven entrepreneurship. An fsQCA approach. J. Bus. Res. **115**, 393–402 (2020)

T. Ferguson, C.G. Roofe, SDG 4 in higher education: Challenges and opportunities. Int. J. Sustain. High. Educ. **8**(22), 959–975 (2020)

I.B. Franco, E. Derbyshire, T. Chatterji, J. Tracey, SDG 4 Quality Education, in *Actioning the Global Goals for Local Impact Towards Sustainability Science, Policy, Education and Practice*, ed. by I. B. Franco, E. Derbyshire, (Springer, Singapore, 2020), pp. 57–68

F. Glen, W. Cédric, *UNESCO Meta-Survey on the Use of Technologies in Education in Asia and the Pacific* (UNSECO, Bangkok, 2003)

U. Hanemann, *Examining the Application of the Lifelong Learning Principle to the Literacy Target in The Fourth Sustainable Development Goal (SDG 4)* (Springer, Hamburg, 2019)

Joint SDG Fund, Goal 4 Quality Education. [Online] (2021). Available at: https://www.jointsdgfund.org/sustainable-development-goals/goal-4-quality-education. Accessed 15 Aug 2021

D. Kekić, S. Miladinovic, *Functioning of Educational System During an Outbreak of Acute Infectious Diseases* (Belgrade, Academy of Criminalistic and Police Studies, 2013)

J. Koßmann, SDG 4: Why Does Education Matter? [Online] (2019). Available at: https://aiesec.at/2019/04/05/sdg-4-education-matter/. Accessed 16 Agu 2021

C. Latchem, Chapter 12: Gender issues in online learning, in *Culture and Online Learning: Global Perspectives and Research*, ed. by I. Jung, C. N. Gunawardena, (Stylus Publishing, LLC, Sterling, 2014)

J. Margolis, A. Fisher, *Unlocking the Clubhouse: Women in Computing*, 1st edn. (The MIT Press, Reno, 2002)

Nerdy Mates, Glimpse at How Education Will Possibly Look Like in 2050. [Online] (2021). Available at: https://nerdymates.com/blog/education-future. Accessed 22 Aug 2021

D. Patel20, Why SDG 4 Quality Education Is Important for Poverty Reduction. [Online] (2019). Available at: https://blogs.lse.ac.uk/internationaldevelopment/2019/11/25/why-sdg-4-quality-education-is-important-for-poverty-reduction/. Accessed 14 Aug 2021

UN Environment Programme, Goal 4: Quality Education. [Online] (2021). Available at: https://www.unep.org/explore-topics/sustainable-development-goals/why-do-sustainable-development-goals-matter/goal-4. Accessed 2 Aug 2021

UNEP, Goal 4: Quality Education. [Online] (2021). Available at: https://www.unep.org/explore-topics/sustainable-development-goals/why-do-sustainable-development-goals-matter/goal-4. Accessed 14 Aug 2021

UNESCO, Leading SDG 4 – Education 2030. [Online] (2021). Available at: https://en.unesco.org/themes/education2030-sdg4. Accessed 1 Aug 2021

United Nations, *Policy Brief: The Impact of Covid-19 on Children* (United Nations, New York, 2020)

United Nations, Quality Education: Why It Matters. [Online] (2021a). Available at: https://www.un.org/

sustainabledevelopment/wp-content/uploads/2018/09/Goal-4.pdf. Accessed 14 Aug 2021

United Nations, The Global Goals for Sustainable Development. [Online] (2021b). Available at: https://www.globalgoals.org/3-good-health-and-well-being. Accessed 17 Aug 2021

United Nations, United Nations Sustainable Development Goals. [Online] (2021c). Available at: https://www.un.org/sustainabledevelopment/wp-content/uploads/2017/03/3_Why-It-Matters-2020.pdf. Accessed 31 July 2021

United Nations Development Programme, Goal 4: Quality Education. [Online] (2021). Available at: https://www.tr.undp.org/content/turkey/en/home/sustainable-development-goals/goal-4-quality-education.html. Accessed 1 Aug 2021

E. Unterhalter, The many meanings of quality education: Politics of targets and indicators in SDG4. Glob. Policy **10**(1), 39–51 (2019)

T. Unwin, M. Weber, M. Brugha, D. Hollow, *The Future of Learning and Technology in Deprived Contexts* (Save the Children, London, 2017)

UN 2015. Transforming our world [WWW Document] https://sustainabledevelopment.un.org/post2015/transformingourworld/publication

SDG-5 Gender Equality

7

Abstract

Gender equality, the fifth of the sustainable development goals of the UN, is a base element for creating a comfortable, sustainable and wealthy world and being a fundamental human right. While achieving the goals for a sustainable future, SDG-5, Gender Equality, will be one of the building blocks of this path. So, taking actions to accomplish the goals of SDG-5 is not only crucial for the related SDG itself, but also it helps to proceed in other SDGs as well. This chapter presents the business models of 16 companies and use cases that employ emerging technologies and create value in SDG-5. We should highlight that one use case can be related to more than one SDG and it can make use of multiple emerging technologies.

Keywords

Sustainable Development Goals · Business models · Gender Equality · Sustainability

The author would like to acknowledge the help and contributions of İlke Burçak, Fatma Balık, Aleyna Yıldız, Fatmanur Babacan, Handan Öner, Enejan Allajova and Batuhan Özcan in completing this chapter. They also contributed to Chapter 2's Artificial Intelligence, Cloud Computing, Distributed Computing, Edge Computing and Quantum Computing sections.

Gender equality, the fifth of the sustainable development goals of the UN, is a base element for creating a comfortable, sustainable and wealthy world and being a fundamental human right. The word "gender" is not the same concept as the word "sex". Sex refers to the biological distinction between men and women, and gender means the social status that is attributed to men and women. Gender roles can change according to religion, ethnicity, age and environment (Kumar Pathania 2017). The concept of gender equality is the name given to ensuring equal rights and freedoms regardless of gender in all social events such as gender ratio in companies, salary ratio, psychological and physical violence, right to vote and gender ratio in education (Shastri 2014). An important step for ensuring gender equality is women's empowerment. Women's empowerment refers to the essentiality of a woman's ability to have more authority in her life independently from her sex (Kumar Pathania 2017).

SDG-5 aims to end all forms of discrimination against women and girls, including ending violence, ensuring access to sexual health and reproductive rights and ending child marriage and sexual exploitation. Also, it aimed to increase the visibility of women in society and encourage women to appear in all spheres of life (Stuart and Woodroffe 2016). Even though gender inequality is decreasing, women still face various difficulties. Several examples can be given (United Nations 2018):

S. Küfeoğlu, *Emerging Technologies*, Sustainable Development Goals Series, https://doi.org/10.1007/978-3-031-07127-0_7

- In 56 countries, 20% of 15–19-aged girls that have been in a sexual relationship have faced violence by their partners between the years 2015 and 2016. This violence was physical and/or sexual.
- Around 2017, 21% of women aged 20–24 were married under 18 or were in an unauthorised union. In other words, approximately 650 million women were married as children.
- In 30 countries, 1 in 3 15–19-year-old girls had been exposed to genital mutilation around 2017. In these 30 countries, mutilation implementation was typical.
- In approximately 90 countries, women worked in nursing and as home labour without getting paid about three times more than men between 2000 and 2016.
- On average, 46% of people from 34 countries said that men's lives are better in their countries, while 15% said women's lives are better. The majority of individuals polled in several countries from Europe and America believe that men have a better life than women in their nation. On the other hand, 75% of those polled think that women in their country will ultimately have the same rights as men, while 5% say that equality has already been reached (Horowitz and Fetterolf 2020).
- One out of every three women over the age of 15 in Europe is exposed to physical and sexual violence. And one in two women is sexually harassed (European Institute for Gender Equality 2021).
- Women are paid 16% less than men (United Nations Women 2021).

The examples given above are only a tiny part of the inequality experienced. These inequalities have driven women to seek equality and effectively include gender equality among sustainable development goals. Taking actions to achieve the goals of SDG-5 is not only crucial for the related SDG itself, but also it helps to proceed in other SDGs. Although it does not seem so, progressing in gender-related issues serves, facilitates and expedites the improvements in some of the remaining main goals (IISD 2017). In this perspective, to achieve gender equality and women

empowerment, six targets were determined by the UN. Figure 7.1 summarises the targets and indicators of SDG-5.

Environmental policy is inextricably related to sustainable development goals, and it's hard to conceive sustainable development goals without modern digital technology. Emerging technologies are increasingly essential instruments for achieving a balance of low environmental impact and high performance (Мачкасова 2020). Sustainable development cannot be accomplished unless gender equality is ensured. However, gender inequality is a reality that exists almost everywhere globally. Therefore, it is a possible and sad event that today's people are exposed to injustice and discrimination because their genders are different. In such a situation, gender equality, which aims to end discrimination between the sexes, is very important to us. Since women and girls make up approximately half of the world's population, they are potentially half its total potential. However, gender inequality still exists today and undermines the development of society (United Nations 2021).

First of all, gender equality is a human right, and all women and girls should have the same rights and lives as men. Gender equality is a basic human right and indispensable for a peaceful and sustainable world. Despite such goals, sexism and discrimination persist. For example, one in five women and girls between the ages of 15 and 49 are exposed to sexual or physical violence by someone else (United Nations 2021). Also, in many countries, the sexes are segregated, such that women are denied inheritance and land ownership (Shi et al. 2019). In addition to this, women and girls are exposed to all kinds of violence and are subjected to all sorts of injustices in business and social life. Considering all these, ending gender inequality has a crucial place for the development of humanity. Therefore, gender equality, the fifth goal of sustainable development, representing equality for women and men everywhere and in all fields according to their needs, is the number one method of eliminating tyranny against people's gender (Murat 2017). During the last 20 years, gender equality has been the main study area for UNDP. According to statistics of UNDP,

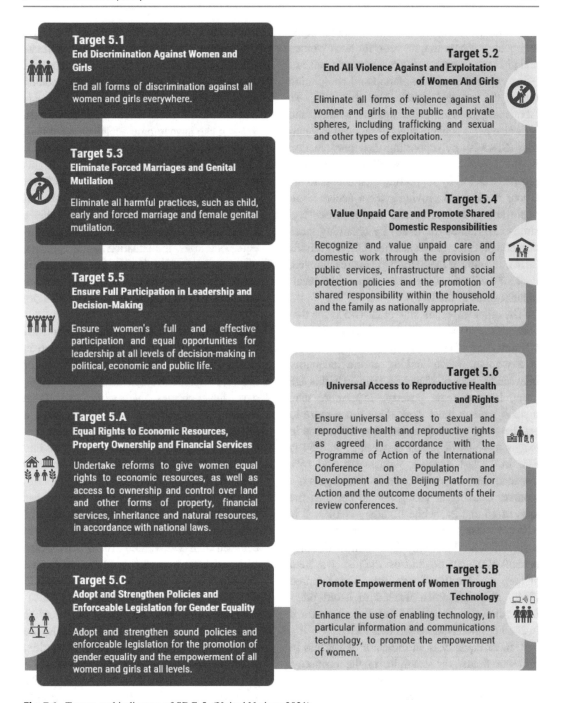

Fig. 7.1 Targets and indicators of SDG-5. (United Nations 2021)

today more girls continue to a school than 15 years ago. However, in some areas, this sexual violence and discrimination are still in progress even by the governments. In addition to all these discriminations (salary inequality, job inequality or others), climate change, natural disasters and migrations affect women and children negatively much more in many ways as they are more prone to adverse conditions (United Nations Development Programme 2021).

Achieving SDG-5's goal will lead to equal and quality access to education for both women and girls who cannot reach education because of being exposed to discrimination and completing one of the goals of quality education (SDG-4). Abolishing discrimination against women and girls also helps achieve goals of reduced inequalities (SDG-10) by supporting their participation in the elections, decent work and economic growth (SDG-8) by supporting their access to the labour market and providing them decent work opportunities to survive and peace, justice and strong institutions (SDG-16) by obtaining more peaceful and connected societies (United Nations 2018). In short, if the SDG-5 goals are achieved, then they will affect the other sustainable development goals of the UN.

The other important side of SDG-5 is the discrimination of women in labour. According to UN data, one of every four people in parliament is a woman. When looking at the inequality between the ages of 25 and 34, women in this age group have extreme poverty, 25% more than men. A woman must work three times more to get the same salary as a man. The difference in labour force participation between 25 and 54 is 31% in the last 20 years. Women are paid 16% less than men, and only one in four managers is a woman. Looking at young people between the ages of 15 and 24, while illiteracy is 14%, it is 31% for women. While 39% of women work as agricultural workers, only 14% own land. Forty percent of women do not trust the justice system and are afraid of its vulnerabilities. One hundred ninety million women wanted to avoid pregnancy; however, they could not find a way for it in 2019 (United Nations Women 2021).

Gender equality is an indicator of success in different fields, for example, in the economy, health, safety, business, racial equality, reducing poverty and bringing peace. Therefore, SDG-5 is a method to ensure gender equality in this regard. There are several specific field examples that SDG-5 affects:

- Moreover, gender inequality has severe consequences during natural disasters. Experts explored how gender disparity contributes to

death and damage at the 2005 World Conference on Disaster Reduction. Climate change, which increases the severity of natural disasters, also affects gender equality, looking at the results of other studies, and puts female individuals in even more poor conditions. Women can have a more decisive part in their own protection when a gender viewpoint is included in discussions (Human Rights Careers 2020). Unequal societies have fewer social and emotional bonds. High rates of non-social harmful behaviour and violence are observed in these countries. In gender-equal countries, the situation is the opposite, where people are cohesive. People in these countries are healthier and have better life conditions (Victorian Government Directory 2021).

- Race equality and gender equality are inextricably related. Race plays a significant part in issues such as the gender pay gap. Women of colour, Hispanic women and native women have fewer wages than white and Asian women. Black women in the USA have a higher death rate for pregnancy-related reasons. When gender equality takes race into account, it also enhances racial equality (Victorian Government Directory 2021).

- The destitution level of young females has been detected to be the highest. As the ages of men and women get older, the poverty gap between men and women also widens. Some of the apparent reasons behind this are the fact that women quit their jobs after marriage and that women are not given the same educational opportunities and career prospects as men. Due to gender inequality, girls are trapped in poverty. However, better education, health and career opportunities can enable a girl to evolve into the best version of herself. As a result, focusing on decreasing gender inequality is a long-term, high-impact method to diminish poverty (Human Rights Careers 2020).

While achieving the goals for a sustainable future, SDG-5 will be one of the building blocks of this path. So, taking actions to accomplish the goals of SDG-5 is not only crucial for the related SDG itself, but also it helps to proceed in other SDGs as mentioned above. Although it does not seem so, progressing in gender-related issues serves, facilitates and expedites the improvements in some of the remaining main goals (IISD 2017). According to the report published by the European Institute for Gender Equality, abbreviated as EIGE, even in the continent that contains some of the most developed countries in the world, that is, Europe, SDG-5 could not progress appreciably (Barbieri et al. 2020). If we look at the larger framework, it is expected that, by 2030, 169 targets which correspond to 17 goals will not be successfully met. Prioritising certain goals, in particular SDG-5, is one possible attitude to take one step forward in speeding up completing the targets at the regional, national and mondial levels. As it can be seen from the previously mentioned topic above, there is no doubt that one of the most inclusive and universal goals is gender equality. Therefore, transforming the SDG-5 into the focal point of the progress plans, both intellectually and officially, can help move forward in the 2030 agenda and shape the new route map after 2030 accordingly (Hepp et al. 2019). To exemplify the bond of SDG-5 with other SDG, combinations between SDG-5 (Gender Equality) and SDG-16 (Peace and Justice, Strong Institutions) have emerged in recent years, and achieving gender equality is considered to be instrumental in progress as countries improve their ability to provide stable governments or vice versa. In any case, such efforts will have to be greatly increased in the future in terms of achieving the SDGs, including the Paris Climate Accord and lengthy gender equality, in these specific situations (Kroll et al. 2019).

In addition, ensuring gender equality, especially in education, will also contribute to the future economically. According to a study that predicts economic growth, if women receive education in science, technology, engineering and mathematics (STEM), the employment rate is expected to grow by 0.5–0.8% until 2030. This growth is expected to increase to 2.5% by 2050 ("Economic Benefits of Gender Equality in the EU", 2021). However, unfortunately, in countries like Nigeria awareness of gender equality is not provided in education. Thus, there can be no development for the future. Since teachers do not have sufficient knowledge about gender equality, they cannot inform people about it both inside and outside the school. It will be tough to build a country based on gender equality if this gap is not filled with a significant gender equality gap. It is challenging to integrate gender equality projects in an environment with no infrastructure. In this respect, insufficient knowledge about gender equality will cause future generations to be unaware of this issue.

There are a lot of differences between genders, and these differences reflect consumption and production activities (Roushdy 2004). Women's reliance on money in the family increases spending and development, which is beneficial for all family members (Schady and Rosero 2008; Rubalcava et al. 2009). A woman can be seen in the business world and be in the role of a worker or an entrepreneur. However, inequalities can create gaps in women's lives, such as entering a new workplace, preventing them from gaining professional competence (BarNir 2012) or having a role in work. Even in the same job, women receive 86% of men's salary in public institutions and 76% in the private sector, which causes difficulties in their careers (Shi et al. 2019). Even for women entrepreneurs, these situations are not different (Fairlie and Robb 2009). When looking at finance performance, the ideas and products of women entrepreneurs are not worse than male entrepreneurs. However, establishments where women entrepreneurs run businesses raise fewer financial resources (Demartini 2019). This leads to the gender employment gap, and if this gap is closed, GDP will increase by 11% (Victorian Government Directory 2021).

Women's influence extends beyond businesses and organisations. As stated in the studies, the economy benefits from improving women's economic engagement. If the average wages of women living in OECD countries were raised to the current Swedish level, GDP would increase by about $6 trillion. Thus, it is obvious that the pay gap between genders is quite harmful to the economy (Human Rights Careers 2020). As it was mentioned as an example above, in Australia's GDP, enterprises, where at least three out of ten people in the management of the company staff are women, are 15% more profitable. If the number of men and women entering the workforce from higher education is equalised, the Australian economy will benefit eight billion dollars. In Victoria, police spend 2 out of 5 h dealing with issues related to family violence, costing more than $3.4 billion annually in financial terms. In Australia, the budget for unpaid care is six times higher than the budget for paid care, and the majority of those who do these unpaid care jobs are women (Victorian Government Directory 2021).

In short, the goal of SDG-5 is about empowering women's economic status and finance. Steps were taken to achieve this goal and accelerate the progress in no poverty (SDG-1) to reduce the number of women who are having a hard time finding a job to live by promoting a prominent endorsement for their economic freedom and rights. Thus, this ensures the ideas that were mentioned above about poverty. Additionally, creating economic opportunities and amenities for women results in sustainable industrial development, which is one of the targets of industry, innovation and infrastructure (SDG-9) (UN Women 2018). Thus, it is concluded that gender equality affects every field, and it will lead to development, especially in the economy.

7.1   Companies and Use Cases

Table 7.1 presents the business models of 16 companies and use cases that employ emerging technologies and create value in SDG-5. We should highlight that one use case can be related to more than one SDG and it can make use of multiple emerging technologies. In the left column, we present the company name, the origin country, related SDGs and emerging technologies that are included. The companies and use cases are listed alphabetically.[1]

---

[1] For reference, you may click on the hyperlinks on the company names or follow the websites here (Accessed Online – 2.1.2022):

http://www.equilo.io/;   http://www.pipelineequity.com/; https://encourage.fi/;   https://equalista.com/;   https:// equileap.com/;   https://safeandthecity.com/about-us; https://salvatio.dk/;   https://solve.mit.edu/challenges/ frontlines-of-health/solutions/4506;   https://www.ava-women.com/;   https://www.bodyguard.ai/individuals; https://www.developdiverse.com/product/;   https://www. euphoria.lgbt/;   https://www.metta-space.com/about; https://www.shimmy.io/;   https://www.sisterwave.com/; https://www.someturva.fi/vaikuttajat

Table 7.1 Companies and use cases in SDG-5

No	Company info	Value proposal (what?)	Value creation (how?)	Value capture
1	Ava Women Switzerland 3, 5 AI, flexible electronics and wearables	It is an AI-assisted end-user product that informs women about their reproductive system and health.	It provides a bracelet that is wearable at night to track certain changes in the body. The data gathered from the bracelet is processed using AI techniques to determine the most fertile days. It also tracks the period cycle and pregnancy progress of women.	By recognising the optimal time to conceive, it helps prevent one of the most common causes of infertility: trying to conceive during infertile times. As a result, women keep track of their own health and save money on treatments that aren't necessary.
2	Bodyguard France 5, 10 AI, natural language processing	It is an AI-based mobile application that protects individuals and businesses against cyberbullying on social media. Through comprehending text, the solution helped by NLP acts as a real-time moderator, detecting toxic comments even when emojis, abbreviations, censored words and typos are used.	Insults, threats, trolling, bodily shaming, racism, homophobia, misogyny, hatred, sexual and moral harassment are all considered when determining comment protection levels. With AI and NLP, the app reads, understands and analyses text. It also ensures that the comments comply with the access permissions that users have granted and that they are removed.	It is aimed to protect mental health for people and to provide a higher-quality social media experience by overcoming cyberbullying. Social and ethical values are captured by protecting people from any form of verbal harm, sexist comments, sexual harassment or any form of discriminatory behaviour on social media.
3	Develop Diverse Denmark 5, 8, 10, 16 AI, natural language processing	It is a startup that creates equal opportunities by fostering a working environment where everyone's opinion counts and is appreciated for adding value, regardless of gender, sexual orientation and identity, physical and mental ability, age, ethnicity or religion.	Natural language processing (NLP) and machine learning (ML) are used by the company to detect implicit and explicit stereotypes in texts and provide inclusive alternatives.	As a social value, the startup helps avoid discrimination in the initial stages of hiring for top positions and increases the quality and diversity of the candidate pool.
4	Equalista Germany 5, 10 Big data	It is a mobile application that analyses people's gender biases with big data and provides training to eliminate these biases.	After the prejudices of the people are analysed with big data, the gender equality learning app lets users discover why inequality exists, what it looks like and how to talk about it, test users' own biases and trains them to take action in real life.	Social value is captured by providing the collection and development of incomplete information about sexism from the social point of view. Thus, by aiming for a life free of biases, it is desired to open the doors to a more equal world.

(continued)

Table 7.1 (continued)

No	Company info	Value proposal (what?)	Value creation (how?)	Value capture
5	Equileap The Netherlands 5 Big data, datahubs	It is a global database on gender equality, with metrics including female representation at all levels of a company, the gender pay gap, parental leave and anti-sexual harassment policies.	It evaluates the businesses based upon their gender equality criteria. It helps investors make their investment decisions according to the data analysed with big data. They provide gender metrics that financial institutions need to assess the risk and opportunities presented by gender inequalities in global markets.	Social value is created by improving the role of women in workplaces by paying attention to inequality. Furthermore, Equileap's database for gender equality enables investors to make better investment decisions.
6	Equilo USA 5 AI, big data	It is a web-based application that provides analysis, recommendations and action plans to customers regarding gender equality and social inclusion using big data and AI.	Machine learning, deep analytics and big data are used to give tailored results for each project and investment based on sector and geography, with a range of advanced analytics tools to support use cases in the development, humanitarian, financial and private sectors.	Social value is created through the use of the data and analysis by aid agencies and humanitarian organisations to incorporate customised best practices for gender equality and social inclusion into their project planning, investment decisions or implementation.
7	Euphoria USA 5, 10 AI, big data	It is a big data-based mobile app provider focusing on accurately planning individuals' health goals and helping transgender people to do savings by its investing algorithms.	Solace is an AI-mediated encrypted mobile app that allows individuals to plan and monitor the lifestyle, medical and legal objectives of their gender transition. With big data, the app analyses collected data of users and informs them of accomplishing their goals. By its investing algorithms, Bliss helps transgender people manage their finances and making savings.	Social value is created by providing solutions to the problems that transgender individuals face such as legal, medical, social and career decisions by offering a solution. The revenue stream is from helping transgender individuals plan their finances and making savings.
8	MediCapt USA 3, 5 Cloud computing	It allows clinicians to capture forensic evidence of sexual violence and securely forward it to police and justice authorities for investigations and prosecutions. Forwarding evidence to police, lawyers and judges use cloud data storage.	The MediCapt application transforms a standard medical entry form into a digital platform for forensic documentation and is combined with a secure mobile camera to facilitate forensic photography, digitisation of medical documentation is achieved and this data is stored with cloud technology, combining the fields of law and health.	Social and ethical value is created by storing, revealing and preventing the loss of evidence of violence against women and femicide. The revenue streams from more efficient, cheaper and faster data storage.

No	Company info	Value proposal (what?)	Value creation (how?)	Value capture
9	Metta Space Spain 5 AI, natural language processing	It is a platform for helping companies tackle and prevent sexual harassment, through AI- and NLP-based reporting apps for their employees.	Using a case management system for the company and an NLP algorithm, the application collects and processes data with AI and provides assistance by using the voice help button or the written form method.	Social and ethical values are created by detecting inequality and harassment from the employees' reports and the words they use when they mention the situation. NLP and AI promise objective assessments of abuse reporting.
10	Pipeline Equity USA 5 AI, cloud computing	It is a software-as-a-service (SaaS) company using AI to prevent gender bias and improve business performance in workplaces.	Pipeline's patented SaaS technology employs artificial intelligence to identify, address and combat gender bias. It may evaluate data based on a sequence of triggered events and offer recommendations that support greater financial performance for the organisation as well as individual growth through direct interaction with organisations' cloud-based human capital management system.	Social value is created through fostering gender equality in workplaces by recommendations made to organisations. In this way, a more equal environment within the institution is initiated, and equality among people is ensured.
11	Safe and the City UK 5 AI	It is a platform created to extract real-time information about location-specific risks and emergencies with a machine learning algorithm scanning news, social and official data sources to accurately detect severe weather, protests, terrorism and public health outbreaks.	i3 Intelligence accesses various open, shared and private data sources to detect personal and public safety concerns with its machine learning algorithm. i3 React provides an immediate way to relay information relevant to where the user is. The user will be the first to know about a crisis, help inform others and follow any official guidance on what to do next.	As an effect of the platform offered, social value is created by making people aware of bad events. In this case, people could feel free from discrimination and safer.
12	Salvatio Push Denmark 3, 5 AI, flexible electronics and wearables, Internet of Things	It is an emergency button and social rescue app that connects users to a community of responders with the aid of a wearable push button.	The company has developed a wearable device that consists of emergency and panic buttons. The IoT systems may inform users about possible dangers (physical attacks, sexual assaults or serious injury) in the areas, by collaboration with API and AI which can prevent death from falls or missing help.	With an IoT and AI-based mechanism, the wearable device offers a better quality of life by helping people who feel in danger in social life and helps women participate in life in comfort and to be comfortable on the streets.

(continued)

Table 7.1 (continued)

No	Company info	Value proposal (what?)	Value creation (how?)	Value capture
13	Shimmy Technologies USA 5, 10 AI	It is a company that uses AI and gamification to teach required skills to the garment industry employees, where most of them are women.	They developed an AI-based training software game, "Shimmy Upskill". It prepares workers for automation in the sector by developing workers' skills such as digital pattern making and 3D modelling.	Social value is created by developing workers' skills and abilities and preparing them for automation in the workforce. Thus, workers' right to livelihood is protected.
14	Sisterwave Brazil 5 Big data	It is a big data-based application that solves women's travel challenges.	The Sisterwave platform introduces "travellers" to "hostesses" and provides a network. Big data technology enables users to navigate the application better.	Social value is provided by women feeling more secure while travelling and earning income by providing employment. The revenue stream is provided by the app's monthly and annual membership fees.
15	Someturva Finland 5, 10 AI, cybersecurity, natural language processing	The company develops software that allows influencers to report incidents of online sexual harassment through an anonymous application. Following that, the solution assigns a specialised lawyer and social psychologist to review the case. The user then receives a personal assessment and necessary tools to resolve the situation. The software also enables them to discuss their situation and get solution proposals over a remote meeting.	The application is downloaded, dangerous situations in the Finnish application are detected, and if there is a dangerous situation, this is reported and some recommendations are received from the application.	A better-quality environment is provided by offering people a way to navigate online platforms more safely. Also, there are different usage packages at different prices according to various groupings, and income is formed in this way.
16	We Encourage Finland 3, 5 AI	The AI-powered platform supports and empowers the victims of intimate partner violence (IPV) and/or gender-based violence.	It offers an AI-powered conversation platform called "AINO Chatbot" which informs the victim about his/her rights and offers suggestions to guide the victim. The data collected from IPV survivors, professionals and strategic partners are processed via AI and used by a boot to inform and support the user.	Social value is created by providing psycho-social support and guidance to the victims of gender-based violence. They are also informed and guided about their reproductive and sexual health.

References

D. Barbieri, G. Lanfredi, B. Mollard, V. Peciukonis, M.B.P. La Horz, J. Reingardé, L. Salanauskaité, Gender Equality Index 2020 – Digitalisation and the future of work **182**, 1–30 (2020)

A. BarNir, Starting technologically innovative ventures: Reasons, human capital, and gender. Manag. Decis. **50**, 399–419 (2012). https://doi.org/10.1108/00251741211216205

P. Demartini, Why and how women in business can make innovations in light of the sustainable development goals. Adm. Sci. **9**, 64 (2019). https://doi.org/10.3390/admsci9030064

Economic Benefits of Gender Equality in the EU [WWW Document]. Eur. Inst. Gend. Equal (2021). https://eige.europa.eu/gender-mainstreaming/policy-areas/economic-and-financial-affairs/economic-benefits-gender-equality. Accessed 8 Nov 2021

European Institute for Gender Equality, What is gender-based violence? [WWW Document]. Eur. Inst. Gend. Equal (2021). https://eige.europa.eu/gender-based-violence/what-is-gender-based-violence. Accessed 8 Oct 2021

R.W. Fairlie, A.M. Robb, Gender differences in business performance: Evidence from the characteristics of business owners survey. Small Bus. Econ. **33**, 375–395 (2009). https://doi.org/10.1007/s11187-009-9207-5

P. Hepp, C. Somerville, B. Borisch, Accelerating the United Nation's 2030 Global Agenda: Why prioritization of the gender goal is essential. Global Pol. **10**, 677–685 (2019). https://doi.org/10.1111/1758-5899.12721

J.M. Horowitz, J. Fetterolf, Worldwide optimism about future of gender equality, even as many see advantages for men. Pew Res. Cent. Glob. Attitudes Proj. (2020). https://www.pewresearch.org/global/2020/04/30/worldwide-optimism-about-future-of-gender-equality-even-as-many-see-advantages-for-men/. Accessed 8 Oct 2021

Human Rights Careers, 10 reasons why gender equality is important. Hum. Rights Careers (2020). https://www.humanrightscareers.com/issues/10-reasons-why-gender-equality-is-important/. Accessed 8 Oct 2021

IISD, S.K.H., Achieve gender equality to deliver the SDGs (2017). http://sdg.iisd.org/commentary/policy-briefs/achieve-gender-equality-to-deliver-the-sdgs/. Accessed 8 Oct 2021

C. Kroll, A. Warchold, P. Pradhan, Sustainable Development Goals (SDGs): Are we successful in turning trade-offs into synergies? Palgrave Commun. **5**, 140 (2019). https://doi.org/10.1057/s41599-019-0335-5

S. Kumar Pathania, Sustainable development goal: Gender equality for women's empowerment and human rights. Int. J. Res. GRANTHAALAYAH **5**, 72–82 (2017). https://doi.org/10.29121/granthaalayah.v5.i4.2017.1797

G. Murat, Sürdürülebilir Kalkınma 2030 Gündemi Bağlamında Çalışma Hayatında Cinsiyete Dayalı Ayrımcılık. Karadeniz Tek. Üniversitesi Sos. Bilim. Enstitüsü Sos. Bilim. Derg. **7**, 7–36 (2017)

R. Roushdy, Intrahousehold resource allocation in Egypt: Does women's empowerment lead to greater investments in children? **18**, 2–5 (2004)

L. Rubalcava, G. Teruel, D. Thomas, Investments, time preferences, and public transfers paid to women. Econ. Dev. Cult. Change **57**, 507–538 (2009). https://doi.org/10.1086/596617

N. Schady, J. Rosero, Are cash transfers made to women spent like other sources of income? Econ. Lett. **101**, 246–248 (2008). https://doi.org/10.1016/j.econlet.2008.08.015

A. Shastri, Gender inequality and women discrimination. IOSR J. Humanit. Soc. Sci. **19**, 27–30 (2014). https://doi.org/10.9790/0837-191172730

R. Shi, K. Kay, R. Somani, Five facts about gender equality in the public sector [WWW Document] (2019). https://blogs.worldbank.org/governance/five-facts-about-gender-equality-public-sector. Accessed 8 Oct 2021

E. Stuart, J. Woodroffe, Leaving no-one behind: Can the Sustainable Development Goals succeed where the Millennium Development Goals lacked? Gend. Dev. **24**, 69–81 (2016). https://doi.org/10.1080/13552074.2016.1142206

UN Women (ed.), *Turning Promises into Action: Gender Equality in the 2030 Agenda for Sustainable Development* (UN Women, New York, 2018)

United Nations, *The Sustainable Development Goals Report 2018* (United Nations, New York, 2018)

United Nations, Gender equality and women's empowerment – United Nations Sustainable Development [WWW Document] (2021). https://www.un.org/sustainabledevelopment/gender-equality/. Accessed 8 Oct 2021

United Nations Development Programme, *Goal 5: Gender Equality [WWW Document]* (UNDP, 2021). https://www.tr.undp.org/content/turkey/en/home/sustainable-development-goals/goal-5-gender-equality.html. Accessed 8 Nov 2021

United Nations Women, *Sustainable Development Goal 5: Gender Equality [WWW Document]* (UN Women, 2021). https://www.unwomen.org/en/news/in-focus/women-and-the-sdgs/sdg-5-gender-equality. Accessed 8 Oct 2021

Victorian Government Directory, The benefits of gender equality | Victorian Government [WWW Document] (2021). http://www.vic.gov.au/benefits-gender-equality. Accessed 8 Oct 2021

Я.К. Мачкасова, Экологическая политика Европейского Союза в период с 2015 года и по настоящее время. Связь с Целями устойчивого развития ООН.: магистерская диссертация по направлению подготовки: 41.04.05 – Международные отношения, 2020

SDG-6 Clean Water and Sanitation

8

Abstract

The sixth sustainable development goal, Clean Water and Sanitation, is to ensure that everyone has access to safe, clean water. Everyone has the right to healthy, adequate, physically accessible and affordable water for household use under the right to water security. Acknowledging that millions of people lack access to clean water for sanitation, there is an urgent need for major investments in infrastructure and governance of water provisioning to ensure public health and increase resilience for transmissible diseases and virus outbreaks. This chapter presents the business models of 36 companies and use cases that employ emerging technologies and create value in SDG-6. We should highlight that one use case can be related to more than one SDG and it can make use of multiple emerging technologies.

Keywords

Sustainable Development Goals · Business models · Clean Water and Sanitation · Sustainability.

The sustainable Development Goals (SDGs) were formally accepted by the United Nations (UN) General Assembly on September 5, 2015, paving the way for a sustained, unified development effort on a global scale, leaving the millennium development goals (MDGs) in the dust. They are a collection of 17 goals that are anticipated to affect global social, economic and environmental policy through 2030. The sixth Sustainable Development Goal (SDG) is to ensure that everyone has access to safe, clean water. Everyone has the right to healthy, adequate, physically accessible and affordable water for household use under the right to water security (UN 2015). Although the MDGs have made progress, the goal of improving basic sanitation through access to latrines and sanitary waste collection remains unmet. In addition, the population predictions of nine billion people by 2050 imply that more work remains to be done. SDG-6 performance and its implications on other SDGs are influenced by a variety of factors at various geographical and temporal dimensions. Significantly, the actuality of SDG-6 is defined by natural limits, regulations and ethnic identities (UN 2015).

Water is a limited resource, and increased demand causes water stress, resulting from water accessibility, need and water quality. These challenges are caused by expanding human population and per capita water consumption, increasing urbanisation, the consequences of climate change, the need for additional irrigation water to boost food production and environmental needs

S. Küfeoğlu, *Emerging Technologies*, Sustainable Development Goals Series,
https://doi.org/10.1007/978-3-031-07127-0_8

for environmental preservation and biodiversity. While climate change impacts water ecosystems and water resource accessibility, socioeconomic factors increase water demand and degrade water sources (Komarulzaman et al. 2017). Briefly, SDG-6 includes local water supply objectives and governance and technologically focused objectives. Achieving the objectives of SDG-6 is essential not just for water-related concerns but also for other SDGs such as SDG-2 on zero waste and SDG-14 on life below water, as well as for the future of the Earth. Clean water is vital not just for humanity but also for flora, fauna and other associated sustainable development initiatives. Sustainable development necessitates the reduction of waste and the recycling of as much water as feasible through the use of a circular system (Gulseven and Mostert 2017). The agenda recognised the need for clean water and proper sanitation for human rights. Clean water is linked to all aspects of life, including food, nutrition, illnesses and poverty reduction. It contributes to promoting sustainable economic growth and the preservation of the planet's biosphere. SDG-6 defines eight global targets. These are all essential elements that are included in SDG-6. These targets are universally accepted, but all governments ensure the implementation of the targets according to their national liabilities (Alshomali and Gulseven 2020). In line with "Transforming Our World: The 2030 Agenda for Sustainable Development" by the UN, the eight main targets of SDG-6 are illustrated in Fig. 8.1. These targets are categorised under two headings: main targets and implementation targets. The main targets of SDG-6 are illustrated from 6.1 to 6.6, whereas the implementation targets are 6.A and 6.B.

So far, the millennium development goals have aided in mobilising the globe to enhance access to clean water and sanitation. By 2015, hundreds of thousands of people have acquired better water and sanitation access. From 2000 to 2015, the percentage of the world's population that used better sanitation climbed from 59% to 68%. This indicates that in 2015, 4.9 billion people worldwide had access to better sanitation (UN 2016). Notwithstanding, there is still a long way to go for hundreds of thousands of people

who do not have access. In 2020, slightly more than half (54%) of the world's population will access adequately managed sanitation. Yet, it's alarming that nearly one-half of the population does not. Around 6% of the population does not have access to sanitation and must practice open defecation (Ritchie and Roser 2021). SDG-6 substantially boosts the degree of expectation for the water sector, asking for universal access to safe water and sanitation while addressing challenges of water quality and shortage concerns over the next 15 years to balance the demands of the environment, energy, communities, agriculture and industry (Leigland et al. 2016).

On the one hand, poor sanitation results in financial damages due to the direct expenses of curing sanitation-related diseases and lost money due to diminished or lost production. Furthermore, poor sanitation costs time and effort owing to inaccessible or inadequate sanitation facilities, reduced product quality due to poor water quality, lower tourism income and more clean-up expenses. As a result, it is undeniable that increased sanitation significantly influences people's health and the overall economy (Van Minh and Hung 2011). On the other hand, the world will not fulfil the SDGs unless the international financial system undergoes significant change. Fulfilling the targets of SDG-6 by 2030 is not possible with the current financing. Meeting SDG goals 6.1 (i.e. ensuring that everybody has access to clean and affordable drinking water by 2030) and 6.2 (i.e. achieving access to adequate and equitable sanitation and hygiene for all and end open defecation, paying special attention to the needs of women and girls and those in vulnerable situations by 2030) is expected to cost around US$150 billion every year (sanitation and hygiene for all in a fair manner). Additional SDG-6 objectives such as protecting water-related ecosystems, minimising water pollution and adopting integrated water resource management will cost significantly more; total global WSS infrastructure development needs are expected to reach US$6.7 and US$22.6 trillion by 2030 and 2050 respectively. Although the most immediate requirements are in the Global South, high-income nations are also suffering

Target 6.1
Access to Clean Water

By 2030, ensure that everyone has access to clean and affordable drinking water.

Target 6.2
Access to Adequate Hygiene

By 2030, all people will have access to adequate and equitable sanitation and hygiene, and open defecation will be eradicated, with a specific focus on the needs of women and girls, as well as those in vulnerable situations.

Target 6.3
Reduce Pollution and Increase Recycling

By 2030, reduce pollution, eliminate dumping, and minimize the discharge of hazardous chemicals and materials, halve the share of untreated wastewater, and significantly increase recycling and safe reuse globally.

Target 6.4
Guarantee Sustainable Freshwater

By 2030, significantly improve water-use efficiency across all sectors and guarantee sustainable freshwater withdrawals and supply to solve water scarcity and minimize the number of people affected by it.

Target 6.5
Effective Management of Water Resources

By 2030, effective management of integrated water resources at all levels, including cooperation between countries with common water resources as possible.

Target 6.6
Mobilize Resources to Implement Policies to End Poverty

By 2020, In order to protect and restore Water-related ecosystems safety measurements needs to be taken such ecosystems are lakes, rivers, wetlands, aquifers ,mountains, and forests.

Target 6.a
International Collaboration for Sanitation-Related Activities

Expand international collaboration and capacity-building assistance to developing countries in water and sanitation-related activities and programs, such as water harvesting, desalination, water efficiency, wastewater treatment, recycling, and reuse technologies, by 2030.

Target 6.b
Increase Local Engagement for Water Management

Encourage and increase local community engagement in improving water and sanitation management.
(sustainable development goals: their impacts on forests and people)

Fig. 8.1 Eight targets of SDG-6. (United Nations 2021)

from severe deficits; the USA, for example, is expected to require US$1 trillion in water supply and sanitation (WSS) investment over the next 20 years (McDonald et al. 2021).

Goal 6 of the Sustainable Development Goals is to guarantee that everyone has access to clean drinking water and sanitation by 2030, at the cost of around $150 billion annually. There exist a few funding sources. One alternative is private finance in the form of direct equity investment from pri-

vate water firms and commercial bank credit. According to a study, private investments in water and sanitation have not materialised as expected due to the industry's risk-return profile. Private investors regard water and sanitation as "too complex", with inadequately attractive returns. An undiscovered resource of public funds, public banks, is one option for filling the financial gap for water supply and sanitation. Even though there are about 900 public banks globally, with assets totalling $49 trillion, academic studies and mainstream policy institutions such as the World Bank have been widely ignored as a key source of water and sanitation funding (McDonald et al. 2021). Therefore, SDG-6 has its challenges as it has its targets. SDG-6 must overcome its challenges to achieve its goal to "ensure the availability and sustainable management of water and sanitation for all" (Katila et al. 2019). Sadly, the water industry is failing to fulfil its targets, and studies imply that SDG-6 will face three main challenges (Alshomali and Gulseven 2020):

1. Finance: Since finance is the key enabler in project implementation, it is the most crucial challenge among the others. Unfortunately, the water development industry requires a vast number of financial resources and financial stability for sustainability. The lack of financial resources in underdeveloped countries makes it even more important to be careful while using existing financial resources to achieve rapid growth in the water industry. Additionally, investments in the water development industry have social and environmental advantages as well as economic ones.

2. Capacity Building: Progressively, successful governments create formal and informal institutions to achieve their goals. Unfortunately, in countries with problems with accessing and managing water resources, in Sub-Saharan Africa, Asia and Southeast Asia, there is a severe lack of capability and execution in water-related problems. These parts of the world are at the top of the human shortages of essential needs such as agriculture, safe drinking water, sanitation and risks related to water impacts, water waste and recycling. Sadly, these countries have been experiencing these concerns for decades.

3. Governance: In many industries, successful and fair governance is critical, but it is especially critical in the water development industry. Many of the developing countries are still coping with internal problems such as conflicts and public health-related problems. Also, their governance structures are not stable. Efficient water management requires

political stability, institutional rules, administrative management, wise decision-making and related implementations. These gaps can be solved by the government using data accountability.

SDG-6 is critical to achieving sustainable development, as access to safe drinking water and adequate sanitation are human rights (UN-Water 2021). The availability of these services, especially water and soap for handwashing, is critical to human health and well-being. They are necessary for improving nutrition, preventing disease and providing healthcare, as well as guaranteeing the smooth operation of schools, workplaces and political institutions, as well as disadvantaged and marginalised groups' full involvement in society. The evidence for the negative health effects of inadequate water and sanitation is overwhelming. Poor water and sanitation can lead to a wide range of severe diseases, such as diarrhoea (Howard et al. 2016). In 2017, approximately 1.2 million people died as a result of contaminated water sources, equivalent to 2.2% of all deaths worldwide (Ritchie and Roser 2021).

As mentioned in the recent Summary Progress Update for SDG-6 (UN-Water 2021), acceleration in future action depends on several factors. One of the main bottlenecks is that due to policy and institutional disintegration between different levels, sectors and actors, decisions are taken in one area, or the sector usually falls short of considering the impacts on water quality and availability in other areas. Such fragmentation along with funding gaps and lack of data and information sharing across sectors and borders result in problems in informed decision-making. Furthermore, the implementation of SDG-6 targets is slowed by institutional and human capacity deficiencies, particularly at the local government and water and sanitation provider levels, as well as inadequate infrastructure and governance models. Five accelerators to drive action at a larger scale were proposed to overcome the problems. Optimising finance, improving data and information, capacity development, fostering innovation and effective governance are all needed for delivering SDG-6 results for the future.

There is also a change needed in three professional perspectives that guide water policy. These are economics, management and engineering (Sadoff et al. 2020). Economics should not treat water resources as abundant resources to minimise the costs of its provision. Economics as a discipline should acknowledge the value of water as the scarce key resource and not the capital that is required for its provision. Water engineering also needs to be revised in a way that needs and goals-based perspectives replace the linear and centralised approach to water engineering. This means that by using the existing technology, water engineers should design wastewater recycling systems and differentiate between the sources of water, its costs and qualities to utilise each for certain needs and goals better. "Water engineering needs to move beyond the concepts of reliability and optimality, which evaluate designs over a narrow set of objectives and possible future conditions, to focus on robustness and flexibility in the face of uncertainty" (Sadoff et al. 2020). Finally, water management must increase its capacity to deal with complexity and trade-offs. Adaptive and integrated water management is required in an uncertain environment to account for interconnections, changes and potential surprises. Integrated techniques help identify and minimise trade-offs, as well as the unravelling of unforeseen consequences. They also help promote inclusiveness in water management, as different stakeholders from different sectors at all scales are brought together.

Finally, in line with the impacts of the recent COVID-19 outbreak, acknowledging that millions of people lack access to clean water for sanitation, there is an urgent need for major investments in infrastructure and governance of water provisioning to ensure public health and increase resilience for transmissible diseases and virus outbreaks.

8.1   Companies and Use Cases

Table 8.1 presents the business models of 36 companies and use cases that employ emerging technologies and create value in SDG-6. We should highlight that one use case can be related to more than one SDG and it can make use of multiple emerging technologies. In the left column, we present the company name, the origin country, related SDGs and emerging technologies that are included. The companies and use cases are listed alphabetically.[1]

---

[1]For reference, you may click on the hyperlinks on the company names or follow the websites here (Accessed Online – 2.1.2022):

http://worldswaterfund.com/; http://www.fluidrobotics.com/; https://4lifesolutions.com/; https://aquarobur.com/; https://asterra.io/; https://boxedwaterisbetter.com/; https://cropx.com/; https://enbiorganic.com/; https://fredsense.com/; https://orbital-systems.com/; https://lifestraw.com/; https://metropolder.com/en/; https://puralytics.com/; https://rentricity.com/; https://robonext.eu/case-studies/water-link/; https://uptraded.com/; https://wasserdreinull.de/en/; https://watly.co/; https://www.aguardio.com/; https://www.bluetap.co.uk/; https://www.ecoworth-tech.com/; https://www.sarastear.com/en/; https://www.hibot.co.jp/; https://www.hydraloop.com/; https://www.idrica.com/goaigua/drinking-water/#; https://www.innovyze.com/en-us; https://www.lishtot.com/tech.php; https://www.mwater.co/; https://www.oceowater.online/oceo/website/index; https://www.oleanetworks.com/; https://www.ranmarine.io/; https://www.semillasanitationhubs.com/; https://www.solarscaremosquito.com/; https://www.takadu.com/; https://www.zte.com.cn/global/; https://zwitterco.com/

Table 8.1 Companies and use cases in SDG-6

No	Company info	Value proposal (what?)	Value creation (how?)	Value capture
1	4Life Solutions Denmark 6 Biotech and biomanufacturing	4Life solutions provides a product-based solution for safe and affordable drinking water for low-income communities where sanitation is costly and access to safe water is problematic	SaWa, the startup's product, is a bag that holds up to 4 litres of water and uses ultraviolet (UV) rays and heat from the sun to kill microorganisms in water. After being exposed to the sun for 4 h, the water is safe to drink. The technology only requires the power of the sun to remove dangerous pathogens making water safe to drink without burning fossil fuels	The SaWa bag is reusable for 1 year. The solution it offers for sanitation of water is cheaper than using charcoal, and at the same time, it reduces carbon emissions by displacing emission-intensive processes. The company was able to provide more than 45,000 people with clean water while reaching more than 7000 households. The company also partners with a healthcare project in Tanzania where the product is provided as an additional benefit to women during their pregnancy and children
2	Aguardio Denmark 6, 11 AI, internet of things	It is an IoT- and AI-based device that guides and motivates hotel guests to reduce their shower time to save water, energy and money	Aguardio G2 (product of Aguardio) starts monitoring water usage as soon as a person enters the shower cabin and turns the water on or off by using the combination of IoT and big data. IoT sensors gather the data which is used for predicting the water usage habits of the customers with the help of AI	Revenue creation is achieved by savings in terms of water, energy and money. The company claims that reducing 20% of shower time results in savings from €60 to €90 per hotel room/year, which is also a 15-month ROI
3	Aqua Robur Technologies Sweden 6 Big data, datahubs, internet of things	It is a hub (also known as Fenix hub) that enables monitoring pipelines by digitalising the network with the use of IoT and big data	With quick and cost-efficient installations, the Fenix hub enables getting data easily and managing assets more effectively by IoT sensors. The final data is shared via IoT and utilises big data to analyse for better water management	Revenue streams come from savings. The company provides extra data through IoT sensors to monitor pipelines in order to prevent water losses due to leaks and to control water distribution. This allows saving valuable time and utilising the information more efficiently
4	ASTERRA USA 6, 12 Big data	It's a satellite-based monitoring system that uses computer vision technology and big data technologies to find and analyse moisture seeping from subterranean pipelines	The system determines leaking points with an "eye-print" of more than 3500 sq. km, collects and analyses geographical image data via satellite deploying ground-penetrating synthetic aperture radar and turns it into predictive insights by using ML algorithms	Revenue creation is achieved with the monitoring system that saves the world over 9200 million gallons of drinkable water, 21,800 MWh of energy annually, and ground crews can also be more efficiently used by utilising big data

(continued)

Table 8.1 (continued)

No	Company info	Value proposal (what?)	Value creation (how?)	Value capture
5	Blue Tap UK 1, 6 3D printing	It is a company that uses 3D printing to improve access to high-quality drinking water in low-resource settings	By using 3D printing, the company manufactures filters to make water drinkable	By the use of 3D printing, the company offers cheaper and faster solutions, unlike conventional methods. It also protects against recontamination which is very common during water collection and storage
6	Boxed Water USA 6 Recycling	The firm provides an alternative to aluminium cans and plastic water bottles via its sustainably produced purified water in 92% plant-based sustainable packaging	The firm provides an alternative to aluminium cans and plastic water bottles via its sustainably produced purified water in 92% plant-based sustainable packaging	The products are 92% renewable with its plant-based sustainable packaging and 100% recyclable and refillable with some common sense care. Replacing conventional harmful materials with ones that are reusable has a positive environmental impact
7	CropX Israel 6, 15 AI, cloud computing, internet of things	The company's services are in optimising and automising farm management with the help of AI, using a scalable cloud platform to integrate data from both above and below ground making use of IoT	Their cloud platform receives data from IoT sensors, and by using ML algorithms, they provide insights. Also, by factoring in thousands of data points, AI helps predict water-uptake patterns and detect faulty irrigation systems and burst pipes	Using their AI-powered adaptive irrigation service that automatically optimises irrigation, they provide crop yield increase and water and energy cost savings to farms
8	EcoWorth Tech Singapore 4, 6, 14, 16 Advanced materials	Using carbon fibre aerogel technology, this Cleantech startup focuses on waste to value-creating applications in industrial wastewater treatment and oil and gas decontamination	Carbon fibre aerogel (CFA) technology is a highly absorbent material that is non-toxic, natural and recyclable. CFA is applied by incorporating the material into industrial-grade cartridges and used as filters	Revenue is captured via product solutions that ensure consumers' financial savings by providing at least twice the absorbency rate compared with other competitors, translating to at least twice the savings on material cost
9	EnBiorganic Technologies USA 6 Biotech and biomanufacturing	Through microbiology, the company purifies water and provides waste solutions for municipalities and lagoons	Their microbiology consists of several strains of naturally occurring soil bacteria that replace traditional microbiology for wastewater treatment solutions. This microbiology is used in water cleaning and preventing wastewater	They create sustainable wastewater treatment solutions by using natural microbiology. It reduces energy consumption and increases efficiency via this biotechnology

(continued)

Table 8.1 (continued)

No	Company info	Value proposal (what?)	Value creation (how?)	Value capture
10	Fluid Robotics India 6 Robotics	Through the use of a robotic system and information technology, the company is assisting cities in India to have better control over water pollution caused by untreated wastewater mixed in rivers, lakes and groundwater	A robotic system is used for mapping and digitising underground pipelines in cities to detect structural defects and to gather information about the capacity of pipelines. The robotic system is able to survey lines with a minimum calibre of 150 mm	The design of the company's robotic equipment enables access to difficult-to-reach and blocked spaces. Structural analysis of the pipelines helps prevent water leakages, sanitary sewer overflows and monsoon flooding. This provides relevant authorities with useful information to take necessary measures to prevent damage and with an opportunity to increase resiliency against the impacts of climate change in the region
11	FREDsense Canada 6, 14 Big data, internet of things	It is a platform that helps detect chemicals in water such as arsenic, iron and magnesium by using technologies such as IoT sensors and big data analytics	It builds fast, sensitive and easy-to-use IoT sensors to detect chemicals in the water which help to spend less time acquiring data and making decisions. The data gathered from the sensors are used for the platform. The platform has a built-in multifaceted way for the rapid customisation of sensors for field use	Revenue creation is achieved through savings in terms of time and money. By using IoT and big data analytics, the company offers a cheaper, faster and more accurate solution for monitoring chemicals in the water
12	Hibot Japan 6 Robotics	The company uses integrated robotic systems to assess water pipes' conditions to help secure the underground infrastructures of a city	It uses algorithms and robots to figure out which pipes in a city or town are most likely to need replacement by assessing pipes' conditions	Revenue creation is achieved by decreasing maintenance fees. By the use of robotics, the company claims that the system can predict future failure rates within 80–90% accuracy rate. This allows companies and municipalities to better allocate their resources and target the pipes that are most likely to cause problems soon
13	Hydraloop The Netherlands 6 Recycling	The company has created a product that collects, cleans and reuses water from home appliances and showers through its water recycling system	The product collects used water from water pipes and purifies wastewater through its water recycling system. It is possible to use the water in toilet flushing and washing machines and for irrigation	The company has created a product that collects, cleans and reuses water from home appliances and showers through its water recycling system

(continued)

Table 8.1 (continued)

No	Company info	Value proposal (what?)	Value creation (how?)	Value capture
14	Idrica Spain 6, 9 Digital twins	It is a digital transformation company in the water industry, delivering services such as digital twins and the technological solution "GoAigua" to manage the entire water cycle	The company creates a digital twin of the water management system to analyse the system better. All the leaks and consumption data can be monitored via this system	Throughout the implementation of digital twin technology for the water management system, handling any problem that occurs during the process becomes easier and cheaper
15	Innovyze USA 6 AI, digital twins	Emagin, which is a product of the company, is a digital twin-based AI-powered platform for water utility operations and process control. It aims to help operators with water operations	The product builds a digital twin of your system from process flow diagrams and is trained with historical SCADA data. Operators can review system conditions in real time and choose an optimal operating mode based on chosen targets. Machine learning is used to enable prescriptive actions that are intelligently delivered to operators	The digital twin of the process helps the operators by saving time and supporting them during their decision-making process. ML helps to lower operating expenses and increase operational efficiency while reducing the carbon footprint of the factory
16	LifeStraw USA 6 Advanced materials, biotech and biomanufacturing	The company is a producer and distributor of straw-like personal water filters, family purification units, portable units and large-scale purification units for institutional use in schools and health clinics	Via advanced materials such as membrane microfilters, membrane ultrafilters and purifiers and activated carbon and ion exchange filters, the products can remove bacteria and parasites, remove harmful chemicals and viruses, filter microplastics and reduce lead	Consumers with limited access to clean water will benefit from the development of a system to readily clean water
17	Lishtot Israel 6 Big data, internet of things	The company manufactures a smart water metre that helps track water use. Water metres connect to WiFi and provide information, such as the usage patterns and leaks, via mobile applications	Water metres measure water utilisation and provide real-time statistics to users' phones by using artificial intelligence and IoT technology. AI detects patterns of water use to ensure optimisation of consumption with the assistance of IoT sensors	Through water metres, they offer their users the ability to prevent waste and excessive bill costs. Also with the options provided in the mobile app, users are able to set consumption goals that could help enable sustainable use of water
18	MetroPolder The Netherlands 6 Big data, internet of things	The company brings IoT-based solutions for flooding in cities by storing all rain that falls on the roof and disposing of or using them at a later time	A buffer system, an internet-connected weir and an online dashboard make up the polder roof. The smart drop weir monitors and controls water levels and run-off. The customised dashboard provides access to information about the roof and operating system	The company creates an environmental value by reviving cities through smart water management and leveraging stored rainwater for cooling, growing and fostering nature within the city

(continued)

Table 8.1 (continued)

No	Company info	Value proposal (what?)	Value creation (how?)	Value capture
19	mWater USA 6 AI, big data	The company offers an operating system that collects and monitors data using AI and big data on water provisioning, sanitation and population. The business is women-owned and engages in nonprofit projects with international organisations and for-profit projects with the US government and private sector	It uses AI and big data to give access to water sanitation and hygiene monitoring, open-source population density data, mappable data and satellite imagery for preventative actions against cyclones and floods. The business has combined open-source population density data with AI and big data to create a layer that anybody can use when planning or monitoring WASH (water, sanitation and hygiene) projects	It creates a significant value by deriving census data using AI from satellite imagery which is critical support for projects undertaken by the private sector and governments. The business enables reliable data systems for water service providers which encourages investment in different regions of the world. The data platform has over 100,000 accounts that belong to users from the water sanitation and hygiene sector in a total of 184 countries
20	Nymphea Labs France 3, 6 Healthcare analytics, internet of things	It is an IoT-based solar-powered platform that enables monitoring and controlling of mosquito-caused diseases such as Zika and dengue virus and malaria	The platform can be placed to any water sources such as rivers, lakes and dams and enables mosquitos to lay eggs. Then, intermittent ripples caused by the water source suffocate and kill the larvae. The company is developing a new version of the product which is capable of identifying the species, sex and potential viruses. The system will also be capable of alarming the population and the authorities via mobile app. The data collected will be shared openly to encourage research and innovation	The product enables early detection of potential epidemics which allows for preventing deaths. Only in 2020, 627,000 people died because of malaria, and 80% of the deaths were children under 5. Considering that the African region suffers the most from the health impacts and spends around 12 billion dollars a year for fighting malaria, the product can mitigate the economic costs of fighting epidemics, while contributing to health improvements and the Well-being of the region
21	OCEO Water India 6, 9, 12 Internet of things	It is an IoT-enabled smart water purifier integrated with AI that provides customers with healthy water in customers' homes without purchasing a device or paying a maintenance fee	Each device has intelligent sensors that gather data and are also connected to live servers on a real-time basis, allowing for engaged interaction between devices and users. The data flow helps monitor devices both on a micro- (individual households) and macro-level (town/city). ML is used while remotely monitoring to assure safety in every water drop	It creates social value by making it possible to reach clean water for everyone via distributing ocean water cleaning devices for free. By use of IoT sensors, the company monitors water consumption and its quality. It also helps to improve remote asset maintenance and filter life health checks through machine learning

Table 8.1 (continued)

No	Company info	Value proposal (what?)	Value creation (how?)	Value capture
22	Olea Edge Analytics USA 6, 11 Edge computing	It is an edge computing platform for the water utility industry	After basic metre data given by the official water department is investigated, a team surveys the metres, deploys sensors and edge computers and reviews the information. A user interface with real-time information, recommendations and reports is set up. Lastly, actions are provided to the water department	Edge computing was used to detect probable malfunctioning water metres, which resulted in a 60% accuracy in water use
23	Orbital Sweden 6 Internet of things, recycling	They offer a water recycling technology to be used in domestic appliances, especially for showering	Water quality is monitored 20 times per second by the sensors. Filtered water that is too unclean to filter is replaced with new, clean water. The water passes through the purification loop the rest of the time, screening away any contaminants	Thanks to orbital shower, one can save up to 90% of the showering water, thus decreasing the domestic water consumption
24	Puralytics USA 6 Advanced materials, biotech and biomanufacturing	The company utilises a water purification technology that uses a light-activated nanotechnology-coated mesh that can both disinfect and detoxify water. There are no chemical additives and 100% of the water is purified	The light and nanomaterials combine five photochemical processes, which are photodisinfection, photolysis, photoabsorption, photocatalytic oxidation and reduction. Puralytics treatment modules or stand-alone systems are scalable in treatment levels and inflow, from millions of gallons per day to less than 1 gallon per minute	After use, the nanomaterial inside the product is not consumed or broken down. When compared to conventional items that degrade after a year, the company provides environmental value through nanomaterials
25	RanMarine Technology The Netherlands 6, 8, 11, 13, 14 Datahubs, drones, recycling	It is an aquatic drone that removes biomass from water, collects wastes like plastics through contributing to recycling and enables analysing the quality of water via machine learning statistics	By monitoring nitrogen accumulation, dissolved oxygen and pH levels through its sensors, it removes biomass and collects wastes from water. It also measures the quality of water with the data collected that is available for reporting and analysis through the datahub	It removes the biomass from the surface, which helps preserve nature with minimal disruption and eliminates operational risks. It also removes half a ton of waste from water per day, offers a way to recycle through recycling facilities and makes water knowledge accessible by sharing water quality with users in real time

(continued)

Table 8.1 (continued)

No	Company info	Value proposal (what?)	Value creation (how?)	Value capture
26	Rentricity USA 6 Energy storage, recycling	It is a renewable electricity generator that targets water utility companies and other large fluid conveyors that install pressure-reducing valves (PRVs) within their pipelines to maintain pre-set pressure ranges. The system harnesses excess pressure within water operations and uses it to generate clean electric power. Rentricity also offers custom information services	Electricity is generated by hydrokinetic power generation applications which are basically generated by harnessing energy from flowing water, such as the water that flows through municipal drinking water pipes. Continuously flowing water through mains offers the capability to generate electricity year-round, 24 h per day. Each Rentricity system continuously produces between 5 and 350 kilowatts of clean and reliable energy. The system is installed in parallel with existing pressure reduction valves to recover energy that these valves would otherwise waste	Besides producing green and renewable energy, water utilities currently use about one-third of their operating budget to pump water, and if this pressure is not used, it is completely wasted. Rentricity systems are located within existing water systems reducing permitting, installation time and environmental issues. The system produces enough to satisfy the demand of between 30 and 300 average homes or a portion of an industrial facility
27	RoboNext Belgium 6 AI, robotic process automation	It is a company that is implementing and managing robotic process automation (RPA) and AI for the customer companies to improve their services. Its project with water-link which is a complete water and sanitation facility includes AI and RPA	RoboNext examines the facility to identify the greatest use cases to automate by appropriately engaging the RPA platform, in which the technological leader is UiPath. RoboNext installed invoicing and master data management in UiPath, resulting in the automation of ten LAC systems	Water-link facilitates the complete water and sanitation needs of over 610,000 consumers and produces 150 million m^3 of drinking water per year. After RPA and AI applications, the productivity is increased, and several processes focused on invoicing and payment follow-up have become more efficient
28	Sarastear Japan 6 Internet of things	The company's water generating equipment converts ambient humidity into drinkable water	Following the removal of dirt and dust from the air via air filters, the air is converted into water droplets by a heat exchanger that condenses moisture to make water. Clean water is generated after the water has been filtered further. The final product of repeating the filtering circulation is clean and drinking water	The Sarastear water server project, which aims to manufacture drinking water from air, employs the vision of a future in which everyone in the globe may simply consume safe and secure water. Individuals and communities that currently have a limited access to clean water would benefit greatly from this technology. Furthermore, in the case that water becomes a much more scarce resource, the development of this technology would greatly benefit everyone

(continued)

Table 8.1 (continued)

No	Company info	Value proposal (what?)	Value creation (how?)	Value capture
29	SEMILLA Sanitation The Netherlands 6, 7, 12, 14 Recycling	It is a startup that develops sanitary wastewater treatment units. It uses biological or physical treatments to channel wastewater streams into safe drinking water, irrigation water and compost. Flexible modules of the company can also provide biogas and electricity	Modules of the company form a closed wastewater treatment unit, using technologies imported from advanced space technology and recycling to convert sanitary wastewater into clean water and nutrients for food production	Revenue is created by enabling the reuse and recovery of water as well as reducing the overall water consumption. After the sanitation process, it also provides nutrients for food production. The company also brings biogas and electricity to underdeveloped countries along with toilets and washing and showering facilities which result in added social value
30	TaKaDu Israel 6 AI, big data, internet of things	The company assists water utilities to manage their networks more efficiently by detecting, analysing and managing the usage patterns through extensive data gathering and processing via AI and machine learning	The company minimises financial damage by exposing hidden leaks and water loss with the data from IoT sensors. Information regarding anomalies is collected in operations through machine learning algorithms via sensors that receive data from water utilities	They offer water utilities the chance to increase savings from water loss reduction, efficient use of human resources and savings in operational costs. The company has helped avoid 1000ML in non-revenue water loss and over $1.3 M in rebates for non-supply, along with 200,000 property impacts. In a 1-year period, the customer utilities on average are able to save close to one million dollars
31	Uptraded Germany 6, 12 AI, recycling	The platform offers the cheapest and most sustainable ways to keep wardrobes up-to-date by an AI-based recommendation engine	It is a novel online clothing marketplace where your own unused garments become the currency. An intelligent matching system based on a swipe mechanism identifies exchange options for your uploaded garments. An algorithm evaluates the user behaviour and can therefore suggest the best possible exchange options (e.g. suitable style, the similar market value of the items to be exchanged, correct size, etc.). Users can then change their clothes and win-win situations arise. Through this form of transaction, the customers achieve the maximum added value for their clothing from their subjective point of view	Customers achieve the maximum added value for their clothing by reusing them. The excessive amount of water spent in the textile sector, together with the reuse of clothes, reduced the individual water consumption of the users of the application

(continued)

Table 8.1   (continued)

No	Company info	Value proposal (what?)	Value creation (how?)	Value capture
32	Wasser 3.0 Germany 6 Advanced materials, bioplastics, biotech and biomanufacturing	The company focuses on the removal of micropollutants and microplastics such as pharmaceuticals, PFAS, heavy metals and pesticides from various water sources	The combination of high-tech materials, low-tech processes and a systemic perspective enables a new level of environmental and health protection in (waste) water treatment. By applying a non-toxic hybrid silica gel, they create table tennis ball-sized balls that float on the water surface and are easily removed from there	The introduction of microplastics and micropollutants into the entire ecosystem is now uncontrollable. Microplastics and micropollutants are found in human and animal bodies, which is extremely unhealthy. All users profit from the company's removal of these toxins
33	Waterfund USA 6 AI, crowdfunding	It is a worldwide water investment and trading business focused on purchasing and managing infrastructure assets that produce and supply clean water to the highest-value end-users using desalination and wastewater treatment technologies	They provide funds via forecasting and modelling tools on market solutions through crowdfunding to various water treatment and desalination projects in order to rationalise consumption and promote sustainable water consumption models	They implement projects for cleaning water, separating sea salt from water and wastewater management at low cost. They make clean and drinkable water accessible
34	Watly Italy 6, 12 Advanced materials	It is a solar-based water purifying system that uses graphene technology	Through cutting-edge galvanisation and graphene-based technology, water is separated from soaps, solvents and hydrocarbons. Water, in an arch shape, is evaporated with the heat power generated by solar panels	A significant impact of the value creation can be demonstrated as the ratio of water input and water output which is 99.5%, when the water is not just contaminated, but also scarce (reverse osmosis ratio is 30–50%)
35	ZTE China 6 5G, autonomous vehicles, spatial computing	The company offers 5G and spatial computing-based water management solutions provided by the company and China telecom to provide intelligent management and three-dimensional monitoring of water. It aims to increase operational efficiency in water operations	For inspection, a UAV (unmanned aerial vehicle) is utilised, and for real-time video checks, an unmanned ship patrol is used. This enables three-dimensional monitoring of the water source location, as well as data from the attached water quality sensor. Because of the 5G network's high bandwidth and low latency, UAVs can conduct visual inspections of the lake surface. The goal is to detect foreign things (such as solid trash) on the water's surface. Those objects will be removed if they are found unsuitable	Revenue creation is achieved through operational efficiency. Compared with the conventional water management systems, the solution allows three major innovations. Firstly, 5G allows monitoring water resources with 4 K quality videos since it uses high bandwidth and low time delay. Secondly, AR/VR is combined with 5G to carry out and control the video monitoring. Also, remote control of UAVs is achieved by AR. Lastly, based on the ZTE IoT platform, AI and big data technologies are used to establish a water quality model

(continued)

Table 8.1 (continued)

No	Company info	Value proposal (what?)	Value creation (how?)	Value capture
36	ZwitterCo USA 6 Advanced materials	The company uses zwitterionic (a chemical molecule known as inner salt) copolymer membranes which prevent organic components from clogging membranes' pores during filtration	A zwitterion molecule includes both negatively and positively charged groups in close proximity. During filtration, these groups pull water to the membrane made of zwitterion while pulling organic components which handicaps filtration	By increasing water reuse capabilities, ZwitterCo enables filtration to be more material-efficient and energy-efficient and therefore cost-efficient compared to conventional treatment methods. The use of membrane technology in bioprocessing, agriculture and food and beverage sectors contribute to the reuse of wastewater, clean water generation and higher-yield concentration in agriculture. Therefore, it allows for both economic and environmental gains

References

I. Alshomali, O. Gulseven, A note on SDG 6 -clean water and sanitation for all. (2020). https://doi.org/10.13140/RG.2.2.16461.38881

O. Gulseven, J. Mostert, Application of circular economy for sustainable resource Management in Kuwait. Int. J. Soc. Ecol. Sustain. Dev **8**, 87–99 (2017). https://doi.org/10.4018/IJSESD.2017070106

G. Howard et al., Climate change and water and sanitation: Likely impacts and emerging trends for action. Annu. Rev. Environ. Resour. **41**(1), 253–276 (2016). https://doi.org/10.1146/annurev-environ-110615-085856

P. Katila et al. (eds.), *Sustainable Development Goals: Their Impacts on Forests and People*, 1st edn. (Cambridge University Press, 2019). https://doi.org/10.1017/9781108765015

A. Komarulzaman, J. Smits, E. de Jong, Clean water, sanitation and diarrhoea in Indonesia: Effects of household and community factors. Glob. Public Health **12**(9), 1141–1155 (2017). https://doi.org/10.1080/17441692.2015.1127985

J. Leigland, S. Trémolet, J. Ikeda, *Achieving Universal Access to Water and Sanitation by 2030 The Role of Blended Finance* (2016), p. 20

D.A. McDonald, T. Marois, S. Spronk, Public Banks + Public Water = SDG 6? **14**(1), 18 (2021)

H. Ritchie, M. Roser, *Clean Water and Sanitation* (Our World in Data, 2021) Available at: https://ourworldindata.org/sanitation. Accessed 11 Aug 2021

C.W. Sadoff, E. Borgomeo, S. Uhlenbrook, Rethinking water for SDG 6. Nat. Sustain. **3**(5), 346–347 (2020). https://doi.org/10.1038/s41893-020-0530-9

UN, *Transforming our world: the 2030 Agenda for Sustainable Development* (Department of Economic and Social Affairs, 2015) Available at: https://sdgs.un.org/2030agenda. Accessed 28 July 2021

UN (2016) Goal 6: Ensure availability and sustainable management of water and sanitation for all — SDG Indicators. Available at: https://unstats.un.org/sdgs/report/2016/goal-06/. Accessed 11 Aug 2021

UN-Water, *Summary Progress Update 2021: SDG 6 — Water and Sanitation for all* (Geneva, Switzerland, 2021), pp. 1–54

H. Van Minh, N.V. Hung, Economic aspects of sanitation in developing countries. Environ. Health Insights **5**, EHI.S8199 (2011). https://doi.org/10.4137/EHI.S8199

SDG-7 Affordable and Clean Energy

Abstract

Reaching affordable, clean, sustainable, modern and reliable energy is the main aim of the Sustainable Development Goal 7. Energy is placed at the centre of environmental and economic issues. Despite this significance, 20% of people living worldwide cannot access electricity in 2021. Adaptation towards SDG-7, Affordable and Clean Energy, brings in new investments and creates a significant economy around it. While private investments and government spending in developed countries concentrate on achieving efficiency and renewable energy production, developing countries focus on obtaining access to electricity and clean energy sources. This chapter presents the business models of 60 companies and use cases that employ emerging technologies and create value in SDG-7. We should highlight that one use case can be related to more than one SDG and it can make use of multiple emerging technologies.

Keywords

Sustainable development goals · Business models · Affordable and clean energy · Sustainability

Reaching affordable, clean, sustainable, modern and reliable energy is the main aim of the Sustainable Development Goal 7. Energy is placed at the centre of environmental and economic issues. Despite this significance, 20% of people living worldwide cannot access electricity in 2021. Also, the utilisation of renewable energy sources must increase because of the high-level demand for energy. For example, the ratio of people reaching for clean energy for cooking has increased from 50% in 2010 to 66% in 2019 (United Nations 2021a). The renewable and non-renewable energy sources are obtained from the environment. Examples of these energy sources are coal, natural gas, petroleum, hydropower, solar wind, etc. Energy sources, including fossil fuels and renewables, are in high demand worldwide. Despite excellent progress in expanding access to power, increasing the use of renewable energy in the electrical sector and improving energy efficiency during the past decade, the world remains far from obtaining cheap, dependable, sustainable and contemporary energy for all. With an average yearly electrification rate of 0.876 percentage points, the worldwide power

The author would like to acknowledge the help and contributions of Berk Ürkmez, Görkem Balyalıgil, Ali Emir Güzey, Samet Özyiğit, Kerem Şen and Esra Kılıç in completing this chapter. They also contributed to Chapter 2's Advanced Materials, Blockchain, Energy Storage, Hydrogen and Internet of Things sections.

access rate improved from 83% in 2010 to 90% in 2019. The worldwide access deficit has shrunk from 1.22 billion in 2010 to 759 million in 2019. Despite significant efforts, there may still be 660 million people without access to electricity in 2030 (United Nations 2021b). Whereas bioenergy is the most common renewable energy source, approximately three billion are utilising the energy sources such as wood, coal, animal waste and so on (United Nations 2021a).

Since 2010, the number of people who have gained electricity access has exceeded a billion, making it possible for 90% of the world population to be linked in 2019. Despite this, 759 million people lack access to electricity, and the majority of these people live in areas that are unstable or affected by conflict. While regional differences continue to exist, the global electricity access deficit concentration is located in Sub-Saharan Africa, accounting for 75% of the gap on a global scale. The number of people without access to electricity is less than 2% of the population in Eastern Asia, Southeast Asia, Latin America and the Caribbean. In Sub-Saharan Africa, this number is around 50%. Within countries with major access deficits, Bangladesh, Kenya and Uganda have shown the largest improvements since 2010. This is attributable to more than three percentage points annual electrification growth rates, largely due to an integrated approach that combines grid, mini-grid and on-grid solar electrification (IRENA 2021a).

According to new data released this year, the number of individuals without access to low-carbon cooking equipment has steadily decreased. Since 2010, around 450 million individuals in India and China have acquired access to clean cooking due to clean air policies and liquefied petroleum gas (LPG) transition programmes. The problem in Sub-Saharan Africa is very severe, and the situation is deteriorating. Only 17% of the population has access to safe cooking water. Over 2.6 billion people in the world still lack access, and household air pollution, caused by cooking smoke, is responsible for roughly 2.5 million early graves each year (IEA 2020).

From 63 (56–68)% in 2018, the percentage of people in the world who have access to clean cooking fuels and technology rose to 66% (confidence ranges of 59–71%) in 2019. There were 2.6 (2.2–3.1) billion people in the world who did not have access to the Internet. Clean fuels and technologies were only nine percentage points more accessible in 2018 than in 2010 when 57% (52–62%) of the world population had access. According to current trends, the world will not be up to the 2030 universal access objective by about 30%, reaching only 72% of the population. To accomplish the aim of universal access to sustainable fuels and technology by 2030, annual increases of more than three percentage points would be required (*The Energy Progress Report 2021* 2021).

Renewable energy comes from naturally renewing but flow-limited sources; renewable resources are nearly limitless in terms of duration but have a limited amount of energy per unit of time (EIA 2020). Technological advancements have increased the consumption and interest in renewable energy sources, owing to rising pollution and rapid fossil fuel usage. Many nations debate energy, energy security and global warming, and rules are being developed in this context. The United Nations (UN) is responsible for the most important pioneering study in this subject. The energy sources listed below are the most well-known renewable energy sources.

The energy produced by the use of sunlight and heat is called solar energy. Solar energy has different uses, and these can be generating electricity from solar energy or heating air and water. The use of solar energy in the world is increasing day by day, and by the end of 2020, more than 700 GW of energy will be produced from the sun, which will equal approximately 3% of the energy consumed in the world (ARENA 2021).

Winds occur in various regions of the world depending on atmospheric events. Winds, moving air, cause kinetic energy, which can be converted into electrical energy with the help of wind turbines (IRENA 2021a, b). The use of both onshore and offshore wind turbines has increased significantly in the last 20 years for electricity generation from wind energy. While there was a production capacity of 7.5 gigawatts (GW) in 1997, this figure increased approximately 75 times in 2018 and reached 564 GW. Electricity generation using wind energy has more than doubled from 2009 to 2013 and equalled 16% of the energy produced using renewable energy sources in 2016 (IRENA 2021b).

One of the commercially developed renewable energy technologies is hydroelectric energy technologies. In this technology, a reservoir for water is created, usually using a large area. The water accumulated in this reservoir is released when energy is needed, and it turns the turbines into the dam to generate energy. The difference between this energy source and other renewable energy sources is that it provides continuity. While solar energy and wind energy can sometimes cause fluctuations in production in energy systems, energy can be produced when needed in hydroelectric energy systems (EDF 2021).

Geothermal energy originates from the heat underneath the Earth's surface. The causes of this heat are the actions that take place inside Earth and the planet's structure. Although this heat is practically endless, it is not allocated evenly under every part of the surface (Barbier, 2002). Places where geysers or fault lines are located are more suitable for geothermal energy production. Produced energy can be utilised in heating and electricity. Worldwide total installed geothermal energy capacity was 14,013 MW in 2020 (IRENA 2021a).

Biomass mostly derives from organic waste generated by plants and animals. It can be used to produce liquid, gaseous or solid biofuel by biological, chemical or thermochemical conversion processes and to produce heat by directly burning it (EIA 2021). Worldwide total installed bioenergy capacity as liquid, gaseous or solid biofuels was over 130,000 MW in 2020 (IRENA 2021a).

The ratio of the amount of consumed energy and created specific output is called energy efficiency. In final usage areas such as manufacturing, buildings, construction, agriculture and transportation, highly efficient technological instruments are key to achieving efficiency goals. Emission savings and air quality are directly associated with energy production and efficiency levels, so any effort to improve efficiency in final

usage areas and electricity production would contribute massively against the GHG problem. Governmental energy policies are major impulses for efficiency and consumption rates, but private efforts such as digitalisation in industry and residences are similarly effective only when they are practised collectively. For instance, it is predicted that digitalisation in the building sector could cut CO_2 emissions by 10% in 2030 (IEA 2019). The building sector includes the construction industry and residential usage and represented 36% of total energy consumption in the world in 2018 (Global Alliance for Buildings and Construction 2018). A conceptual residential digitalisation includes smart grid infrastructure, IoT-enabled learning algorithms, management and edge computing systems (Küfeoğlu 2021). Industrial digitalisation advanced rapidly due to high energy usage density per facility, but in comparison, single houses have much lower energy usage density. Thus, only a collective switch to smart houses and smart grid systems would make a difference in global energy efficiency.

Other high energy density industries such as manufacturing and agriculture are quite open fields for automatisation and digitalisation, and at some point, energy-saving actions in these fields must be taken by machines and intelligent systems instead of humans to get rid of human intervention and error (Küfeoğlu 2021). Even though intelligent systems have a cost at the beginning, long-term effects include, first and foremost, energy efficiency, positive budget changes, air quality and individuals' well-being.

One of the goals in SDG-7 is to increase access to clean energy through international cooperation. This direction supports research and development projects on clean energy, energy efficiency and cleaner fossil fuel technologies (United Nations 2021a). In addition, incentives are provided for investments in the energy infrastructure with the studies to be carried out. To achieve this goal, investments to be made by developing countries in clean energy, including hybrid systems, are supported financially. Although the financial flow in this area continues, the amount of financing transferred to the least developed countries remains below the

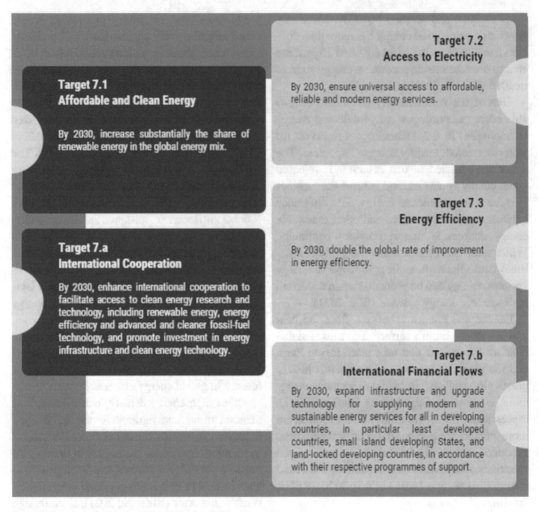

Fig. 9.1 Targets of SDG-7. (United Nations 2021a, b)

desired level. Although there are financial flows in this area today, studies need to be done especially on the countries that are left behind. In this direction, among the future targets under the title of international financial flow are strengthening international connections and increasing the financial flow in general, especially in lagging countries. Although governments and various development finance institutions have announced support in line with the targets set today, the desired levels have not been reached yet, and studies on this issue need to be increased (*The Energy Progress Report 2021* 2021). Figure 9.1 summarises the targets of SDG-7.

With the targets set by the United Nations in 2015, sustainable development has been discussed in three dimensions. Holistic development

is aimed at considering economic, social and environmental perspectives. In this study, plans were made by considering different countries' development levels and national realities (Calzadilla and Mauger 2017). Efforts within the scope of SDG-7 are aimed at affordable, reliable, modern and sustainable energy for everyone (Sustainable Development Goal on Energy (SDG-7) and the World Bank Group 2016). One of the important steps to be taken in line with this purpose is to realise financial flow. Thanks to this financial flow that has been realised and will be realised, especially developing countries will be able to fulfil the 2030 targets set by the UN. With the studies in the field of renewable energy, important steps have been taken, especially in decentralised electricity generation. Studies have been carried out on the problem of not delivering electricity to underdeveloped regions with the contribution of renewable energy. While 1.2 billion people did not have access to electricity in 2010, this figure decreased to 759 million in 2019. Although this figure is desired to be reduced even more for 2030, the target set for 2030 is to reduce the number of people who cannot reach electricity to 660 million, with the effect of the pandemic in the world. Progress in achieving clean cooking conditions is as stable as it gets. By 2019, 2.6 billion people still do not have clean cooking facilities. The smoke produced during cooking still endangers the lives of many people. The target set for 2030 is to reduce this figure by 30%. To achieve this goal, radical steps must be taken (Sustainable Energy for All 2021a). Another target set is energy efficiency. The United Nations also carried out various studies to increase energy efficiency, starting with the motto of "the cleanest energy, is unconsumed energy." In line with this target, efforts are underway to reduce primary energy intensity from 5.6 megajoules (MJ) per USD in 2010 to 3.4 by 2030. In 2018, this figure decreased to 4.75 MJ/USD, and studies in this direction continue (Sustainable Energy for All 2021b). When these targets are achieved by 2030, significant progress will be achieved in clean and affordable energy, although not enough. Although there are various regressions in the studies carried out, especially with the effects of the pandemic period, studies continue to reach these targets with the right targets to be set.

In addition to these targets set for 2030 by the United Nations, countries also have their plans for renewable energy and energy efficiency. For example, while the United States provides 21% of its electricity production from renewable sources in 2020, the target set for 2050 is to increase this figure to 42% (Dubin 2021). European countries are particularly ambitious in this regard. Sweden aims to eliminate fossil fuels in electricity generation by 2040 and is investing in this direction. Denmark, which met more than half of its energy from solar and wind in 2017, has set a fossil fuel-free future target for 2050. Germany aims to meet 65% of its electricity production from renewable sources in its 2030 target. Studies on this subject are not only carried out in Europe and America (Climate Council 2019). In addition, while China aims to produce 35% of its electricity from renewable sources in 2030, Chile has set a 70% renewable energy target for 2050. Indonesia, one of the Asian countries, aims to increase the share of renewable energy from 2% in the country to 23% by 2025, in line with its decisions (Siahaan 2014). Many countries have similar renewable energy future plans. By 2050, it is foreseen that 100% of the energy of 139 countries will be met by renewable energy (McKenna 2017).

Like the investments made in renewable energy, many countries are also working on energy efficiency, as it is the cheapest and cleanest means of energy source. Besides, with the advancements in energy efficiency, there is a decrease in carbon emission rates. Each country has its own goals and its own specific goals in its unions, such as the European Union. By 2030, the EU aims to achieve 32.5% progress in energy efficiency (Malinauskaite et al. 2020). For example, Germany is working to reduce its primary energy consumption by 50% compared to its consumption in 2008 as a target for 2050 (Federal Ministry for Economic Affairs and Energy 2021). Chile has carried out studies to support the renewable energy target it set in 2015 regarding energy efficiency and aims to significantly improve energy efficiency by 2050 by setting new standards (Siahaan 2014). The UK aims to reduce greenhouse gas emissions by 78% from 1990 levels by 2035 and zero emissions by 2050 (GOV. UK 2021). To achieve this goal, they turn to renewable energy sources and work on energy efficiency to reduce energy consumption. All these investments and studies carried out in these areas emerge as a result of certain financial planning and create many new job opportunities. In achieving these goals, it is important to manage the financial flow correctly.

The energy that is both affordable and clean guarantees that modern civilisation can function smoothly and productively. Industries have thrived, thanks to the existing energy system based on fossil fuels. However, it has led to severe climate change and exponential environmental deterioration. The private sector accounts for 23% of global electricity consumption, mostly met by damaging fossil fuels. Businesses may help speed up the transition to sustainable energy systems by investing in renewable energy technologies, research and development, focusing on energy efficiency and incorporating clean energy into their daily operations. They may go beyond their practices and invest in renewable energy sources for their communities, ensuring that their employees and customers have access to it and even providing renewable technology to developing countries (Bureau 2021).

Adaptation towards SDG-7 brings in new investments and creates a significant economy around it. While private investments and government spending in developed countries concentrate on achieving efficiency and renewable energy production, developing countries focus on obtaining access to electricity and clean energy sources. As an example, in 2020, the government of the United Kingdom introduced a £350 million package for the transition from fossil fuels to environment-friendly sources ("PM commits £350 million to fuel green recovery," 2020). Additionally, the United States and Germany raised funding equal to US$7 billion and US$45 billion, respectively, and South Korea declared a 5-year package of US$63 billion to increase the portion of clean energy (Beyer and Vandermosten 2021). Furthermore, international financial flow to developing countries rose to US$14 billion in 2018 from US$10.1 billion in 2010 (IRENA 2021a). However, African countries struggle to find private investment, and governments' spending on energy is not strong enough to achieve the goals. According to recent surveys, very few nations in Sub-Saharan Africa spend higher than 4% of their GDP on infrastructure, which pre-vents access to electricity and clean energy sources by all (Baumli and Jamasb 2020).

Moreover, the energy sector presents many employment opportunities. Global renewable energy employment is expected to reach 11.5 million in 2019 from 11 million in 2018. Although a few countries dominate the market, for instance, the top 10 biofuel producer countries account for 90% of the present jobs, the profitability of new technologies expands the current network to a wider range of countries (IRENA 2020). From 2015 to 2019, the energy efficiency sector has created more than 400,000 new jobs solely in the United States. It grew at a rate of 20%, roughly three times faster than the whole US economy (NASEO and EFI 2020). Furthermore, expansion of the renewable energy sector has a beneficial impact on jobs, as more positions are being provided in the power, agriculture and construction industries. Even though it is causing job losses in the fossil fuel distribution industry and in high energy-demanding businesses as rising energy prices reduce their profitability, the effect of the expansion in the renewable energy sector to employment is still positive in different scenarios. In 2050, 1.3% of EU employment is expected to be redistributed among several sectors with the low-carbon transition in the EU (Fragkos and Paroussos 2018).

9.1   Companies and Use Cases

Table 9.1 presents the business models of 60 companies and use cases that employ emerging technologies and create value in SDG-7. We should highlight that one use case can be related

Table 9.1 Companies and use cases in SDG-7

No	Company info	Value proposal (what?)	Value creation (how?)	Value capture
1	AeroShield	The company is an advanced and energy-efficient window materials producer.	Based on silica aerogel technology, they produce a super-insulating glass sheet. This solid new substance is made up of 95% gas trapped inside pores so microscopic that they are invisible to the human eye.	Energy consumption costs will decrease at the rate of decreasing consumption. From an environmental point of view, since these products are more efficient and reduce energy consumption, they also play an active role in reducing carbon emissions.
	USA			
	7, 13			
	Advanced materials			
2	Ambri	The liquid metal battery systems include a liquid calcium-alloy anode, a molten salt electrolyte and a cathode consisting of antimony particles, providing the use of low-cost advanced materials and a small number of steps in the cell assembly.	Wind and solar energy are used to generate electricity, which is then stored. Grid-scale energy storage devices operate as a "buffer," smoothing out the output from wind and solar farms and allowing them to replace fossil fuel generators on a large scale.	Smarter, quicker, longer-lasting, low-degradation and affordable renewable energy-generating and energy storage systems create additional revenue. When compared to developing traditional power plants, boosting renewable energy sources using innovative materials lowers electricity prices.
	USA			
	7, 13			
	Advanced materials, energy storage			
3	Ambyint	It is an AI-based software service providing plunger and rod lift optimisations.	By incorporating advanced physics and subject matter expert knowledge with artificial intelligence and machine learning to automate operations and production optimisation workflows throughout all well types and artificial lift systems, AI-powered improvement for the oil and gas industry provides step-change improvements to E&P (exploration and production) production results and profitability.	Thanks to increased efficiency via AI, users can increase production by 7%, as well as reduce operational losses by 30% and reduce error rates by 50%.
	USA			
	7, 13			
	AI			

(continued)

Table 9.1 (continued)

No	Company info	Value proposal (what?)	Value creation (how?)	Value capture
4	ANNEA Germany 7 AI, digital twins	It is a company that offers customised AI-based automated predictive and prescriptive maintenance platforms for renewable energy companies' power plants.	End-to-end solutions create digital twins of each component of customers' power plants through AI, physical modelling and normal behaviour modelling. Failure predictions, energy production and performance optimisation forecasts are made for wind turbines, solar farms and hydropower plants of customer companies.	With condition monitoring and automated failure prediction, customers could act against plant failures up to 365 days before, cut operating expenses by 50% and increase energy output by 15%.
5	Aurora Solar USA 7, 13 Big data	It is a software-as-a-service (SaaS) company that uses aerial imagery and big data to assess solar installations.	The programme has a comprehensive information centre with all the data needed to complete the solar project such as the azimuth of the project, slope of roofs, yield calculation and many simulations.	Monthly membership sales generate revenue, which is used to plot solar projects utilising all available data.
6	Bandora Systems Portugal 7 AI, big data, cloud computing, Internet of things	It is an AI-powered, cloud-based facility manager aiming to increase energy usage efficiency and detect and notify anomalies in commercial buildings.	Bandora.OM is a non-intrusive cloud-based platform that collects data from all of a facility's systems. It uses machine learning algorithms to anticipate energy usage, interior temperature, comfort level, occupancy and malfunctions, among other things. Because it is a cloud-based system, it allows one building to benefit from all other buildings that are connected to it.	The artificial intelligence engine may be used to streamline building management, decrease the number of full time equivalents (FTEs) necessary, conduct predictive maintenance and lower energy use.
7	BeFC France 7, 12, 13 Biotech and biomanufacturing, flexible electronics and wearables, Internet of things	It is a fuel cell provider that produces bio-enzymatic fuel cells that can replace conventional batteries in wearable electronics and IoT devices.	To convert natural substrates like glucose and oxygen into electricity, fuel cells use biological catalysts rather than chemical or expensive noble metal catalysts. Enzymes, carbon electrodes and paper microfluidics are all used in their products. BeFC fuel cells can run on a drop of tap water or biological fluids, making them ideal for wearable electronics and IoT devices that collect data on a semi-continuous basis.	The affordability of BeFC fuel cells' production materials compared to conventional batteries makes them cheaper, and also being paper-based makes them environmentally friendly. Their potential use in health monitoring and IoT devices can make these devices more affordable and sustainable and thus more widely used.

8	BlueSky Energy	It is a company that manufactures large-scale non-toxic batteries with only saltwater as battery electrolyte.	Saltwater electrolyte, non-toxic, non-flammable and non-explosive specifications provide users with a completely safe and environmentally friendly energy storage solution. Being maintenance-free and durable against overcharging and undercharging is also a relief for the user's budget.
	Austria		
	7	With their GREENROCK products, green energy storage solutions are provided for residences. Battery cells are modular, clusterable and maintenance-free, and energy management system electronics and software are provided. Approximately 5000 charge-discharge cycles and 15 years of lifespan are promised.	
	Energy storage		
9	BluWave-ai	It is an AI-based platform optimising grid energy for distribution utilities, microgrids and fleet electrification with the help of big data, edge computing and cloud computing.	Savings are achieved through AI-based grid energy optimisation to adapt to the fluctuations. Environmental value is captured through the adoption of renewables with the capability to predict power fluctuations, thanks to collecting big data and analysing it through AI algorithms.
	Canada		
	7, 9	They balance the grid energy according to changes in wind and solar energy production to optimise the cost, availability and carbon footprint of diverse energy sources – both renewable and non-renewable – with energy demand in real time to encourage the use of renewable energy sources. Their BluWave-ai Edge AI-assisted software takes data from the grid and applies AI algorithms to improve the grid by forecasting renewable variations. Their BluWave-ai Center software, which is hosted in the cloud, trains and develops AI algorithms as well as updates distributed Edge software.	
	AI, big data, cloud computing, edge computing		

(continued)

Table 9.1 (continued)

No	Company info	Value proposal (what?)	Value creation (how?)	Value capture
10	Brightmark USA 7,13 Biotech and biomanufacturing, recycling	It is an anaerobic digestion company that uses organic waste materials to produce renewable natural gas.	Natural gas produced from organic waste materials such as food waste and animal- and plant-based materials is known as renewable natural gas (RNG). Landfills, animal manure and solid waste recovered during wastewater treatment are the main sources of RNG. Because it is made from trash that is continually produced and occurs naturally as part of the decomposition process, the gas is called "renewable." Raw biogas is captured, cleaned, upgraded and compressed using anaerobic digestion technology.	Customers using this product can produce energy using the garbage they have. In this way, these customers generate income through energy sales and agricultural by-product sales. In addition, from an environmental benefit perspective, methane is captured and converted as biogas during the production process, rather than being released directly into the atmosphere. In addition, when this product is used in vehicles, it emits 6% to 11% lower levels of greenhouse gases than gasoline over the fuel life cycle.
11	Brightmerge Israel 7 AI	It is an AI-based platform focused on the operational and financial optimisation of microgrids that include renewable energy sources.	The AI-driven performance prediction model provides a fast, accurate and cost-effective optimisation environment for microgrid managers, energy developers and financiers.	Real-time data analysing microgrid management software is supplied by the company. As a result, a cost-saving, reliable and efficient microgrid operation is provided to customers.
12	Cadenza Innovation USA 7 Advanced materials, energy storage	The firm offers cell designs that combine cylindrical jelly rolls and massive prismatic cells to enable energy storage that is dependable, high in energy density, low in cost and safe.	The design improves efficiency, while ceramic-based housing materials prevent combustion while allowing for a compact construction.	Value is created through enabling broad adoption of electric vehicles through the development of safe, inexpensive and high-performing battery technology.
13	CrowdCharge UK 7 Energy storage	It is a company that provides V2G (vehicle-to-grid) charging.	Optimised charging sessions are provided using machine learning and AI. The vehicle-to-grid scheme provides grid balancing and real-time reporting services.	Revenue is generated by supplying energy stored in electric vehicle batteries at peak times for use in grid services.

14	Data Gumbo USA 7, 9 Blockchain	It's a blockchain network that allows industrial customers and suppliers to combine transactional data with automated smart contracts.	Pulling specified data into an encrypted distributed ledger based on blockchain creates a third-party verifiable record that each transaction is accurate and definite.
15	Drift USA 7 AI, blockchain	It is a blockchain-based renewable energy retail utility platform that incorporates AI and machine learning.	GumboNet™ is a usage-based subscription service for their network. It enables organisations to communicate relevant field data with counterparties that require it, resulting in automatic service confirmation and payment execution as the service is performed. The platform allows users to buy and sell renewable energy, both as consumers and as suppliers. It forecasts energy usage and offers wholesale energy costs that are lower. The residential user has the option of selecting their preferred power mix. It provides local renewable energy to customers and invoices them on a monthly basis.	Customers would save between 10% and 20% on electricity due to the platform's elimination of intermediaries.
16	enercast Germany 7 AI, big data	It is an AI-based platform for forecasting power levels for wind turbines and solar plants.	Its self-learning software-as-a-service solutions provide precise power estimates for wind and solar facilities, allowing renewable energy to be integrated into power networks. Its AI and big data integration platform aids its clients' decision-making processes in the industrial sector.	The AI capabilities of the platform enable the user to predict the amount of production to be made by their power generation facility. In addition, with the production estimates made, renewable energy systems are connected considering the potential production when connected to the grid. In this case, grid stabilisation is achieved more effectively.
17	Energy Vault Switzerland 7 Advanced materials, energy storage	It's a large-scale gravity-based energy storage device that doesn't rely on surface topography or subterranean geology. They provide a long-term solution that does not degrade over time and has competitive performance, with round-trip efficiency ranging from 80 to 90%.	The system combines advanced material science and proprietary machine-vision AI software to autonomously orchestrate the charging and discharging of electricity using ultra-low-cost composite bricks and innovative mechanical crane systems, using traditional physics fundamentals of gravity and potential energy.	Revenue is made by recycling waste materials to create moving masses and establishing a circular economy.

(continued)

Table 9.1 (continued)

No	Company info	Value proposal (what?)	Value creation (how?)	Value capture
18	Enexor USA 7, 11, 12, 13 Recycling	It is a company that produces clean onsite bioenergy by recycling organic waste.	Their Bio-CHP™ solution generates clean onsite energy using free (or low-cost) locally generated organics or plastic waste. It provides microgrids with controlled, continuous renewable energy. By reducing methane emissions from landfills, offsetting fossil fuel-based electricity generation and reducing waste disposal transportation emissions, each system can reduce up to 2200 metric tonnes of CO_2-equivalent emissions yearly.	Revenue is captured through reduced energy costs, thanks to recycling the waste as an energy source and eliminating transportation. Also, reducing landfills creates an environmental value since it decreases GHG emissions.
19	EQuota Energy China 7 AI, big data	The company is an AI- and big data-based energy-intelligent management service provider. The main services are energy savings and carbon management.	Insight™ and InsightLite are software solutions. Through AI and big data, Insight™ is an energy management solution that is used for power monitoring, energy efficiency improvement, equipment management and predictive operation and maintenance. Energy reporting, real-time warnings, optimisation ideas and analysis tools are part of InsightLite.	Revenue creation is achieved through energy efficiency, energy savings and carbon emissions control management and improved control. By utilising the EQuota service, a steel factory in East China saved $1 million in energy expenses and saved 5000 tonnes of coal in just 6 months.
20	ESS USA 7 Energy storage	The company manufactures environmentally safe, cost-effective and high cycle life iron flow batteries containing water as electrolyte, iron and salt for commercial usage.	The energy storage is carried out by redox reactions. Long cycle life and durability are obtained via electrode design and control system. Flexibility in terms of adjusting power and amount of energy exactly.	Environmental value is captured by earth minerals not used in battery systems.
21	Fluence USA 7, 13 Energy storage	The firm creates a comprehensive basis for safe and repeatable energy storage systems by combining hardware, controls, software and intelligence.	The Sunstack solar energy storage technology was created with sunlight capture and distribution in mind. Its system architecture integrates batteries and PV on the same side of the DC bus to benefit from larger PV-to-inverter ratios, increase solar production and simplify the connectivity procedure.	Revenue is created via cost-effective, safe and scalable energy storage systems working with new storage constituents.

22	Form Energy USA 7, 13 Energy storage	The company manufactures battery systems consisting of iron and air electrodes and water-based non-flammable electrolytes for energy storage applications.	Thousands of battery modules are joined together in modular megawatt-scale power blocks, which are housed in an environmentally protected container. Tens to hundreds of these power blocks will be linked to the energy grid, depending on the scale of the energy storage system.	Revenue is created through low-cost, scalable, optimisable, reliable, safe battery systems. The batteries include cheap and abundant components and exhibit multiple-day energy storage.
23	Grid4C USA 7, 12 AI, big data	The company utilises AI and big data in their software to detect anomalies and predict future conditions in energy consumption.	By examining smart metre data, their algorithms can identify, diagnose and anticipate problems, inefficiencies and behavioural abnormalities in a wide range of household equipment. HVAC systems, water heaters, pool pumps, the house envelope and refrigerators are all classified as having these problems. Customers may see projected daily prices and power use for the future week. Daily and monthly power demand forecasts are also generated at 1-hour intervals by the models.	Savings are captured through AI-based energy monitoring and forecasting, increased safety through meter data collection and AI algorithms' malfunction detection and prediction capabilities.
24	GridCure Canada 7 AI, big data	It is a predictive analytics software provider for the energy industry that utilises big data and machine learning to optimise their strategy and operations.	GridCure uses a patented machine learning approach to sort, categorise, assess and transform data into an industry-standard common information model. Most significantly, data is protected by the same role-based access constraints that protect the client utility.	Faster fault management and self-healing, as well as better grid dependability, provide value.
25	HYDROGRID Austria 7 AI	It is a company that optimises hydroelectric power plants using machine learning software.	The developed machine learning technology helps to make the flow rate estimation. In addition, it determines the production capacity based on the price estimations and thus realises the optimum production planning.	Hydroelectric power plant owners can generate up to 18% more energy using this software. It enables flow estimates and reduces the work that needs to be done manually, ensuring both more accurate production and lowering the operational costs of the production facility.

(continued)

Table 9.1 (continued)

No	Company info	Value proposition (what?)	Value creation (how?)	Value capture
26	LevelTen Energy USA 7 Big data, blockchain	It is a market model for P2P energy trading with renewable energy that combines big data with blockchain.	Their CFO-Ready platform offers solutions for energy data analysis. The data and analytics required to fulfil the power purchase agreement (PPA) provide real-time value and risk assessment.	The main revenue source is the P2P trading platform. The company provides an extra value added by request for proposal (RFP) automation, performance monitoring and energy data analysis solution.
27	Meyer Burger Switzerland 7, 13 Advanced materials	It is a company that produces sustainable-type solar modules by using advanced materials.	The SWCT technology used in modules enhances cell stability and makes modules less vulnerable to so-called micro-cracks, which are one of the most prevalent causes of solar module energy losses. The cells are further protected from moisture and other external impacts by the ultra-stable unique backsheet.	Solar modules are able to produce more efficient energy by reducing losses. The rate of loss in production over time is also lower in solar modules and offers the user the opportunity to produce more efficient energy for a long time. In terms of the environment, solar modules are designed as recyclable as possible, toxic heavy metals are not used in their production, and the energy used during production is provided from renewable energy sources.
28	Nafion USA 7, 13 Advanced materials, energy storage	The company produces advanced membranes, resins, dispersion materials for energy storage, hydrogen production, chemical processing and transportation.	Nafion membranes keep separate electrolytes from each other allowing conduction of the desired number of charge carriers with high mechanical properties and high ionic conductivity, selectivity and chemical stability for flow battery energy storage systems.	Extra value is produced by manufacturing economical, safer and scalable energy storage.

29	NDB USA 7, 13 Advanced materials, energy storage	It's a nanotechnology firm that specialises in the development and production of semiconductors, energy and battery solutions.	NDB was among the first companies to use atomic voltaic cells in mid- and high-power applications and was one of the first to produce them. Throughout its lifespan, which is often several years, the self-charging battery delivers steady power by transforming the energy produced from radioactive decay into usable energy. Nuclear batteries are optimised for high-power applications by NDB. NDB suggests that nuclear fuel be recycled to extract radioisotopes, allowing it to be reused. Isotopes derived from recycled nuclear waste or a reactor emit a high concentration of energy particles that can be converted into usable energy.	Recycling nuclear waste generated by nuclear power plants and using these wastes for energy generation prevents environmental damage that may arise from nuclear waste.
30	NEARTHLAB South Korea 7 AI, big data	They use autonomous drones to generate visuals of wind energy assets and AI algorithms to identify and analyse defects.	An artificial intelligence algorithm uses big data to classify the types and severity of faults and make early maintenance suggestions.	They allow managers of onshore and offshore wind farms to automate comprehensive and spot checks, enhancing safety at work.
31	New Holland USA 2, 7, 13 Hydrogen	They are an agricultural machinery manufacturing company. They also have hydrogen-powered tractors in their product mix.	During the piston intake stroke, hydrogen is injected and combined with air. Diesel is injected at top dead centre of the compression stroke, as with other diesel engines, and compression heat ignites the mixed fuels. The tractor was created in collaboration with Blue Fuel Solutions, which uses a hybrid of H_2 and diesel fuel.	Producers using this tractor can engage in sustainable agriculture, market their products and earn additional income from their products. In terms of the environmental value created, the amount of fossil fuel in the fuel used decreases up to 50% in terms of CO_2 emissions.

(continued)

Table 9.1 (continued)

No	Company info	Value proposition (what?)	Value creation (how?)	Value capture
32	Nozomi Networks USA 7 Cybersecurity, Internet of things	It's a start-up that creates solutions to secure electricity generation, transmission and distribution networks while also increasing operational visibility and reliability.	Industrial network vulnerability assessment systems uncover vulnerabilities in operational technology (OT) and Internet of things (IoT) devices, allowing energy utilities to address security concerns they may face. This technology detects utility system vulnerabilities automatically and organises them by vendor and severity. As a result, cybersecurity teams can prioritise high-risk exposures as a result of this identification.	By eliminating most of the security threats that energy companies may encounter, energy interruptions are prevented, and thus a value is created. Thus, both energy and economic losses are prevented.
33	ORE Catapult UK 7, 13 Robotics	They integrate robotic systems into the offshore wind energy sector.	The maintenance of energy-producing parts can be streamlined with the aid of robotic systems.	Wind turbine parts that are well-controlled and have a long lifespan generate revenue.
34	Orison USA 7 Energy storage	It's an all-in-one energy storage system that's suitable for residential, light commercial and industrial use.	A self-installable, all-in-one household battery solution. Battery storage, power conversion electronics and a smart networking control system are all included in Orison units.	Value is created by providing a domestic battery system with a domestic consumption monitoring instrument panel and a smartphone application.
35	Oxford PV UK 7 Advanced materials	It is a company that produces advanced solar cells.	Silicon solar cell and module producers will be able to break past the performance barrier, thanks to perovskite solar cell technology. Significantly enhancing the performance of silicon photovoltaics will result in cost reductions, transforming the economics of solar energy and accelerating its worldwide expansion.	Solar cells made of perovskite provide electricity with a high efficiency and at a low cost. This technology adds value to the consumer by allowing them to produce significantly more energy for a lower cost. More energy is generated in small areas due to increased efficiency, and environmental value is captured. This helps to reduce the amount of area needed for energy generation.

36	PHYSEE	The company is an IoT-compatible sensor and smart window manufacturing company.	Proposed glass coating systems include windows with custom sensors, a high-resolution data hub that is coherent with IoT networks and automated operation settings through central IoT software. Temperature and natural illumination control is made possible and could be automated by the central IoT management system.	Customers could gain up to 15% energy savings in their buildings through the IoT automation capabilities of the smart windows system. An additional data hub that allows the system to communicate with the IoT automation network and dedicated UI software is also provided to customers.
	Netherlands			
	7			
	Internet of things			
37	Plug Power	A hydrogen energy solution-based company providing comprehensive hydrogen fuel cell turnkey solutions.	The ProGen (hydrogen engine) fuel cell engine suite is intended to be used in both transportation and stationary power applications.	OEM firms producing power solutions for a variety of industries can benefit from the company's versatility.
	USA			
	7			
	Hydrogen			
38	PONTON	A B2B IT firm that specialises in energy trading, grid management and customer-related procedures. The firm offers energy trading (power, gas and CO_2 certificates) as well as grid operating solutions.	The Enerchain project includes PONTON. It's an over-the-counter wholesale energy market decentralised energy trading platform. Prosumers, DSOs and aggregators can exchange locally produced energy using this platform. Gridchain is a blockchain-based initiative that focuses on the integration of TSO/DSO processes. It streamlines and standardises the grid management process' collaboration and communication among market players (TSOs, DSOs, aggregators, suppliers and generators)	The P2P trading tool, which will serve as the basis for the first European energy exchange on the blockchain, will be the main source of revenue. The participants' charging procedure has not yet been made public.
	Germany			
	7			
	Blockchain			

(continued)

Table 9.1 (continued)

No	Company info	Value proposal (what?)	Value creation (how?)	Value capture
39	Powerledger Australia 7 Big data, blockchain	The company provides blockchain-based solutions including P2P trading, microgrid trading, carbon product trading, electric vehicle settlement, asset germination and virtual power plants.	Consumers of electricity in an energy microgrid or embedded network are empowered by comprehensive visibility of their electricity usage and transactions, which is provided simply, securely, transparently and in real time. xGrid is a peer-to-peer electricity trading platform that operates on the regulated electricity grid. It makes it possible for neighbours to swap low-carbon electricity. µGrid allows for unparalleled granularity in electricity metering, large data collecting, quick micro-transactions and microgrid management.	Customers that generate surplus power can profit from their renewable energy investments and increase their return. Prospective customers include residential strata complexes such as apartment buildings or flats, shopping malls, retirement villages, caravan/holiday parks and industrial parks.
40	Qubit Engineering USA 7, 12, 13 Quantum computing	The company produces quantum-based optimisation software for pre-construction wind farm planning.	Within minutes, quantum-enhanced optimisation algorithms provide a bird's eye view of the plan site (all prospective building sites, wind data and land features) and automatically identify the best placements for the wind farm's turbines. The programme also enables interactive and easy manual modification of turbine positions, allowing for real-time estimates of predicted annual energy production (AEP) for the present turbine design.	The return on investment of a wind farm planned with quantum-based algorithms is improved since the wind farm generates more electricity and asset fatigue due to wake turbulence is reduced.
41	Raptor Maps USA 7 AI, drones	A company that uses machine learning and drone technologies to monitor (PV) photovoltaic production facilities, identify anomalies and eliminate losses due to this.	Raptor Solar is a software solution that uses machine learning to assist solar enterprises save expenses and enhance energy output. The user can use software to analyse thermal and colour imagery obtained from drone and/or manned aircraft aerial solar site inspections. Every image is evaluated by software, which detects, classifies and locates abnormalities.	Raptor Solar enables the user of the product to observe the problems in the solar power generation facility and quickly solve the problem. In this way, possible production losses in the system can be avoided in a short time, and the production amount of the production facility can be increased.

#	Company	Description	Impact
42	RCAM Technologies USA 7, 9 3D printing, energy storage	The company's main activities are 3D concrete printing for wind power facilities and energy storage.	Using 3D printing processes, they design and build suction anchors or permanent foundations for offshore wind turbines. They also design onsite-printable concrete 3D-printed wind turbine towers. Furthermore, their energy storage technology, Marine Pumped HydroElectric Storage (M-PHES), mimics traditional pumped hydro energy storage in that it employs many enormous hollow concrete spheres on the seafloor to store mechanical energy in the form of pressure.	Savings are captured through reduced transportation needs and reduced construction costs, thanks to the use of onsite 3D-printed concrete in wind power facilities and energy storage.
43	Samsung SmartThings South Korea 7 Internet of things	The company provides comprehensive IoT software for home appliance automation and monitoring to compatible smart households for residential users.	SmartThings IoT software supports a large number of smart homes, sensors and other firms' smart gadgets, and the company allows users to establish personal IoT networks in their homes. They also enable access to a variety of devices using approved communication protocols, as well as software that allows consumers to track and regulate home energy usage.	Energy monitoring and automation options allow users to monitor and act against energy waste. IoT control software also suggests better utilisation options for energy-saving to the resident without any disturbance to the comfort of the inhabitant.
44	Schlumberger New Energy USA 7, 12, 13 Carbon capture and storage	They are a sub-company of Schlumberger, which is a technology provider for the energy sector, that focuses on making collaborations to execute clean energy projects.	The BECCS facility will transform agricultural biomass waste, such as almond trees, into a renewable synthesis gas that will be combined with oxygen in a combustor to create power. By injecting carbon dioxide (CO_2) underground into neighbouring subsurface geologic formations, more than 99% of the carbon from the BECCS process is predicted to be collected for permanent preservation.	The procedure is supposed to result in net-negative carbon emissions, effectively eliminating greenhouse gas from the environment, by using biomass fuel that consumes CO_2 during its lifespan to create electricity and then securely and permanently storing the produced CO_2. When completed, the facility is planned to remove around 300,000 tonnes of CO_2 per year, which is similar to the emissions from more than 65,000 American homes.

(continued)

Table 9.1 (continued)

No	Company info	Value proposal (what?)	Value creation (how?)	Value capture
45	Schneider Electric France 7 Internet of things	The company is an electricity production, storage, management and industrial automation company providing efficiency and sustainability in energy storage and microgrids.	IoT-enabled and digitised architecture software EcoStruxure pledges to increase energy efficiency and reduce maintenance costs, carbon footprint, engineering costs and required time in nearly every type of industry.	Revenue streams are optimised through increased efficiency of industrial appliances, reduced energy and maintenance costs, increased productivity and reduced carbon footprint.
46	Sekab Sweden 7, 13 Biotech and biomanufacturing	The company produces biofuel and ethanol from cellulose.	In the manufacturing of ethanol, the firm employs wastes from forestry and agriculture, such as wood chips and sugarcane haulm. The growth of large-scale ethanol production and the breaking of the transportation industry's dependency on oil will need the manufacturing of ethanol from cheap raw materials that are easily accessible in big amounts and do not compete with food production.	Revenue is captured from ethanol production, which is a plant-based biofuel. Environmental value is also generated by reducing petroleum-based fuel consumption.
47	SGH2 Energy USA 7, 13 Hydrogen, recycling	It is a green hydrogen producer that produces hydrogen from any kinds of waste ranging from paper to plastics, tires to textiles.	The gasification process at SGH2 is based on a plasma-enhanced thermal catalytic conversion method that is tuned for oxygen-enriched gas. The ultimate product is high-purity hydrogen and a little quantity of biogenic carbon dioxide, neither of which contributes to greenhouse gas emissions.	Solena Plasma Enhanced Gasification (SPEG) technology enables affordable green hydrogen production. The use of green hydrogen contributes to the reduction of carbon dioxide emissions.
48	Siemens Energy Germany 7 Cybersecurity	The company provides cybersecurity solutions to oil and gas industries.	Cyberspace, Vulnerability and Basic Compliance Evaluation is carried out, and protection methods are developed in accordance with the results of these evaluations. Secure architecture, device hardening, malware pattern updates and application whitelisting are used as protection methods.	Value capture by ensuring energy security and uninterrupted continuation of energy production.

49	Sierra Energy USA 7, 11, 12, 13 Recycling	It is a company that recycles organic waste through waste gasification for energy generation.	Their FastOx® method breaks down waste at the molecular level using heat, steam and oxygen. Organic molecules decompose into a high-energy syngas. Metals and inorganics melt together to form a non-leaching stone. Waste is completely transformed into high-value goods, with no emissions produced in the process. There are no harmful by-products or emissions from the procedure.	Savings are captured through reduced energy costs with clean energy production from zero-cost waste and environmental value capture through reduced landfills, thanks to the recycling of otherwise dumped waste.
50	sonnen Germany 7 Energy storage	The company produces battery storage systems for energy storage applications.	With a PV system plus a sonnenBatterie, around 75% of a household's annual energy needs with self-produced, renewable energy will be met.	The users reduce the energy cost by producing their energy independently by themselves.
51	Trine Sweden 7, 13 Crowdfunding	It is a crowdfunding platform for solar energy project investments.	They provide capital as loans to solar panel companies that produce and lease home-applicable solar panel systems. Thus, companies can continue to lease solar systems although the previous leasers did not complete their payments yet. Investors earn money as interest when companies pay their loans after they get payments from leasers.	More wide use of renewables is achieved due to the increased leasing capability of solar companies, thanks to loans offered by crowdfunding. Revenue is generated to crowd investors through the interest in loans.
52	TRION Energy USA 7 Advanced materials, energy storage	The company develops and offers advanced silicon–graphite anode material for lithium-ion battery manufacturers.	SiMoGraph, an advanced silicon–graphite anode material, replaces conventional anode materials in lithium-ion batteries, diminishes the risk of explosion and increases the capacity of batteries. Availability in multiple Li-ion applications increases battery efficiency in a wide range of energy storage options.	Customers achieve 35% higher cell performance despite the low-cost assembly of the proposed material. Also, extra equipment is not needed in the battery manufacturing process. High-capacity batteries cause less waste or recycle load in energy storage facilities or individual usages.

(continued)

Table 9.1 (continued)

No	Company info	Value proposal (what?)	Value creation (how?)	Value capture
53	Tvinn Sweden 7 AI, big data, cloud computing	It's a cloud-based data-driven optimisation software company that integrates with clients' energy-related activities and products.	They optimise energy and power utilisation by combining grid electricity, locally produced energy and energy storage to fit the present power demand. The energy may then be balanced, prioritised and scheduled to charge automobiles, power a nearby home or even sell energy or electricity back to the grid via their platform.	Revenue is captured through energy generation, storage and trading locally. The company maximises the use of locally produced energy, reduces peak power and, most importantly, allows you to charge your electric fleet.
54	Utilight Israel 7 3D printing	The company is a laser 3D printing provider for producing solar cells and electronics.	The Pattern Transfer Printing (PTP) method is used for solar PV cell high-volume production. Using light source deposition technology in this non-contact and high-efficiency production method provides an increase in solar cell conversion efficiency when using standard industry pastes and processes. Besides, thanks to this printing method, contactless, correctly aligned printing can be performed. In this way, the consumption of materials used in production is reduced.	PV cell manufacturers can increase the efficiency of their products. It also reduces production costs by reducing the materials used during production. Increasing the efficiency of solar energy conversion also has a positive effect on the environment, because the same amount of energy produced with other systems can be produced in a much smaller area, depending on the increase in efficiency. This contributes to the reduction of the space required for energy production.
55	V-Labs Switzerland 7, 9 Spatial computing	The company is an augmented reality glasses producer that provides accurate geospatial data.	In outdoor settings, geospatial data (GIS, BIM) is shown with a <10 cm accuracy. Users may attain great precision even near to structures using computer vision technologies paired with GNSS/RTK. Geospatial data and its associated properties may be precisely measured and rectified in the field and then saved in industry-standard GIS data formats. The user may carry out their normal tasks while seeing virtual data about cables and pipelines, 3D building blueprints and/or maintenance procedures with the use of a headset.	Value is captured through increased efficiency and ease of work for the surveyors by using the geospatial data provided with augmented reality glasses.

#	Company	Description	Value
56	Vodafone UK 7 5G	It is a mobile network company that offers faster connections to utilities with 5G.	Vodafone 5G enables energy and utility businesses to better track usage and enhance customer service by connecting smart grids, surveillance drones, predictive maintenance and smart metres.	Savings are captured through improved monitoring, thanks to the frequent data transfer capacity among monitoring devices with the increased bandwidth of 5G.
57	Wartsila Finland 7 AI, big data, energy storage	Using machine learning and historic and real-time data analytics to optimise the asset mix, GEMS enables customers to remotely monitor, operate, identify and diagnose equipment with unrivalled safety, reliability and flexibility.	It offers automated decision-making based on real-time and forecasted data (device status, weather, grid measurements and market data). This intelligence optimises system performance while reducing costs and offers flexible warranty management of multiple battery technologies.	GEMS IntelliBidder provides the following capabilities to evaluate bid optimisation functions and strategy performance: electricity market price and renewable forecasting, schedule commitment, automatic bid generation, manual bid entry, integration with ISOs and APIs for bid submission and key performance indicator monitoring.
58	WiBotic USA 7, 13 Wireless power transfer	It is a firm that develops technology that allows robots and unmanned aerial and undersea vehicles to charge wirelessly.	The firm specialises on wireless charging hardware, as well as a wireless battery management system and fleet-level power management software. The business has created its own adaptive wireless charging technology, which has a longer range (up to three times the diameter of the transmitting coil) and is more efficient.	The company creates value by enabling autonomous robots to operate mobile, aerial and in space with a wireless battery management system.
59	WiTricity USA 7, 11 Wireless power transfer	It is a company that develops wireless power transfer solutions in car parks for electrical vehicles.	Resonant coupling occurs when the natural frequencies of a source and a receiver are nearly identical. The company's power sources and receivers are magnetic resonators with designs that efficiently transfer power over long distances using the magnetic near-field.	Value is captured by providing wireless energy solutions which reduce outage risk caused by wire damage and save place and time.

(continued)

Table 9.1 (continued)

No	Company info	Value proposal (what?)	Value creation (how?)	Value capture
60	ZeroAvia	It is a company that produces hydrogen-powered powertrains for the aviation sector.	By using the hydrogen-powered powertrain produced by the company, the transformation process in aeroplanes is achieved. In this way, aircrafts are loaded with hydrogen as fuel, and hydrogen is used during flight. The flight range of the converted aircraft is between 300 and 500 miles, and these flights do not cause carbon emissions.	A hydrogen-powered powertrain contributes to the reduction of fuel and maintenance costs by 75% on the user's side. This ensures that the total travel expenses are reduced by 50%.
	USA			
	7, 13			
	Hydrogen			

to more than one SDG and it can make use of multiple emerging technologies. In the left column, we present the company name, the origin country, related SDGs and emerging technologies that are included. The companies and use cases are listed alphabetically.[1]

References

ARENA, Solar energy [WWW Document]. Aust. Renew. Energy Agency (2021), https://arena.gov.au/renewable-energy/solar/. Accessed 25 Oct 2021

E. Barbier, Geothermal energy technology and current status: An overview. Renew. Sustain. Energy

---

[1]For reference, you may click on the hyperlinks on the company names or follow the websites here (Accessed Online – 2.1.2022):

http://befc.global/; http://bluwave-ai.com/; http://datagumbo.com/; http://enexor.com/; http://equotaenergy.com/en/; http://grid4c.com/; http://rcamtechnologies.com/; http://sierraenergy.com/; http://v-labs.ch/; https://agriculture.newholland.com/eu/en-uk; https://ambri.com/; https://annea.ai/; https://cadenzainnovation.com/technology/; https://crowd-charge.com/; https://energyvault.com/#about-us; https://essinc.com/energy-center/; https://fluenceenergy.com/energy-storage-technology/; https://formenergy.com/team;https://gridcure.com/; https://hydrogrid.eu/tr/home_tr/; https://ndb.technology/; https://newenergy.slb.com/; https://ore.catapult.org.uk/; https://orison.com/; https://qubitengineering.com/; https://raptormaps.com/; https://sonnengroup.com/sonnenbatterie/; https://storage.wartsila.com/; https://trionbattery.com; https://tvinn.se/; https://witricity.com/; https://www.aeroshield.tech/; https://www.ambyint.com/; https://www.aurorasolar.com/; https://www.bandorasystems.com/; https://www.bluesky-energy.eu/en/home-2/; https://www.brightmark.com/; https://www.brightmerge.com/; https://www.drift-trader.com/; https://www.enercast.de/; https://www.leveltenenergy.com/; https://www.meyerburger.com/en/; https://www.nafion.com/en/products; https://www.nearthlab.com/; https://www.nozominetworks.com/; https://www.oxfordpv.com/; https://www.physee.eu/real-estate-solutions/smartskin; https://www.plugpower.com/; https://www.ponton.de/; https://www.powerledger.io/; https://www.se.com/ww/en/about-us/sustainability/; https://www.sekab.com/en/products-services/biofuel/; https://www.sgh2energy.com/; https://www.siemens-energy.com/global/en/offerings/industrial-applications/oil-gas/cybersecurity.html; https://www.smartthings.com; https://www.trine.com/; https://www.utilight.com/; https://www.vodafone.co.uk/business/5g-for-business/5g-energy-and-utilities; https://www.wibotic.com/; https://www.zeroavia.com/

Rev. 6, 3–65 (2002). https://doi.org/10.1016/S1364-0321(02)00002-3

K. Baumli, T. Jamasb, Assessing private investment in African renewable energy infrastructure: A multi-criteria decision analysis approach. Sustainability 12, 9425 (2020). https://doi.org/10.3390/su12229425

J. Beyer, A. Vandermosten, Greenness of Stimulus Index. Vivid Economics (2021)

G.B. Bureau, SDG 7 Affordable and clean energy: How to adopt it into your business. Green Bus. Bur. (2021), https://greenbusinessbureau.com/topics/sdg/sdg-7-affordable-and-clean-energy-how-to-adopt-it-into-your-business/. Accessed 7 Aug 2021

P.V. Calzadilla, R. Mauger, The UN's new sustainable development agenda and renewable energy: The challenge to reach SDG7 while achieving energy justice. J. Energy Nat. Resour. Law 36, 233–254 (2017)

Climate Council, 11 countries leading the change on renewable energy [WWW Document]. Clim. Counc. (2019), https://www.climatecouncil.org.au/11-countries-leading-the-charge-on-renewable-energy/. Accessed 12 Aug 2021

K. Dubin, EIA projects renewables share of U.S. electricity generation mix will double by 2050 [WWW Document]. US Energy Inf. Adm. (2021), https://www.eia.gov/todayinenergy/detail.php?id=46676. Accessed 1 Aug 2021

EDF, Types of renewable energy [WWW Document]. EDF (2021), https://www.edfenergy.com/for-home/energywise/renewable-energy-sources. Accessed 25 Oct 2021

EIA, Renewable energy explained – U.S. Energy Information Administration (EIA) [WWW Document] (2020), https://www.eia.gov/energyexplained/renewable-sources/. Accessed 25 Oct 2021

EIA, Biomass explained – U.S. Energy Information Administration (EIA) [WWW Document] (2021), https://www.eia.gov/energyexplained/biomass/ (accessed 10.31.21).

Federal Ministry for Economic Affairs and Energy, Germany makes it efficient [WWW Document]. Fed. Minist. Econ. Aff. Energy. (2021), https://www.bmwi.de/Redaktion/EN/Dossier/energy-efficiency.html. Accessed 8 Aug 2021

P. Fragkos, L. Paroussos, Employment creation in EU related to renewables expansion. Appl. Energy 230, 935–945 (2018). https://doi.org/10.1016/j.apenergy.2018.09.032

Global Alliance for Buildings and Construction, 2018 Global Status Report. Global Alliance for Buildings and Construction (2018)

GOV.UK, UK enshrines new target in law to slash emissions by 78% by 2035 [WWW Document]. GOV.UK (2021), https://www.gov.uk/government/news/uk-enshrines-new-target-in-law-to-slash-emissions-by-78-by-2035. Accessed 11 Aug 2021

IEA, Multiple Benefits of Energy Efficiency (IEA, Paris, 2019)

IEA, SDG7: Data and Projections [WWW Document].
    IEA (2020), https://www.iea.org/reports/sdg7-data-
    and-projections. Accessed 31 July 2021

IRENA, Renewable energy and jobs – Annual review
    2020 (2020)

IRENA, Tracking SDG 7: The Energy Progress Report
    (2021) [WWW Document]. Publ.-SDG-7-2021
    (2021a), https://www.irena.org/publications/2021/
    Jun/Tracking-SDG-7-2021. Accessed 2 Aug 2021

IRENA, Wind energy [WWW Document]. /wind (2021b),
    https://www.irena.org/wind. Accessed 25 Oct 2021

S. Küfeoğlu, *The Home of the Future: Digitalization and
    Resource Management* (Springer, Cambridge, 2021)

J. Malinauskaite, H. Jouhara, B. Egilegor, F. Al-Mansour,
    L. Ahmad, M. Pusnik, Energy efficiency in the indus-
    trial sector in the EU, Slovenia, and Spain. Energy 208
    (2020)

J. McKenna, Most of the world's countries could run on
    100% renewable energy by 2050, says study [WWW
    Document]. World Econ. Forum (2017), https://
    www.weforum.org/agenda/2017/09/countries-100-
    renewable-energy-by-2050/. Accessed 7 Feb 2021

NASEO, EFI, U.S. Energy and Employment Report 2020
    (2020)

PM commits £350 million to fuel green recovery [WWW
    Document] (GOV.UK, 2020), https://www.gov.uk/
    government/news/pm-commits-350-million-to-fuel-
    green-recovery. Accessed 15 Aug 2021

T.S. Siahaan, Indonesia passes national energy regulation,
    set to raise price of fuel, power [WWW Document].

Amcham Indones (2014), http://www.amcham.or.id/
    en/news/detail/indonesia-passes-national-energy-
    regulation-set-to-raise-price-of-fuel-power. Accessed
    7 Aug 2021

Sustainable Development Goal on Energy (SDG7) and
    the World Bank Group, Sustainable Development
    Goal on Energy (SDG7) and the World Bank Group
    [WWW Document] (World Bank, 2016). https://www.
    worldbank.org/en/topic/energy/brief/sustainable-
    development-goal-on-energy-sdg7-and-the-world-
    bank-group. Accessed 8 Oct 21

Sustainable Energy for All, Understanding Sustainable
    Development Goal 7 (SDG7) [WWW Document].
    Sustain. Energy All (2021a), https://www.seforall.org/
    data-and-evidence/understanding-sdg7. Accessed 19
    Aug 2021

Sustainable Energy for All, SDG 7.3 – Energy efficiency
    [WWW Document]. Sustain. Energy All (2021b),
    https://www.seforall.org/goal-7-targets/energy-
    efficiency. Accessed 8 Aug 2021

The Energy Progress Report 2021, Washington, DC
    (2021)

United Nations, Goal 7 | Department of Economic and
    Social Affairs [WWW Document] (2021a), https://
    sdgs.un.org/goals/goal7. Accessed 29 July 2021

United Nations, The Sustainable Development Goals
    Report: 2021 [WWW Document]. U. N. (2021b),
    https://unstats.un.org/sdgs/report/2021/goal-13/.
    Accessed 31 July 2021

SDG-8: Decent Work and Economic Growth

10

Abstract

Economic growth can be defined by increasing consumption due to the increase in population and reaching production amounts to meet consumption with technological developments and governmental incentives. The correct placement of people in the production and consumption equation can be expressed as decent work. Although decent work and economic growth may seem like different terms at first glance, they are inseparable terms for each other. SDG-8, Decent Work and Economic Growth, aims to attain full and productive employment, as well as respectable work, for all women and men by 2030. This chapter presents the business models of 37 companies and use cases that employ emerging technologies and create value in SDG-8. We should highlight that one use case can be related to more than one SDG and it can make use of multiple emerging technologies.

Keywords

Sustainable development goals · Business models · Decent work and economic growth · Sustainability

The author would like to acknowledge the help and contributions of Berkay İspir, Ekrem Gümüş, Zeynel Salman, Begüm Nur Okur, Canberk Özemek, Elif Berra Aktaş and Beyza Özerdem in completing of this chapter.

Economic growth can be defined by increasing consumption due to the increase in population and reaching production amounts to meet consumption with technological developments and governmental incentives. Economic activity has been boosted by various governmental initiatives, ranging from tax optimisation to free-market protection, infrastructure and education investment (Bleys and Whitby 2015). On the other hand, the correct placement of people in the production and consumption equation can be expressed as decent work. Although decent work and economic growth may seem like different terms at first glance, they are inseparable terms for each other. Decent work includes equal opportunities for all people, without gender, social and opportunity discrimination. According to International Labour Organization (2015), decent work encompasses opportunities for productive work that pays a fair wage; workplace security and social protection for families; improved prospects for personal development and social integration; freedom for people to express their concerns and organise and participate in decisions that affect their lives; and equal advantages for all women and men. Although slavery has been abolished in the past centuries in human history, there is still a lot of employment in inhumane work that can be defined as modern slavery. The increase in decent employment is directly related to economic growth. The work done will be finalised in less time with more quality, waste will be reduced, and this optimised framework will directly contribute to the economy. Decent work is made possible by rights, security, conditions, remuneration, being organised and represented and equality in every sense (Rodgers 2009).

Before 1990, until the millennium development goals (MDGs), poverty was one of the leading indicators affecting decent work and economic growth, while in developing countries, the rate of extreme poverty was 47%, and the rate of work over $4 a day was only 18% (United Nations 2015). While not the exact MDG counterpart of SDG-8, it is directly related to more than one MDG, which are MDGs 1 (eradicate extreme poverty and hunger), 3 (promote gender equality and empower women) and 8 (develop a global partnership for development). According to the research and observation, it illustrates significant changes in poverty. For instance, in a quarter of the years between 1990 and 2015, the rate of poverty decreased by 33% in developing countries. From the gender discrimination perspective, women make up 41% of paid workers except those who work in the agricultural sector. Also, there was an increment in 1990, which was about 35%. In the last 20 years, women gained a right to represent themselves in 90% of 174 countries' parliaments. The research also states that between 2000 and 2004, from developed countries, the official development assistance enhanced 66%, and it reached $135.2 billion (United Nations 2015).

Based on these developments, in the environment created by the goals achieved, it was decided to continue these studies with more comprehensive goals in September 2015 at a historic UN Summit. The sustainable development goals (SDGs) seek to promote long-term economic growth by increasing productivity and fostering technological innovation. Promoting policies that stimulate entrepreneurship and job development and effective efforts to eliminate forced labour, slavery and human trafficking is critical. Keeping these goals in mind, the objective is to attain full and productive employment, as well as respectable work, for all women and men by 2030

("Goal 8: Decent Work and Economic Growth", 2016). The SDG-8 is among the top five, with the most progress between 2018 and 2019 based on target achievement (UNDESA 2020).

Global economic growth, which has already slowed down for many reasons in recent years, has slowed down even more with the effect of the pandemic. The COVID-19 crisis has disrupted many global economic activities. According to projections, real GDP (gross domestic product) is expected to decrease by 4.6% in 2020 compared to 2019. Considering these studies, it is expected to remain well below the 7% target envisioned by the 2030 Agenda for Sustainable Development (United Nations 2021). As a result, SDG-8 is measured in a general framework, with which indicators, the sub-goals and the potentials of decent work and economic growth are mentioned, and how these studies are carried out on a global scale. In 2020, after the global economy and employment were affected by the COVID-19 pandemic that affected the world, the projections have changed, and it will be seen how it is affected by 2030.

The United Nations' SDG-8 promotes inclusive and sustainable economic growth, full and productive employment and decent work for all (United Nations 2015). In his study, Sachs (2012) mentions that as MDGs fulfilled their schedule and appeared to be insufficient, SDGs were suggested to take their place, which was more detailed and comprehensive, although some of the SDGs do not have a corresponding MDG (2012). Due to the lack of existence for this development goal in the MDGs, economic development was included as an important item together with the SDGs. This goal involves some business themes: employment, economic inclusion, non-discrimination, capacity building, availability of a skilled workforce and elimination of forced or compulsory labour ("Goal 8: Decent Work and Economic Growth", 2016). The targets of SDG-8 are including economic growth sustainability per head for eligibility of national state of affairs and especially minimum 7% GDP growth in the countries with the least developed economies; diversification, technological upgrading and innovation, particularly an emphasis on high-value-added and labour-intensive sectors, will help countries achieve better levels of productivity. Also, targets of SDG-8 include micro, small and medium-sized businesses that encourage formalisation and expansion, especially via access to financial services through development-oriented policies that support constructive things, fair job creation, entrepreneurship, creativity and innovation. Progressively improvement of global resource consumption and production is also aimed. Additionally, achieving full and efficient employment, decent work and the same pay for work of the same value for all women and men and decreasing the youth rate not in training, education or employment is crucial. Lastly, SDG-8 targets design and applications for new policies to promote sustainable tourism; supporting local financial institutions capacity; aid for trade support for developing nations; creating and operationalising a worldwide youth employment strategy and putting it into action are the goals. The targets of SDG-8 are shown in Fig. 10.1.

The continuous increase in the population in the world causes concern about the use of natural resources and the fact that people can settle in suitable jobs to live at a certain level of welfare. On the other hand, technology disparities between countries, sectors and businesses are generally acknowledged as the major drivers of productivity inequalities (Acemoglu 2012). However, with the effect of technological developments, economic growth supports meeting human needs and keeping welfare high. Lastly, the models of endogenous innovation and technology have advanced significantly.

In this chapter, the importance and aim of SDG-8, some sub-targets, and the effects of COVID-19 on the urgency of SDG-8 will be explained along with statistical examples. SDG-8, decent work and economic growth aim to help build a sufficient and healthy economy and pro-vide a job with satisfying work conditions for every person in need. Economic power is a key contributor to any other SDGs.

Economic growth depends on the increase in the total production in an economy. This production may be of goods or services required to meet human needs. One of the most critical targets of SDG-8 is "sustainable economic growth", which represents building up the economy while caring for social and environmental issues for current and future generations (de MELLO et al. 2020). Target 8.1, the first target of SDG-8, focuses on sustainable economic growth, which is usually not easy to predict, as it depends on many different parameters. Some of these parameters are usual (government policies, technological developments, new enterprises, global and local trends, etc.), while some are unpredictable scenarios. The pandemic of COVID-19 is such an example. In January 2020 (before the pandemic), the total GDP of the world was expected to grow by 2.5%. In June 2020, the forecasts had changed to a decrease of 5.2% (World Bank 2021). According to the World Bank (2021) data, the effects were enormous: In 2020, world trade volume had decreased roughly by 8.3% compared to the previous year.

SDG-8 contains economic growth and work opportunities with decent conditions for all. According to Embrapa (2020), Brazilian Agricultural Research Corporation, one of the main indicators of economic growth is the employment level. If economic growth sets back and fails to provide jobs to everyone in need, the policy environment will be considered insufficiently business-friendly (Frey 2017). Therefore, government and institutional policies for business will be subject to questioning. As a recent example, the effect of COVID-19 on people's lives can be given. The disease has impacted economic growth and trade volume and employment opportunities. With the virus becoming a pandemic, many companies have downsized. Therefore, employment opportunities have gone down.

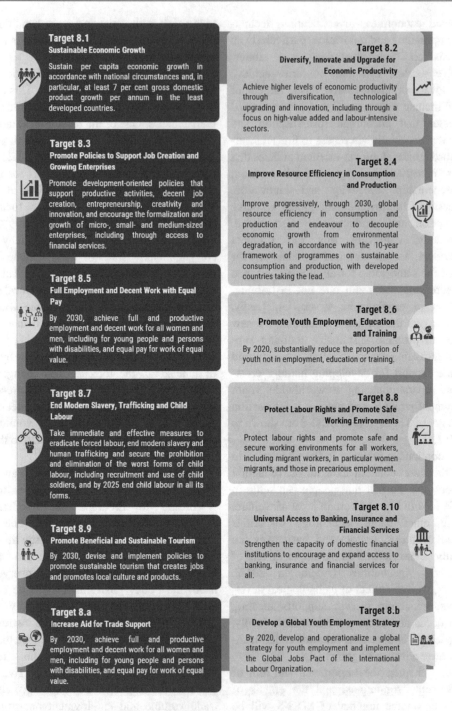

Fig. 10.1 Targets of SDG-8 "Goal 8: Decent Work and Economic Growth" (2016)

SDG-8 also aims to take immediate and effective steps of precaution to protect labour rights and end forced labour, modern slavery and human trafficking (de MELLO et al. 2020). Decent work conditions are also major factors in people's lives as they affect their physical and psychological well-being, income and social status. Therefore, SDG-8 is a matter to be considered vital for the development of society. According to the International Labour Organization (ILO 2006), "decent work" includes providing everyone access to "productive and quality work in freedom, equity, security and human dignity conditions". Therefore, SDG-8 is also an essential contributor to maintaining human rights. The ILO framework for the measurement of decent work includes four main titles: "International labour standards and fundamental principles and rights at work", "Employment creation", "Social protection" and "Social dialogue and tripartism" (Stoian et al. 2019).

The COVID-19 pandemic has had a tremendous influence on job potential in nations across the globe. According to the United Nations Economic and Social Council (2021), following an average of 2% growth from 2014 to 2018, real global GDP per capita expanded by only 1.3% in 2019 and is expected to decline by 5.3% in 2020 related to the pandemic's damage. For the near future, post-pandemic global GDP per capita is expected to rise by 3.6% in 2021 and 2.6% in 2022. The economic growth is steadily rebounding; however, it may continue to lag behind pre-pandemic levels for some time. It has had devastating consequences not only for job opportunities but for many other SDG-8 objectives, such as working hours and income. To reverse this trajectory in the future, economies need to be transformed in ways that promote high productivity, such as shifting from low-yield agriculture to high-yield agriculture (Lambrechts and Stacy 2020). While the global economy improves, and workers' living circumstances improve, inequities persist, and there aren't enough jobs to keep up with the world's expanding population (van den Breul et al. 2018).

10.1 Companies and Use Cases

Table 10.1 presents the business models of 37 companies and use cases that employ emerging technologies and create value in SDG-8. We should highlight that one use case can be related to more than one SDG and it can make use of multiple emerging technologies. In the left column, we present the company name, the origin country, related SDGs and emerging technologies that are included. The companies and use cases are listed alphabetically.[1]

---

[1]For reference, you may click on the hyperlinks on the company names or follow the websites here (Accessed Online – 2.1.2022):

http://passiontoprofit.co/; https://aeinnova.com/; https://anscer.com/; https://aws.amazon.com/what-is-aws/?nc1=h_ls; https://azure.microsoft.com/en-us/overview/; https://enosi.energy/; https://feedzai.com/; https://joonko.co/; https://jos-quantum.de/; https://movandi.com/; https://pezesha.com/; https://smartforest.world/;https://store-h.com/; https://www.accenture.com/us-en/about/technology-index;https://www.aquafarmsafrica.com/; https://www.bitpesa.co/;https://www.bobemploi.fr/; https://www.dataroid.com/; https://www.diversely.io/; https://www.eyelight.tech/; https://www.geminus.ai; https://www.google.com/intl/en_ie/business/; https://www.ibm.com/thought-leadership/institute-business-value/report/ar-vr-workplace; https://www.impactterra.com/; https://www.marklabs.co/; https://www.morpher.com/; https://www.oureye.ai; https://www.ovamba.com/; https://www.poketapp.com/; https://www.qcware.com; https://www.realblocks.com/home; https://www.samsara.com/; https://www.simularge.com; https://www.springboard.com/; https://www.uber.com/global/en/cities/; https://www.upcyclingplastic.com/en/; https://www.yeself.com/

Table 10.1 Companies and use cases in SDG-8

No	Company info	Value proposal (what?)	Value creation (how?)	Value capture
1	Accenture Ireland 8, 9 AI, big data, cloud computing, Datahubs, Internet of Things	It is a company that provides consultancy and information technology services in technology, digital marketing, strategy and operational processes and is a solution partner of AI, big data and cloud computing.	Data becomes the ultimate competitive asset and differentiator with data-led transformation (the company's project), and by scaling AI with cloud, enterprises can reposition their offers, extend capabilities and increase data and AI maturity to generate new sources of value and sustainable development. Customers may set key objectives and develop their governance approach with responsible AI (company project), resulting in solutions that enable AI and their business to thrive.	Economic value is captured by optimising AI systems of clients and transforming clients' business models into sustainable models, thanks to the increase in data access, computing power and cloud speed realised with the services they offer. Additionally, they add social value by supporting cultural differentiation for the workforce to better utilise data and AI.
2	AEInnova USA 8, 9 Edge Computing, Internet of Things	They digitise manufacturing industries with IoT to become more profitable and environmentally friendly.	INDU-EYE is an industrial IoT Condition Monitoring solution that allows clients to remotely monitor the condition of their machines or processes in order to determine the best time for repair. It's a long-range, remotely operated software-as-a-service (SaaS) that uses cutting-edge computer technology to process all data in the appropriate location (Edge Computing).	Revenue capturing is based on enabling smart digitalisation, monitoring and management of equipment and operations and lower maintenance costs.
3	Amazon Web Services USA 8, 9 AI, big data, cloud computing, Datahubs, Internet of Things	It is a cloud computing platform that aims to save companies from high data storage costs with its cloud service, offers them agile and innovative data management and provides an environment for them to easily use technologies such as AI, data lakes and analytics and IoT.	Users may deploy their application in numerous physical sites with only a few clicks because the organisation has infrastructure all over the world. It provides a variety of databases that are purpose-built for various sorts of applications, allowing customers to select the best tool for the task in terms of cost and performance. From infrastructure services like computation, storage and databases to IoT, machine learning, data lakes and analytics and much more, users may spin up resources as needed.	Customers can get economic advantage by using serverless computing services, which allow them to pay only for what they need in information technology operations rather than keeping all data on physical servers and data centres. It captures social value by assisting public and private sector organisations in developing sustainable business models for managing public safety and public health services in the cloud.

#	Company / Country / SDGs / Technology		Value capture
4	Anscer Robotics India 8, 11 AI, autonomous vehicles, robotics	They produce autonomous robotic vehicles managed by AI, which aim to minimise the task of the worker in physical works involving safety problems.	The company designs and manufactures multifunctional robots capable of moving without any human intervention or assistance. Their autonomous mobile robots contribute to the human-machine interaction in the work environment.	As an economic value, by the aid of AI algorithms, more simple executions of physical tasks are accomplished. Preventing humans from hazardous tasks also captures social value.
5	AquaFarms Africa Guinea 1 ,2. 8, 10 Soilless farming	The agri-tech firm uses aquaponics farming techniques (a combination of fish farming and soilless agriculture) to produce fish and speciality vegetables and fruits locally and sustainably that would otherwise be imported from outside the continent.	The promoters focus on production, while the company manages other areas of the operational value chain such as procurement, quality control, marketing, sales and logistics. Their proprietary app provides their franchisees to manage their sites as well as enables them to manage quality control and logistics at lower costs.	By making urban agriculture widely available and accessible through the company's locally built and cost-effective aquaponics system, women and youth will have lucrative prospects in the food space while lowering carbon emissions from food importation. They provide timely farm-to-table delivery of high-value vegetables and fish goods allowing young urban farmers to earn more money and repatriate funds spent in overseas markets.
6	AR and VR in the workplace - IBM USA 8 AI, spatial computing	The company manufactures and distributes computer hardware and software, as well as hosting and consulting services in a variety of fields. In this project, the company expands the use of augmented reality and virtual reality in the workplace.	The company utilises AR/VR and AI to generate XR applications in its business processes to increase work efficiency and improve remote working experience after the necessity of working remotely.	Value capture through enhancing employee productivity and reducing costs with the help of AI and XR and transforming workers' training, workflows and engagement. In addition, XR solutions are delivering measurable returns on investment (ROI) to the business.
7	Bitpesa Kenya 8 Blockchain	They are a blockchain-based transaction company that enables its users to transfer money to and from African countries with their own currencies.	It makes use of blockchain to significantly reduce costs and speed up money transactions. Therefore, it provides easier and cheaper transactions for investors planning to use currency in African countries. In addition, it uses several cybersecurity methods to maintain reliability.	Value is captured by savings and encouraging foreign currency transfers into African economies, enabling easier and faster money transactions.

(continued)

Table 10.1 (continued)

No	Company info	Value proposal (what?)	Value creation (how?)	Value capture
8	Bob Emploi France 1, 8, 10 AI	The platform assists organisations in training their new employees on how to use their corporate software faster by using AI-based analytics.	The app gathers and analyses user behaviour data. Based on the data collected, it delivers an individual and interactive guide of the onboarding, engagement and adoption processes following the correct use of the software. By combining AI and user behaviour analysis, the guidance increases users' experience by providing the personalisation of their requirements in the direction of their specific goals.	Value capture via the use of AI (AI) to improve employees' technological expertise and assist them in adapting to new software without spending extra money or working longer hours.
9	Dataroid Turkey 8, 9 AI, big data	It's an enterprise-level digital analytics and customer engagement platform that uses data-driven insights to assist consumers improve their encounters and offer a better service.	It offers interesting insights and interactive dashboards that go beyond vanity stats to help you make better business decisions. Organisations can follow each interaction step by step owing to its big data architecture and make it intelligible and actionable utilising reporting capabilities, according to its data and analytics capabilities. It assists clients in making better and quicker decisions by recognising abnormalities and outliers in user behavioural data using machine learning techniques.	With its data analytics perspective and machine learning solutions, value is created by better campaign results with personalised interactions; building long-lasting, meaningful relationships with customers; and shipping digital products with better user experiences.
10	Diversely UK 8 AI	It is a data-driven recruitment AI-based platform increasing diversity and inclusion (D&I) in the workplace.	They provide companies with analytics, guidance, tips on recruitment and diversity reports, identifying and prioritising D&I gaps (or hot spots) across the organisation and areas to focus on.	Social and ethical values are captured through enhanced diversity and inclusion in workplaces.

11	Enosi Australia 8, 9, 11, 12, 13 Blockchain	The company delivers a solution to enterprises of all scales and provides solutions regarding clean energy and zero carbon for all households and businesses.	Through its platforms, the firm facilitates sustainable energy tracking and provides blockchain-based payment solutions that are quick, safe and dependable. Customers can choose whether they wish to buy solar or wind generators. They'll be able to see exactly when each of these providers delivers electricity to you, as well as how much money you've saved by going direct.	The value is captured by tracing clean energy and retailing via blockchain. In addition, the environmental value obtained from carbon zero is provided by the supply of electricity directly from solar and wind farms.
12	Eyelight Israel 8 AI, flexible electronics and wearables	It is a wearables company that develops a product for partially sighted and blind people, based on the substitution of the sense of sight through hearing and touch.	The gadget is powered by a 3D wearable camera that records the environment in real time in front of the user. Following the collecting of environmental data, AI algorithms are used to analyse the data and convert it into a usable format. Transformed data, such as traffic lights, GPS directions and so on, are transmitted to the user through Bluetooth via the company's 3D dynamic surface.	Social value is captured by supporting blind people to sustain their functional activities in life and business. Real-time navigation with AI algorithms prevents possible dangers and creates a secure environment for users.
13	Feedzai Portugal 1, 8 AI, big data, cybersecurity	The company is fighting financial crime with AI. Their management platform is powered by big data and machine learning.	They protect businesses against financial risks on their AI-based platforms. By using these technologies, they ensure that decisions can be made very quickly and that fraud can be detected and prevented. They provide a safer environment for businesses, traders and buyers by preventing money laundering and transaction fraud.	Revenue capture models are based on services that analyse money laundering, fraud and similar transactions faster using a machine learning algorithm built by examining big data in end-to-end encrypted databases.

(continued)

Table 10.1 (continued)

No	Company info	Value proposal (what?)	Value creation (how?)	Value capture
14	Geminus USA 8, 9 AI, digital twins	Digital Twins based on physics-constrained AI are used to generate self-optimised industrial process designs.	Operators and engineering service providers may benefit from their industrial process-focused digital twin offering, which provides self-optimised design, predictive operational intelligence and improved asset performance. Their solution uses physics-constrained AI to provide high-fidelity O&M (Operating & Maintenance) intelligence and constantly improve asset and process designs.	The company develops value by offering profits and digital transformation by optimising the economic value-generating phases of factories and enterprises using products based on digital twin and AI.
15	Google My Business USA 8 AI, big data, cloud computing	The company is helping local businesses establish an online presence to increase their revenues. In India, Google helped SMEs by introducing the mobile app "Google My Business". The app allows businesses to create and manage their content for free on Google products in both Hindi and English by using big data, AI and cloud computing.	With the app, the business can stay up to date on Google Search and Google Maps to stand out among competitors. The business shows up when customers search for it on Google or Google Maps. This technology makes it easier to find and easier for customers to visit offices and respond to reviews, messages or questions.	Interacting with customers by responding to reviews, updating working hours and sharing photos of products and services and using the app will help the business to stand out from the competition and increase revenue.
16	Impact Terra Netherlands 1, 2, 8, 13 AI, big data, Internet of Things	The company offers an insights platform for agricultural stakeholders. Their application offers farming tips and weather and disaster predictions for farmworkers. Farm owners and traders can benefit from market information collected from various data sources that the application combines.	The company has a data pool that includes several public institutions along with their data. They use ML to process this data to inform their customers. The reports include weather predictions, disaster predictions and market analyses.	Value is captured by using big data, AI and data processing to provide meaningful predictions. It also creates social value by helping farmers build up their businesses and environmental value by providing useful instructions about agriculture to prevent hazardous farming activities for nature and the environment.

17	Joonko USA 8 AI	It is an AI-based diversity recruitment software that enhances workforce diversity and helps companies achieve their equality and diversity goals.	By analysing the data, they optimise the recruitment processes of the companies and make the processes efficient by giving feedback. They also manage recruitment and position efficiency by performing key performance indicator (KPI) management.	Social and ethical value is captured by making the human resources processes of companies more efficient and supporting diversity.
18	Joss Quantum Germany 8, 9 Quantum computing	The company offers tailored solutions using quantum computing.	The quantum computing company provides solutions for finance, insurance and energy. The company also guides its customers in their transition to quantum computing.	Value is captured by helping to solve the problems of large companies in several industries.
19	Mark Labs USA 1, 8 AI, cloud computing, Datahub	The company focuses on shareholder engagement and management by using AI to assess environmental, social and governance (ESG) risks and opportunities for sustainable investing	Their software-as-a-service (SaaS) platform helps asset managers and asset owners drive environmental and social transformation. Ongoing corporate engagements across customers' portfolios or market-wide with automatically log and track engagements are identified. It has been building solutions to link financial capital with AI technology to achieve real-world sustainability outcomes by database using a combination of natural and AI. All data are accessible via data feeds and APIs.	Value creation through data visualisation and better engagement management and improved social, financial and environmental outcomes.

(continued)

Table 10.1 (continued)

No	Company info	Value proposal (what?)	Value creation (how?)	Value capture
20	Microsoft Azure USA 8, 9 AI, big data, cloud computing, Datahubs, Internet of Things	It is a public serverless computing platform including infrastructure, software and platform services for big data analytics, network, virtual computing, storage, AI and IoT solutions and much more.	Users may create, operate and launch apps from anywhere, using their favourite languages, frameworks and infrastructure – including their own data centres and other clouds – to tackle problems big and small. They may use tools and services built for hybrid clouds to integrate and manage their environments. They can get real-time information from live streaming events and data from IoT devices.	Economic value is captured by providing technical guidance to the customers for choosing short- or long-term cloud preferences based on their needs, and customers can pay per use and prevent physical data server costs. Environmental value is captured by reducing the data centre footprint through minimising physical server use for storage with cloud infrastructures. The company also assists clients in understanding, tracking, reporting, analysing and reducing carbon emissions related to cloud usage.
21	Morpher Austria 8 Blockchain	The blockchain-based trading platform allows its users to access the financial markets, removing barriers that prevent people from entering the cryptocurrency market.	They operate with their ERC20-based tokens and offer smart contracts, which are also based on Ethereum. As blockchain trading is much cheaper (thanks to lower fees), safer and faster than traditional methods, users can trade beyond pre-market and intraday, act on breaking news over the weekend, sell to save their gains or buy a breakout stock before markets open.	Value is captured by savings and improved security. Social value is captured by creating the possibility for all traders to easily use the platform.
22	Movandi USA 8, 9 5G, AI, edge computing, Internet of Things	It is a company that offers 5G millimetre wave solutions to enterprises in their software and hardware systems for the connection of IoT devices into AI applications in processes.	Movandi is tackling real-world 5G mmWave deployments with distinctiveness and high-performance core technologies in 5G integrated circuits, antennas, systems, algorithms and design disciplines, allowing 5G to fulfil its full potential.	Economic value is captured through time and bandwidth savings by integrating 5G into IoT devices, making computation and data storage closer to the source of data with edge computing. Providing low latency wireless connection and extended range between existing designs also contributes to its economic value. Environmental value is achieved through reduced footprint and optimised power consumption.

23	OurEye.ai India 8 AI	It is an AI company that offers solutions by monitoring security cameras and analysing data for its customers.	It is a real-time video analytics platform that enables businesses to use video analytics to supercharge their operations. The firm uses pre-existing CCTV-IP (Closed-Circuit Television Cameras) infrastructure and keeps track of several standard operating procedures throughout our industry checklists. During COVID-19, the company also presented a new solution in which they scanned CCTV-IP records and issued warnings to business owners for recognising and analysing social distance and mask usage.	The value is captured by using AI for real-time video processing and analytics via using pre-existing CCTV-IP infrastructure. Also, the company creates social value by tracking social distancing and mask usage for a healthier community.
24	Ovamba USA 8 AI, big data	The company has developed data analytics-based growth and performance systems to provide customers with interest-free (NIB), risk-reduced capital for trade, imports, manufacturing and business growth by partner banks.	A program is in place that provides central banks with data analytics and technology to assist policy efforts that reduce non-performing loans (NPLs), improve capital markets and maintain a secure and inclusive banking system for all.	It helps institutions deliver financial inclusion, access capital and promote economic growth for small and medium businesses in these regions on a risk-mitigated basis.
25	Passion Profit Kenya 1, 8 big data	The company serves entrepreneurs the mentorship, resources and access to the capital they need to establish profitable businesses that create employment and decrease poverty.	They provide profit creation, business advisory, trade and export analysis services to their users by using big data in the market. Budget and credit forecasting, cash flow management and recommendations for relevant and profitable value propositions are made with the dependence of this analysis.	To capture economic value, they enable their users to get insights and improvement opportunities in the decisions they make on business strategy and financial growth by big data analysis made on the research market.

(continued)

Table 10.1 (continued)

No	Company info	Value proposal (what?)	Value creation (how?)	Value capture
26	Pezesha Kenya 8 Cloud computing	It built a digital financial infrastructure intending to become the leading enabler platform and marketplace with the use of cloud computing.	It connects small and medium-sized businesses with working capital through a collaborative approach in which banks, microfinance institutions and other financial institutions or networks can connect on our platform to be matched with quality SMEs (small- and medium-sized enterprises), resulting in meaningful financial inclusion and lowering the cost of capital.	Economic value is captured by scales and ensures the longevity of customers' loan operations by utilising proprietary credit scoring services and cloud-based financial end-to-end infrastructure.
27	Poket Canada 8, 9 Big data	The company is an all-in-one customer rewards platform that helps businesses attract, grow and retain customers.	The business offers a means to personalise a set of mobile data crowdsourcing tools. Self-reported data, citizen-generated data, high-frequency data collecting and participatory mapping are all good use cases.	The value captured by big data in the processing of crowdsourcing data enables these companies to develop location-specific insights by empowering residents in a community. The spotlight is on programs that can benefit from self-reported data, data created by people or community-driven insights.
28	Qc Ware France 8, 9 AI, quantum computing	The company is a quantum computing software firm that develops corporate solutions that operate on quantum hardware.	It is a quantum computing firm that offers data-related solutions to organisations. They handle challenging issues and their applications with the use of quantum computing and AI. Their focus also includes making quantum computing accessible to classically educated data scientists and providing performance speed-ups on near-term hardware to their customers.	The company creates value by utilising machine learning and mathematical data modelling techniques, in addition to quantum computing processes.

29	RealBlocks USA 8, 17 Blockchain, crowdfunding	The fintech company provides its customers a way to reach advisors and investors to get the best alternative investment managers in its SaaS (software-as-a-service) platform. Through its blockchain-enabled platform, shares of alternative investment products are digitised and made tradable.	The platform uses crowdfunding to help investors raise capital for an enterprise. The blockchain-enabled feature allows any fund distributor to invest in a safer, cheaper and faster way.	The value is captured through the use of blockchain in that trusted shares of alternative investment products are digitised and made tradable in the company's platform. Withal, social value is created by crowdfunding with enabling increased liquidity and improving increased access to prevent barriers to entering global markets.
30	Samsara USA 8 AI, big data, Internet of Things	It is an integrated platform to increase safety, efficiency and sustainability and to digitally transform the world of physical operations.	They capture rich data from sensors, cameras and OEM integrations and centralise all operational data on one unified platform. IoT sensors and AI-powered cameras are used.	Value is created by managing equipment, fleet and sites all from a single platform digitally.
31	Simularge USA 8, 9 AI, cloud computing, digital twins	They provide industrial applications and a development platform that includes data analytics, machine learning, mechanical simulation and app creation features that function in combination.	Machine learning and digital twin applications produced on the company's solution platforms lower industrial enterprises' production costs, estimate maintenance time and identify quality problems before they arise.	The company creates value based on digital transformation for factories and facilities. They also use AI, cloud computing and Digital twin for growth for businesses.
32	Smart Forest UK 8, 13, 15 Blockchain, Internet of Things	It is a platform created to make it easier for planters to find funds and support in their business model. It is also a beneficial investment area for people who want to invest their money with the advantages of blockchain security.	Planters/farmers are supported in their business strategy as well as the installation of IoT devices by the company. The firm builds a one-of-a-kind tokenised tree system based on the ERC-721 standard to provide investors with blockchain-secured investment regions. Each tokenised tree has a complete description that includes the tree's planting date, kind, GPS coordinates, series name, CO_2 offset and anticipated wood price.	It creates environmental value with the financial support system it provides to the endeavour of planting trees. Its economic value is also secured by blockchain compared to other investments, and the tree provides stable financial value as an invariable requirement. "Save and earn sustainably." can be said to be the best summary.

(continued)

Table 10.1 (continued)

No	Company info	Value proposal (what?)	Value creation (how?)	Value capture
33	Springboard USA 4, 8 AI	The company helps job seekers find employment in the field of information technology with AI-based matchmaking and accurate education recommendations.	The company provides education in data, design, cybersecurity and coding fields. The website allows regular users to access specialised education and training to find jobs easily and quickly.	They capture value by providing education for people to find jobs. Social value is created by providing people with accessible education, and the student must only pay the course fee once they have found a job.
34	Store H Italy 7, 8, 9, 12 Energy storage, hydrogen	For its customers, the blockchain firm keeps track of on-chain emissions and energy use.	The start-up will support all major dApps and provide a decentralised and automated approach for dApp developers and users to be climate-positive. It is built to be as modular and pluggable as feasible. The company's objective is to become the industry's go-to multi-chain sustainability middleware for DeFi and NFT.	Value is captured by automatically tracking blockchain transactions. The company aims to enable any user or dApp to be 100% sustainable. In addition, the environmental value is captured by sourcing the highest quality green tokens from different marketplaces and their customers. This process helps create emission-free smart contracts and transactions.
35	Uber USA 8 AI, big data, cloud computing, distributed computing	The ride-hailing company offers a service in which one can submit a trip request that is automatically sent to a driver nearby.	The company's platform automatically finds the navigational route for the driver and calculates the distance and fare using AI and big data. Also, for accurate matching the company uses distributed computing.	The value is captured by using AI, distributed computing and big data during the process of ride-hailing for customers and drivers. Also, the company creates environmental value by decreasing the carbon footprint on transportation by using big data in traffic.
36	Upp Netherlands 8, 13, 16 Recycling	The company supports communities and governments to become plastic waste-free. They create long-lasting, recyclable goods from your discarded plastic.	They offer the loop on plastic waste upcycling locally and in a circular way.	The value is captured by plastic waste recycling. Also, the company creates environmental value by helping reduce plastic waste and its recycling.
37	YesElf Slovakia 8 AI	The company has a software tool that automatically analyses stereotypic content for gender, age and ethnicity in texts and proposes non-stereotypic alternatives. The aim is to empower companies to make diversity the new norm in every company.	The DD-scan is an AI-based software tool that promotes inclusive writing, which enables the user to reach a wider and more diverse talent pool. The DD-scan detects and colour highlights stereotypic language in the texts and suggests neutral alternatives in real time. This way, the customers attract qualified candidates up to threefold, saving time on the recruitment process.	This technology focuses on removing stereotypes in recruitment and communication content to make all the language inclusive and thus enabling recruiters to hire talent efficiently.

References

D. Acemoglu, Introduction to economic growth. J. Econ. Theory **147**, 545–550 (2012). https://doi.org/10.1016/j.jet.2012.01.023

B. Bleys, A. Whitby, Barriers and opportunities for alternative measures of economic welfare. Ecol. Econ. **117**, 162–172 (2015). https://doi.org/10.1016/j.ecolecon.2015.06.021

de MELLO, L. M. R., NSS BASSI, L. A. dos SANTOS, and AFA de A. GERUM. "Decent work and economic growth: contributions of Embrapa." Área de Informação da Sede-Livro científico (ALICE) (2020).

D.F. Frey, Economic growth, full employment and decent work: The means and ends in SDG 8. Int. J. Hum. Rights **21**, 1164–1184 (2017). https://doi.org/10.1080/13642987.2017.1348709

Goal 8: Decent Work and Economic Growth [WWW Document], 2016. Goal 8 Decent Work Econ. Growth Jt. SDG Fund. URL https://www.jointsdgfund.org/sustainable-development-goals/goal-8-decent-work-and-economic-growth. Accessed 15 Aug 2021

M. Lambrechts, B. Stacy, *8 Decent Work and Economic Growth: Increasing Productivity and Reducing Vulnerable Employment* [WWW Document]. (2020). URL https://datatopics.worldbank.org/sdgatlas/goal-8-decent-work-and-economic-growth/. Accessed 15 Aug 21

ILO 2006. Decent Work FAQ: Making decent work a global goal. [WWW Documment] https://www.ilo.org/global/about-the-ilo/newsroom/news/WCMS_071241/lang--en/index.htm

G. Rodgers, The goal of decent work. IDS Bull. **39**, 63–68 (2009). https://doi.org/10.1111/j.1759-5436.2008.tb00446.x

UNDESA, SDG Good Practices, A compilation of success stories and lessons learned in SDG implementation (2020)

United Nations, The Millennium Development Goals Report 2015 (2015)

United Nations, The Sustainable Development Goals Report 2021 (2021)

L. van den Breul, R.-J. Haar, J. Korver, E. Kostense-Smit, J. Muller, J. Olivier, H. Sonneveld, M. van der Valk, Sustainable Development Goals A business perspective, in *Deloitte*, (2018)

Sachs, J. D. (2012). From millennium development goals to sustainable development goals. The Lancet, 379(9832), 2206–2211. https://doi.org/10.1016/S0140-6736(12)60685-0

D. Stoian,, M. Iliana, C. Dean, SDG 8: Decent work and economic growth–potential impacts on forests and forest-dependent livelihoods. Sustainable Development Goals: Their Impact on Forests and People (2019): 237–78.

World Bank 2020. [WWW Document] https://www.worldbank.org/en/news/feature/2020/06/08/the-global-economic-outlook-during-the-covid-19-pandemic-a-changed-world

SDG-9: Industry, Innovation and Infrastructure

11

Abstract

Developing and underdeveloped countries need durable infrastructure investments, sustainable industrial breakthroughs and innovative approaches to achieve sustainable economic growth and social and grassroots development and combat climate change. SDG-9, Industry, Innovation and Infrastructure, is based on three main themes. To provide transportation, information and communication infrastructures, which are an important part of development in line with these goals, the key to sustainable economic growth and raising the welfare level of the society is to develop industrialisation, and new technological developments and new skills in line with innovation are discussed. This chapter presents the business models of 59 companies and use cases that employ emerging technologies and create value in SDG-9. We should highlight that one use case can be related to more than one SDG and it can make use of multiple emerging technologies.

Keywords

Sustainable development goals · Business models · Industry · innovation and infrastructure · Sustainability

Economic welfare is decreasing day by day; meanwhile, inequalities are increasing. People have problems in reaching their basic needs. One out of every three people in the world does not have access to clean drinking water ("1 in 3 people globally do not have access to safe drinking water – UNICEF, WHO"); 940 million people (13% of the world) live without the miracle of electricity (Ritchie and Roser 2020). According to WHO projections, five billion people will be deprived of health services by 2030 (World Health Organization 2017). This situation started to become much more dangerous, especially in underdeveloped and developing countries. Therefore, world leaders adopted SDG-9 specifically for infrastructure and industrial investments within the 2030 Agenda for Sustainable Development to cope with these inequalities and combat climate change through the UN in 2015. In this agenda, developed countries have committed to providing development assistance to developing and underdeveloped countries. Developing and underdeveloped countries need durable infrastructure investments, sustainable industrial

The author would like to acknowledge the help and contributions of Ulaş Özen, Eren Fidan, Uğur Dursun, Büşra Öztürk, Asya Nur Sunmaz and Muhammed Emir Gücer in completing of this chapter. They also contributed to Chapter 2's Carbon Capture and Storage, Cellular Agriculture, Crowdfunding and Flexible Electronics and Wearables sections.

S. Küfcüoğlu, *Emerging Technologies*, Sustainable Development Goals Series, https://doi.org/10.1007/978-3-031-07127-0_11

breakthroughs and innovative approaches to achieve sustainable economic growth and social and grassroots development and combat climate change within the scope of this SDG. In this direction, governments, non-governmental organisations, the private sector and universities need to find solutions to these problems together.

SDG-9 infrastructure is primarily based on environmental considerations and global commitments and is driven by scientific research and innovation. In 2015, Sweden had an ambitious aim to ramp up investments in solar and wind and clean transport and eliminate fossil fuel within its boundaries. Then the ambition evolved among the European Union, where the member states could provision resources. In 2016, Canada took this ambition a step further by lessening traffic congestion to reduce fuel consumption and air pollution and modernising the workplace to use its office places better. After the initiatives of countries such as Australia, China and India in order, countries worldwide have been triggered to make efforts to build resilient infrastructure and sustainable industries and foster innovation (Saxena 2019).

Access to financial services and markets is crucial for developing countries. These countries need loans and credit for their growth. Surveys covering from 2010 to now show that 34% of small-scale industries in developing countries receive loans or credit, which is for competing in the global market. But, in sub-Saharan Africa, only 22% of small-scale industries benefit from loans or credit (United Nations 2020). According to the World Bank's data, individuals using the Internet rate of all populations have increased from 20.412% to 48.997%. The average of OECD members is 83%. Even though individuals using the Internet rate is high for OECD members, the average of the least developed countries, which the UN classifies, is 17%. Access to the Internet is still meagre for the least developed countries (International Telecommunication Union Database 2020). Accessing a mobile network is another main parameter. Almost all of the world's population is covered by mobile networks. It is estimated that 96.5% of the world population is covered by at least 2G communication, and 81% of the world population is covered by the Long-term Evolution (LTE) network (United Nations 2020).

The 5th target is enhancing research and upgrading industrial technologies with the indicators of new product development or established technology and infrastructure and increasing expenditure on R&D (Källqvist 2021). Regarding SDG 9.5, worldwide R&D spending has risen, with R&D investments totalling 1.7% of global GDP in 2014, as stated by the SDG Progress report (United Nations 2021b). For example, R&D expenditure in wealthy nations amounted to 2.4% of GDP, in developing countries 1.2% and in the least developed countries (LDCs) 0.3%. The number of researchers per million is similarly divisive: 1098 researchers worldwide in 2014, 3739 in the millions in the rich nations and just 63 in the millions in the LDCs (International Telecommunication Union Database 2020; United Nations 2021b). According to a UNCTAD media statement released, the research calls for focused investment policies in developing countries to increase connectivity infrastructure, encourage digital enterprises and assist the larger economy's digitisation (UNCTAD 2021). So, this SDG-9 aims to trigger new action plans according to innovative movements, thanks to contributing to the vast financial resources in most countries, especially developing countries.

SDG-9 is based on three main themes. To provide transportation, information and communication infrastructures, which are an important part of development in line with these goals, the key to sustainable economic growth and raising the welfare level of the society is to develop industrialisation, and new technological developments and new skills in line with innovation are discussed. Figure 11.1 summarises targets of SDG-9.

Industry-innovation cooperation in the case of developing sustainable infrastructure becomes crucial over the sustainable development goals agenda. Achieving these objective goals complies with the execution of infrastructural enhancement. Economy, environment and society are the three main pillars in focusing on the vital step of implementing innovation and indus-

Target 9.1
DEVELOP SUSTAINABLE, RESILIENT, AND INCLUSIVE INFRASTRUCTURES
Develop quality, reliable, sustainable and resilient infrastructure, including regional and transborder infrastructure, to support economic development and human well-being, with a focus on affordable and equitable access for all

Target 9.2
PROMOTE INCLUSIVE AND SUSTAINABLE INDUSTRIALIZATION
Promote inclusive and sustainable industrialization and, by 2030, significantly raise industry's share of employment and gross domestic product, in line with national circumstances, and double its share in least developed countries

Target 9.3
INCREASE ACCESS TO FINANCIAL SERVICES AND MARKETS
Increase the access of small-scale industrial and other enterprises, in particular in developing countries, to financial services, including affordable credit, and their integration into value chains and markets. Promote inclusive and sustainable industrialization and, by 2030, significantly raise industry's share of employment and gross domestic product, in line with national circumstances, and double its share in least developed countries Promote the rule of law at the national and international levels and ensure equal access to justice for all

Target 9.4
UPGRADE ALL INDUSTRIES AND INFRASTRUCTURES FOR SUSTAINABILITY
By 2030, upgrade infrastructure and retrofit industries to make them sustainable, with increased resource-use efficiency and greater adoption of clean and environmentally sound technologies and industrial processes, with all countries taking action in accordance with their respective capabilities.

Target 9.5
ENHANCE RESEARCH AND UPGRADE INDUSTRIAL TECHNOLOGIES
Enhance scientific research, upgrade the technological capabilities of industrial sectors in all countries, in particular developing countries, including, by 2030, encouraging innovation and substantially increasing the number of research and development workers per 1 million people and public and private research and development spending.

Target 9.A
FACILITATE SUSTAINABLE INFRASTRUCTURE DEVELOPMENT FOR DEVELOPING COUNTRIES
Facilitate sustainable and resilient infrastructure development in developing countries through enhanced financial, technological and technical support to African countries, least developed countries, landlocked developing countries and small island developing States.

Target 9.B
SUPPORT DOMESTIC TECHNOLOGY DEVELOPMENT AND INDUSTRIAL DIVERSIFICATION
Support domestic technology development, research and innovation in developing countries, including by ensuring a conducive policy environment for, inter alia, industrial diversification and value addition to commodities

Target 9.C
UNIVERSAL ACCESS TO INFORMATION AND COMMUNICATIONS TECHNOLOGY
Significantly increase access to information and communications technology and strive to provide universal and affordable access to the Internet in least developed countries by 2020.

Fig. 11.1 The targets of SDG-9 (United Nations 2021a)

trial development into societal development (The Economist Intelligence Unit 2019).

In the economic aspect, cases of generating infrastructure linked to the innovative industry are profitable in practical steps that vary from the first area of job creation to producing active industrial links "such as a bridge that links a rural village to urban markets" (The Economist Intelligence Unit 2019). The basis of a healthy economy lies on the ground of sustainable devel-

opment of creating values. "Concrete, steel and fibre-optic cable are the essential building blocks of the economy" (Puentes 2015). Therefore, generating infrastructure by investing in energy projects, telecommunication systems, pipelines, parks and water systems keep that ground fruitful. While pointing out that economic growth is visibly linked with infrastructural progress, it enables many other goals that depend on it to be actualised and should not be left out unspoken.

"[…] sustainable infrastructure enables governments and the private sector to provide […] broader economic growth while improving quality of life and enhancing human dignity" (The Economist Intelligence Unit 2019).

A senior policy analyst in the OECD's Environment Directorate, Virginie Marchal, says "Infrastructure is really at the centre of the delivery of the SDGs" (The Economist Intelligence Unit 2019). Looking at the main circles of SDGs which are not only focusing on materialistic development but also the reduction of every kind of inequality within nations, providing access through infrastructural development plays an undeniable role. "Infrastructure is a tool in increasing social mobility" (The Economist Intelligence Unit 2019). Access to affordable and fair clean water, food, sanitation, education and employment and gender equality cannot be separated with the pipeline installation, innovative and efficient agriculture developments, construction of education centres in safe walking distances and providing those who are out of employment and boosting new job areas and transportation options.

Through the environmental aspect, with innovative methods, industries are now ready to follow a greener path in which they reduce the harmful impact of their work and, in some cases, even neutralise it. In this case, being the most used by people in industrial outcome, green-focused provided infrastructure enables millions if not billions of people to contribute with or without knowing. To give an example, the Economist Intelligence Unit published data from the USA in which they say that it is estimated that for a person travelling to work back and forth, switching to use public transportation from private may have the power in the reduction of carbon footprint close to 2177.243 km per year (The Economist Intelligence Unit 2019). By actively contributing, green infrastructure installations can also provide quite a beneficial improvement for the environment and directly to the city populations' life quality. Other data from The Economist Intelligence Unit based on a simulation suggests that, in downtown Toronto, a reduction by 2 °C in the temperature would have been achieved if half of the suitable exteriors were installed with a green roof (The Economist, 2019, p. 8). Innovative developments and simple cases of planting trees and greens also have a huge impact on many lives by being effective protection in the case of natural disasters such as floods and soil erosion.

SDG-9 proposes that industrialisation and technological progress are the basis for growth and that all countries should industrialise sustainably. Investment in infrastructure and innovation are important factors of economic development. Cities now accommodate more than half of the global population. It is now more important than ever to establish new enterprises and develop public transportation and renewable energy and information and communication technologies. Long-term approaches to financial and ecological concerns, such as energy efficiency and employment generation, need innovative transformation. Sustainable development may be fuelled by encouraging sustainable industries, technological research and innovation investment. Several countries across the world are working on SDG-9. Germany's Chemnitz University of Technology won the German Excellence Initiative with its Merge Technologies for Multifunctional Lightweight Structures research centre (MERGE). In this cluster, seven of Chemnitz University of Technology's eight faculties, two local Fraunhofer-Institutes as extramural research partners and several industrial partners collaborate in a trans-disciplinary approach to develop new lightweight materials that will allow cars, for example, to lose weight and consume less fuel while conserving natural resources (United Nations 2018).

Another example is the 10th Annual Longjiang Cup, an innovation competition for advanced mapping technology and product information modelling students, held at the Harbin Institute of Technology (HIT) in China. The competition's goal is to apply the spirit of the government of Heilongjiang Province's Proposal to Encourage University Graduates' Innovation and Entrepreneurship. With the involvement of 20 teams comprising of 150 undergraduates from 16 institutions, including HIT, it also seeks to promote students' inventive abilities (United Nations 2018).

The HLFP (High-level Political Forum) thematic review, released in 2017, emphasises that funding SDG-9 implementation would need large investments in infrastructure and innovation and instances of how SDG-9 expenditures may help other SDGs develop. According to the report, more than 1.1 billion people lack access to electricity, 663 million do not have access to safe drinking water, 2.4 billion do not have adequate sanitation, and one-third of the world's population does not use all-weather roads. Even though closing these gaps requires infrastructure, innovation and roughly US\$1–1.5 trillion per year in developing countries, official development assistance (ODA) to developing countries for economic infrastructure, particularly transportation, totalled around USD 57 billion in 2015 (SDG Progress Report), up 32% from 2010 (United Nations 2017) (International Institute for Sustainable Development (IISD) 2017).

As a result of investments in local infrastructure and technology (such as water pumps, power, clean cookstoves and mini-grids), local growth may be accelerated and more inclusive. At the same time, efficiency is increased, and repair and maintenance time is reduced. As explained in the review, financing at this level is likewise fraught with difficulties. For example, growing financial services in the agri-food and rural sectors, commonly comprising small-scale operations, present obstacles due to a lack of credit histories and collateral.

11.1 Companies and Use Cases

Table 11.1 presents the business models of 59 companies and use cases that employ emerging technologies and create value in SDG-9. We should highlight that one use case can be related to more than one SDG and it can make use of multiple emerging technologies. In the left column, we present the company name, the origin country, related SDGs and emerging technologies that are included. The companies and use cases are listed alphabetically.[1]

---

[1] For reference, you may click on the hyperlinks on the company names or follow the websites here (Accessed Online – 2.1.2022):

http://envelio.com/; http://futuresisens.com/en/home/; http://greyscale.ai/; https://appliedvr.io/; https://awakesecurity.com/product/; https://carbonengineering.com/; https://carrier.huawei.com/en/; https://cattleeye.com/; https://climeworks.com/; https://credosemi.com/; https://deepmind.com/; https://eigentech.com/; https://epigamiastore.com/; https://gravity.co/; https://www.avantmeats.com/; https://integriculture.jp/?locale=en; https://intellistruct.com/; https://iomoto.io/en/; https://jobs.netflix.com/; https://mirreco.com/; https://new-home.superpedestrian.com/; https://nextwavesafety.com/; https://northvolt.com/; https://samsunghealthcare.com/en; https://saphium.eu/?lang=en; https://spatial.io/; https://steeltrace.co/; https://surgicaltheater.net/; https://verily.com/; https://www.apple.com/healthcare/apple-watch/; https://www.armis.com/; https://www.ayfie.com/; https://www.beebryte.com/; https://www.bluebeam.com/uk/solutions/revu; https://www.bostondynamics.com/spot; https://www.dow.com/en-us; https://www.ebayinc.com/company/; https://www.fbr.com.au/; https://www.gene.com/; https://www.hybirdtech.com/; https://www.ibm.com/quantum-computing/; https://www.intellisense.io/; https://www.microsoft.com/en-us/hololens; https://www.mobiusbionics.com/; https://www.multiplylabs.com/; https://www.nvidia.com/en-us/geforce-now/; https://www.perceptiveautomata.com/; https://www.redrockbiometrics.com/; https://www.rubberconversion.com/; https://www.seadronix.com/; https://www.sensefly.com/; https://www.starlink.com/; https://www.terracycle.com/en-GB/; https://www.track160.com/; https://www.tsmc.com/english; https://www.ubiqsecurity.com/; https://www.uipath.com/; https://www.valerann.com/; https://www.waste2wear.com/

Table 11.1 Companies and use cases in SDG-9

No	Company info	Value proposal (what?)	Value creation (how?)	Value capture
1	Apple USA 3, 9 AI, flexible electronics & wearables, healthcare analytics	They make smartwatches that can identify cardiologic abnormalities, detect an epilepsy crisis and alert medical personnel and the patient's family.	Apple Watch has an optical heart sensor, electrical heart sensor, accelerometer and gyroscope. It can measure heartbeats per minute, detect irregular rhythm and tachycardia and record ECG by using an optical and electrical sensor on it. It also notices epilepsy crises by using an accelerometer, gyroscope and call emergency services. It records all medical data of the patient and reports to the patient about its risks using the healthcare analytics database.	Value is captured by providing early diagnosis of cardiological diseases, detection of epilepsy crises and referring the user to emergency services.
2	Applied VR USA 3, 9 Spatial omputing	It is a virtual reality content firm that focuses on assisting individuals with pain and anxiety.	With VR glasses that the patients wear on their heads, the patients feel themselves in a different environment, and the interactive structured environment offered by VR makes it an effective tool for the designed attention modulation.	Value is captured by a de facto digital pain management platform and by establishing a new treatment paradigm for pain management in the hospital, in the clinic and at home.
3	Armis USA 9 Cybersecurity, Internet of Things	It provides a cybersecurity solution that enables businesses to control their networks and devices with IoT.	It identifies unmanaged devices with IoT without intermediaries and allows compromised devices to report their security status. Cybersecurity enables visibility and control by keeping the environments with systems secure.	It protects business devices and networks against cyber threats by removing security agents that are highly likely to be hacked against cybersecurity threats arising from the increase of connected devices.
4	Avant Meats China 2, 9, 14 Biotech & biomanufacturing, cellular agriculture	They offer lab-grown fish by using cultivation processes that are sustainable and scalable.	They isolate small samples of cells from healthy fish. Then they feed them nutrients and give them a nourishing environment. And they allow them time to grow in perfectly nutrient-rich conditions.	Environmental value is captured by producing lab-grown fish by avoiding fish farming and reducing pollutants in the fish.

5	Awake Security USA 9 AI, cybersecurity	It is a cybersecurity firm that combines data with in-depth security knowledge utilising artificial intelligence.	The Awake Security Platform analyses the network traffic and autonomously identifies, assesses and processes threats – giving customers actionable insight to respond effectively. It identifies the true attack surface, isolates threats that go undetected and enables rapid response.	Value is captured through enhancing digital security and keeping data at the point of collection to ensure privacy and compliance.
6	Ayfie Norway 9, 16 AI, blockchain	It provides search and text analytics solutions such as efficiently summarising, filtering and improving business intelligence, by language analysis and indexing unstructured data to extract meaning from data.	With AI, it detects meaningful data sets in the text and reviews extensive information such as a person, organisation, date, etc., with semantic analysis. It prepares reports with AI insight-based predictions and research based on user needs determined by machine learning with detected data.	By providing aggregated analysis of textual data, it enables users to understand what is mentioned, evaluate prepared reports and cluster documents without wasting time and energy. Thus, it enables faster and more accurate decisions to be taken with less labour and time.
7	BeeBryte Singapore 9, 11 AI, Internet of Things	It develops energy-saving software as a service for businesses, industries and electric vehicle charging stations.	Internet of Energy is the notion. Automatic control of heating-cooling equipment (e.g. HVAC), pumps, EV charging points and/or batteries, using AI, reduces utility expenditures. If solar power is available, they aim to maximise self-consumption.	Up to 40% savings can be achieved based on weather forecasts, occupancy, use and energy price signals.
8	Bluebeam USA 9 Autonomous vehicles, cloud computing	It is a collaboration platform that provides smart solutions to keep architects, engineers, construction firms and government agencies in sync; interact anytime, anywhere; and complete construction projects faster.	It uses cloud computing to store and manage entire projects so that global data infrastructure is brought together and expanded. Smart annotations, hyperlinking and automation tools are presented, thanks to autonomous things technology for providing organised work.	It provides increased value for clients in terms of reducing design conflicts and for design teams/consultants in terms of efficient planning and schematic design. Easier collaboration and document interactivity are served for team members' value.

(continued)

Table 11.1 (continued)

No	Company info	Value proposal (what?)	Value creation (how?)	Value capture
9	Boston Dynamics USA 9 AI, autonomous vehicles, edge computing, robotics	It is a robotics company that creates robots that mimic the mobility, dexterity and agility of humans and animals in order to improve people's lives.	AI allows its robot Spot, to walk, climb stairs, avoid obstacles, traverse terrain and autonomously follow predetermined routes with little or no input from users. With dynamic motion technology supported by Edge CPUs, Spot can navigate tough, unstructured, unknown and antagonistic terrain with ease.	It provides robots with the technical requirements to reach terrains inaccessible to others and perform automated tasks in unstructured environments. Spot prevents casualties caused by dangerous encounters and potentially dangerous situations. It provides data to the authorised person by scanning the environment without the need for the authorities to enter threatening environments.
10	Carbon Engineering Canada 9 Carbon capture & storage	It is a sustainable energy company that absorbs CO2 from the atmosphere and converts it into a safe and inexpensive transportation fuel.	The company uses direct air capture (DAC) technology which pulls in atmospheric air and then, through a series of chemical reactions, extracts the carbon dioxide (CO2) from it while returning the rest of the air to the environment.	As an environmental value, it does photosynthesise much faster, with a smaller land footprint, and also delivers climate-relevant quantities of permanent carbon removal and delivers clean, affordable energy to the world.
11	CattleEye UK 9 AI	It is a firm that watches dairy cows without the use of hardware using a deep learning video analytics platform.	A simple low-cost security camera mounted at the entrance or exit of the milking parlour begins to learn how to uniquely identify cows and monitor welfare and an increasing number of other behaviours using AI algorithms in the cloud.	Value is captured by providing evidence that farmers are working to the highest level of stock management.
12	Climeworks Switzerland 9 Carbon capture & storage	The firm creates, manufactures and runs direct air capture equipment that capture carbon dioxide from the atmosphere and permanently store it underground.	The company's capturing system works in two stages: At first, the air is drawn into the collector with a fan. Secondly, the bonds are released under the high temperature of between 80 and 100 °C. The adopted storage method is storing in the underground mineralisation of carbon dioxide.	The environmental value which makes direct air capture a unique approach is that it has the smallest land and water usage of all carbon dioxide removal approaches.

#	Company	Description	Use case	Value capture
13	Credo USA 9 Advanced materials, cybersecurity	For hyperscale, cloud, corporate and edge data centres, 5G wireless and telecom service providers and high-performance Artificial Intelligence industries, they allow secure, high-speed, low-power networking with high signal integrity.	Data security between Ethernet-connected devices is provided by the firm. The OWL-800 MACsec chip is a high-end security chip. It encrypts and decrypts data at a rate of 800 gigabits per second and supports a variety of protocols.	Secure data transport across Ethernet-connected devices generates value.
14	DeepMind UK 9 AI	It is an AI company that works on inventions and advancements in machine learning, neuroscience, engineering, mathematics, simulation and computing infrastructure, along with new ways of organising scientific endeavours.	AlphaFold, which is accepted as a solution to the 50-year-old protein folding problem through an open platform of DeepMind, uses AI to accurately predict the shape of a protein at scale and in minutes, down to atomic accuracy.	Value is captured by providing an open platform for everyone contributing to modern biological research. This research has the potential in helping tackle diseases and quickly find new medicine to unlock the mysteries of how life itself works.
15	Dow USA 3, 6, 7, 9, 11, 12, 13, 14 Advanced materials	The company provides services in essential sectors such as textile, chemistry, construction, food and health. They produce advanced materials for buildings.	V Plus Perform solutions use panel insulation technology to guarantee that buildings are both sustainable and energy-efficient. The chemical characteristics of the advanced insulating material utilised have enhanced its performance, resulting in higher recyclability and a reduction in CO_2 emissions. These assist building users in evaluating environmental performance and increasing energy efficiency by up to 20%.	In the chemical industry, advanced materials science is used to produce environmental and cost-effective building standards.
16	eBay USA 9 AI, distributed computing	It is an online shopping site that enables AI-driven sales assistance and auction-vice shopping.	It creates an environment where both buyers and sellers are analysed by detail to connect each of them with the right fit. As buyers search for listings, their queries are analysed and broken down into structured signals allowing downstream systems to capture buyer demand. These signals are tracked and stored in a data warehouse and subsequently processed as aggregated demand over time.	Financial value is captured by a variety of fees for services provided such as branding solutions and advertising business, allowing individuals or businesses to list items for auction.

(continued)

Table 11.1 (continued)

No	Company info	Value proposal (what?)	Value creation (how?)	Value capture
17	Eigen Technologies UK 9 AI, natural language processing	It is a research-driven AI firm that specialises in natural language processing (NLP) for companies in the banking, legal and professional services industries.	It enables organisations to unlock the value of their qualitative data by automating the extraction of text from documents using machine learning and NLP.	It meets the highest standards in information security and offers a full suite of APIs and system integrations with a client success team that will support customers every step of the way.
18	envelio Germany 7, 9 AI, big data	It is a big data and AI-powered digital power-grid management platform.	By utilising AI-based algorithms and gathering big data from the grid, the platform assists grid operators with grid planning, optimisation, forecasting, automatisation and operation.	Revenues are generated by using AI algorithms to automate operations and ease grid design and management.
19	Epigamia India 9 Internet of Behaviours	It is a start-up that produces chocolate syrup and spreads.	It understands the needs and wants of the customers and non-customers by using the technology of the Internet of Behaviour.	By using IoB it can detect the behaviour of its customer base. It tracks down their geographical locations and the restaurants they visit, to customise their products that cater to customers' needs.
20	FBR Australia 9, 13 Robotics	They manufacture robotic block laying devices and give system solutions, allowing them to perform safely and accurately in uncontrolled conditions outside.	To increase home building efficiency, an optimisation programme turns wall designs into block placements and reduces handling and waste of block products. Hadrian X works with accuracy in outdoor conditions, thanks to FBR's dynamic stabilisation technology. In real time, this system corrects for dynamic interference and vibration in the boom and layhead and precisely lays blocks.	Revenue generation via faster and more efficient construction. Environmental value is captured by reducing waste in construction processes.
21	FutureSiSens Spain 3, 7, 9, 11 Advanced materials, autonomous vehicles, big data, healthcare analytics, Internet of Things	It detects changes in the environment through thermoelectric sensors it produces from silicon nanoparticles and uses this data to develop products and services in the fields of health and energy infrastructure with big Data.	It measures the change of gas flows in pipes with autonomous microsensors, collects and shares data with IoT, performs maintenance and hazard detection in the field of energy and produces healthcare analytics products by detecting respiratory problems in the field of health services.	Instant monitoring of difficult-to-measure and controllable works ensures the pre-detection of critical problems, thus saving maintenance costs and resources.

22	Genentech USA 3, 9 AI, biotech & biomanufacturing	They transform cancer patient data into cancer treatment.	They transform cancer patient data into computer models to be used in cancer treatment.	The company facilitates the healing process of patients by ensuring early and rapid cancer treatment and diagnosis.
23	Gravity USA 9 Flexible electronics & wearables	It designs, builds and flies jet suits that can be wearable.	Their jet suits are in backpack form and fly with jet fuel. The weight of jet suits is 300 pounds, but it is designed not to transfer high force to the pilot. Pilots who are less than 200 pounds can fly about 7 minutes by dint of jet fuel. The controller of jet suits is made of flexible electronics and attached to the hand of the pilot to be controlled ergonomically on the flight.	It saves time by swiftly traversing short distances. It provides a wide range of military and entertainment-related solutions.
24	GreyScale.AI USA 9 AI	The firm uses cutting-edge technology to test pouches, bags and rigid containers' seals.	By combining advanced imaging solutions with AI/machine learning software, sealing defects are detected and also real-time insight is provided into sealing efficiency to ensure a tightly controlled sealing process.	AI improves overall sealing quality, detects seal faults and provides 100% inspection traceability as well as strong cloud analytics.
25	Huawei Carrier China 9, 11 5G, cloud computing, cybersecurity	The company provides innovative infrastructure solutions and digital transformation services to industries, especially the telecommunications sector, using 5G, cloud computing and cybersecurity technologies.	5GtoB solution provides low-latency communication and advanced mobile broadband with 5G, while cloud computing digitises the infrastructure.	By increasing the productivity of applications and devices in the industry chain and digitising the industry, it will simplify business users' operations, increase revenues and make their workflows more efficient.
26	HyBird Technologies UK 9 AI, cloud computing, digital twins, drones	It facilitates asset management and saves resources and time by making failure predictions before critical decisions for industrial companies with its software tools and technologies.	Clarity creates a model of reality with the drones, sensors and cameras it uses and its digital twin. The created model makes failure predictions with AI and transfers it to project portfolio management via the SaaS platform.	Value is captured by cost reduction and fast and remote control of assets by optimising information management and workflow with predictive models.

(continued)

Table 11.1 (continued)

No	Company info	Value proposition (what?)	Value creation (how?)	Value capture
27	IBM Quantum USA 9 Quantum computing	They provide quantum hardware and quantum software and adapt them to industrial solutions. The technology is also used to understand subatomic particles, determine the behaviour of proteins and find optimal routes for cruises.	Their quantum computer has 65 qubits and 128 quantum volumes, and it can deliver tasks 128 times faster than supercomputers. It can also carry out complex tasks with its quantum circuits which cannot be done by classical computers.	Value is captured by solving complex tasks much faster which cannot be solved by classical computers.
28	IntegriCulture Inc. Japan 9 Cellular agriculture	Cell-based meat is a sustainable protein developed by the cellular agriculture platform firm.	The technology of cellular agriculture is used for large-scale cell cultivation during the production of meat. The system creates the best growth conditions for cells in vivo so that any cells and products produced by living things become possible in principle cultures.	Revenue is generated by allowing a significant reduction in the cost of cell culture. As a benefit for individuals, IntegriCulture's cultivated cells produce supplements that may be depleted due to age.
29	Intelligent Structures USA 9 AI, autonomous vehicles, Internet of Things	It is a bridge performance asset management system that gives bridge owners data and decision assistance based on IoT, analytics and machine learning to help them make creative decisions that reduce risk and increase productivity across the bridge life cycle.	The firm delivers data and decision assistance to bridge owners based on IoT, analytics and machine learning, allowing them to make creative decisions that reduce risk and increase productivity in bridge asset management.	It unlocks wasted infrastructure value and creates high-performing structures for a safer and more productive world by decreasing traffic congestion, enhancing public safety, communicating with autonomous vehicles and improving air quality.
30	IntelliSense UK 9, 12 AI, cloud computing, digital twins, Internet of Things	The software firm focuses on the mining sector, with the goal of using AI to make mining operations more efficient, sustainable and safe.	The operation as a service infrastructure platform provides process optimisation in mining operations with AI-powered operational decision-making software and digital twin simulation technologies. With IoT adapted to sensors, mining-related data are obtained remotely.	It ensures the efficient use of data in making critical decisions in the mining industry, making critical decisions less costly and more reliably.
31	iomoto Germany 9 Blockchain	It is a platform that transforms the car into a business entity capable of paying for things like parking and charging on its own.	It is based on blockchain and provides a low-cost transaction platform with no creditor risk and complete security. Smart contracts, which are kept as a chain of data blocks, are used to conduct all transactions.	Revenue is captured by removing third-party intermediaries for charging and parking purposes. The service allows customers to use all charging stations.

32	Microsoft HoloLens USA 3, 4, 9, 11 Spatial computing	With the help of mixed reality, it is a Microsoft service that has multiple use cases in manufacturing, engineering and construction, healthcare and education to boost productivity, user accuracy and output.	HoloLens 2 can be used in safely completing tasks error-free with hand tracking, built-in voice commands, eye tracking, spatial mapping and a large field of view. One can also access a robust ecosystem of applications that are supported by the security, reliability and scalability of Microsoft Azure.	Value is captured by reducing downtime and speeding up onboarding and upskilling, accelerating the pace of construction and mitigating risks earlier in the construction cycle, enhancing the delivery of patient treatment at the point of care and improving student outcomes and teaching from anywhere with experiential learning.
33	Mirreco USA 9 Blockchain, carbon capture & storage	Mirreco Cast Blockchain envisages a "seed to solution" CO2 framework; from initial carbon sequester to tracking system validating CO2 contained.	Blockchain is used in a Cast unit to measure the value/volume of CO_2 from the location the hemp was grown, and this will allow a value to be placed, creating a tradeable asset across the entire ecosystem.	Environmental value is captured by offering traceability of CO_2 that allows assisting industry, government, growers and communities to make conscious choices that deliver on their lower carbon responsibilities as well as reducing industrial CO_2.
34	Mobius Bionics USA 9, 3 Flexible electronics & wearables, robotics, wireless power transfer	It is a medical device firm that focuses on introducing sophisticated prosthetic technology to amputees who have lost their upper limbs.	Their prosthesis, LUKE Arm, is connected to neurons with electrodes and can move with its 10 powered joints. The arm includes software and can perform the daily activities of amputees' robotic arm. In addition to the many control input options, the LUKE arm may be controlled with Inertial Measurement Units (IMUs) that are typically worn on the user's shoes. They read the tilt of the user's foot and interpret each movement like a joystick to control the arm. This unit can be charged with wireless charging by using wireless power transfer.	Social and ethical value is captured by providing the possibility of performing hand actions and daily activities for amputee persons.

(continued)

Table 11.1 (continued)

No	Company info	Value proposal (what?)	Value creation (how?)	Value capture
35	Multiply Labs USA 3, 9 3D printing	The customised pharmaceutical firm creates 3D printed capsules that include all of a patient's prescribed medications.	The capsule is manufactured autonomously by using 3D printing and programmed to release pharmaceutical ingredients in a controlled way throughout the day. The capsule is produced from a specific pharmaceutical polymer called hydroxypropyl cellulose, or HPC.	The company allows patients who consistently are on certain prescriptions to take their pills in a controlled and systematic manner. The patient does not over-use nor miss medication and saves time as well. By using only one capsule, the company reduces the production and usage waste and capture environmental value.
36	Netflix USA 9 Distributed computing	It is a streaming entertainment service that produces TV series, documentaries and feature films across a wide variety of genres and languages.	It takes advantage of distributed systems that have a large number of clusters deployed over numerous AWS EC2 instances, each of which has a large number of Memcached nodes and cache clients. The data is shared across the cluster within the same zone, and sharded nodes store several copies of the cache.	Thanks to distributed computing, the site generally doesn't experience any disruptions if a single machine fails. It is fault-tolerant as it can be made up of hundreds of nodes that work together. It is efficient because workloads can be broken up and sent to multiple machines.
37	NextWave Safety USA 9 Spatial computing	It is a platform that gives safety training to employees as part of their job orientation for conducting safety investigations and operating heavy equipment.	Virtual reality (VR) technology is used for enhancing safety training that reproduces practical scenarios in a virtual environment.	As a social value, the company provides health and safety for the workers. The revenue stream is optimised by providing a notable decrease in labour training costs.
38	Northvolt Sweden 9 Energy storage, recycling	The firm is a battery maker with a reduced carbon impact and a battery recycling service.	Cell production lines enable deep cell traceability and energy storage, more efficient manufacturing of recycling materials and enhanced cell performance through lifetime performance and degradation prediction.	Environmental value is created by decreasing the carbon footprint and providing greener energy storage.

39	NVIDIA	GeForce NOW is a cloud gaming service that lets users play games hosted on remote servers and broadcast to one of the compatible devices via the Internet.	Data centres through partnerships around the world provide servers utilising Nvidia Tesla graphics cards and can stream games at up to 1080p resolution at 60 frames per second. With an Internet connection, using adaptive bitrate streaming, cloud computing and 5G technologies to scale the quality based on bandwidth, users reach quality gaming experiences.	It provides gamers with low-cost hardware and a chance to reach the latest gaming experience.
	USA			
	9			
	5G, cloud computing, distributed computing			
40	Perceptive Automata	It is a firm that develops human behaviour prediction technologies for the safe deployment of highly automated (L2/3) and autonomous (L4/5) vehicles on a broad scale, particularly in cities. The business makes it possible for these cars to predict what people will do next, allowing them to safely manoeuvre around humans such as pedestrians, cyclists and other drivers.	The product is an AI inference model that integrates into autonomous driving stacks. The core software module is wrapped in a C++ API that enables a seamless integration. It detects pedestrians and other related moving things as they are and not as black boxes. It analyses their certain body movements and velocity while deciding on how to act upon possible crises.	The company provides a safer and more consistent ride for both the drivers and pedestrians. By eliminating the risks of accidents, it helps to save money as well as save lives.
	USA			
	9			
	AI, robotics			
41	Redrock Biometrics	It provides a technology named PalmID, a user-friendly biometric authentication that brings security for digital devices with a camera (either RGB or IR).	PalmID Capture Module uses sophisticated machine vision techniques to turn the RGB video input stream into an authentication-ready palm image. The PalmID Matching Module matches in real time the captured palm image against cloud stored references. This process uses proprietary algorithms extensively tested against large datasets of palm images that ensure no false positives.	It makes the authentication process fast and easy. It provides VR and AR gamers non-touched passwords not reliant on keyboards, avoiding the need to create usernames and passwords for devices workers and consumers use every day. It facilitates P2P transactions, offering greater security than PIN and chip solutions, increasing transaction speed and data protections while paying.
	USA			
	9			
	Biometrics, cloud computing			

(continued)

Table 11.1 (continued)

No	Company info	Value proposal (what?)	Value creation (how?)	Value capture
42	Rubber Conversion Italy 9 Recycling	Rubber devulcanisation company that produces high-quality compounds by releasing the potential of valuable raw materials from post-production and post-consumer rubber waste.	Behind the concept of recycling, the company employs a mechanical process initiated by an eco-sustainable powdered devulcanised agent at room temperature and pressure that selectively dissolves sulphur bonds, thereby keeping intact the mechanical qualities of the compound.	By reducing raw material costs and improving carbon and material impact, recycling and upcycling materials help to improve product sustainability.
43	Samsung Healthcare South Korea 3, 9 AI, flexible electronics & wearables, healthcare analytics, Internet of Things	The company produces solutions to users' problems with human-oriented innovations using flexible electronics and wearables and healthcare analytics technologies.	They analyse the data collected from digital products by using wearable technology products and IoT, personalising daily activities such as health and sports using AI, with healthcare analytics technologies.	It optimises the health of the individual, such as early diagnosis in the fields of individual health and disease detection, by analysing the personal data of users.
44	Saphium Germany 9 Bioplastics, hydrogen, recycling	They develop natural, non-toxic and compostable bioplastics with the aid of microorganisms.	Polyhydroxyalkanoates (PHAs)-based bioplastics are waterproof, durable in sterile and dry environments and degraded in the soil after 180 days. They use hydrogen and carbon dioxide as building blocks. They extract carbon dioxide from the air or industrial exhaust streams and produce hydrogen with electricity from renewable energy sources via electrolysis.	Environmental value is captured by decreasing plastic waste and CO_2 from the atmosphere. The production of 1 kg of plastic according to this process removes 2 kg of carbon dioxide from the air.
45	Seadronix South Korea 9 5G, AI	It is a deep tech maritime solution provider that uses 5G to give an AI-powered autonomous aid navigation system.	Its AI technology-based ship control system allows the devices to work together in real time to avoid obstacles and make necessary adjustments using 5G to avoid crashing.	Value is captured by decreasing the losses that can occur in terms of human life, the environment and money when accidents take place.
46	senseFly Switzerland 2, 9, 11, 15 AI, drones	senseFly is an AI-powered drone solutions provider. The collection and analysis of geospatial data allow surveying, agriculture, engineering and humanitarian aid to make better and faster decisions.	Through the use of high-tech cameras on their AI tech drones, the company provides mapping and surveying solutions by analysing geospatial data. Solutions are offered by senseFly, for example GeoBase, for those customers who don't own a base station without Virtual Reference Station (VRS) access.	Value is captured by more efficient mapping and surveying operations with drones. Revenue is gained through time and money-saving via creating more efficient drones. Data analysis through AI helps solve agricultural problems. It offers the reduction of fertiliser applications by 20% using its drones.

47	Spatial USA 9 Spatial computing	It is a 3D collaborative platform that lets users use avatars to represent themselves in a virtual workspace from anywhere in the world.	The user uses any device including VR/AR headsets, desktop or phone use. The user joins his teammates by creating a lifelike avatar and creates an environment just as if the same environment were working together.	The company enables teams that collaborate closely and visually on physical products, especially from remote locations, to get their work done.
48	Starlink USA 9 5G	It is a satellite Internet constellation service operated by SpaceX providing satellite Internet access to a wide range of populations worldwide.	It has more than 1600 satellites in orbit. They use 5G technology to provide consumers with high-speed, low-latency broadband and internet for video chats and online gaming without the need for wired connections.	It has low latency, faster Internet connection and the ability to provide Internet access to rural areas and communities with low socioeconomic standards.
49	SteelTrace Netherlands 9 Blockchain	It is a digital platform that makes the steel supply chain for the chemical and petrochemical industries more transparent, traceable and efficient.	Within the blockchain, every action or data input is recorded and track of the data related to certification in the steel supply chain is kept instantly so that full traceability of who did what and when is recorded, making all users fully accountable for their actions or data inputs are available.	The platform allows companies to collaborate in a standardised way and faster exchange of certificates and fewer supply chain delays resulting in cost savings.
50	Superpedestrian USA 3, 9, 11, 13 AI, autonomous vehicles, robotics	It is a transportation robotics firm that develops technology to improve the safety and performance of micro-electric cars. The business is mass-producing the semi-autonomous robotic bicycle wheel by focusing on urban mobility.	The semi-autonomous robotic bicycle wheel learns how the user rides, amplifies pedal power, has regenerative braking abilities and connects to smartphones for ride analysis. The wheel's battery is charged via an external cord that fits a standard wall outlet. Electronic Braking Assistance while riding will partially recharge the wheel when coasting or backpedalling.	Environmental value is captured by promoting urban mobility and sustainable transportation in order to decrease the carbon emissions of both private and traditional public transport. It provides an innovative cycling experience, by creating a faster and more stable solution.

(continued)

Table 11.1 (continued)

No	Company info	Value proposition (what?)	Value creation (how?)	Value capture
51	Surgical Theater USA 3, 9 Spatial computing	It is a firm that delivers virtual and augmented reality-based healthcare services.	The company's VR software allows surgeons to walk through and visualise difficult procedures, including brain surgery, in virtual reality before ever touching the patient.by combining simulation technology with a patient's own anatomy scans, using medical imaging such as MRI, CT and DTI, to create a 360-degree virtual reality reconstruction of the patient's own anatomy and pathology.	Patients can see, in VR, exactly what a surgeon intends to do to their body before they give their consent. Precision VR prevents patient engagement opportunities in the clinic while providing surgical planning and navigation capabilities in the operating room, as well as for medical education and collaboration.
52	TerraCycle UK 9 Recycling	To eliminate waste, the company specialises in the recycling of materials that are difficult to recycle such as pet food packets, bread bags, crisp packets and nitrile gloves.	For the recycling phase, the R&D team of the company analyses the materials to determine the mechanical recycling or the chemical recycling processes to apply it into new materials that include how to break down the waste, separate it into its building blocks and then recycle those materials for new applications.	The environmental value is captured by providing recycling services to non-recycling materials. The company eliminates waste and creates a more sustainable environment.
53	Track 160 Israel 9 AI	It is a software-as-a-service platform (SaaS) that enables precise monitoring and performance statistics as well as automated solutions, giving clubs and academies access to extremely accurate data, video and events all in one place.	Using a single camera set, deep-learning algorithms and groundbreaking optical tracking technology, Track160's platform provides teams at any level with an accurate 360° match and player analysis via AI. It's able to accurately identify the location of each joint in the body of a player and track its motion.	Revenue through an improved scouting system. It lets clubs find young skilled potential players. With system tracking the movement of the players of both teams and the ball across the entire field at all times using deep learning in its software, it provides team coaches to find their optimal tactics and formations.
54	TSMC Taiwan 9 Advanced materials	It is a semiconductor technology business that makes components for automotive electronics, high-performance computing, the Internet of Things and mobile devices.	Their manufacturing technologies include integrated circuits, MEMS, CMOS image sensors, embedded NVM, radio frequency, analogue, high voltage and BCD power processes.	High-performance computing devices enable more advanced, capable, intelligent, energy-efficient and safer products and allow to greatly increase the quality of life and move towards a sustainable society for the common good.

#	Company	Description	Value
55	Ubiq Security USA 9 AI, cybersecurity	They enable developers to quickly integrate encryption directly into their applications – without requiring any prior encryption knowledge or expertise.	It protects sensitive data with its cybersecurity system on IoT devices and networks – by enabling them to integrate data encryption directly into devices through rapid, lightweight and scalable APIs.	The revenue model is based on providing solutions to customers by saving time and money, reducing data theft and avoiding a multifaceted extortion event that causes data to be sold online.
56	UiPath Romania 9 AI, robotic process automation	It offers an end-to-end platform for automation, combining the robotic process automation (RPA) solutions and AI tech to enable every organisation to rapidly scale digital business operations.	The company builds, deploys and manages software robots that simulate human actions interacting with digital systems and software using RPA and AI.	Robotic process automation streamlines workflows, which makes organisations more profitable, flexible and responsive. It automates the workflow.
57	Valerann Israel 9 AI, big data, Internet of Things	A traffic management data platform, The Smart Road System, is a proprietary solution for sensing technologies into roads, transforming them into a data-generating, connected infrastructure.	The Smart Road System platform uses all available sources of information, to provide real-time, high-resolution insights and predictions by leveraging AI and Big Data algorithms. In addition, IoT wireless sensing technologies are integrated into roads, turning them into a connected infrastructure that generates data.	Revenue generation through improving incident management, optimising road operations, making roads ready for the future of smart mobility. Social value is captured by increased traffic management and reducing accidents.
58	Verily USA 3, 9, 10 Big data, flexible electronics & wearables, healthcare analytics	The company produces research and solutions such as hardware, software and clinical expertise to improve healthcare.	The Onduo project provides virtual care, health coaching, daily activity tracking and telemedicine opportunities by personalising instant data from the individual with big data technology and flexible wearable products by using Healthcare Analytics technologies such as the digital phenotyping method.	It increases public health by making health data more usable and contributing to individual health-enhancing solutions and research in the field of health. It reduces social inequality by lowering healthcare costs and making vulnerable populations more accessible to healthcare.
59	Waste2Wear China 9 Blockchain, recycling	For fashion firms, promotional stuff, home décor and other sectors, the company provides traceable textiles and products made from recycled materials such as recycled polyester (RPET).	The fabric's value chain is entirely transparent and supported by Blockchain technology. The Blockchain records the recycling process step by step, allowing each stage in the value chain from plastic garbage to textiles to be tracked. The QR codes on the Blockchain-verified items provide access to product information and recycling.	Environmental value is captured by reducing the plastic waste that ends up in landfills, rivers and the ocean. Furthermore, fewer carbon emissions are emitted and less water is used in the production.

References

International Institute for Sustainable Development (IISD), Policy brief: How can progress on infrastructure, industry and innovation contribute to achieving the SDGs? | SDG knowledge hub (2017). Available at: https://sdg.iisd.org/commentary/policy-briefs/how-can-progress-on-infrastructure-industry-and-innovation-contribute-to-achieving-the-sdgs/. Accessed 2 Nov 2021

International Telecommunication Union Database, Individuals using the Internet (% of the population) | Data (2020). Available at: https://data.worldbank.org/indicator/IT.NET.USER.ZS. Accessed 1 Nov 2021

T. Källqvist, The Sustainable Development Goals in the EU budget (2021). Available at: https://ec.europa.eu/clima/policies/international/negotiations/paris_en. Accessed 1 Nov 2021

R. Puentes, Why infrastructure matters: Rotten roads, Bum economy, Brookings (2015). Available at: https://www.brookings.edu/opinions/why-infrastructure-matters-rotten-roads-bum-economy/. Accessed 2 Nov 2021

H. Ritchie, M. Roser, Energy. Our world in data (2020). Available at: https://ourworldindata.org/energy. Accessed 1 Nov 2021

S. Saxena, Sustainable Development Goal 9: Building resilient infrastructure, sustainable industrialization and fostering innovation. ABS Int. J. Manag. (2019) Available at: https://absjournal.abs.edu.in/ABS-Journal-Volume-7-isuue-2-December-2019.pdf#page=29. Accessed 1 Nov 2021

The Economist Intelligence Unit, The 2019 infrascope. The Economist (2019). Available at: https://infrascope.eiu.com/wp-content/uploads/2019/04/EIU_2019-IDB-Infrascope-Report_FINAL-1.pdf. Accessed 2 Nov 2021

UNCTAD, Digital economy report 2021 (2021). Available at: https://unctad.org/system/files/official-document/der2021_en.pdf. Accessed 2 Nov 2021

United Nations, 2017 HLFP thematic review of SDG-9: Build resilient infrastructure, promote inclusive and sustainable industrialization and foster innovation 1 A global perspective on SDG-9 (2017). Available at: https://sustainabledevelopment.un.org/content/documents/10786Chapter2_GSDR2016.pdf. Accessed 2 Nov 2021

United Nations, #SDGsinAcademia: Goal 9. United Nations (2018). Available at: https://www.un.org/en/academic-impact/sdgsinacademia-goal-9. Accessed 2 Nov 2021

United Nations, Economic and Social Council (2020). Available at: https://undocs.org/en/E/2020/57. Accessed 1 Nov 2021

United Nations, Goal 9 | Department of Economic and Social Affairs (2021a). Available at: https://sdgs.un.org/goals/goal9. Accessed 2 Nov 2021

United Nations, The Sustainable Development Goals report (2021b). Available at: https://unstats.un.org/sdgs/report/2021/The-Sustainable-Development-Goals-Report-2021.pdf. Accessed 1 Nov 2021

World Health Organization, World Bank and WHO: Half the world lacks access to essential health services, 100 million still pushed into extreme poverty because of health expenses (2017). Available at: https://www.who.int/news/item/13-12-2017-world-bank-and-who-half-the-world-lacks-access-to-essential-health-services-100-million-still-pushed-into-extreme-poverty-because-of-health-expenses. Accessed 1 Nov 2021

SDG-10: Reduced Inequalities

12

Abstract

The concept of inequality is that two different people or two different societies do not have equal rights and freedoms on the same event, depending on certain factors. Inequality is a situation that prioritises one segment and excludes the other segment. These inequalities can be mainly age, gender, disability, race, ethnicity, origin, religion and economic situation. Ensuring SDG-10, Reduced Inequalities, is an important step in the path of achieving a more sustainable world. This chapter presents the business models of 21 companies and use cases that employ emerging technologies and create value in SDG-10. We should highlight that one use case can be related to more than one SDG and it can make use of multiple emerging technologies.

Keywords

Sustainable development goals · Business models · Reduced inequalities · Sustainability.

The author would like to acknowledge the help and contributions of İlke Burçak, Fatma Balık, Aleyna Yıldız, Fatmanur Babacan, Handan Öner, Enejan Allajova and Batuhan Özcan in completing of this chapter.

The concept of inequality is that two different people or two different societies do not have equal rights and freedoms on the same event, depending on certain factors. Inequality is a situation that prioritises one segment and excludes the other segment. These inequalities can be mainly age, gender, disability, race, ethnicity, origin, religion and economic situation. Also, these inequalities begin to increase due to people's place in society and class differentiation. The tenth sustainable development goal (SDG-10) "reduced inequalities", which is expected to be reduced by 2030, is very important for all developed or developing countries (European Commission 2021). Because inequality occurs in many areas, it should be examined in different groups as it has many cultural, regional and religious layers. Since these factors have led to various inequalities in the individual and society term, the European Commission has divided inequalities into two groups so that "reducing inequalities" can be achieved and studies can be carried out on this issue (European Commission 2021). So, when the main problems that cause inequality are considered, some of them are seen to be more personal problems, while others are more social. For example, it is stated that while poverty caused by economic inequality is a more personal problem, gender inequality is more social. Thus, some inequalities that can be grouped based on individual (European Commission 2021) include:

S. Küfeoğlu, *Emerging Technologies*, Sustainable Development Goals Series,
https://doi.org/10.1007/978-3-031-07127-0_12

1. **Economic inequality** is expressed as a result of differences such as income, consumption and wealth of individuals in society.
2. **Social inequality** can be expressed as inequalities experienced in social areas such as education or employment. Some differences in social status and position among individuals in society, which lead to economic inequalities, are strongly associated with social inequality.
3. **Political inequality** is expressed as the unequal consequences of decisions taken by political authorities. This can lead to different groups being prevented from participating in political processes.
4. **Environmental inequality** can be expressed in the unequal distribution of environmental risks such as air and water pollution and inequality in their access to ecosystem services such as land, parks and fresh water in the surrounding area. Changes in the social status of individuals and their positions in society cause environmental inequality to be closely related to social and economic inequality.

Some other inequalities can be grouped based on society (Stewart et al. 2009):

1. **Vertical inequality** is expressed as some differences between individuals in a particular country, region or the whole world.
2. **Horizontal inequality** is characterised as inequalities that arise due to culturalal differences between groups such as ethnicity, religion and race within a specific country or region or in the world.

According to some studies carried out by the European Commission, it has been determined that "horizontal inequality", which also includes gender inequality, is the group where violence and conflict are seen the most and which causes social peace, social order and democracy to be affected (European Commission 2021).

Inequalities are not only between people. There are inequalities between countries, such as those connected to representation, migration and

development assistance, which are also targeted by the goal (UN 2015). It is also important to implement the laws to prevent the economic system and the current global wealth inequality against the inequalities between countries. Thus, thanks to the targets and indicators established under SDG-10, it also acknowledges that it helps to alleviate inequalities within the country (Ofir et al. 2016). SDG-10 aims to improve developing nations' representation in international markets, regulate migration and increase the flow of finance to developing countries via foreign direct investment and government development assistance. In this approach, eliminating inter-country inequalities serves as both a goal and a means of lowering inequalities inside countries (Katila et al. 2020). In addition, reducing income inequality will lead to positive economic and sociocultural results (Pickett and Wilkinson 2010). Also, another area of inequality is in education. The increasing level of education will increase development worldwide and reduce economic inequality. According to a study, increasing the early child education rate to 25% in all countries has a benefit of $10.6 billion, and increasing it to 50% has a benefit of $33.7 billion (Engle et al., 2011).

Socioeconomic injustice manifests itself in the form of exploitation, economic marginalisation and denial of essential living conditions, resulting in unequal concepts of justice. Thus, socioeconomic injustice can lead to disasters. For example, thousands of migrants lost their lives on their journeys due to high migrant movements in 2020. 4186 deaths and disappearances were seen in 2020 (United Nations Economic and Social Council 2021). Remedies against distributional injustice can be realised through factors such as changing the division of labour, renewing incomes and transforming economic structures. The transformation of economic structures can also occur with the restructuring of political-economic policies (Katila et al. 2020). So, political decisions are vital in tackling these inequalities. It has been said that legally protecting rights will also reduce social and cultural inequality (Guha-Khasnobis and Vivek 2007).

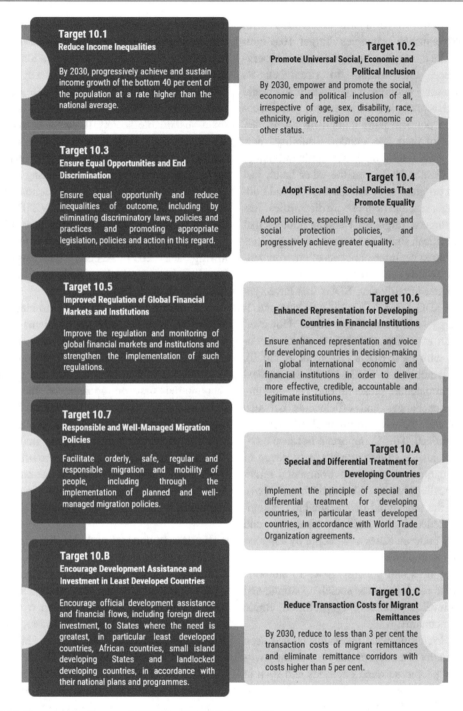

Target 10.1
Reduce Income Inequalities

By 2030, progressively achieve and sustain income growth of the bottom 40 per cent of the population at a rate higher than the national average.

Target 10.2
Promote Universal Social, Economic and Political Inclusion

By 2030, empower and promote the social, economic and political inclusion of all, irrespective of age, sex, disability, race, ethnicity, origin, religion or economic or other status.

Target 10.3
Ensure Equal Opportunities and End Discrimination

Ensure equal opportunity and reduce inequalities of outcome, including by eliminating discriminatory laws, policies and practices and promoting appropriate legislation, policies and action in this regard.

Target 10.4
Adopt Fiscal and Social Policies That Promote Equality

Adopt policies, especially fiscal, wage and social protection policies, and progressively achieve greater equality.

Target 10.5
Improved Regulation of Global Financial Markets and Institutions

Improve the regulation and monitoring of global financial markets and institutions and strengthen the implementation of such regulations.

Target 10.6
Enhanced Representation for Developing Countries in Financial Institutions

Ensure enhanced representation and voice for developing countries in decision-making in global international economic and financial institutions in order to deliver more effective, credible, accountable and legitimate institutions.

Target 10.7
Responsible and Well-Managed Migration Policies

Facilitate orderly, safe, regular and responsible migration and mobility of people, including through the implementation of planned and well-managed migration policies.

Target 10.A
Special and Differential Treatment for Developing Countries

Implement the principle of special and differential treatment for developing countries, in particular least developed countries, in accordance with World Trade Organization agreements.

Target 10.B
Encourage Development Assistance and Investment in Least Developed Countries

Encourage official development assistance and financial flows, including foreign direct investment, to States where the need is greatest, in particular least developed countries, African countries, small island developing States and landlocked developing countries, in accordance with their national plans and programmes.

Target 10.C
Reduce Transaction Costs for Migrant Remittances

By 2030, reduce to less than 3 per cent the transaction costs of migrant remittances and eliminate remittance corridors with costs higher than 5 per cent.

Fig. 12.1 Targets and indicators of SDG-10. (United Nations 2021)

SDG-10 has 10 sub-targets. As shown in Fig. 12.1, Targets 10.1-4 of SDG-10 are illustrated to understand how they are distributed among social groups, minorities and multiple groups across variables such as political, social and economic (Kabeer et al. 2017). Target 10.1 recognises economic differences within a country, whereas Target 10.3 recognises potential distribution; both have a lot of overlap with distributive justice principles. Target 10.2 conforms to the concepts of recognising and observing justice, thus aiming to strengthen its inclusiveness economically, politically and socially. However, it aims to ensure social, economic and political participation for all people, regardless of their age, gender, disability, race, ethnicity, origin, religion or economic or another status. Indicator 10.2.1, on the other hand, just evaluates progress in economic aspects and contexts of age, sex, and people with disabilities. Just as other targets, Target 10.4 adopts policies that review political, social and economic inequalities by adhering to the three concepts of environmental justice (Katila et al. 2020).

There are, in particular, SDGs that have synergy, like SDG 1-8-5. SDG 5 is a unique one. It has synergy with most of the SDGs. However, there is an exception as SDG-10, which reconciles not with most SDGs (Hegre et al. 2020). SDG-10 is not tied to SDGs 12–15, which deal with environmental preservation. The possible conflicts, exchanges and efficiencies between SDG-10 and these environmental SDGs have gone unnoticed. This is important because environmental justice study is particularly aware of the implications of global environmental remedies on localised battles (Sikor and Newell 2014). In short, the tenth goal of sustainable development, "Reduced Inequality", aims to end inequality between states and people by supporting policies implemented to eliminate discriminatory practices and policies. SDG-10 supports everyone to reach the same social, economic and political level regardless of age, religion, ethnic origin or economy.

After discussing what inequality is, it is seen that inequality is a concept that is constantly witnessed by people between countries and even in all areas of society. Among the types of inequality, social inequality, which affects society and humanity the most, still makes a name for itself in today's age. Even in today's conditions, inequality is a problem in the world. Disadvantaged ethnic groups or religious communities with a bad image may experience difficulties while benefiting from public services such as health and education. The deprivation of some segments of such public services in a sustainable life is a major obstacle to sustainability for that society because public services have a key value for a sustainable society (Ogwezi et al. 2020). Therefore, states have demonstrated their political intention and desire to decrease inequality within and across countries by adopting SDG-10. Each country can choose which direction it wants to take to meet this lofty target by 2030. With this, in September 2015, 193 states have affirmed that they have the political willpower to act following the 2030 Agenda for Sustainable Development, which includes SDG-10 (Kaltenborn et al. 2020).

Although the approach of states to inequalities is positive, societies still face problems in many areas due to inequalities. In this case, to better comprehend the significance of preventing inequality, it is necessary to state the problems encountered first on account of the present inequalities. In this way, the importance of adopting this purpose can be better explained by revealing how many problems are imposed on life. In the course of daily life, it is possible not to pay enough attention to whether one can recognise the problems that arise from these inequalities.

- **Health:** In unequal societies, the expectation of long life is lower as the healthcare services are not provided equally. At the same time, in this kind of society, psychological health problems, child death rates and overweight issues are also higher. Furthermore, the percentage of HIV infection is higher in unequally developed and still developing countries (The World Economic Forum 2015). If we consider the reverse case, healthy individuals mean more efficient work, happier people and a prosperous society. Therefore, more people must achieve affordable and professional healthcare services (ESCAP 2019).

- **Social life:** It is claimed that inequality affects mental health, drug use, obesity, education, youth/birth, high crime rate and anti-social/fearful behaviour (Pickett and Wilkinson 2010). Also, reducing the existing inequality between population groups triggers the reduction of the conflict rate (Stewart et al. 2009).
- **Work-life:** The importance of equality in access to decent work can be seen in economic aspects and social aspects. To avoid poverty, working without social protection in dangerous works while taking insufficient money is named vulnerable employment. The more people work as vulnerable employees, the more economic growth slows, and the more society gets uneasy (ESCAP 2019).
- **Human capital development:** Scores in UNICEF's child welfare index appear to be worse in some countries with higher inequality levels. In these countries, the number of people who leave their education and jobs or become mothers at a young age increases as inequality increases. As a natural consequence of all these, social mobility is restricted and innovation rates are lower. Due to better social mobility, equal countries are prone to use higher ratios of innovation (The World Economic Forum 2015).
- **Economic progress and stability:** Economic progress and stability: The International Monetary Fund (IMF) states that decreasing inequality will support economic growth in the long run. This expression can mean that poverty reduction is achieved through income equality (The World Economic Forum 2015). The probability of crashing global markets due to drastic wealth disparity between and among the countries was reported by the most prominent economists globally, including the IMF itself (Berg et al. 2018). Inequality is associated with economies becoming more frequent and severe in their ups and downs. This makes economies even more volatile and vulnerable to crises (The World Economic Forum 2015). The 2018 World Inequality

Report aims to measure income and property inequality systematically. Earnings in some countries can be several times higher than in other countries. In this case, the increasing political influence in the economy makes it difficult to trust the parliaments and the state. Today, even in countries that have survived the crisis, economic inequality can arise (United Nations 2020a). From another perspective, economic growth will be slow in societies with high-income inequality. While the cost of ignoring inequality is high, countries with high-income inequalities will experience an economic recession, and it will be difficult for the society to get out of poverty (United Nations Economic and Social Council 2021).

- **Ecological:** Drought, rising sea levels, hurricanes, landslides, low-income people living in villages and large cities and developing countries have disproportionately affected the impacts. Their location and serious socio-economic damage affect these countries more (Scholz 2020). As all know, some countries maintain their existence with agriculture-livestock, and some countries survive with widespread infectious and respiratory diseases or find it difficult to find food. Because these conditions cannot be equalised, poor or disadvantaged countries suffer more from climate change and natural events. Because, for example, in countries that live on agriculture and animal husbandry, people have to stay connected to the land and the country, they must adapt to situations immediately to survive. However, the disadvantages do not adapt easily (United Nations 2020a). Additionally, when there is a lack of clean and economical alternatives, people burn crop wastes or coal for lighting, heating or cooking. That's why equality in accessing clean energy is important for a healthier and cleaner world (ESCAP 2019).
- **Technological:** The development of automation and artificial intelligence has made some occupational groups unemployed, and the

impacts of these technologies have contributed to income inequality. On the other hand, however, some countries that cannot reach health services have started to reach them, thanks to the developing technologies. If a system that can facilitate learning is invented at the educational level, the people in rich countries have more advantages than developing countries because of transportation/communication costs (United Nations 2020a).

The problems caused by inequality in the areas encountered in life have been explained above. SDG-10 has an important role in solving problems that are so integrated into human life. It is necessary to solve by considering the specific areas and situations of the 2030 targets and the problems after.

The action plan of sustainable development goals is aimed to be completed by all states in the world by 2030. However, there are some debates on whether the SDGs can be achieved by the target date or not. According to the "UN's 2020 report on the SDGs", it has been stated that due to the negative impact of the recent pandemics and regional wars, the work towards the successful completion of the SDGs has slowed down (United Nations 2020b). If SDG-10 is achieved, studies have been carried out on what the "imaginary" world would be like. For example, according to a study in the UK West Midlands Combined Authority Area, city issues have been solved under cover of SDGs. The estimates of future oriented to SDG-10 (Bonsu et al. 2020):

- Reducing health disparities and enhancing population health and well-being, including mental health difficulties, air pollution and the global climate catastrophe.
- Improving life prospects for everyone, including those who face particular challenges or disadvantages.

- Spreading the word about a movement for inclusive leadership/inclusivity, wealth and fairness.
- Building the correct skills leads to increased productivity and prosperity, allowing people from all walks of life to find work. Improving skill levels so that people have the knowledge and certificates they need to find work.
- Addressing climate change through improving location, infrastructure, air quality and the environment.
- As a result of these efforts, the WMCA has been recognised as one of the best achievers on local sustainability programs nationally in yearly league tables.

As can be seen from the study, ensuring SDG-10 is an important step in the path of achieving a more sustainable world. However, whether or not these SDGs can be completed by the targeted date has also become important for the future. When the United Nations' SDG report of 2020 is examined, it has been observed that the pandemic in 2020 has some consequences that may also affect the SDGs. Considering that the course of the SDGs depends on some developments, evaluations of improvement/worsening made based on areas become important. While there has been improvement in some areas, other areas such as food insecurity and the increase in natural disasters have caused many inequalities to emerge. These developments show that the pandemic has produced some unpredictable effects in 2020, and it is not exactly known what effect it will have on the course of the SDGs. According to the report, it becomes more difficult to achieve these targets until the targeted date (United Nations 2020b). In other words, it is thought that there is not enough data about whether the targets can be achieved or not in the future and that the effect of the pandemic can be reversed as a result of the ongoing studies.

The matter of inequality is a popular research area in economics. Inequality has risen over the world. Countless studies relate inequality and economic growth (Galor 2011). The 2018 World Inequality Report, co-authored by Alvaredo, Piketty and Zucman, likewise strikes a fresh and distinct tone, warning that if growing inequality is not adequately tracked and tackled, it might lead to a wide range of political, economic and social disasters (World Inequality Lab 2018). Especially, policy circles have become more interested in the economic growth rates affected by income inequality. Although the effect may vary based on the wealth of the corresponding and some other variables, it is stated that the changes in income inequality affect the gross domestic product (GDP) per capita (Hossen and Khondker 2020). On the other hand, the World Bank blazed the way, demonstrating that there are policies that can reduce inequality while also increasing growth and productivity (World Bank 2016).

Furthermore, the 2018 World Inequality Report has some latest studies: Economic inequality exists in all parts of the world. For example, in Europe, it is the least, whereas, in the Middle East, it is the highest. The disparity has expanded in nearly all nations in recent years. However, at varying rates, since 1980, economic inequality has climbed fast in North America, China, India and Russia, while it has increased moderately in Europe (Kaltenborn et al. 2020). The expansion of democracy into economic institutions, in addition to international and national initiatives to develop progressive tax systems and combat tax evasion and tax havens, can have a significant influence on decreasing inequality. Nations may adopt some basic, doable steps to increase equality alongside economic democracy.

12.1 Companies and Use Cases

Table 12.1 presents the business models of 21 companies and use cases that employ emerging technologies and create value in SDG-10. We should highlight that one use case can be related to more than one SDG and it can make use of multiple emerging technologies. In the left col-

[1]For reference, you may click on the hyperlinks on the company names or follow the websites here (Accessed Online – 2.1.2022):

http://www.almacenaplatform.com/; http://youbenefit. spaceflight.esa.int/3d-printing-for-refugee-camps/; https://ceretai.com/; https://lexmachina.com/; https://mouse4all.com/en/; https://quadfi.com/index.html; https://skilllab.io/en-us; https://tykn.tech/about/; https://voiceitt.com/; https://www.ava.me/; https://www.dotincorp.com/; https://www.gapsquare.com/; https://www.knockri.com/ethical-ai/; https://www.letsenvision.com/; https://www.marinusanalytics.com/; https://www.microsoft.com/en-us/ai/seeing-ai; https://www.scewo.com/en/; https://www.sevaexchange.com/; https://skilllab.io/en-us; https://www.skillhus.no/about-us; https://www.visualfy.com/; https://www.wandercraft.eu/fr/accueil-2

Table 12.1   Companies and use cases in SDG-10

No	Company info	Value proposal (what?)	Value creation (how?)	Value capture
1	Almacena Platform Netherlands 1, 5, 8, 10, 12 AI, blockchain	It is a marketplace that uses AI and blockchain to allow African farmers to maximise their potential and gain access to EU markets.	The company ensures more efficient and sustainable trade for both sides through fully digitising the supply chain, from the small-scale producer to the final buyer. They use Blockchain to create digital identities of coffees; cloud computing to create coffee origins, routes and data records; and NFC for product traceability and tracking.	Revenue generation is captured by empowering small-scale farmers. They ensure risk reduction for both sides of trades.
2	Ava USA 10 AI	It is an AI-based speech recognition technology that converts speech into writing.	Ava Closed Captions and Ava App provide subtitles for screen, images or presentations using AI. It is also capable of providing what is spoken around the user in written form when everybody speaking is connected to the app.	The app reaches out to those who require hearing aids and assists them in living a more intelligible life. It allows these persons to integrate into society without any restrictions.
3	Ceretai Sweden 10 AI	It's an AI-based automated programme that identifies and dispels preconceptions and standards in popular culture, such as movies and music.	The start-up's products utilise Artificial Intelligence to sift through vast datasets or archives of client information. The Diversity Dashboard then visualises the outcomes, allowing media companies to keep track of content diversity. It monitors new material on a regular basis to check if it corresponds with the customer's diversity and inclusion plan.	Social and ethical values are created by increasing awareness and eliminating discriminatory patterns by controlling ethical behaviours. The behaviours that cause discrimination in the entertainment industry can be blocked.
4	Dot South Korea 10 AI, Big Data, Cloud Computing, Flexible Electronics & Wearables, Internet of Things	The company develops wearable technological items for visually impaired people	Dot Watch and Dot Mini gadgets use AI, IoT, Big Data, and Cloud Computing to enable the vision-handicapped to "read" a whole book, similar to the Kindle	Social value is created by making a fair environment for disabled people to be part of life. They aim to make devices more affordable for underdeveloped countries and help vulnerable people participate in education

5	Envision Netherlands 10 AI, Big Data, Flexible Electronics and Wearables	It develops AI-powered products for visually impaired and low vision people. AI technology is used to extract information from images received by the products.	The product Envision Glasses is glasses with a camera, Wi-fi, Bluetooth and audio system built-in. It is able to convert an image to an audio description with the purpose of providing a better understanding of the scenarios such as face recognition, object detection and scene description to help visually impaired people. Additionally, the product Envision App can be used to read any kind of text, scan people or things or visualise scenes and colours with audio descriptions.	Social value is captured by achieving empowerment and engagement of visually disabled people in daily life. The product helps people hear what they want to see.
6	Gapsquare UK 5, 10 AI, big data, cloud computing	Gapsquare is a software platform that makes use of AI, Big Data, and Cloud Computing to reveal the wage gap in workplaces to ensure fair payment for everyone.	The FairPay PRO platform exposes insights about the wage gap taking into account gender, race, sexual orientation and disability for the corresponding organisation. HR and payroll data entered are processed using AI and Big Data via a cloud-based environment (SaaS).	Social value is captured by identifying unfair payment resulting from employees' demographics in order to provide fair wages for every employee in the organisation. Personal preferences are welcomed to embrace diverse talent ecosystems with the use of this platform to form better payment insights for a workplace.
7	Knockri Canada 5, 10 AI, Natural Language Processing	The company produces software used in the hiring process that is free of human bias to increase diversity using an AI and NLP-based technique.	It makes use of video interviews, audio and written assessments during the hiring process. It analyses the content of those behavioural assessments using NLP and AI algorithms. Their algorithm evaluates the merit of a candidate based on the customer's skills, so it is free of race, gender, age, ethnicity, accent, appearance or sexual preference discrimination.	Social and ethical value creation through the removal of bias factors in the hiring process is eventually helping to increase racial and gender diversity. Since it is a merit-oriented process in job recruitment, it increases the work efficiency and the vision of the company.

(continued)

Table 12.1 (continued)

No	Company info	Value proposal (what?)	Value creation (how?)	Value capture
8	Lex Machina USA 10, 16 AI, natural language processing	It is a legal data analytics platform that sifts through millions of pages of legal data to give strategic insights into courts, judges, attorneys, law firms and parties, allowing lawyers to predict the outcomes of various legal strategies.	Every 24 h, Lex Machina collects data by browsing many judicial databases. Lex Machina then uses Lexpressions®, a proprietary Natural Language Processing and Machine learning tool, to clean, code and tag all of the data. Lex Machina extracts all of the actors in each case, including attorneys, law firms, parties and judges.	The firm offers data on damages, findings, resolutions and remedies, all of which are important in determining what transpired in a case and who won
9	Marinus Analytics USA 10 AI, Cloud Computing, Cybersecurity	Some crimes, such as human trafficking and colonisation, that are concealed within adverts are stopped by employing AI, and they are preserved with the use of Cloud Computing.	AI analyses adverts with illegal purpose, and the data is saved in a cloud system so that others who try a similar strategy may quickly locate it. After that, the authorities are alerted.	The firm produces societal benefit by eliminating discriminatory behaviours such as people trafficking and sex terrorism. As a result, society becomes more aware.
10	Microsoft Seeing AI USA 3, 10 AI, Cloud Computing	It is a Microsoft-developed programme that can narrate various sceneries, objects, documents or a human description to assist visually impaired persons using AI and Cloud Computing.	The App "seeing AI" avails for providing audible information to a visually impaired person. It can be used to describe a person, convert a text into audio, describe a perceived colour, identify a product using its barcode and describe a scene, etc. with an AI-based algorithm. Additionally, the app supports different languages and is available on iOS devices	Social and ethical value is captured by providing a partially assistive tool for the visually impaired. Their dependence on other people can be reduced, enabling them to further participate in daily life
11	Mouse4all Spain 10 Robotics	It is a start-up that is working on a robotic gadget that allows persons with physical limitations like cerebral palsy and Parkinson's disease to use an Android tablet or smartphone completely hands-free.	The robotic device can be used with one or two switches, a trackball or a joystick option, and smart device control can be provided by connecting to the app via Bluetooth.	The designed robotic device helps to eliminate the accessibility problem of physically disabled individuals who cannot touch the screen of the device while using a smart device. Thus, the daily life of these individuals becomes easier.

#	Company	Description	Social value
12	Quad-Fi Canada 4, 8, 10, 17 AI, Big Data	Using Machine Learning and data analytics, the start-up develops an online lending platform for student loans and reduces borrowing costs for young people and immigrants.	The app collects and combines insurance and academic data from students and recent graduates. The AI then makes an accurate credit assessment in terms of financial and academic success and provides better financial ratios.	The company provides low-interest loans to students and recent graduates to contribute to better quality education. Also, social value is created by supporting them to establish a business and family life.
13	Scewo Germany 10 Robotics	The company has a new generation robotics technology-based wheelchair that people with reduced mobility can reach their desired location independently and flexibly.	Scewo BRO is an electric wheelchair that is equipped with robotics technologies, combining riding on two wheels with climbing stairs.	By making the vehicle used by individuals with reduced mobility smarter, the vehicle can continue on its way even in unsafe areas such as stairs and rough/slippery surfaces, providing an extremely comfortable and safe use for disadvantaged individuals.
14	Seva Exchange USA 10 AI	The firm is an AI-based platform, which pairs volunteers' skills with the people that need help.	The platform uses Artificial Intelligence to link people with talents, interests, and resources with others who require assistance in those areas, while also recording the value of each talent that can be passed forward.	Social value is built on promoting the types of jobs that are underrated by some economies, creating new avenues and reducing inequality. They prevent social inequality by strengthening social relations between volunteers and people who need help
15	Skillhus Norway 1, 3, 8, 10 Big Data	The company is a recruitment and consultancy agency that bridges the gap between skilled migrants and refugees and the Norwegian job market.	By collecting and processing data, the start-up provides diversity training for the companies to improve diversity leadership. They aim to reduce unconscious recruiting prejudice.	The social value is created by filling the gap between skilled migrants and refugees. The start-up aims to raise awareness on issues such as diversity, equality and belonging. It applies a comprehensive approach to organisations and leadership development.
16	SkillLab Netherlands 1, 4, 5, 8, 10, 17 AI	It is a startup that uses artificial intelligence to assist refugees and migrants in finding work.	The company helps people express their experiences and skills and explore and apply for career opportunities by using AI. Users can match with job opportunities according to their skills and generate job applications.	Social value is captured by ensuring refugees and migrants use their skills to find a job. They help people have a livelihood and reduce inequalities in society.

(continued)

Table 12.1 (continued)

No	Company info	Value proposition (what?)	Value creation (how?)	Value capture
17	Tykn Netherlands 10 Blockchain	It is a platform that stores the identities of refugees and immigrants digitally by using blockchain.	The application "my wallet" offers a blockchain-based verifiable credential wallet that provides a higher trustworthy digital identity.	It integrates refugees into society who do not have identity documents. Thus, it ensures that such people have fewer problems in matters such as healthcare, education, banking or finding a job.
18	Visualfy Spain 10 AI	For persons with hearing loss, the business makes an AI-based gadget with sight and vibration sensors that gives convenience.	The device Visually Home can be trained with AI. It brings the desired sounds into the desired warning form. These notifications can be changed individually and can send notifications to every device in the house connected via bluetooth.	Value is captured by enabling the hearing-impaired people to adapt to normal life. The AI-based device transforms sounds into colour and vibration, equalising living conditions for people with disabilities.
19	Voiceitt Israel 10 AI	The firm develops products to assist persons with speech impairments.	The application, "Voiceitt", is trained with the phrases used daily by the users like how they say their name. Using AI and statistical modelling, it learns the speaking styles of users and improves itself as they use it.	Social value is generated by helping people with speech problems. They facilitate their adaptation to daily life and make speech accessible to all.
20	Wandercraft France 10 Flexible electronics & wearables, robotics	The company supplies a robotic walking exoskeleton that allows simulating the permanent imbalance in human walking with dynamic walking algorithms created and supported by flexible electronics and wearables and robotic technologies.	Atalante is a robotic wearable product that facilitates motor movements of people with walking disabilities and helps the upper limbs to move. Robotic dynamic walk algorithms are integrated into a lower limb exoskeleton capable of emulating human self-balanced walk.	Social value is created by allowing people with a permanent imbalance or disorder in walking to walk without help from other people, thus increasing their living standards and further integrating them into society.
21	You Benefit Netherlands 10 3D printing, AI, robotics	By using 3D printing, AI and robotics technologies, new buildings and roads for emergency shelters in refugee camps are constructed.	New structures are created by 3D printing from sand while the structures are being built. The vision of this space innovator comprises an autonomous rover capable of building structures using free energy from the sun and local resources. It will use sintering as an additive manufacturing technique to turn terrestrial sand into solid rock, without the need for any other binders.	Social and environmental values are captured by providing refugees housing in a faster, cheaper and more efficient way. These goals create awareness about the hard circumstances of refugees.

umn, we present the company name, the origin country, related SDGs and emerging technologies that are included. The companies and use cases are listed alphabetically.[1]

References

A. Berg, J.D. Ostry, C.G. Tsangarides, Y. Yakhshilikov, Redistribution, inequality, and growth: New evidence. J. Econ. Growth **23**, 259–305 (2018). https://doi.org/10.1007/s10887-017-9150-2

N.O. Bonsu, J. TyreeHageman, J. Kele, Beyond agenda 2030: Future-oriented mechanisms in Localising the sustainable development goals (SDGs). Sustainability **12**, 9797 (2020). https://doi.org/10.3390/su12239797

P.L. Engle, L.C. Fernald, H. Alderman, J. Behrman, C. O'Gara, A. Yousafzai, M.C. de Mello, M. Hidrobo, N. Ulkuer, I. Ertem, S. Iltus, Strategies for reducing inequalities and improving developmental outcomes for young children in low-income and middle-income countries. Lancet **378**, 1339–1353 (2011). https://doi.org/10.1016/S0140-6736(11)60889-1

ESCAP, *A Guide to Inequality and the SDGs* (ESCAP, 2019)

European Commission, *Reducing Inequality [WWW Document]* (International Partnerships | European Commission, 2021). https://ec.europa.eu/international-partnerships/sdg/reducing-inequality_en. Accessed 22 Aug 2021

O. Galor, Inequality, human capital formation, and the process of development, in *Handbook of the Economics of Education*, (Elsevier, 2011), pp. 441–493

B. Guha-Khasnobis, S. Vivek, The rights-based approach to development: Lessons from the right to food movement in India, in *Food Insecurity, Vulnerability and Human Rights Failure*, ed. by B. Guha-Khasnobis, S. S. Acharya, B. Davis, (Palgrave Macmillan, London, 2007), pp. 308–327. https://doi.org/10.1057/9780230589506_13

H. Hegre, K. Petrova, N. von Uexkull, Synergies and trade-offs in reaching the sustainable development goals. Sustainability **12**, 8729 (2020). https://doi.org/10.3390/su12208729

Z. Hossen, B.H. Khondker, How is India flaring in achieving SDG 10 on reduced inequality? in *Sustainable Development Goals an Indian Perspective*, vol. 2020, (Springer Nature Switzerland AG, 2020), pp. 153–165

N. Kabeer, R. Santos, UNU-WIDER, *Intersecting Inequalities and the Sustainable Development Goals: Insights from Brazil*, WIDER Working Paper, 167th edn. (UNU-WIDER, 2017). https://doi.org/10.35188/UNU-WIDER/2017/393-6

M. Kaltenborn, M. Krajewski, H. Kuhn (eds.), *Sustainable Development Goals and Human Rights, Interdisciplinary Studies in Human Rights* (Springer International Publishing, Cham, 2020). https://doi.org/10.1007/978-3-030-30469-0

P. Katila, C.J.P. Colfer, W.D. Jong, G. Galloway, P. Pacheco, G. Winkel, *Sustainable Development Goals: Their Impacts on Forests and People*, vol 654 (Cambridge University Press, 2020). https://doi.org/10.1017/9781108765015

Z. Ofir, T. Schwandt, S. D'Errico, K. El-Saddick, D. Lucks, *Five Considerations for National Evaluation Agendas Informed by the SDGs* (Public Library, 2016)

J. Ogwezi, C. Okeke, B. Uzochukwu, W. Mitullah, K. Saffron, U. Bhojani, G. Mir, T. Mirzoev, B.E. Ebenso, N. Dracup, G. Dymski, D. Doan, S. Ouma, P. Trust, F. Onibon, *Achieving SDG 10: A Global Review of Public Service Inclusion Strategies for Ethnic and Religious Minorities* (UNRISD, Geneva, 2020). https://doi.org/10.13140/RG.2.2.17100.36486

K. Pickett, R. Wilkinson, *The Spirit Level: Why Greater Equality Makes Societies Stronger* (Bloomsbury Press, 2010)

I. Scholz, *Reflecting on the Right to Development from the Perspective of Global Environmental Change and the 2030 Agenda for Sustainable Development* (Springer Open, 2020), pp. 191–206

T. Sikor, P. Newell, *Globalizing environmental justice?* (Elsevier Enhanced Reader, 2014). https://doi.org/10.1016/j.geoforum.2014.04.009

F. Stewart, G. Brown, A. Cobham, *The Implications of Horizontal and Vertical Inequalities for Tax and Expenditure Policies* (Centre for Research on Inequality, Human Security and Ethnicity, Oxford, 2009), p. 46

The World Economic Forum, *5 Reasons why we Need to Reduce Global Inequality* [WWW Document] (World Economic Forum, 2015). https://www.weforum.org/agenda/2015/09/5-reasons-why-we-need-to-reduce-global-inequality/. Accessed 22 Aug 2021

UN, *Transforming Our World: The 2030 Agenda for Sustainable Development.* [WWW Document] (UN, 2015). https://undocs.org/Home/Mobile?FinalSymbol=A%2FRES%2F70%2F1&Language=E&DeviceType=Mobile. Accessed 20 Aug 2021

United Nations, *World Social Report 2020: Inequality in a Rapidly Changing World* (UN, 2020a). https://doi.org/10.18356/7f5d0efc-en

United Nations, *The Sustainable Development Goals Report 2020* [WWW Document] (UN, 2020b). https://unstats.un.org/sdgs/report/2020/The-Sustainable-Development-Goals-Report-2020.pdf. Accessed 23 Aug 2021

SDG-11: Sustainable Cities and Communities

13

Abstract

City governance is vital for sustainable development goals and resource management and allocation as well as urban climate-related initiatives, as it is estimated that more people will reside in the urban areas in further years. As more people migrate to cities, the world steadily becomes more urbanised. The population of the cities accounts for 55% of the total population, and cities generate 85% of global gross domestic product and emit 75% of greenhouse gas emissions. SDG-11, Sustainable Cities and Communities, aims to ensure inclusive, safe, resilient, sustainable urban and human settlements by providing inexpensive transit solutions, decreasing urban sprawl, enhancing urban governance involvement, improving the protection of cultural assets and addressing urban resilience and climate change issues. This chapter presents the business models of 50 companies and use cases that employ emerging technologies and create value in SDG-11. We should highlight that one use case can be related to more than one SDG and it can make use of multiple emerging technologies.

Keywords

Sustainable development goals · Business models · Sustainable cities and communities · Sustainability

The author would like to acknowledge the help and contributions of Alaattin Canpolat, Zeynep Kaya, Zafer Güray Gündüz, Uğur Cem Yılmaz, Selenay Sonay Tufan, Yalkın Kızılkan, Zeynep Yaren Dabak, Buse Gönül Bostancı and Emre Koç in completing of this chapter. They also contributed to Chapter 2's 3D Printing, 5G, Biometrics and Biotechnology & Biomanufacturing sections.

City governance is vital for sustainable development goals and resource management and allocation as well as urban climate-related initiatives, as it is estimated that more people will reside in the urban areas in further years. As more people migrate to cities, the world steadily becomes more urbanised. The population of the cities accounts for 55% of the total population, and cities generate 85% of global gross domestic product (GDP) and emit 75% of greenhouse gas emissions. It is forecasted that by 2050, the total city population will be equal to 6.5 billion people. If the urban areas are going to be designed and managed as now, sustainable development will not be achieved. Additionally, due to the rising populations and migration, rapid urbanisation has resulted in a surge of populated cities, particularly in developing nations, and slums have become a critical issue of urban life. The challenges of global sustainability cannot be solved without a significant focus on urban sustainability. Furthermore, making cities sustainable requires the establishment of jobs and economic opportunities, as well as safe and affordable housing, resilient communities and strong economies (UNDP 2020; Vaidya and Chatterji 2020). Further to that, there is a high potential of collaboration and coordination across various industries at the city scale, as well as the vital potential for policymakers in governments to recognise the interconnections and the need for interoperability among the stakeholders responsible for planning and designing sustainable development plans (Radovic 2019).

Perceiving the importance of cities, the United Nations General Assembly (UNGA) voted in 2015 to make "sustainable cities and communities" another target within the 2030 Agenda for Sustainable Development. Data obtained from 911 cities in 114 countries in 2020 shows that spatial urbanisation has been substantially quicker than population increase throughout the 1990–2019 period, and smaller cities are urbanising faster than larger cities (United Nations, 2021). In particular, from 2000 to 2018, the percentage of people living in slums fell from 39.66% to 29.25% among the global urban population. However, this percentile decrease is equivalent to an almost 80 million people increase (The World Bank 2021). This fact is a sign of the need for taking precautions to avoid devastating results.

"Sustainable Cities and Communities", which is within the "Sustainable Development Goals of United Nations as Goal 11", aims to "ensure inclusive, safe, resilient, sustainable urban and human settlements" by removing slum-like situations, providing inexpensive transit solutions, decreasing urban sprawl, enhancing urban governance involvement, improving the protection of cultural assets, addressing urban resilience and climate change issues, improving urban management (pollution and waste management), ensuring access for all to secure public places and enhancing urban management through improved urban rules and regulations (Franco et al. 2020).

SDG-11 and prospective innovations and efficient solutions to enhance city policy coherence include several major sectoral interlinkages and urban synergies. Despite the worldwide progress to lead and drive all processes on sustainable development, there are still significant information gaps and difficulties that might stymie SDG-11 implementation. The New Urban Agenda of UN-Habitat presented by "The United Nations Human Settlements Programme" emphasises the importance of a concentrated emphasis at the city and neighbourhood levels. It also has direct, tangible benefits for people's quality of life and the achievement of long-term developmental goals. To provide successful implementation and make concrete improvements in people's daily lives, the global goals laid forth in SDG-11 must be integrated with local development agendas (Franco et al. 2020). As shown in Fig. 13.1, there are ten targets within the context of SDG-11.

SDG-11 and the subject of sustainable urbanisation are important for most countries, given the high rates of urbanisation and the expected future share of the urban population (Koch and Krellenberg 2018). For instance, nearly three-quarters (320 million people) of the European Union's (EU) population reside in urban regions such as cities, towns and suburbs. Europe's urban population is predicted to rise to just over 80% by 2050. As a result, sustainable cities, towns and suburbs are vital for their residents' well-being and quality of life (Eurostat 2021). Another critical fact that should be stated is, while occupying only 3% of the Earth's territory, cities account for 60–80% of global energy consumption and 75% of global carbon emissions (United Nations 2021). Thus, the results of related regulations in the cities could impact the entire earth.

When creating sustainable smart cities that focus on SDG-11, several factors are to consider. The growth of information communication technologies (ICT) has significantly influenced the way people live their lives and how they arrange work, leisure and society. A variety of innovative products, services and business models have been facilitated by a drop-in computer capacity costs and size. Two significant developments could be stated for the

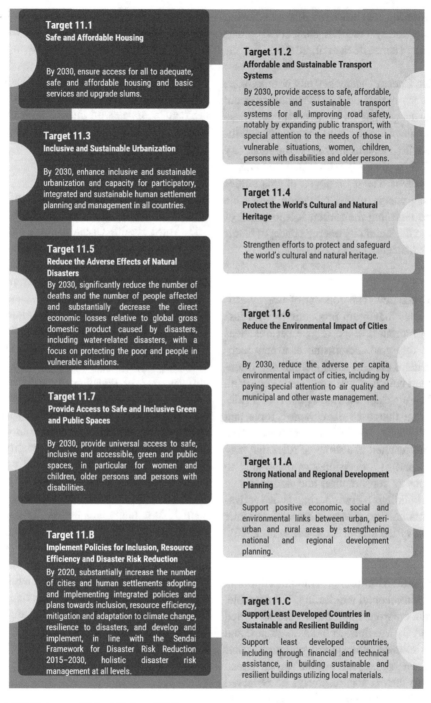

Fig. 13.1   SDG-11 targets. (United Nations, 2021)

worldwide growth of ICT and to make cities smart. The first is the transition from cables to wireless services, including telephones and the Internet. The second trend is related to the rising number of devices linked to the Internet and the change to the "Internet of Things" (Townsend, cited in Höjer and Wangel 2015). Furthermore, the impact of smart cities on sustainability cannot be underestimated. Renewable and green energy, energy efficiency, air quality, environment monitoring and water quality monitoring are all noteworthy research subjects in smart city planning (Ismagilova et al. 2019):

Renewable Energy Many key city entities, such as wireless sensor networks and water distribution, require power systems for basic operation. These have to be adapted into being optimised, intelligent and environmentally friendly in the smart city concept. This is possible with renewable energy and ICT systems. The main targets of smart cities are reducing energy usage, providing renewable energy and lessening the carbon footprint. All of this leads to the smart city energy concept (Aamir et al. 2014; Ismagilova et al. 2019).

Energy Efficiency The concept of energy efficiency enables maximum productivity with less energy consumption. Experts give several ideas to achieve this goal. For instance, a new technique that helps prevent energy efficiency anomalies in smart buildings was presented (Peña et al. 2016). The suggested method is built on a rule-based system that uses data mining tools and energy efficiency specialists' expertise. This research has resulted in a series of rules that may be used as part of a decision support system to optimise power consumption and anomalies in intelligent buildings by monitoring device activation and minimising power consumption while considering varied user needs (Peña et al. 2016).

Environmental Monitoring Another important focus is environmental monitoring. For example, six different environmental factors are identified for "Smart City Mission" in India: landscape and geography, climate, atmospheric pollution, water resources, energy resources and urban green areas. These factors should always be observed and accessible through online platforms to achieve public participation for problem-solving. This was achieved in Pisa, Italy, where the system gathered, processed and disseminated data

on air quality using a low-cost, distributed and efficient sensor network. Fixed and mobile sensor nodes were included in the system. Moreover, the data from the citizens were stored and later converted into indices such as Air Quality Index, Traffic Index, etc. All parties interested in obtaining regular updates on the city's air quality can access this information (Bacco et al. 2017; Dwivedi et al. 2019).

Air Quality Air pollution is one of the most serious concerns for industrialised societies. The World Health Organization (WHO) states that pollution is the prominent reason for mortality among children under the age of 5. A case study in the context of air quality monitoring was implemented in Christchurch, New Zealand, after the earthquake with a magnitude of 6.2. The research focused on near-real-time monitoring of fine-scale air pollution and connections to respiratory illnesses. The project's purpose was to create a citywide continuous real-time air pollution surface and provide the data in the form of an interactive dynamic map and raw data stream. A grid of four dust mote devices and low-cost IoT air quality sensors were used to collect the data. All people and interested parties were given access to data on air quality in a variety of formats, including main forms, maps and tables. Its goal was to encourage individuals to check air quality information simply and understandably. Also, citizens could collect information about their exposure (Marek et al. 2017). Identifying the city's most polluted and cleanest regions can help to enhance the environment and citizens' quality of life. Illnesses such as cerebral stroke can be minimised by reducing air pollution (Zaree and Honarvar 2018).

Water Quality Monitoring Managing the quality of water and providing safe drinking water are challenging in crowded cities. Nowadays, cities confront difficulties such as ageing water infra-structure, high maintenance costs, new contaminants and increased water use as a result of the rising population. Therefore, an effective water management system is needed by sustainable cities (Hrudey et al. 2011; Hou et al. 2013; Polenghi-Gross et al. 2014). In particular, a study has been released that improved ICT may enhance drinking water quality throughout the world. In the study, wireless communication, data processing, storage and redistribution have been suggested for Bristol's quality monitoring system. Data collection, transfer, storage and visualisation are parts of the system which is based on cloud computing (Chen and Han 2018).

Cities will have to reconsider their systems and their environmental consequences as more people migrate into urban areas and environmental concerns become more urgent. Many cities across the world have already started to embrace more environmentally friendly practices (mostly in America and Europe), and certain patterns are emerging (Martin et al. 2018). Sustainable cities will build on these foundations, going beyond today's environmental standards. Cities have vital roles in sustainable development and are thus critical for both regional and global destinies. However, there is no one-size-fits-all solution for creating a sustainable city due to the climate, geography and law differences. Long-term planning is required for the most drastic changes aimed at creating a sustainable city, and future studies can lead to further discussions and decision-making processes. Future studies should focus on improving one's understanding of future opportunities for adapting to or avoiding future influences and consequences (Phdungsilp 2011).

Along with the developing sustainability industry, thanks to increasing investment ratios from companies around the world, many new business areas are emerging and will continue to emerge in the future. Investing in SDG-11 can bring many benefits to the company. Companies may benefit from a better brand image, a greater staff retention rate and increased financial performance by investing in the sustainability of their communities. They will be able to keep up with changing laws and avoid penalties under their state's environmental legislation (Valuer | SDG 11 Forecast, p. 32).

Let's assume the appropriate policies are put in place. In that case, 24 million new jobs will be created by adopting sustainable energy practices and shifting to a greener economy, such as increasing electric vehicles usage and energy efficiency in existing and future buildings (International Labor Organization 2011). For instance, South Korea will invest USD 61 billion to raise renewable energy capacity from 12.7 GW to 42.7 GW by 2025 and increase its green mobility fleet to 1.33 million electric and hydrogen-powered vehicles. The plan will efficiently renovate public rental housing and schools to become more energy-efficient and transform urban areas into smart green cities (European Commission 2019). Moreover, the global electric vehicle market is estimated to reach 34,756 thousand units by 2030, up from an estimated 4093 thousand units in 2021 (Research and Markets 2021). Furthermore, a united effort to improve communities' sustainability will require investments in various sectors such as transport, waste management and construction (Valuer | SDG 11 Forecast, p. 32). Two sectors, which are indispensable for sustainable cities, will continue their development in the future; by 2023, the smart transportation industry will be worth $149.2 billion (MarketsandMarkets 2020a, b), while the worldwide waste management market will be worth $530 billion in 2025 (*Waste management market value worldwide 2027* 2020).

The built environment is one of the major causes of environmental degradation. Excessive energy and resource consumption are caused by the embodied energy of the built environment during construction and the energy needs of structures during use (Wieser et al. 2019). The construction industry will be the most demanded market in the future. For instance, the global construction industry will be worth \$15 trillion by 2025 (Deloitte-Marketing & Brand Department 2021). Meanwhile, the global modular construction market is expected to reach \$157.19 billion by 2023, up from \$106.15 billion in 2017, with a CAGR of 6.9% (MarketsandMarkets 2020a, b). Additionally, the global construction sustainable materials market is expected to be worth \$523.7 billion by 2026 to provide a more environmentally friendly solution (BIS Research 2017). Well-managed cities will make efficient use of natural resources and technology, resulting in a beneficial and crucial impact on society, the environment and the economy (Revi and Rosenzweig 2013). By 2050, smart cities will have saved \$22 trillion through initiatives such as public transit and energy-efficient buildings (Smart City Futures 2017). Mobility as a service (MaaS) solutions are expected to increase in popularity as technological infrastructure improves, and data becomes more accessible worldwide. The global MaaS market will grow from \$38.76 billion to \$358.35 billion by 2025 (The Insight Partners 2018).

13.1   Companies and Use Cases

Table 13.1 presents the business models of 50 companies and use cases that employ emerging technologies and create value in SDG-11. We should highlight that one use case can be related to more than one SDG and it can make use of multiple emerging technologies. In the left column, we present the company name, the origin country, related SDGs and emerging technologies that are included. The companies and use cases are listed alphabetically.[1]

---

[1] For reference, you may click on the hyperlinks on the company names or follow the websites here (Accessed Online – 2.1.2022):

http://neer.ai/; http://www.fingerprints.com/; http://www.intel.com/; https://carge.co/; https://emsol.io/; https://enviosystems.com/; https://evreka.co/; https://nordsense.com/; https://numina.co/; https://phantom.ai/; https://restado.de/; https://skycatch.com/; https://ucomposites.com/; https://urbanfootprint.com/; https://view.com/; https://waymo.com/; https://www.actility.com/smart-building-facility-management/; https://www.altaeros.com; https://www.betolar.com/; https://www.brighterbins.com/; https://www.cepton.com/; https://www.cyvision.com/; https://www.ekodenge.com/; https://www.five.ai/; https://www.foam.space/; https://www.fuelcellenergy.com/; https://www.gofar.co/; https://www.hayden.ai/; https://www.iberdrola.com/home; https://www.interactions.com/; https://www.interstellarlab.com/; https://www.latitudo40.com/; https://www.oneclicklca.com/; https://www.optibus.com/; https://www.ourcrowd.com/; https://www.pirelli.com/global/en-ww/homepage; https://www.printyour.city/; https://www.quantafuel.com; https://www.sigfox.com/en; https://www.skeletontech.com/; https://www.smartcultiva.com/; https://www.smart-enspaces.com/; https://www.spacemakerai.com/; https://www.ubicquia.com/simply-connected-simply-smart; https://www.urbansdk.com/; https://www.visionful.ai/; https://www.weride.ai/en/; https://zeleros.com/

Table 13.1 Companies and use cases in SDG-11

No	Company info	Value proposal (what?)	Value creation (how?)	Value capture
1	Actility France 8, 9, 11 Autonomous vehicles, big data, Internet of Things	The company deploys unified, scalable, multipurpose IoT network infrastructure for utilities and cities, providing well-defined points of interoperability between systems.	By building a long-range low-power network for IoT, city authorities can connect countless battery-powered things into a single network, enabling data collection through parking sensors, environmental monitoring equipment or flood detection sensors. The data are then fed to centralised intelligent systems which can make recommendations to optimise city service management.	Value is produced by providing sustainable solutions that assist consumers in becoming more energy-efficient by reducing their energy consumption habit's environmental effect.
2	Altaeros USA 11 5G, AI, autonomous vehicles, Internet of Things	It is a firm that manufactures aerostats in a variety of sizes and with permanent or mobile bases to meet the demands of clients. These aerostats are intended to make network deployments for industrial IoT, rural connectivity and a variety of other data-intensive applications easier and faster.	"ST-Flex" aerostats offer mobile carriers and internet service providers (ISP) a way to provide service to rural and remote areas by enlarging coverage zone. ST-Flex expands customers' 4G footprint with wide-area coverage from an aerial cell tower. It upgrades to 5G by swapping the payload equipment from the work platform on the ground. Also, this process completely covers a remote worksite with uninterrupted coverage, thanks to IoT.	Value creation is achieved by simplified deployments, shorter build-out times and less money required to keep the network up and running.
3	Betolar Finland 11, 13 Advanced materials	The company focuses on turning different industrial side streams from the energy, mining, steel and forestry industry into low-carbon, cement-free construction materials that perform the same as concrete in terms of qualities such as strength.	The company uses Advanced Materials named "Geoprime": geopolymer-based low-carbon construction material and a sustainable alternative to cement. Its durability and strength are comparable to cement-based products.	They create sustainable building solutions in the construction sector by using advanced Materials and offering dmcreased carbon emissions. They produce cement-free and cost-effective materials.

(continued)

Table 13.1 (continued)

No	Company info	Value proposal (what?)	Value creation (how?)	Value capture
4	BrighterBins Belgium 11, 12 AI, Internet of Things	They develop complete sensor and route optimisation solutions for waste management and hauliers by using IoT and AI.	Their sensors collect data on fill levels and garbage disposal, which they communicate to the firm or a third-party IoT platform. The software optimises garbage collections and city planning using data from sensors and AI algorithms.	Revenue is generated by preventing overflowing bins, keeping communities clean and optimising trash pickup truck routes. Their sensors and IoT solutions improve traffic, reduce CO_2 emissions and make cities greener by using AI.
5	Carge Greece 7, 9, 11 Cloud computing	It is a cloud computing-based mobility as a service (MaaS) mobile application that allows electric vehicle drivers to connect to any public charging networks.	It includes thousands of charging stations across Europe and provides information on the closest charging points. Moreover, it helps users to reserve a place at the demanded charging point and gives them the fastest and safest route to the stations. They utilise smart algorithms to balance the grid directly with each vehicle utilising cloud computing.	Revenue is captured by energy savings. Users can maintain the battery life of their electric vehicles by no longer waiting at the charging stations.
6	Cepton Technologies USA 9, 11 Spatial computing	They develop lidar-based solutions for a variety of areas, including automotive (Advanced Driver Assistance Systems (ADAS) /AV (autonomous vehicle), smart cities, smart spaces and smart industrial applications by using spatial computing.	They process the data that they collect via lidars by applying spatial computing. They provide intelligent and safe roads and rail, as well as pedestrian and traffic analytics, in the context of smart cities. Their Micro Motion Technology (MMT)-based lidar solutions can be used in a variety of intelligent sensing applications, such as ADAS, AV and smart infrastructure.	Their proprietary MMT-based lidars provide reliable, scalable and cost-effective solutions – thanks to spatial computing – for smart applications that require long-range, high-resolution 3D perception.
7	CY Vision USA 11 Spatial computing	The company is a 3D augmented reality head-up display manufacturer for future vehicles.	From broad sunshine to dark stormy situations, and with continuous depth from in-cabin to infinity, their AR head-up displayers provide 3D capabilities with binocular disparity, full-motion parallax and focus blur cues.	Economic value is captured by bringing better vehicular experiences and providing safer rides by using augmented reality.

8	Ekodenge	It is a company that specialises in digitalisation and sustainability, targeting buildings and industry. They have a project named "Ekobina+", which is developed as an IoT-enabled, intelligent energy monitoring and management platform for public and private buildings.	The platform can anticipate energy demand through energy simulations, thanks to an IoT-enabled database; it can also provide decision support for energy management through demand forecasting, cost optimisation and digital twin methodology, as well as calibrated energy simulation capability using live and historical IoT data.	Revenue is captured through the reduction of energy consumption, cost and emissions while improving indoor comfort and air quality conditions. Another benefit is the reduction of environmental and climatic impact by decreased consumption.
	Turkey			
	11			
	Big data, digital twins, Internet of Things			
9	EMSOL	It is a cloud-based software-as-a-service (SaaS) platform that links to a network of vehicle tracking, air and noise pollution sensors to offer real-time monitoring and analytics by using IoT.	Air and noise pollution is measured every minute by using calibrated sensors, and retrieved data are processed with the help of cloud computing. By using IoT, exact sources of air and noise pollution are identified, thanks to their dashboards.	Revenue is gained through the mitigation of air and noise pollution, especially for supply chain operations and construction activities by using their dashboards supported by cloud computing and IoT.
	UK			
	11			
	Cloud computing, Internet of Things			
10	Envio	The company develops smart building solutions for existing commercial buildings. Through IoT and cloud-based management, the system learns, predicts and optimises operations.	Envio Systems creates cloud-based technology for commercial buildings that are already up and running. The primary IoT product suite is a complete end-to-end solution for any current commercial building, regardless of its size, age or infrastructure, that helps boost comfort and save energy costs while allowing you to monitor and manage your building from any web-enabled device.	A smart building system for existing commercial facilities that helps increase comfort and reduce energy cost while enabling you to monitor and manage your building from any web-enabled device.
	Germany			
	11			
	Energy storage, Internet of Things			

(continued)

Table 13.1 (continued)

No	Company info	Value proposal (what?)	Value creation (how?)	Value capture
11	Evreka Turkey 11 AI, Internet of Things	The waste management company offers software and hardware solutions to both public and private institutions. Evreka promotes smart and sustainable cities by optimising waste management processes across many industries. The company utilises sensors and machine-to-machine communication.	Evreka spans the entire waste process and delivers highly innovative technology and environmentally friendly solutions from the collecting phase through treatment and recycling with the seamless integration of cutting-edge equipment and unique software. Evreka uses AI and Machine Learning to digitise these procedures in order to save money and time and enhance efficiency while still ensuring citizen happiness.	Evreka is a SaaS company and makes profits by providing software and hardware solutions tailored to customers' needs. By providing efficient waste management solutions, Evreka saves local governments and private institutions' time and resources. Their efforts also result in environmental benefits. Efficient waste management processes divert more waste from landfills, reduce operational carbon emissions and mitigate health and safety risks.
12	Fingerprints Sweden 11 AI, biometrics, natural language processing	The company develops Distinct Area Detection (DAD) algorithm, which is a feature-based algorithm that looks for things that are unique in its surroundings within the touch platform.	The DAD algorithm works to locate distinct areas in the three-dimensional fingerprint image derived from the capacitive sensor. It consists of enrolment and verification/identification steps. Firstly, a number of distinct areas are extracted from the fingerprint image. The areas together with their geometric relationships form a template that is unique to each fingerprint. Secondly, the template is used as an operator acting on the fresh fingerprint image. If the match is approved, the authentication of the person is completed.	The values offered by the technology include efficient image quality, low power consumption, cost efficiency and biometric systems with features for a satisfying user experience with quick confirming a secure biometric payment.
13	Five AI UK 11 AI, autonomous vehicles	It is a platform that builds self-driving software components and development platforms to help autonomy programs solve the industry's challenges by using AI	They supply AI algorithms for creating/choosing different traffic situations, performing various simulations, analysing performances and enhancing the stack by applying machine learning. Their algorithms enable an Autonomous Driving System (ADS) to comprehend the environment around it by combining data from cameras, RADAR, LIDAR, IMU, GPS and other sensors.	Delivering sophisticated online components that answer the issues in self-driving systems and building a modular platform that enables performance and scale for system development and assurance utilising AI capture value.

14	FOAM USA 11 Blockchain, Internet of Things	The company provides a permissionless and autonomous network of radios. It offers secure location services through time synchronisation, independent of centralised sources like GPS.	Thanks to blockchain, FOAM is transparent and censorship-resistant and can provide secure location data for other applications.	Social value is captured through secure and transparent location services, especially for cities.
15	FuelCell Energy USA 11 Carbon capture and storage, energy storage, hydrogen	The company provides solutions for power generation, carbon capture, local hydrogen production for transportation and industry and long-duration energy storage for utilities, industrial and large municipal power customers.	The company provides environmental solutions for various applications, including long-duration energy storage and fuel cell power plants. The energy storage system utilises solid oxide electrolysis cells to affordably and efficiently convert excess power into hydrogen, an energy carrier, for long-duration storage applications.	Revenue is captured through SureSource™ products, which deliver efficient, affordable and clean solutions to enable a world developed by clean energy. The SureSource product facilitates clean air permitting, reducing permitting costs and approval time.
16	GoFAR Australia 11 Internet of Things	The company registers business expenditure miles, analyses your car's health using IoT and measures driver performance.	The adapter plugs into the vehicle's diagnostic port and constantly monitors the car's health. Then it instantly alerts the driver about any problems it detects. Then, the car's data are sent from the adapter to the mobile app.	Value is created by reducing fuel consumption and increasing automobile operational lifetime. On average, it can save up to 9.8% fuel.
17	Hayden AI USA 11 AI, digital twins	It is a fully integrated urban mobility/traffic management platform that enhances road safety and traffic flow by utilising sources from AI and digital twin.	It uses AI-powered perception systems and a citizen app to turn public and private cars into street sensors, creating a real-time digital twin of the city by merging mobile sensors with citizen-driven data, and allows community and stakeholder involvement through easily accessible digital tools.	Economic and social value is captured through enabling safer and more sustainable cities by improving the traffic flow in terms of efficiency.

(continued)

Table 13.1 (continued)

No	Company info	Value proposal (what?)	Value creation (how?)	Value capture
18	Iberdrola Spain 7, 11, 13 Internet of Things	With the help of building automation systems, the company works on projects that focus on producing cost-cutting and energy-efficient solutions.	The company is implementing energy-saving measures in its facilities around the world to contribute to sustainability. In this implementation, IoT-based remote-controlled precautions are mostly used in lamps, electronic devices, etc. in the building.	Revenue is captured through informing and training users and providing sustainable solutions to help customers become more energy-efficient and reduce the environmental impact of their energy habits and consumption. Besides, they support women's and Paralympic sports to accelerate social development.
19	Imagine Intelligent Materials Finland 11 Advanced materials, Internet of Things	The company develops sensing solutions that deliver data from large surface areas in buildings, infrastructure and logistics and can be manufactured at scale.	Functionalised graphene transforms ordinary materials into electrically conductive, intelligent materials by using the Internet of Materials (IoM). These materials sense and report real-time changes in stress, temperature and moisture. In this process, signal processing expertise in hardware and software is combined in the nanotechnology sector.	The system provides real-time actionable insights on structural health, stress, pressure, leaks and fire detection. Its primary capabilities include asset monitoring of controlled surfaces and improving risk management, safety and productivity making the environment better and safer.
20	Interactions USA 11 AI, biometrics, natural language processing	It's a firm that makes intelligent virtual assistants, which mix AI and human understanding to help organisations and consumers have effective discussions.	Full-stack AI technologies with a unified technology suite enabling automatic speech recognition (ASR), natural language processing (NLP), dialog management and voice biometrics are included in the adaptable Platform. NLP offers the foundation for actionable intelligence applications and corporate processes by recognising the persons, places, subjects and intents present in any piece of voice or text.	Value is captured through Conversational AI which is used to enhance the patient's or member's experience by providing more efficient, convenient and accessible interactions between health systems, medical professionals, insurers and patients in the healthcare industry.

21	Interstellar Lab France 2, 6, 7, 11, 15 AI, advanced materials, 3D printing	The company helps humans live in harmony with the environment both on Earth and in the future on Mars by developing living pods with optimal conditions for humans and plants.	The company has developed a closed-loop, environmentally controlled and modular station by using a space exploration design approach, 3D printed material systems, advanced biosystems technologies and AI-based monitoring and control. A station is a combination of sealed modules; each of them can generate and recycle food, air and water to support life for a group of people.	The company aims to preserve biodiversity and sustain human life. They accelerate the transition to environmental regenerative solutions on Earth by creating integrated food production and water and waste recycling system.
22	Karamba Security USA 11 Cybersecurity	It is an embedded cybersecurity solutions provider for connected systems in the automotive industry, IoT and the enterprise edge.	The XGuard suite offers unrivalled self-protection against the loss of device control. Karamba XGuard can identify and stop buffer overflows, which can lead to external malicious code or code reuse attacks (such as return-oriented programming). Manufacturers may use these solutions to apply this technology to seal systems automatically during manufacturing.	With negligible network overhead, value is captured by concealing network traffic by verifying the sender of every communication. Unauthorised communications and over-the-air (OTA) malware are prevented via authenticated encryption (AE).
23	Latitudo 40 Italy 2, 3, 6, 7, 9, 11 AI, big data, cloud computing, Internet of Things	They convert satellite photos into geospatial data to ease and improve decision-making processes. They use AI and machine learning algorithms to filter and elaborate the photos.	It's a cloud-based Unified Data Analytics tool that extracts data from satellite pictures and IoT data automatically. They use a back-end link to their partners' planning systems to accomplish automated tasking on a large number of satellites. They sift through petabytes of data from a variety of sensors and constellations. Their image capture engine can locate the best image and trim away the area required for any given investigation.	Value is captured by providing more efficient object detection, better visualisation of land use, urban heat islands and coastline erosion detection services and more precise farming. Moreover, the inventory and asset management, building density analysis and environmental hazard and risk assessment can be enhanced as well.

(continued)

Table 13.1 (continued)

No	Company info	Value proposition (what?)	Value creation (how?)	Value capture
24	Neer Technologies USA 6, 11, 12 AI, Internet of Things	It is a fully integrated real-time water management platform powered by AI and machine learning.	They simulate and evaluate the risk status of drinking water distribution mains, sewage treatment plants and stormwater collection systems. They employ machine learning to detect leaks and anticipate system breakdowns in the water, sewage and stormwater collecting systems.	Revenue is earned through managing and planning drinking water, wastewater and stormwater systems more effectively and creating safer and cost-effective water systems in the long term.
25	Nordsense Denmark 11, 13 AI, autonomous vehicles, big data, Internet of Things, robotic process automation	The company provides services in the waste management industry by using sensors and data. They aim to halve waste collections by picking bins up once the sensors go on with punctuality.	It provides real-time data on customers' bins and waste generation patterns with Smart Bin Sensors. Benefiting from IoT and Autonomous Things, they keep track of the location and movements of waste containers, secure containers against theft, and increase the transparency of daily waste operations with asset tracking, which is ensured by using intelligent route trucks using AI. Smart sensors help robotic process automation to detect fill rates.	Environmental revenue is captured by reducing CO_2 emissions, lessening noise pollution and decreasing traffic congestion while preventing unsanitary bin overflows. Trucks use shorter routes by navigation combined with AI.
26	Numina USA 9, 11 AI, big data, Internet of Things	The company's computer vision-enforced sensors collect mobility data from open spaces. The processed data provide important insights into how people move within urban spaces. These insights allow urban planners to design cities that are inclusive, liveable and sustainable.	Numina's goal is to use data to help cities become more adaptive and egalitarian. Numina employs computer vision to track all types of activities on the street and provides new insight to city planners, facility managers, mobility operators and real estate developers. Numina displays the volumes, pathways, dwell durations and other behaviours of people, cyclists, various types of cars, pets, garbage bags and more – all in digital formats that are easily consumable.	The data they provide allow designers to make better-informed choices about people's needs which results in more inclusive and equitable cities. Their mission makes use of the mass amounts of data generated by people's activities in public spaces and provides economic, environmental and societal benefits to everyone involved.

#	Company	Description	Use case	Value
27	One Click LCA Finland 11, 13 Big data	The company produces software that helps calculate and reduce the environmental impacts of any building, thanks to stored data in its database.	To be able to make life cycle assessment (LCA) interpretations, known data are stored by using Big Data technology. For LCA, benefiting Big Data provides an advantage by finding, accessing and reusing relevant data.	Easy and affordable carbon assessment and life cycle assessment (LCA) by using big data.
28	Optibus Israel 11 AI, cloud computing	It is a cloud-native AI platform and software-as-a-service (SaaS) company that plans and schedules complex transit operations and the movement of vehicles and drivers to improve the quality and reliability of transit service.	Customers can use AI and cloud computing to model their transportation network (rules and preferences) at a granular level, allowing schedulers and operations executives to compare scenarios, publish reports and answer complex questions, transforming the related operation into a smart, data-driven business.	Economic, environmental and social value is generated by making effective use of a combination of AI optimisation algorithms and distributed cloud computing to make public transportation smarter and nourish freedom of movement and sustainable cities while reducing costs by modelling the transportation network and creating optimal routes, timetables and vehicle schedules with these technologies.
29	OurCrowd Israel 11 Crowdfunding	The company is a venture capital-crowdfunding hybrid platform for accredited investors facilitating investments in Israeli startups.	OurCrowd finds deals, does due diligence and makes investments available to its members, all while investing its own money. Through board seats and a mentorship programme that matches entrepreneurs with OurCrowd contacts from relevant sectors, it plays an active role in the firms it funds.	They build value for its portfolio companies throughout their life cycles, providing mentorship, recruiting industry advisors, navigating follow-on rounds and creating growth opportunities through its network of multinational partnerships.

(continued)

Table 13.1 (continued)

No	Company info	Value proposal (what?)	Value creation (how?)	Value capture
30	Phantom AI USA 11 AI, autonomous vehicles	It is an autonomous vehicle company that presents a comprehensive autonomous driving platform featuring computer vision, sensor fusion and control capabilities with AI solutions.	They employ a Deep Learning-based computer vision solution to detect vehicles, pedestrians, bicyclists, free-space, traffic signs and traffic lights, as well as optimal data, which includes all types of corner cases for deep learning and novel data augmentation techniques to increase data diversity. Autonomous cars also employ a Deep Learning detection pipeline with powerful tracking techniques.	Revenue is captured through PhantomVision which has real-time detection and target tracking on a bird's-eye view that provides safe driving with an accurate motion estimate of road objects to various Advanced Driver-Assistance Systems functions such as Auto Emergency Braking, Adaptive Cruise Control, Traffic Jam Assist and Lane Keeping Assist System.
31	Pirelli Italy 3, 7, 11, 12 Advanced materials, biotech & biomanufacturing, energy storage	It is a tyre company that has one project named "cycl-e around" aiming for sustainable transportation.	In these bikes, advanced materials are used in bicycle frames and biomaterials used in tires. E-bikes have a 500 Wh battery for energy storage.	Revenue is achieved through convenient, healthy and sustainable mobility for customers .
32	Print Your City Netherlands 11, 12 3D printing, recycling	It is a company that transforms cities' plastic waste into custom urban furniture for public spaces using recycling and 3D printing.	With a robotic arm and recycling facilities, plastic waste that is brought by citizens is custom designed as urban furniture through a plastic waste loop with technologies like 3D printing and recycling at the local level. This technological solution enables a sustainable city and less CO_2 emissions through combining modular repair and mass customisation.	The Pots Plus collection is an environmentally friendly furniture company that creates its collection by using the plastic waste of cities and turns them into furniture which follows ergonomic curvatures to suit the natural geometries of the human body.
33	Quantafuel Norway 11 Advanced materials, recycling	Quantafuel developed a patented technology that converts mixed plastic waste, including non-recyclable plastic waste, into advanced low-carbon synthetic fuel and chemicals.	They developed a technology that purifies the gas formed by the process and alters the molecules. This supplies them with a way to create a chemically recycled attractive product that satisfies a huge demand in the market.	Revenue and environmental value are captured by using waste plastics to create two major products: synthetic diesel fuel and recycled naphtha. Recycled naphtha is used by petrochemical companies to create better quality plastics, while premium-quality synthetic diesel fuel is used by individual consumers.

34	Intel USA 3, 7, 11 AI, edge computing, Internet of Things	Together with many other applications, the company provides Intelligent Transportation Systems for smart cities of the future.	IoT and AI facilitate intelligent transportation systems for land, air, rail and sea. These systems link automobiles, traffic lights, toll booths and other infrastructure to help alleviate traffic congestion, avoid accidents, decrease emissions and improve transportation efficiency. Fleet management, intelligent traffic management, Vehicle-to-Everything (V2X) communication, electric vehicle charging, electronic toll collecting and a variety of other mobility solutions are examples.	Extra value is captured, thanks to edge computing and inference, with which customers can benefit from fast response times, free up bandwidth and help keep sensitive data private.
35	Restado Germany 11,12 AI, recycling	The company transforms architecture, construction, craft and commerce materials and ensures that they can be reused.	Using an AI-based matching algorithm, the platform promotes change by matching the supply of materials from demolition with the overstock from demand in new projects.	Reusing building materials saves energy, emissions and waste while also requiring less new production, less resource-intensive recycling processes and less landfill.
36	Sigfox France 11 Internet of Things, wireless power transfer	The company is a service provider of network and IoT. Its network provides service to billions of devices to connect to the Internet.	The firm is deploying a 0G network to monitor billions of items transmitting data without the need to build and maintain network connections. 0G networks are software-based communication solutions that manage all networks and compute complexity in the Cloud rather than on the devices. They provide wireless communication, with minimal signalling overhead, a small and efficient protocol and devices that are not connected to any network in the globe.	Its network allows devices to connect to the Internet while consuming as little energy as possible and drastically reduces the energy consumption and costs of connected devices.

(continued)

Table 13.1 (continued)

No	Company info	Value proposal (what?)	Value creation (how?)	Value capture
37	Skeleton Technologies Germany 11, 13 Energy storage	The company is the ultracapacitor-based energy storage. They deliver solutions across the industry for customers with needs for high power, high energy and long-life energy storage.	The company provides ultracapacitor technology driven by advances in nanomaterials, the electrification of infrastructure and industry and fuel efficiency and emissions in the automotive and transportation segments. Proprietary raw material, "curved graphene", provides ultracapacitors with an advantage in power and energy density.	Revenue is captured by developing and manufacturing ultracapacitors that have very little internal resistance (down to 0.12 mΩ). In order to increase sustainability in cities, it develops environmentally friendly capacitors for big vehicles such as cars, buses and trains.
38	Skycatch USA 11 AI, drones, edge computing	The data company focuses on indexing and extracting critical information from the physical world using drones and AI.	A cloud-based and edge-based 3D reconstruction engine converts 2D photos into high-precision 3D point clouds and models. To ensure data is matched with current data, it fully automates the transfer of global coordinates to project/local coordinate systems. Then it pulls many sorts of measures from 3D data automatically. It enables users to execute 3D data analysis, such as parallel volumetric measurements and continuous time output for cloud platforms, among other things.	Revenue generation is achieved by reducing the amount of time to capture and process data.
39	SmartCultiva USA 11 Big data, cloud computing, Internet of Things	The company delivers a set of sensors, connected devices and software for farm management. They design and manufacture nano-sensing devices that offer real-time monitoring and control solutions essential to the new generation of farmers.	Nano CL Series B measures humidity, air, water temperature, light intensity, carbon dioxide levels and soil moisture before transferring this data to cloud-based IoT applications using various network protocols. The information is accessible on both mobile and web applications. The product is operated by solar energy and an ion lithium battery.	Revenue is captured through producing sensing devices to help the farmer with full real-time data and metrics, to monitor the entire farming business.

40	Smarten Spaces Singapore 11 AI	The company aims to help deliver experiences in the spaces of the future with AI for workplace management, space management and tenant engagement.	They create a revolution in working experience by utilising AI technology and bringing the workplace and employees together. They provide an end-to-end SaaS solution for the digital workplace, as well as an AI-powered tenant management system with adjustable features such as contactless access, safety protocols and proximity alerts.	Revenue is captured by enabling optimal future-ready spaces for the users. Their AI-based solutions allow users to make informed financial decisions based on how their spaces are being utilised while also enhancing the experiences with the spaces.
41	Spacemaker Norway 11 AI, cloud computing	It is a cloud-based AI software for real estate sites that empowers teams to collaborate, analyse and design sustainable cities.	They work with cloud-based AI to perform feasibility studies and optimise site proposals for density and living qualities and contribute to reducing the speed of site assessments using ready-to-use data sets, a digital 3D model and a quick simulation of alternative planning scenarios.	Revenue is captured through designing smart proposals and solving site constraints and density requirements using real-time analyses and generative design.
42	Ubicquia USA 9, 11 AI, Internet of Things	It is an IoT smart city platform that offers vital services such as lighting management, video AI, and public Wi-Fi.	Its plug-and-play network can transform any existing street light into a multi-function router by using AI, supporting a growing range of services such as sophisticated lighting controls, public Wi-Fi and the connection to third-party sensors such as air quality monitors and surveillance cameras with IoT.	Revenue is captured by providing a cost-effective and extensible IoT platform for smart city services to municipalities, utilities, broadband service providers and lighting manufacturers. Their products are intended to assist utilities and communities in detecting, anticipating and utilising AI in order to prevent outages before they occur.
43	Ucomposites Denmark 9, 11, 12, 14 Advanced materials, recycling	The firm focuses on recycling composite materials and glass fibres, as well as supplying virgin-quality recycled fibres (advanced materials) to important industry sectors including automotive, building materials and oil and gas.	The company accumulates and recycles previously used glass fibre materials. They work in the advanced material technology field which makes it possible to develop the same quality as the virgin alternatives.	Value is captured by reducing the environmental impact of waste management. Life cycle assessment (LCA) is used to document the carbon footprint of recycled materials.

(continued)

Table 13.1 (continued)

No	Company info	Value proposal (what?)	Value creation (how?)	Value capture
44	Urban Footprint USA 1, 2, 4, 7, 12 Big data	The software company's urban Intelligence platform helps public and private-sector organisations prioritise investments and resources where they are most needed, collecting data and applying it to make long-term recommendations.	It is aimed to store data in the housing, energy and finance sectors of various cities and to benefit the public and private sectors when necessary. The firm uses cloud-based software and web-based software. With the software it uses, it regularly transfers the necessary data from many complex and intensive data to the user.	Value is created by sustainable and profitable suggestions obtained using Big Data. They create value by saving the customer's time by serving them certain data about a specified region or neighbourhood before they take an action.
45	Urban SDK USA 11 Big data, cloud computing, Internet of Things	It is an IoT-based planning tool that uses real-time location information to help smart cities improve mobility, transit, sustainability and safety operations. They use big data and cloud computing to retrieve real-time data.	They use big data and cloud computing to display traffic counts and trip demand by measuring the number of travel across autos, motorcycles and pedestrians. Furthermore, real-time traffic speeds are used to assess congestion and dependability. By anticipating heat maps at different time periods and route segments, it also helps to predict collisions and assure traffic safety.	Value is produced by delivering better data to government agencies, politicians and the general public, allowing them to make more informed policy and budgetary choices with the IoT platform, thanks to the use of big data and cloud computing.
46	View USA 3, 7, 11, 12 Advanced materials	The company produces sustainable smart windows which are digital and connected and can be controlled from anywhere.	They make smart windows that reject solar radiation and glare while maintaining a comfortable temperature. The electrochromic coating is made up of multiple thin layers of metal oxide, and they use electrochromic technology to predictably adjust tint levels in response to external conditions and user preferences.	Value is achieved through their smart windows which automatically adjust in response to the sun to increase natural light and access to views. This is done to improve people's health and wellness by reducing headaches, eyestrain and drowsiness while simultaneously saving energy.

47	Visionful USA 11 AI, edge computing	It's a platform that uses AI and edge computing to give real-time parking assistance, fully automated parking enforcement and predictive analytics.	They present edge-computing based sensors, employ algorithms for parking space classification and have large and consistent datasets. These features give rise to implement advanced deep learning techniques for parking lot recognition. Their real-time deep learning occupancy recognition is a solution for increasing the parking experience of drivers.	Revenue is collected by using a specialised GPU and AI software to monitor a 360-degree view. These self-contained and portable devices may be powered by solar panels and have their own IoT cellular connections, eliminating the need for power and network infrastructure. Transferring huge files to the cloud is, no longer necessary thanks to edge computation.
48	Waymo USA 11 AI, autonomous vehicles	The company makes smart vehicles using AI and autonomous things technologies by adding features like lane-keeping and fuel tracking.	The company's cars are designed for complete autonomy, with sensors that provide 360-degree views and lasers that can identify objects up to 300 metres away. Radar is used to look around cars and track moving things, while short-range lasers detect and focus on items close to the vehicle, thanks to AI.	They provide a safe and effortless driving experience to customers by transforming their cars into smart cars by implementing AI technology. One of the aims of the company is to decrease the number of deaths inflicted by a car crash.
49	Weride China 11 AI, autonomous vehicles	The autonomous driving (AD) company develops high-driving automation and self-driving technologies by using AI.	They enable their AD vehicles and AI-driven algorithms to plan and execute safe, dependable and human-like driving in complicated dynamic settings by providing real-time HD Map and localisation supported by numerous sensors, including LiDAR, camera, GNSS and INS. Each day, their data is sent to a data processing platform, and labelling data is created.	Value is captured by providing accurate recognition that outperforms human drivers, multi-sensor fusion at multi-levels to improve perception quality, multiple perception paths to achieve redundancy and powerful models and algorithms to recognise long-tail objects by using AI.
50	Zeleros Spain 7, 8, 9, 10, 11, 13 Energy storage	The company designs and develops new technologies for sustainable and efficient transportation, which will allow travelling at 1000 km/h with renewable energies, combining the best from the aeronautics and the railway industries.	The company provides a way to build a cost-efficient and energy-efficient hyperloop system for city-to-city travel at high speeds. A combination of vehicle technologies for the cruise phase integrated into the pod, with infrastructure systems for acceleration and deceleration provides the system with scalability and capacity, radically reducing infrastructure stress.	The company's hyperloop system is up to 5–10 times more energy-efficient than an aeroplane for the same inland route and operates with 0 direct emissions with a fully electric powertrain. These provide safety and reliability for communities.

References

M. Aamir et al., Framework for analysis of power system operation in smart cities. Wireless Pers. Commun. **76**(3), 399–408 (2014). https://doi.org/10.1007/s11277-014-1713-3

M. Bacco et al., Environmental monitoring for smart cities. IEEE Sens. J. **17**(23), 7767–7774 (2017). https://doi.org/10.1109/JSEN.2017.2722819

BIS Research, Global construction sustainable materials market – Analysis & forecast, 2017–2026 (2017), Available at: https://bisresearch.com/industry-report/global-construction-sustainable-materials-market-2026.html. Accessed 17 Aug 2021

Y. Chen, D. Han, Water quality monitoring in smart city: A pilot project. Automat. Construct. **89**, 307–316 (2018). https://doi.org/10.1016/j.autcon.2018.02.008

Deloitte-Marketing & Brand Department, Global powers of construction (2021), Available at: https://www2.deloitte.com/global/en/pages/energy-and-resources/articles/deloitte-global-powers-of-construction.html. Accessed 17 Aug 2021

Y. Dwivedi, et al. (eds), ICT unbounded, social impact of bright ICT adoption: IFIP WG 8.6 international conference on transfer and diffusion of IT, TDIT 2019, Accra, Ghana, June 21–22, 2019, Proceedings (IFIP advances in information and communication technology) (Springer, Cham, 2019). https://doi.org/10.1007/978-3-030-20671-0

European Commission, A European green deal (2019). Available at: https://ec.europa.eu/info/strategy/priorities-2019-2024/european-green-deal_en. Accessed 17 Aug 2021

Eurostat, Sustainable cities and communities in the EU: Overview and key trends (2021), Retrieved from: https://ec.europa.eu/eurostat/statistics-explained/index.php?title=SDG_11_-_Sustainable_cities_and_communities

I. B. Franco, T. Chatterji, E. Derbyshire, J. Tracey (eds.), *Actioning the Global Goals for Local Impact: Towards Sustainability Science, Policy, Education and Practice, Science for Sustainable Societies* (Springer, Singapore, 2020). https://doi.org/10.1007/978-981-32-9927-6

M. Höjer, J. Wangel, Smart sustainable cities: definition and challenges. ICT innovations for sustainability. Springer, Cham, 333–349 (2015).

D. Hou et al., An early warning and control system for urban, drinking water quality protection: China's experience. Environ. Sci. Pollut. Res. **20**(7), 4496–4508 (2013). https://doi.org/10.1007/s11356-012-1406-y

S.E. Hrudey et al., Managing uncertainty in the provision of safe drinking water. Water Supply **11**(6), 675–681 (2011). https://doi.org/10.2166/ws.2011.075

International Labor Organization, Towards a GREENER EConomy: The social dimensions: Sustainable development knowledge platform (2011), Available at: https://sustainabledevelopment.un.org/index.php?page=view&type=400&nr=380&menu=1515. Accessed 17 Aug 2021

Ismagiloiva, E. et al. (2019) 'Role of smart cities in creating sustainable cities and communities: A systematic literature review', in Dwivedi, Y. et al. (eds) ICT Unbounded, Social Impact of Bright ICT Adoption. Cham: Springer (IFIP Advances in Information and Communication Technology), pp. 311–324. doi: https://doi.org/10.1007/978-3-030-20671-0_21.

F. Koch, K. Krellenberg, How to contextualize SDG 11? Looking at indicators for sustainable urban development in Germany. ISPRS Int. J. Geo-Inf. **7**(12), 464 (2018). https://doi.org/10.3390/ijgi7120464

L. Marek, M. Campbell, L. Bui, Shaking for innovation: The (re)building of a (smart) city in a post disaster environment. Cities **63**, 41–50 (2017). https://doi.org/10.1016/j.cities.2016.12.013

MarketsandMarkets, Modular construction market global forecast to 2025 (2020a), Available at: https://www.marketsandmarkets.com/Market-Reports/modular-construction-market-11812894.html. Accessed 26 Aug 2021

MarketsandMarkets, Smart transportation market size, share and global market forecast to 2025 -COVID-19 impact analysis (2020b), Available at: https://www.marketsandmarkets.com/Market-Reports/smart-transportation-market-692.html. Accessed 26 Aug 2021

C.J. Martin, J. Evans, A. Karvonen, Smart and sustainable? Five tensions in the visions and practices of the smart-sustainable city in Europe and North America. Technol. Forecast. Soc. Change **133**, 269–278 (2018). https://doi.org/10.1016/j.techfore.2018.01.005

M. Peña et al., Rule-based system to detect energy efficiency anomalies in smart buildings, a data mining approach. Expert Syst. Appl. **56**, 242–255 (2016). https://doi.org/10.1016/j.eswa.2016.03.002

A. Phdungsilp, Futures studies' backcasting method used for strategic sustainable city planning. Futures **43**(7), 707–714 (2011). https://doi.org/10.1016/j.futures.2011.05.012

I. Polenghi-Gross et al., Water storage and gravity for urban sustainability and climate readiness. Am. Water Works Assoc. **106**(12), E539–E549 (2014). https://doi.org/10.5942/jawwa.2014.106.0151

Research and Markets, The worldwide electric vehicle industry is expected to grow at a CAGR of 26.8% from 2021 to 2030 (2021), Available at: https://www.globenewswire.com/en/news-release/2021/06/11/2245781/28124/en/The-Worldwide-Electric-Vehicle-Industry-is-Expected-to-Grow-at-a-CAGR-of-26-8-from-2021-to-2030.html. Accessed 26 Aug 2021

A. Revi, C. Rosenzweig, The urban opportunity: Enabling transformative and sustainable development. Background research paper submitted to the High-level Panel on the Post-2015 Agenda (2013)

Smart City Futures (2017), Available at: https://www.shapingtomorrow.com/home/alert/4165100-Smart-City-Futures. Accessed 25 Aug 2021

The Insight Partners, Mobility as a Service (MaaS) Market 2025. [online] (2018), Available at: https://www.researchandmarkets.com/reports/4471796/mobility-as-a-service-maas-market-2025-global. Accessed 26 Aug 2021

The World Bank, Population living in slums (% of urban population) (2021), Available at: https://data.worldbank.org/indicator/EN.POP.SLUM.UR.ZS?end=2018&start=2000&view=chart. Accessed 3 Aug 2021

UNDP, Goal 11: Sustainable cities and communities [WWW Document] (2020), https://www.tr.undp.org/content/turkey/en/home/sustainable-development-goals/goal-11-sustainable-cities-and-communities.html

United Nations, Sustainable Development Goals 11: Make cities and human settlements inclusive, safe, resilient and sustainable (2021). Available at:

https://sdgs.un.org/goals/goal11. Accessed 3 Aug 2021

Valuer, SDG 11 Forecast (2021), Available at: https://www.valuer.ai/resources/report/sdg-11-forecast. Accessed 5 Aug 2021

H. Vaidya, T. Chatterji, SDG 11 sustainable cities and communities." Actioning the Global Goals for Local Impact. Springer, Singapore, 173–185 (2020).

M.A.-Z. Vesela Rodovic, *SDG11-Sustainable Cities and Communities: Toward Inclusive, Safe, and Resilient Settlements* (Emerald Publishing Limited, London, 2019)

A.A. Wieser et al., Implementation of Sustainable Development Goals in construction industry – A systemic consideration of synergies and trade-offs. IOP Conf. Ser. Earth Environ. Sci. **323**, 012177 (2019). https://doi.org/10.1088/1755-1315/323/1/012177

T. Zaree, A.R. Honarvar, Improvement of air pollution prediction in a smart city and its correlation with weather conditions using metrological big data (2018), p. 12

SDG-12: Responsible Consumption and Production

14

Abstract

SDG-12, Responsible Consumption and Production, strives to break the current cycle of economic growth, resource usage and environmental degradation, which has fuelled unsustainable global development for decades. While producing countries bear responsibility for natural resource depletion, pollution and other negative consequences of their production, wealthy countries' practical and legal responsibilities are significantly high due to their high consumption levels. An increase in consumption is often associated with an improved quality of life, which creates a conflict between the pillars of sustainable development and the environmental well-being of the planet. This issue becomes more complicated since cross-border resource management methods are more controversial than cooperative. This chapter presents the business models of 46 companies and use cases that employ emerging technologies and create value in SDG-12. We should highlight that one use case can be related to more than one SDG and it can make use of multiple emerging technologies.

Keywords

Sustainable development goals · Business models · Responsible consumption and production · Sustainability

The sustainable development goals (SDGs) are offered as a blueprint for transforming the activity on the planet towards sustainable development. The prominence of sustainable development in international environmental conferences and policies has aided in its global adoption as a conceptual framework for addressing environmental issues at a variety of policy levels. SDG-12 strives to break the current cycle of economic growth, resource usage and environmental degradation, which has fuelled unsustainable global development for decades. Additionally, the divide between developed and developing countries in terms of consumption and production widens. Hence, while producing countries bear responsibility for natural resource depletion, pollution and other negative consequences of their production, wealthy countries' practical and legal responsibilities are significantly high due to their high consumption levels. An increase in consumption is often associated with an improved quality of life, which creates a conflict between the pillars of sustainable development and the environmental well-being of the planet. This

The author would like to acknowledge the help and contributions of Alaattin Canpolat, Zeynep Kaya, Zafer Güray Gündüz, Uğur Cem Yılmaz, Selenay Sonay Tufan, Yalkın Kızılkan, Zeynep Yaren Dabak, Buse Gönül Bostancı and Emre Koç in completing of this chapter.

© The Author(s) 2022
S. Küfeoğlu, *Emerging Technologies*, Sustainable Development Goals Series,
https://doi.org/10.1007/978-3-031-07127-0_14

issue becomes more complicated since cross-border resource management methods are more controversial than cooperative. Thus, sustainable consumption and production are one of the most cost-effective and successful methods to accomplish economic development, minimise environmental consequences and improve human well-being (Amos and Lydgate 2020).

The foundations of SDG-12 go back to 1972, when the Club of Rome analysed a computer simulation of a planet with limited natural resources. According to this simulation, which examines the effects of economic and population growth on the planet, if there is no growth trend, the planet will show a sudden and irrevocable decline in terms of population and industrial production in 2072. This report, called The Limits to Growth (LTG), validated its data from 1972–2000 with real empirical data (Monnet et al. 2014). The concept of responsible consumption and production gained importance over time and was promoted by international organisations. For example, the report "Our Common Future" (1987) by the Brundtland Commission (formerly World Commission on Environment and Development) underlined the importance of reducing global poverty, as well as the disparity in consumption patterns between wealthy and poor, allowing for debate on consumption levels (Gasper et al. 2019). Similarly, during the UN Conference on Environment and Development (known as the Earth Summit) in 1992, another call was made to "reduce and eliminate unsustainable production and consumption patterns" by The Rio Declaration on Environment and Development. Moreover, during the 2002 World Summit on Sustainable Development in Johannesburg, 10-year frameworks of programmes for the actions were introduced. As it was further elaborated during the 2012 UN Conference, "protection of the natural resources and following sustainable consumption and production" is essential to achieve global sustainable development (Gasper et al. 2019).

Sustainable development goal 12 calls for "responsible consumption and production". Global production and consumption are significant driving factors of the global economy but

have led to the destruction of the planet's natural ecosystem and resources. To reach the objective of sustainable consumption and production patterns, people must quickly transform how societies produce and consume. SDG-12 promotes resource and energy efficiency, the creation of sustainable infrastructure, increased access to green and decent employment and improved quality of life. These goals can be achieved through water management, waste management, sustainable products and services, sustainable supply chains and synergies with circular-based renewable energy systems (Khaw-ngern et al. 2021).

To ensure sustainable economic development for all, humanity must reduce its carbon footprint by changing its production and consumption habits. As highlighted in SDG-12, it is important to use natural resources efficiently and to protect the environment from toxic waste pollutants (UNDP 2021). Encouraging industries, businesses and consumers to recycle and reduce waste is important since both of them have roles in transforming consumption into sustainable patterns by 2030 (UNDP 2021). Industries' water, soil and air pollution concerns can be exceeded by the "Rs" concept, which covers rethinking, reducing, redesigning, reusing, repairing, refurbishing, remanufacturing, recycling and repurposing (Khaw-ngern et al. 2021). Figure 14.1 summarises the targets of SDG-12.

The transformation towards more responsible and sustainable consumption and production is essential to slow the effects of climate change and prevent further irreparable damage to the planet. The production of materials and goods has a range of environmental impacts, and we are already witnessing the profound effects of destructive production patterns. Unregulated production has led to mass deforestation, excessive waste and other ecological destruction. It also promotes an inefficient use of resources. SDG-12 is significantly important for the success of many other goals due to their interdependent nature.

Responsible consumption and production can promote the transformation of a linear economy to a circular economy (CE) through sustainability and continuity. A circular production and con-

Target 12.1
Implement the 10-Year Sustainable Consumption and Production Framework
Implement the 10-year framework of programmes on sustainable consumption and production, all countries taking action, with developed countries taking the lead, taking into account the development and capabilities of developing countries

Target 12.2
Sustainable Management and Use of Natural Resources

By 2030, achieve the sustainable management and efficient use of natural resources.

Target 12.3
Halve Global per Capita Food Waste

By 2030, halve per capita global food waste at the retail and consumer levels and reduce food losses along production and supply chains, including post-harvest losses.

Target 12.4
Responsible Management of Chemicals and Waste

By 2020, achieve the environmentally sound management of chemicals and all wastes throughout their life cycle, in accordance with agreed international frameworks, and significantly reduce their release to air, water and soil in order to minimize their adverse impacts on human health and the environment.

Target 12.5
Substantially Reduce Waste Generation

By 2030, substantially reduce waste generation through prevention, reduction, recycling and reuse.

Target 12.6
Encourage Companies to Adopt Sustainable Practices and Sustainability Reporting

Encourage companies, especially large and transnational companies, to adopt sustainable practices and to integrate sustainability information into their reporting cycle.

Target 12.7
Promote Sustainable Public Procurement Practices

Promote public procurement practices that are sustainable, in accordance with national policies and priorities.

Target 12.A
Support Developing Countries' Scientific and Technological Capacity for Sustainable Consumption and Production

Support developing countries to strengthen their scientific and technological capacity to move towards more sustainable patterns of consumption and production.

Target 12.8
Promote Universal Understanding of Sustainable Lifestyles

By 2030, ensure that people everywhere have the relevant information and awareness for sustainable development and lifestyles in harmony with nature.

Target 12.C
Remove Market Distortions that Encourage Wasteful Consumption

Rationalize inefficient fossil-fuel subsidies that encourage wasteful consumption by removing market distortions, in accordance with national circumstances, including by restructuring taxation and phasing out those harmful subsidies, where they exist, to reflect their environmental impacts, taking fully into account the specific needs and conditions of developing countries and minimizing the possible adverse impacts on their development in a manner that protects the poor and the affected communities.

Target 12.B
Develop and Implement Tools to Monitor Sustainable Tourism

Develop and implement tools to monitor sustainable development impacts for sustainable tourism that creates jobs and promotes local culture and products.

Fig. 14.1 Targets of SDG-12 (UNDP 2021)

sumption system prioritises the optimisation of raw materials to create sustainable products. These products can be easily maintained, reused, repaired, recycled and/or refurbished to extend their lifetimes and even create novel products. A circular system also promotes waste reduction at every phase in the extraction-production-consumption cycle (European Investment Bank 2020).

Consequently, SDG-12 is one of the most effective goals that meet the CE action plan targets. To exemplify, a plan of action which says explicitly that CE is a system-wide solution aimed at fostering sustainable consumption and production methods has been released by the European Union and therefore helping to achieve the goals set by this particular SDG (Rodriguez-Anton et al. 2019; Dantas et al. 2021). Furthermore, excess resource use, energy and waste production are minimised through data collection and surveillance, indicating a convergence with the CE principles on the reduction of raw resource use and waste and pollution design (Andrews 2015; Inoue et al. 2020).

The first function of the CE is smarter product use and manufacturing. Three strategies used to perform this function are reuse, rethink and reduce. A strategy of reuse could be explained as making a product redundant by eliminating or replacing its function with a new (digital) product or service. Rethink can be described as increasing the product usage intensity (e.g. sharing product). In addition, reducing refers to decreasing resource and material consumption by increasing product efficiency. Extending the lifespan of a product and its parts is the second function of the CE. Reuse, repair, refurbish, remanufacture and repurpose are the five strategies to fulfil the function. Reuse strategy is reprocessing of an abandoned product that is still in good shape and performed its original function by another customer. The strategy of repair can be described as the inspection and maintenance of a damaged product so that the original function can be prolonged. Moreover, the strategy of refurbishing is restoring and updating an outdated product. Remanufacturing is the use of scrapped products in the manufacturing of novel products with the same purpose. Repurposing is using the scrapped product's elements in a novel product with different purposes. The last function of the CE is the useful application of materials. Recycle and recover are the two strategies utilised to satisfy this function. The strategy of recycling is recovering waste materials for reprocessing into new goods, materials or substances. However, energy recovery and reprocessing into materials for use as fuels or backfilling processes are not included. Finally, the strategy of recovery refers to incinerating waste materials to obtain energy recovery (Iordachi 2020; Al et al. 2021).

Responsible consumption and production are critical elements of a sustainable future. This is due to the intrinsic environmental impact of production processes and consumption patterns. In making these processes more responsible, the goal is to minimise the environmental impact of production and consumption while also ensuring that everyone has adequate resources. There is an ongoing debate as to whether it is the duty of consumers, businesses or governments to drive the change towards more sustainable consumption and production patterns. Consumers are often expected to drive this change through their individual choices and actions; however, they are extremely limited by several factors, including societal and economic norms and the market incentives of globalised capitalism. Sustainable consumption is likely to become more feasible when sustainable products and consumption trends are more mainstream and incentivised. Even if enough individual consumers could collectively change their consumption patterns, however, the environmental impact of production would continue to devastate (Stevens 2010). Businesses must also be held responsible for the shift to more sustainable production and consumption patterns. The productivity and efficiency of businesses could apply to using the world's scarce resources more sustainably. However, competition between businesses with very limited regulation benefits business models that produce more at lower costs, typically at the expense of the environment (Tukker et al. 2008). With a largely unregulated market regarding environmental impact, the onus of sustainability then falls on governments. Governments can incentivise individuals and groups to consume and produce more sustainably. On the individual level, governments can encourage responsible consumption by subsidising sustainable products, such as electric vehicles and renewable energy sources (Stevens 2010). Governments can also discourage certain unsustainable consumption habits using taxes (Stevens 2010). Governments and international organisations can take similar approaches with corporations, rewarding them with subsidies for sustainable production methods and penalising overconsumption of scarce resources and excessive environmental impact.

By 2050, it is estimated that the natural resources required to sustain existing lifestyles will need the equivalent of nearly three planets at the current rate of population increase and consumption (McNeill 2020). Obviously, change is necessary, and everyone from the manufacturer to the ultimate customer must participate. This essential change will lead to new business areas such as reducing food loss at all stages of the food supply chain, reducing the amount of plastic waste, sustainable construction, decrease in fossil-fuel subsidies (production and consumption), etc. Also, this goal has affected people's daily lives by implementing a new phrase: "Circular Economy". SDG-12 can be named as "Starting Point of Transformation from Linear to Circular Globally". The fact of being a responsible consumer/producer will manage to be more circular from every aspect in the future. Being successful in this direction is not easy at all. This is a process that can be achieved with the whole world acting together since global material consumption is predicted to rise 15% by 2030 and 75% by 2060 to 167 billion metric tonnes. Growth in low- and middle-income economies aiming to equal high-income nations' level of living might be a major driver of material consumption (The Atlas of Sustainable Development Goals 2020, 2020).

A circular economy scenario is particularly essential given the region's economic importance of the extractive industries and poor recycling rates. Production and consumption trends are expected to alter significantly. For instance, in Latin America and the Caribbean, it is expected that by 2030, more than one million jobs would have been created in net terms, owing to an energy transition and efforts to keep global average temperature rise well below 2 °C over pre-industrial levels (Economic Commission for Latin America and the Caribbean 2020). Job creation in industries such as metal reprocessing and wood reprocessing would more than balance the losses associated with the extraction of minerals and other raw materials in a circular economy scenario. This is because reprocessing's value chain is longer and more labour-intensive than mining's, and higher recycling rates would raise demand for waste management services.

The worldwide food waste management industry is one of the most important markets of interest to those who follow the SDG-12 agenda. Grand View Research estimated the market to be worth $34.22 billion in 2019 and expects it to expand at a compound annual growth rate (CAGR) of 4.7% from 2020 to 2027 (Grand View Research 2020). The Food and Agriculture Organization (FAO) has discovered that one-third of all food produced is wasted, necessitating supply chain improvements. Some of the current emerging developments in the industry include gasification for converting food waste into combustible gases, anaerobic digestion for extracting energy and nutrients from produce and AI systems that forecast demand (Food and Agriculture Organization of the United Nations 2021).

Another important industry is the plastic waste industry. Plastic usage has increased as a result of the COVID-19 pandemic. Single-use plastics, including masks, personal protective equipment and sanitiser bottles, have been deemed necessary for controlling the spread of the disease. Disposable face mask sales have grown by 20,650% globally, from 800 million in 2019 to 166 billion in 2020, resulting in a massive increase in plastic waste and is expected to reach 750 billion by 2028 (Grand View Research 2021). Better plastic production and waste management techniques can lead to more responsible plastic consumption and production. Dematerialisation, substitution and improved biodegradability are some of the most popular plastic production techniques. An effective after-use plastics economy with the efficient collection and reprocessing is required to decrease leakage into natural systems for improved plastic waste management. From worst to best, three possible scenarios are discussed (Lebreton and Andrady 2019).

First scenario: Business goes up just like today. Based on growing demand through the years, global mismanaged plastic waste nearly triples by 2060, from 80 per year to 213 per year metric tonnes.

Second scenario: Waste management has improved. By 2060, global unmanaged plastic waste will have decreased to 50 metric tonnes per year, thanks to improvements in waste management infrastructure, particularly in developing countries.

Third scenario: At the same time, less plastic is used, and management is enhanced. Waste management efforts improve in the preceding scenario, and home plastic trash will be reduced to 5% of municipal solid waste by 2040.

Governments and companies were undoubtedly working to decrease plastic trash before the COVID-19 pandemic. In 2019, 188 nations agreed to amend the UN Basel Convention of 1989 to include plastic as a hazardous waste, and companies throughout the globe committed to increasing recycled plastic in packaging to 22% by 2025 (Global Commitment 2019 Progress Report 2019; Global Material Resources Outlook to 2060: Economic Drivers and Environmental Consequences | READ online 2019). Countries have the chance to "reset the clock" (The Atlas of Sustainable Development Goals 2020, 2020) and then continue their commitment to eliminating plastic and food waste as preparations to recover from COVID-19 evolve.

14.1 Companies and Use Cases

Table 14.1 presents the business models of 46 companies and use cases that employ emerging technologies and create value in SDG-12. We should highlight that one use case can be related to more than one SDG and it can make use of multiple emerging technologies. In the left column, we present the company name, the origin country, related SDGs and emerging technologies that are included. The companies and use cases are listed alphabetically.[1]

---

[1] For reference, you may click on the hyperlinks on the company names or follow the websites here (Accessed Online – 2.1.2022):

https://www.toyo-eng.com/jp/en/; https://muratechnology.com/; http://threefold.io/; http://www.fuergy.com/; https://ampliphi.io/; https://bloombiorenewables.com/; https://diwama.com; https://effabrush.com/; https://energenious.eu; https://hexafly.com/; https://insights.sustainability.google/; https://loopworm.in/; https://lumitics.com/; https://pro.hydrao.com/en/; https://pulpoar.com/; https://recyclingtechnologies.co.uk/; https://seenons.com/en/; https://sensoneo.com/; https://wandelbots.com/en/; https://www.acorecycling.com/; https://www.actandsorb.com/; https://www.amprobotics.com/; https://www.aquaai.com/; https://www.bambooder.nl/; https://www.bdwaste.com/; https://www.biopipe.co/; https://www.circulor.com/; https://www.deme-group.com/; https://www.ducky.eco/#; https://www.greentrash.nl/; https://www.greyparrot.ai/?hsLang=en; https://www.lignin.se/; https://www.merckgroup.com/en/sustainability-report/2020/environment/waste-and-recycling.html; https://www.norsepower.com/key-advantages; https://www.olleco.co.uk/; https://www.osram.com/os/applications/biometric-identification/index.jsp; https://www.relectrify.com/; https://www.plenty.ag/; https://www.samson-logic.com/; https://www.seeo2energy.com/; https://www.sensegrass.com/; https://www.smart-farming-system.com/; https://www.thephaseshift.com/; https://www.upprintingfood.com/; https://www.wasteless.com/; https://www.eugene-app.io

Table 14.1 Companies and use cases in SDG-12

No	Company info	Value proposal (what?)	Value creation (how?)	Value capture
1	Aco Recycling Turkey 9, 11, 12 Autonomous vehicles, Internet of Things, recycling	The company develops recycling and trash management applications by using smart solutions. Underground garbage containers and smart waste management are among their ecologically friendly solutions.	Their vending machines can save and collect garbage separately and neatly at the spot where it occurs by depositing recyclable materials such as plastic, glass bottles or aluminium beverage cans in the smart machine. Customers can choose from a variety of awarding methods to receive cash. The machine is ready to be implemented globally in both deposit and non-deposit situations.	Revenue is captured by producing waste collection machines that users can recycle systematically. These products will have a positive impact on both the recycling culture and the environment.
2	Act&Sorb Belgium 11, 12, 13 Biotech & biomanufacturing, recycling	The company has created a sustainable and value-added process for recycling medium-density fibreboard (MDF) and other wood wastes. Wood residue waste is transformed into high-value activated carbon using the carbonisation and activation method, which provides our planet with cleaner air, water and soil.	Non-recyclable wood residue waste is collected from production and end of life; waste is optimised through grinding, sieving and mixing; wood residue waste is converted into char, a coal-like substance; char is converted into activated carbon during a second burning; and carbon is given bespoke treatment based on customer needs.	Value creation is based on the goal of making the world cleaner and greener. By enabling the reuse of products that are difficult or impossible to recycle, a significant contribution is made to the environment and waste management economy.
3	AMP Robotics USA 12 AI, big data, recycling, robotics	Material recycling with robotic devices enhanced by computer vision and deep learning.	They use a robotics system to automate the identification and sorting of recyclables from mixed material streams. Their AI platform continuously trains itself by recognising different colours, textures, shapes, sizes, patterns and even brand labels to identify materials and their recyclability and enables the robotic sorting of material as granular as a type of plastic at a pick rate of upwards of 80 items per minute.	Revenue is captured by increasing the value that can be extracted from recyclable materials through separation, purity enhancement and identification of new end markets for recycling and reuse.

(continued)

Table 14.1 (continued)

No	Company info	Value proposal (what?)	Value creation (how?)	Value capture
4	Ampliphi USA 12, 17 Recycling	It is a smart environmental action platform that allows consumer brands to analyse, decrease and share their plastic impact in a simple yet methodical way.	The data-driven platform takes the guesswork out of plastic footprint management for recycling.	The platform amplifies the environmental movement and speeds the transition to a circular economy by leveraging the power of data to save time, cut through the noise and maximise effect.
5	Aquaai USA 12, 14 AI, robotics	It is a marine robotics B2B service company that has developed an integrated AI-powered web platform and a specialised fishlike autonomous underwater vehicle (AUV) to collect and deliver real-time visual and environmental data.	The company's product is developed by using threefold technology. These folds are data, software platform and a robot fish that combines it all. This bio-inspired fishlike robot is an autonomous underwater vehicle (AUV) and provides Function as a Service (FaaS) to their customers. The fish robot transmits real-time visual and environmental data to an online visual panel with these systems.	The product reduces outgoing costs in underwater monitoring and environmental data collection. It is used to control aquaculture in smart cities, ports, rivers and dams to support first responders in emergencies. It can be used in disaster management as well as illegal fishing and ocean exploration. The product contributes to nine SDGs with its multiple usage areas.
6	Bambooder Netherlands 12 Biotech & biomanufacturing	It is a BioTechnology company that aims to replace fossil raw materials with lightweight bamboo products by using patented technology for a variety of industrial applications.	Long bamboo fibres are extracted to create an endless bamboo thread for high-performance composite applications. Carbon, glass fibre, steel, aluminium, flax and hemp fibre may all be used to substitute Bambooder Biobased Fibres in composites.	Revenue generation is achieved by contributing to sustainable production in terms of decreasing the use of oil-based raw materials with biotechnology.
7	BD Waste Ghana 12 Recycling	The company connects households with local waste collectors and recyclers to make ecologically sound waste disposal and recycling easier.	A pick-up request is submitted for either plastic, paper, organic, e-waste or household to various aggregators and recyclers through an app.	The company is leveraging digital technology to mobilise stakeholder actions for improved recovery and recycling of your waste.

#	Company / Country / Codes / Tech	Description	Outcome
8	Biopipe Turkey 6, 12 Recycling	The company developed a wastewater treatment system that entirely takes place in-pipe which is protected by patents and proprietary processes.	Biopipe is a biological wastewater treatment technology in which the entire process occurs within a network of pipes. Biopipe microorganisms interact with impurities in wastewater and eradicate them. The pressure differential draws air in automatically, creating a perfect environment for bacteria to thrive quickly and aerobic digestion to speed up.	Households, hotels, residential complexes and governments benefit from the revenue provided by the wastewater treatment system. Biopipe may be deployed everywhere wastewater is created, regardless of size or capacity, owing to its compact footprint and adaptability.
9	Bloom Biorenewables Switzerland 7, 12, 13 Bioplastics, biotech & biomanufacturing, recycling	It is a company that provides a sustainable replacement for petroleum in current products. Furthermore, they produce BioPlastics in addition to biodiesel.	By using energy obtained by recycling non-edible biomass, biodiesel is produced. Besides, thanks to cellulose and lignin, polymer-based biomanufacturing companies produce textile goods and cosmetics.	Revenue is captured by cost-competitive outcomes, with a lower environmental cost and contributing transition from a linear, petro-based economy to a circular, plant-based one.
10	Circulor UK 9, 12, 13 14 AI, blockchain, digital twins	Their mission is to optimise industrial supply chains while making them more transparent to minimise the effect of these processes on the environment. The company offers Blockchain-based software solutions.	Their solutions use digital twins to follow the flow of materials through manufacturing processes while measuring the associated emissions at each level of the supply chain. Circulor's solutions can help to identify any possible dangers by supplementing traceability data. Corrective measures may be made in response to these risks with a comprehensive audit trail and immutable recordings on the blockchain.	Among the various causes contributing to the climate problem, the crucial role of industrial supply chains cannot be overlooked. Circulor addresses this important global issue by allowing businesses to get visibility into their supply chains in order to show responsible sourcing, enhance ESG performance, decrease GHG emissions and manage supply chain risks.
11	DEME USA 12, 14 AI, autonomous vehicles, spatial computing	They provide a service where floating and suspended waste is passively collected by a fixed installation, and larger waste is actively collected by a mobile system using virtual reality and AI.	The project incorporates a smart detecting system, an autonomously navigating workboat and a charging station. Smart cameras use AI to identify floating garbage. The workboat intercepts garbage and transports it to a collecting pontoon, where it is transferred into a container by a crane fitted with a grab.	Rivers are protected from harming the ocean and marine life by reducing the amount of macroplastics in them. As a result, virtual reality and Artificial Intelligence are being used to combat the growing problem of river pollution.

(continued)

Table 14.1 (continued)

No	Company info	Value proposal (what?)	Value creation (how?)	Value capture
12	Diwama Lebanon 9, 11, 12, 13 AI, Internet of Things, recycling	It is an AI-based monitoring and waste analysis software platform that enables the detection and auditing of waste streams for waste management companies.	VITRON is an AI and image recognition technology-based platform that automates the detection and auditing of recyclables on conveyor belts to track material composition and quality at various stages of the recovery process.	It collects data from waste streams with IoT, contributes to the recovery of recyclable materials, increases operational efficiency and product value and enables companies to make responsible consumption.
13	Ducky Norway 12, 13 Big data, cloud computing	It is a climate mitigation company that has a platform for people to change their behaviours to reduce their carbon emission footprint.	The SaaS, big data and cleantech-based carbon emission reduction platform measures, educates and mobilises people to take action on carbon sustainability. It is an employee engagement platform where businesses can engage their employees and customers in targeted challenges, where users log everyday actions in a time-bound, team-based competition.	Environmental impact revenue is captured through contributing to a sustainable planet by carbon footprint and CO_2 emission calculations.
14	Effa Ukraine 3, 12 Bioplastics, biotech & biomanufacturing, recycling	The company aims to minimise the plastic used in toothbrush manufacturing and to make its toothbrushes 100% recyclable	The body part of the toothbrush is produced from paper, and the head part consists of cornstarch and Nylon11. Since the head and body parts can be easily separated from each other, they can be thrown into the relevant recycling bins. The bristles and the head part are in one piece. The toothbrush bristles never leave the product.	The company generates revenue with its toothbrush product, which reduces CO_2 by 60% and plastic usage by 70%. With this product, an environmentally friendly user profile is created.
15	energenious Germany 7, 9, 11, 12 Big data, cloud computing	It is a cloud-based online platform that allows decentralised energy system planners and operators to model and optimise energy infrastructure projects and industrial applications.	MicrogridCreator is a cloud-based microgrid modelling and optimisation service that helps developers and end users plan, create and optimise complex sector-coupled decentralised energy systems.	When it comes to improving decentralised systems, the SaaS solution's cross-sectoral approach offers a persuasive additional benefit. Customers benefit from the integration of environmental elements such as CO_2 levels, sector coupling and flexible operation techniques.

16	Eugene France 12 Big data	Customers may use the company's recycling app solutions and barcode scanners. The software and device provide instructions on how to recycle waste materials.	It is a mobile app where users can access detailed information about the composition of the product and the recyclability of the packaging, using big data and IoT technologies.	Packaging recycling performance captures value through customer knowledge and incentive. They reward users for recycling by cooperating with sponsors to give points and rewards that users may claim based on their behaviours.
17	Fuergy Slovakia 7, 12 AI, Internet of Things	The company developed brAIn – a proprietary hardware device and AI software to optimise energy consumption and maximise the efficiency of renewable energy sources.	The AI-powered device helps users optimise energy consumption and maximise the energy efficiency of renewables. The company's proprietary technology connects to the Internet of Things (IoT) devices, such as smart appliances, to make the most effective use of energy consumption.	Revenue is captured through optimising energy consumption and maximising the efficiency of renewable energy sources; solutions rely on existing energy grids. The company also creates a new energy ecosystem which aims to maximise energy savings.
18	Google EIE USA 11, 12, 13 AI	Environmental Insights Explorer (EIE), whose goal is to make the process of setting an emission baseline and identifying reduction opportunities, is founded on the idea that data and technologies enable the world's transition to a low-carbon future.	EIE provides cities with data and a tool to help estimate their GHG inventory. Firstly, it defines a city boundary and the activity sectors. Then, it gathers the city activity data like energy consumed and type of fuels burned or products, etc. within these boundaries and sectors. It calculates total GHG emissions using the right conversion factors which are the average GHG emissions from burning a type of fuel, driving a type of vehicle or generating electricity.	They use advanced Machine Learning to assist cities and local governments around the world in analysing emissions data and identifying climate action strategies, as well as applying regional scaling, efficiency and emissions factors to create rich inventories and robust data analytics that are freely available to cities and local governments.
19	Green Trash Netherlands 12, 13 Internet of Things, recycling	The company produces smart waste bins, sustainable waste solutions, waste reduction and compaction machines.	The company offers a solution that recognises, sorts and compresses the waste automatically through collecting data from machines by using IoT.	Machines provide environmental advantages in the form of disposable waste recycling, as well as minimising the volume of waste by up to 90%. And as a result of the decreased collection frequency, they directly benefit from lower trash expenses.

(continued)

Table 14.1 (continued)

No	Company info	Value proposal (what?)	Value creation (how?)	Value capture
20	Greyparrot UK 12 AI, recycling	The company uses AI-based computer vision to monitor and sort garbage at scale, automating waste composition analysis.	Greyparrot understands the makeup of the waste stream from infeed to end product using AI intelligence, providing clients access to previously untapped data to anticipate issues before they emerge and make smarter business decisions.	The company helps companies to make better choices of materials and accurate prices which ends up in a stronger trust relationship with customers. The AI technology that is used helps companies build up their circular economy. With the hazardous objects reported by AI, the well-being and safety of the employees are also enhanced.
21	Hexafly Ireland 12 Biotech & biomanufacturing, recycling	The company bioconverts food waste into protein, oil and fertiliser by cultivating black flies.	Black soldier fly larvae thrive in small spaces with minimal resources. The fly colony is self-sustaining, and they require very little water making it possible to consistently produce vast amounts of protein and organic fertiliser in small spaces.	The company bioconverts food waste into sustainable, affordable and environmentally friendly products without chemical processes or using antibiotics.
22	Hydrao France 1, 2, 6, 11, 12 Internet of Things	Hydro metres record the volume, flow and temperature of the water circulating in the cold and hot water networks of buildings with IoT databases and LoRA technology.	The micro-turbine, which starts moving as soon as the water flows in the self-powered motor, allows for a long lifespan without requiring a battery. It identifies unusual water usage, such as leaks and improperly closed taps, and analyses water usage in smart buildings.	The firm develops and models technology solutions for water resource preservation that balance ecology and usage in order to reduce costs and consumption. Under the shower, the hydrometer device saves 70% of water.
23	Lignin Sweden 12 Advanced materials, bioplastics, biotech & biomanufacturing and recycling	It is a biotechnology company that converts lignin so it can be blended with virgin and recycled thermoplastics creating new types of bioplastics.	The approach lowers the boiling point of the lignin polymer without adding additional pressure, resulting in a highly energy-efficient process with no toxic emissions or harmful residues. One tonne of lignin may be transformed into one tonne of RENOL bio-granulates in minutes. Then RENOL may be mixed into a variety of bioplastics.	It has an environmental impact by avoiding the burning of lignin for its energy value. Revenue generation is achieved by enhancing the manufacturing of degradable bioplastics

24	Loopworm	The company aims to utilise food waste from landfills by generating protein-rich diets from them using insects that decompose organic materials into useful nutrients. They offer the food generated from waste to fish and poultry farms.	They utilise black army flies to turn organic waste into high-protein meals. BSF is a fly with several unusual characteristics. It isn't a pest like a conventional house fly; thus, it may really be used by people. As vital decomposers, black soldier fly larvae have a similar role as redworms in breaking down organic substrates and returning nutrients to the soil.	In addition to reducing trash in landfills, the firm also delivers protein-rich meals to fish and poultry farms, making their feeding procedures more sustainable. Their avowed mission is to generate sustainable food and feed for tomorrow while also closing the loop on food waste.
	India
	12, 13
	Biotech & biomanufacturing, recycling
25	Lumitics	Insight is a smart food waste tracker that uses sensors and AI image recognition technology to weigh and identify the many types of trash that restaurants and kitchens generate.	Insight is a smart food waste tracker that uses sensors and AI image recognition technology to weigh and identify the types of trash that restaurants and kitchens generate. Steps are being done to avoid food waste based on the data from the food obtained with this device.	Every year, 1.6 billion tonnes of food are wasted worldwide, with 50% of all food made for buffets ending up in the trash. Growing crops, harvesting, transporting and selling food are all energy- and resource-intensive activities that deliver food to our tables. We save resources and energy that could otherwise be used in other ways if we waste less food.
	Singapore
	12
	AI, big data
26	Merck	It is a company that generates or optimises new production techniques to minimise waste production.	A waste scoring system is put in which allows comparison and tracking of generated waste amounts. Waste separation allows raw materials to be recovered and recycled, while non-recyclable waste is disposed of following the tightest waste disposal regulations.	The company creates a long-run value through their core business model while targeting a balance between environmental, socioeconomic and governance concerns for their businesses, stakeholders and society. Their perspective helps solve the biggest challenges of the twenty-first century, such as disease, poverty, hunger and climate change.
	Germany
	12
	Recycling

(continued)

Table 14.1 (continued)

No	Company info	Value proposal (what?)	Value creation (how?)	Value capture
27	Mura Technology UK 12, 13, 14, 15 Recycling	The company recycles waste plastic back into the chemicals and oils from which they were made by using supercritical water.	Mura's HydroPR system is based on the Cat-HT technology, which uses supercritical water, heat and pressure to transform waste plastics into valuable chemicals and oils by breaking down long-chain hydrocarbons and donating hydrogen to yield shorter-chain, stable hydrocarbon products for sale to the petrochemical industry to be used in the manufacturing of various plastics and other materials.	Value is created by reducing plastic waste and recycling reusable materials by offering a solution for all plastic types and creating a circular economy.
28	Norsepower Finland 7, 9, 12 Autonomous vehicles	The company develops products that reduce the environmental impact of the maritime industry by using rotor sail solution technology.	Rotor sail solution technology maximises the fuel efficiency of the ship by utilising wind. In addition, the system works with an automation system, which autonomously detects enough wind power to save fuel and starts the rotors when it detects this power.	Value creation is done by producing sails that can be used in many types of ships and that will reduce carbon emissions to zero and minimise the damage to nature caused by ships. The sails used are ten times more efficient than an ordinary sailboat in terms of fuel consumption.
29	Olleco UK 7, 12 Carbon capture and storage, Internet of Things, recycling	They help the food/beverage industry and service sector to tackle the climate emergency by converting their waste resources into renewable energy and clean fuel at biogas plants.	The company has a system named "OilSense". This system is a specially designed oil handling system that uses telemetry to carry information (IoT). After oil waste is collected, they convert it to biodiesel which has a 0% carbon ratio. Likewise, by recycling waste oil in anaerobic digestion chambers, electricity is generated in combined heat and power plants.	The company contributes to the circular economy and carbon reduction from the beginning to the end of the recycling process. Furthermore, revenue is generated through greenhouse reductions extending over 88%, thanks to carbon capture technology.

30	Osram	It is a broadband emitting infrared LED producer, where the compact emitter allows consumers to analyse food, medicine and more on mobile devices.	Biometric identification uses unique human traits such as facial features, fingerprints or the vein network in the hand. The technology provides a secure and simple way to manage complicated passwords. Special infrared LEDs from Osram Opto Semiconductors are used in biometric recognition systems. These LEDs light up the target region, allowing the devices to capture high-quality photos and ensure accurate identification.	It enhances the level of reliability and security to make sure the most important things in one's life are as safe as they can be.
	Germany			
	9, 12			
	Advanced materials, biometrics			
31	Phaseshift Technologies	A software company that focuses on the development of high-performance alloys for injection moulding and additive manufacturing by using machine learning and quantum computing.	The company enables the modelling of material properties from quantum scale to macroscopic levels by linking different regimes of chemistry simulations together. By combining machine learning approach and computational chemistry, a large design space (with thousands of combination behaviours) is mapped out, and optimal features (top-performing metrics) of the materials are identified.	Revenue generation is done through waste reduction by optimising the design phases with Machine Learning and quantum computing and manufacturing globally optimum materials. The company shortens the timelines of experiments, thus increasing the return on investment (ROI) as the initial investments decrease.
	Canada			
	12			
	AI, quantum computing			
32	Plenty	It is a vertical farming company that aims to produce organic vegetables and salads by optimising the external conditions with machine learning.	The company raises its plants in towering towers with LED lights within a climate-controlled facility. Pesticides, herbicides, synthetic fertilisers and genetically modified organisms are not used. The technology collects data in the farms using hundreds of infrared cameras and sensors, which is then evaluated using machine learning to optimise growth.	Revenue is captured through increased yield of crops, efficient use of farmland, lower water use and higher productivity by soilless farming and Machine Learning. Revenue is also generated through organic and pesticide-free crops.
	USA			
	12			
	AI, soilless farming			

(continued)

Table 14.1 (continued)

No	Company info	Value proposal (what?)	Value creation (how?)	Value capture
33	PulpoAR	It is an augmented shopping company that works on technologies that enable the user to create virtual product experiments in the optical and cosmetic industries through their social media, website, mobile application and smart screens.	For AR/AI beauty, the face identification algorithm uses facial micro-feature tracking to identify facial traits. To correctly change colours, nail colour shifting is employed in nail segmentation Machine Learning systems. On digital platforms, hair segmentation and colour change machine learning algorithms accurately replicate real-world colours.	Virtual try-on adds value by lowering basket drop rates by up to 30%, according to statistics from real-world examples. It also promotes ethical consumption and hygiene by utilising smart mirror solutions that allow clients to virtually try on things in a physical store.
	Turkey			
	12			
	AI, spatial computing			
34	Recycling Technologies	In the production of new polymers, it is a company that utilises thermal cracking to turn leftover plastic trash into a liquid hydrocarbon feedstock using RT7000 machine.	Their technology, the RT7000, converts plastic waste into chemical feedstock for plastic manufacture. The machine employs a method called thermal cracking, which uses heat in the absence of oxygen to break down long polymer chains into shorter ones, reducing wasteful plastic trash transportation and related carbon emissions.	Revenue is captured through Plaxx, which is a valuable chemical liquid hydrocarbon feedstock after refinement, which can be used in the manufacturing of new virgin quality plastics and is produced from the waste plastic processed through the RT7000. It is a valuable building block in the circular economy and the plastics value chain, providing post-consumer recycled content for new plastic products in line with governmental targets.
	UK			
	9, 12, 13			
	Advanced materials, recycling			
35	Relectrify	The company is a battery storage provider with second-hand electric vehicle batteries.	ReVolve is a battery energy storage solution that is both inexpensive and long-lasting. ReVolve is designed for industrial and commercial installations ranging from 120 kWh to 2 MWh and is powered by Relectrify's technology for repurposing high-quality second-life batteries from electric vehicles.	Recycling used electric vehicle batteries adds value to the environment. Providing cheap storage solutions for residences, industry and the service sector captures business value.
	Australia			
	7, 12			
	Energy storage, recycling			

No.	Company	Description	Details	Value creation
36	Samson Logic Israel 12 Internet of Things	The company develops material and equipment management solutions for transportation, as well as on-site storage and delivery.	IoT sensors and material handling software are used to guarantee that materials are transported and lifted securely to the job site. The software programme distributes materials among the modules based on maximal transit capacity, as well as creating packing instructions for the maker.	The company provides smart packages integrated with IoT sensors and software that converts raw material orders into packaging instructions and manages the inventory of materials at construction sites.
37	SeeO2 Energy Canada 7, 12 Advanced materials, carbon capture and storage	The company converts carbon dioxide (CO_2) and/or water into marketable and clean value-added products using reversible fuel cell technology.	The employment of a catalyst material in the reversible solid oxide fuel cells (RSOFCs) technology allows for the electrochemical conversion of water to hydrogen, carbon dioxide (CO_2) to carbon monoxide (CO) and CO_2 and water to form syngas, which is a combination of CO and H_2. When used in the fuel cell mode, the technology electrochemically transforms H_2, CO, syngas and methane to electricity and heat for off-grid/remote consumers, residential and commercial users.	Value creation is achieved by providing reliable costs, competitive and clean gases for chemical and electronic end-users and energy storage for intermittent renewable energy.
38	Seenons Netherlands 7, 11, 12, 13, 17 AI, Internet of Things, recycling	The company provides a platform that pairs wastes and recycling businesses.	The platform matches waste (also called residual flows) with the right processor. The processor turns it into a new product, green electricity or biogas. Making use of an app, the waste is collected periodically from registered locations by environmentally friendly sustainable vehicles. The IT technology selects the most efficient and sustainable route and vehicle. The wastes are taken to companies that transform the waste into new products.	The company aims to achieve a zero-waste environment. Preservation of natural resources, reducing carbon emissions and becoming fully circular is their mission.
39	SenseGrass France 2, 12, 13 AI, drones, Internet of Things	It is a Soil Intelligence Platform that helps companies and farmers reduce the excessive nitrogen and fertilisers from the soil using sensors. It also optimises the crop data through AI.	IoT sensors monitor soil and water status, while drone images help farmers to monitor the field. The AI algorithm allows the farmers to make the right decision and helps them get a better yield.	Environmental value is achieved through reducing excessive fertiliser usage and better soil and water management. Revenue is generated by increased efficiency and productivity.

(continued)

Table 14.1 (continued)

No	Company info	Value proposal (what?)	Value creation (how?)	Value capture
40	Sensoneo Slovakia 11, 12 AI, Internet of Things	The company offers smart, IoT-based waste management solutions for cities and businesses to improve environmental responsibility and cost-efficiency.	The smart sensors use ultrasound technology to measure the fill levels in bins and containers and send the data to the Smart Waste Management System, a cloud-based platform, via IoT providing cities and businesses with data-driven decision-making and optimisation of waste collection routes, frequencies and vehicle loads.	Revenue is captured by smart waste management solutions that enable cities and businesses to manage their waste efficiently, lower their environmental footprint and improve the quality of services. The waste management solution achieves a 30%–63% reduction of waste collection routes and 97% accuracy on actual waste production.
41	Smart Farming System France 2, 12 AI	It is a company that uses AI solutions and their applications to agricultural technology (AgTech) devices to analyse the development of crops, optimise their growth and prevent diseases.	It is a company that uses AI solutions and their applications to agricultural technology (AgTech) devices to analyse the development of crops, optimise their growth and prevent diseases.	Value is captured by the aeroponic system which works with a low-energy consumption model, uses 97% less water than conventional agriculture and produces a pesticide-free and environmentally friendly product. Its intelligent component allows it to continuously optimise the growth process.
42	ThreeFold Belgium 9, 12 Blockchain, cloud computing, edge computing	The company created a Blockchain-based decentralised cloud infrastructure that allows participants to exchange Internet capacity on the grid through hardware that runs on their operating system, ZeroOS. Their product offers an alternative to large and centralised data centres located in only a handful of locations around the world.	Their grid is an open system that lets anybody participate in the burgeoning digital economy by becoming a node. Anyone may join an open and inclusive Internet ecosystem by connecting computation, storage and network resources to the grid.	The Internet accounts for 10% or more of the world's energy usage. Their co-founders, with decades of expertise in the Internet and Cloud industries, created a simple operating system and infrastructure, reducing layers of complexity from conventional and Web3 approaches.

#	Company	Description	Value
43	Toyo Engineering Japan 7, 12, 13 Biotech & biomanufacturing, carbon capture and storage, hydrogen	A hydrocarbon and petrochemical company generating green and blue ammonia with the aid of carbon capture and storage.	Liquid fuels derived from carbon recycling include not only bio-derived fuels synthesised biologically and chemically using biomass as a raw material but also e-fuels created by synthesising CO_2 from Direct Air Capture and fossil fuels emitted from industrial plants, as well as hydrogen produced from renewable energy.	Business and environmental value through monetisation of CO_2 and production of low-carbon fuel.
44	Upprinting Food Netherlands 2, 12 3D printing, recycling	It is a company that provides restaurants with sustainable food printing by turning food waste into tasty food through the use of 3D printing.	They assist high-end restaurants in reducing their leftover food flows by blending and combining the various components from residual food flows, resulting in purees that are subsequently 3D printed by a food printer using additive manufacturing. These prints are baked and dried to give them crunch and durability.	Value is created by closing the ring at the end of the food chain. They contribute to responsible consumption and production through creating new recipes from wasted food products by using 3D printing in a world where bad-looking and too ripe to be sold vegetables and fruit as well as bread are often wasted.
45	Wandelbots Germany 12 AI, robotics	It is a robotics business that provides a no-code robotics platform with the goal of optimising robotic programming/automation procedures.	The user conducts the robot's activity by marking specific spots or pathways with the TracePen, and the movement types/blendings are determined (via collaboration with the platform) using AI. The specified routes are then created in the Wandelbots app, which allows users to update the path and procedures as well as do quick testing. The platform generates safe code when these are executed, and robotic programming is thus done.	Revenue is generated by enhancing robotic automation processes. This is done by enabling maximum efficiency by delivering a new kind of robotic programming technique to manufacture which leads through reducing the cost and time which rise from robotic coding/programming and related training.
46	Wasteless Israel 2, 12 AI	It is a firm that uses Artificial Intelligence to improve merchants' bottom lines by optimising markdowns based on expiry dates, resulting in less waste.	For supermarket management, the firm uses radio-frequency identification (RFID) technology, machine learning, dynamic pricing and real-time shelf monitoring. The price engine uses "reinforcement learning", a type of machine learning that allows it to learn how customers react to dynamic pricing and then discover the best discounting strategy.	It recaptures the full value of their perishable products and reduces food waste through AI-powered dynamic pricing on supermarkets and online grocery stores. Revenue generation is achieved by tackling waste at the food retail level and also increases the consumers' awareness towards making sustainable purchase decisions.

References

K.K. Al et al., The 9Rs strategies for the circular economy 3.0. Psychol. Educ. J. **58**(1), 1440–1446 (2021). https://doi.org/10.17762/pae.v58i1.926

R. Amos, E. Lydgate, Trade, transboundary impacts and the implementation of SDG 12. Sustain. Sci. **15**, 1699–1710 (2020). https://doi.org/10.1007/s11625-019-00713-9

D. Andrews, The circular economy, design thinking and education for sustainability. Local Econ. J. Local Econ. Policy Unit **30**(3), 305–315 (2015). https://doi.org/10.1177/0269094215578226

T.E.T. Dantas et al., How the combination of Circular Economy and Industry 4.0 can contribute towards achieving the Sustainable Development Goals. Sustain. Prod. Consum. **26**, 213–227 (2021). https://doi.org/10.1016/j.spc.2020.10.005

Economic Commission for Latin America and the Caribbean, Responsible consumption and production (2020). Available at: https://www.cepal.org/en/sdg/12-responsible-consumption-and-production. Accessed 23 Aug 2021.

European Investment Bank, The EIB Circular Economy Guide – Supporting the Circular Transition (2020). https://doi.org/10.2867/578286

Food Loss and Food Waste, Food and Agriculture Organization of the United Nations (2021). Available at: http://www.fao.org/food-loss-and-food-waste/en/. Accessed 8 Aug 2021

D. Gasper, A. Shah, S. Tankha, The framing of sustainable consumption and production in SDG 12. Glob. Policy **10**(S1), 83–95 (2019). https://doi.org/10.1111/1758-5899.12592

Global Commitment, Progress Report (2019) New Plastics Economy (en-GB) (2019). Available at: https://www.newplasticseconomy.org/about/publications/global-commitment-2019-progress-report. Accessed 8 Aug 2021

Grand View Research, Food waste management market size report, 2020–2027 (2020). Available at: https://www.grandviewresearch.com/industry-analysis/food-waste-management-market. Accessed 8 Aug 2021

Grand View Research, Plastic market worth $750.1 billion by 2028 | CAGR: 3.4% (2021). Available at: https://www.grandviewresearch.com/press-release/global-plastics-market-analysis. Accessed 8 Aug 2021

M. Inoue et al., A modular design strategy considering sustainability and supplier selection. J. Adv. Mech. Des. Syst. Manuf. **14**(2), JAMDSM0023–JAMDSM0023 (2020). https://doi.org/10.1299/jamdsm.2020jamdsm0023

V. Iordachi, Circular economy as a new industrial paradigm (2020). https://doi.org/10.5281/ZENODO.4296305

K. Khaw-ngern, P. Peuchthonglang, L. Klomkul, C. Khaw-ngern, A digital circular economy for SDG 11 and SDG 12. Psychol. Educ. **58**(1), 1380–1386 (2021)

L. Lebreton, A. Andrady, Future scenarios of global plastic waste generation and disposal. Palgrave Commun. **5**(1), 1–11 (2019). https://doi.org/10.1057/s41599-018-0212-7

L. McNeill, *Transitioning to Responsible Consumption and Production* (MDPI Books, 2020). https://doi.org/10.3390/books978-3-03897-873-2

J. Monnet et al., *Sustainable Development Goals Network Policy Brief Series SDG 12: Responsible Consumption and Production* (2014), pp. 1–6

OECD, Global material resources outlook to 2060: Economic drivers and environmental consequences (2019). Available at: https://read.oecd-ilibrary.org/environment/global-material-resources-outlook-to-2060_9789264307452-en. Accessed 8 Aug 2021

J.M. Rodriguez-Anton et al., Analysis of the relations between circular economy and sustainable development goals. Int. J. Sustain. Dev. World Ecol. **26**(8), 708–720 (2019). https://doi.org/10.1080/13504509.2019.1666754

C. Stevens, Linking sustainable consumption and production: The government role. Nat. Resour. Forum **34**, 16–23 (2010). https://doi.org/10.1111/j.1477-8947.2010.01273.x

The Atlas of Sustainable Development Goals 2020. (2020). Available at: https://datatopics.worldbank.org/sdgatlas/. Accessed 8 Aug 2021

A. Tukker, S. Emmert, M. Charter, C. Vezzoli, E. Sto, M.M. Andersen, T. Geerken, U. Tischner, S. Lahlou, Fostering change to sustainable consumption and production: An evidence based view. J. Clean. Prod., 16 (2008). https://doi.org/10.1016/j.jclepro.2007.08.015

United Nations Development Programme, Goal 12: Responsible consumption and production (2021). Retrieved from: https://www.tr.undp.org/content/turkey/en/home/sustainable-development-goals/goal-12-responsible-consumption-and-production.html

SDG-13: Climate Action

15

Abstract

SDG-13, Climate Action, aims to adapt to climate change by mitigating adverse effects and keeping the temperature rise below 1.5° by the end of this century and prepare low-carbon development plans. Investing in adaptation is critical for limiting the adverse effects of climate change on human society. Every efficient policy for combating climate change, on the other hand, must decrease emissions to prevent future warming while also adapting to the unavoidable effects of climate change. This chapter presents the business models of 52 companies and use cases that employ emerging technologies and create value in SDG-13. We should highlight that one use case can be related to more than one SDG and it can make use of multiple emerging technologies.

Keywords

Sustainable development goals · Business models · Climate action · Sustainability.

The impact of climate change have been felt and experienced by many countries for many years. Today, greenhouse gas (GHG) is 50% higher than the levels in 1990. Global warming has caused climate change for a long time, and these irreversible changes threaten all countries if the countries around the world do not act. Economic losses caused by natural events due to this climate change are at the level of hundreds of billions of dollars. The nations aim to prevent permanent changes to be experienced in the climate system and to prevent economic losses due to this change. In this direction, the United Nations funds developing countries under SDG-13 to adapt to climate change and prepare low-carbon development plans (United Nations 2021a). The work carried out to support sensitive areas within the scope of Goal 13 also helps to achieve other goals. It is aimed to keep the global temperature change below 1.5° with the work to be carried out within the framework of this purpose. In addition, the carbon dioxide emission rate in 2030 should be reduced by 45% compared to 2010, and net-zero is targeted for 2050 (United Nations 2021a).

Climate change manifests itself in every way nature acts. In the middle of 2021, Germany suffered one of the most fatal and devastating natural disasters in its history by receiving record rainfalls in 100, 500 and even 1000 years in some regions ("European Floods Are Latest Sign of a Global Warming Crisis" 2021). Moreover, wildfires worldwide drew attention in 2021, especially in Siberia, Mediterranean countries and Canada. Despite the cold climate of the northern regions of Russia, Yakutia hit 39 °C and experienced the driest weather since 1888 and suffered

S. Küfeoğlu, *Emerging Technologies*, Sustainable Development Goals Series,
https://doi.org/10.1007/978-3-031-07127-0_15

record wildfires, causing massive amounts of smoke and abnormal temperature rise (Magnay 2021). Human losses, property damages, internal displacements, outages of healthy water and electricity, deforestation and high carbon releases indicate that humanity is still not ready to cope with frequent and abnormal climate-related disasters (Die Welt 2021). By 2030, the UN plans to cope with wild disasters in every country adaptively. Deaths, injuries, internal replacements or homelessness due to disasters should be reduced as much as possible (United Nations 2021b).

To achieve future goals of reducing climate change, national governments and international institutions need to establish new policies and plans with climate change measures included. Indicators of this targeted focus on decarbonisation strategies and the amount of greenhouse gas emissions (SDSN 2021). Contributor countries of the Paris Agreement on climate change prepared their nationally determined contributions (NDCs), which declare each country's plan on reducing GHG emissions and climate change impacts. In addition to NDCs, volunteer countries have begun to adopt new plans for protecting from the natural disasters caused by climate change, such as floods or cyclones (United Nations 2021c). However, some countries' declared targets are insufficient, meaning that they can be satisfied without the adoption of new policies. Also, another group of countries adopted policies that cannot satisfy the insufficient targets. This situation complicates the progress for the 2030 and 2050 targets. Therefore, the policies of the countries that make rapid progress should be imitated (Doni et al. 2020). Furthermore, policies must optimise benefits to different SDGs. For example, a regulation that bans fertiliser use totally to help climate action would decrease yields, and thus, it could increase poverty and hunger and contradict SDG 1 (no poverty) and SDG 2 (zero hunger). However, not having a policy that regulates fertiliser use could cause the overuse of fertilisers and damage the climate action (Campbell et al. 2018). Hence, the adop-

tion of optimised policies is required to achieve the future targets of SDG 13. Also, revision of current policies and plans may be needed to strengthen the tools of governments against climate change. Figure 15.1 compiles the targets of SDG-13.

Climate change research in the built environment and related disciplines of transportation and utilities is still in its early stages. To plan effectively for the future, decision-makers require a significantly wider knowledge base. To accomplish this, academics and decision-makers must collaborate to generate this knowledge, which can then be used to develop effective climate change plans (EPSRC 2003). Some approaches, knowledge and skills are required to counteract climate change.

The course of adaptation necessitates adaptive management considering the evolving impacts of climate, the normative nature of risk tolerance and the tipping points between them (European Commission 2012; Wise et al. 2014). As emphasised by the UN, the importance of adaptation can be observed in the number of countries that have national adaptation plans and the financing of such plans. Recent reports identified six least developed countries that have already implemented a national adaptation plan, while a grand majority of developing countries are prioritising and implementing one. Furthermore, the mean annual finance for climate has been reported to be $48.7 billion between 2017 and 2018, which is an increase of 10% from 2015 to 2016. Most notably, the target of financial support for climate has started to shift from mitigation action to adaptation as more and more countries show increased support for adaptation (United Nations 2021c). The importance of creating reliable frameworks as measures of adaptive capacity has been emphasised both by researchers and policymakers (Solomon et al. 2007). However, there is no template for what this should contain due to adaptation's context-specific, process-based nature. Knowing how to adapt is deemed necessary to measure adaptive capacity, as it is highly related

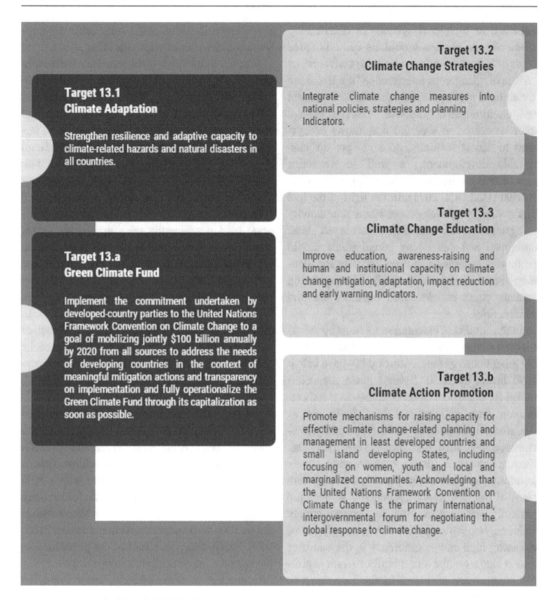

Fig. 15.1 Targets of SDG-13.(UNDP 2021)

to further deciding factors (Klein 2014). Learning is becoming more important in providing insights into what constitutes effective adaptation, which we define as improvements that minimise susceptibility to current and future climate change.

The United Nations Framework Convention on Climate Change (UNFCCC) is the international system for addressing climate change. The convention has been ratified by a large number of developing as well as developed countries such as the USA. The convention aims to "Prevent dangerous human interference in the climate system." In spite of the high amount of support for the convention worldwide, obtaining this goal is contentious (UNFCCC 2021).

The United Nations Framework Convention on Climate Change (UNFCCC) recognises the influence of biological systems in determining when climate change should be ceased. Three fields of influence against which the conventional criteria of "dangerous interference" are tested are agricultural production, sustainable development and environmental response. Climate change must be ceased in a period that allows ecosystems to "adapt naturally, does not prevent sustainable development, as well as maintains agricultural productivity", according to the convention (UNFCCC 2021). In the light of the fact that sustainable development keeps productivity in agriculture and electricity, developed, least developed and developing island states should promote mechanisms to raise capacity for planning and management while incorporating women, youth and localised organisations and communities.

In the context of essential reductions by industrialised nations, the EU is committed to lowering greenhouse gas emissions by 80–95% below 1990 levels by 2050. Several goals are determined for decarbonisation by 2050. In the short-medium term, conventional fossil fuels such as coal and oil are planned to be replaced by low-emission fuels such as natural gas and hydrogen. Nuclear power is also a low-emission power technology and has a huge place in the long-term plans of the EU. Renewable energy sources are primarily preferred in diversified energy supply technologies. With all these energy supply plans, obtaining high-energy efficiency in the end-user area is also a crucial aim. Finally, carbon capturing systems are proposed in the long term to decrease the release of inevitably produced carbon gases into the atmosphere (European Commission 2012).

The time horizon in WEO-2019 is 2050, rather than 2040, as in previous editions, to reflect announcements by various governments to attain carbon neutrality by 2050 and to model the possibility for new technologies (such as hydrogen and renewable gases) to be deployed at scale. As a result of continued CO_2 emissions and advances in climate research, the interpretation of the climatic target included in the sustainable development scenario varies with time (IEA 2020).

Green Recovery's progress in biodiversity conservation, responsible production and climate action will be stronger by 2100 (90%, 94% and 84%, respectively) with a longer timescale and even more ambitious aims, with 12 out of 13 targets on track or improving (IEA 2020; Moallemi et al. 2020). The fossil-fuelled development path achieves the fastest improvements in socioeconomic indices, such as gross world product (GWP) per capita, by 2100 while fulfilling moderate (and occasionally even aggressive) aims. However, in fossil-fuelled development, human and economic development leads to a significant increase in the share of fossil fuels in the energy supply, driven by rising energy demand from high-energy-intensity industries and services. In almost all 10,000 realisations of the fossil-fuelled development route, reliance on fossil fuels results in high climate consequences from energy-related CO_2 emissions by 2100 (Moallemi et al. 2020).

In 2009, the World Energy Outlook published the 450 Scenario, a detailed energy transition scenario. The scenario was named after the CO_2 concentration of 450 parts per million (ppm), which was thought to be consistent with a 50% chance of keeping average global temperature rise below 2 degrees Celsius at the time (assuming that net-zero emissions were reached in 2100) (IEA 2020). More than half of the population in the region's major centres, where 1.2 billion new residents are predicted by 2050, live in low-lying coastal areas. As a result, more than 742 million urban dwellers are today exposed to several dangers caused by climate-related disasters, which pose a threat to infrastructure and communities (UNESCAP 2019).

Through their internal operations and supply chains, businesses are responsible for a large percentage of GHG emissions and resource use. Furthermore, items are frequently incorrectly disposed of and seldom enter a new life for future use due to a lack of end-of-use (EOU) planning.

This contributes to global waste, pollution and a depletion of fresh material supply. As a result, companies play a critical role in collaborating with governments to limit global warming below 1.5 °C and develop resilience to present and future climate change consequences. Top businesses are taking this responsibility seriously today, offering innovative solutions to cut emissions, minimise effects across the value chain, enhance climate resilience and raise climate awareness. This can be accomplished in a variety of ways, including the publication of a sustainability action plan, the procurement of low-carbon materials, the investment in on-site renewable energy, the purchase of renewable energy credits (RECs), the construction of net-zero factories and the return of products at the end of their useful lives (Bureau 2021).

Many of the adjustments required to establish a carbon-free world would have an initial cost on various economic sectors. When zero-carbon economy infrastructure is developed, and legislative incentives are implemented, high-carbon businesses are expected to lose market share throughout the transition. With the adoption of climate policies, a study by Malerba and Wiebe suggests that Germany will experience the highest job increase in the EU. However, some other countries worldwide, such as Japan and the USA, will have a higher increase of available jobs. The study also concluded that there is no correlation between poverty rates and proportional job increases. A country with a high portion of its population living in poverty may witness a high proportional job increase, while another country with similar population characteristics experiences a low proportional job increase. For example, Brazil experiences a high increase of 0.8%, and India experiences a low increase of 0.3% (Malerba and Wiebe 2021). Another study found that if the measures to restrict global warming to 2 °C become in charge, available jobs will increase 0.3% more compared to current measures (Montt et al. 2018). This study also found that Bulgaria, Indonesia and Taiwan will experience the highest proportional job increases of 0.9%. The study suggests that there

will be 4.9 million new jobs created in China, one million in the USA and 1.3 million in India with the climate action measures. Since its economy relies mostly on fossil fuels and the industries that are expected to grow under climate action measures are not developed, the Middle East may experience job losses, unlike the rest of the world (Montt et al. 2018).

To reduce emissions, expenditure that belongs to individuals and the public must change. The government must develop new regulations that encourage individuals to spend their money in new ways, like using decarbonised heating for homes and using alternative modes of transportation such as electric automobiles and motorcycles. Simultaneously, changes in government expenditure and new fiscal incentives are required to spur the development of breakthrough zero-carbon technologies, improve building energy efficiency, modify farming methods and create sustainable energy grids.

Carbon emissions that originate from individuals or companies can be exposed to economic sanctions such as emission trading systems as well as carbon taxes. Even though this system can provide equal pay, low-incomers can be affected disproportionately by these measures (Abdallah et al. 2011). Favourably, this problem may be addressed by rebate programmes that return at least part of the collected earnings to lower-income individuals. Calculations of the precise future costs of adaptation generate a wide range of outcomes, as they are highly dependent on the level of future greenhouse gas emissions, how the climate system will respond to them in various locations and if they are effective. As a result, estimating the economic costs and benefits of adaptation is extremely challenging. However, adapting without attempts to mitigate climate change would be prohibitively expensive (Imperial College 2021).

According to the World Bank, developing countries can face economic damages of approximately 75–100 billion dollars per year because of the 2 °C warming until 2050. However, it is considered that these values are pretty low. In the meantime, the UN Environment Programme (UNEP) estimates this loss can approach 280–500 billion dollars each year until 2050. If global warming continues to rise beyond this point, the costs of adapting will skyrocket (Fankhauser 2019; Olhoff et al. 2015; World Bank 2010).

According to the Intergovernmental Panel on Climate Change (IPCC), reducing global warming to 2 °C would lower yearly per capita global consumption growth by 0.06 percentage points relative to growth in a hypothetical future without climate change (Onencan et al. 2016). The Organisation for Economic Co-operation and Development (OECD), on the other hand, claims that if climate change is taken into consideration in reform and budgetary plans, this consideration will create a 1% and 2.8% increment to GDP in G20 nations for 2021 and 2050, respectively (Organisation for Economic Co-operation and Development 2017).

Investing in adaptation is critical for limiting the adverse effects of climate change on human society. Every efficient policy for combating climate change, on the other hand, must decrease emissions to prevent future warming while also adapting to the unavoidable effects of climate change. According to the Global Commission on Adaptation, climate change will affect people despite the most effective strategies. As a result, reducing emissions is "the best form of adaptation" (Global Commission on Adaptation 2019).

15.1   Companies and Use Cases

Table 15.1 presents the business models of 52 companies and use cases that employ emerging technologies and create value in SDG-13. We should highlight that one use case can be related to more than one SDG and it can make use of multiple emerging technologies. In the left column, we present the company name, the origin country, related SDGs and emerging technologies that are included. The companies and use cases are listed alphabetically.[1]

---

[1]For reference, you may click on the hyperlinks on the company names or follow the websites here (Accessed Online – 2.1.2022):

http://algaepro.no/; http://c-combinator.com/; http://flash-forest.ca/; http://geopard.tech/; http://geyserbatteries.com/; http://mangomaterials.com/; http://spgroup.com.sg/; http://wingtra.com/; https://akercarboncapture.com/; https://biomemakers.com/; https://charmindustrial.com/; https://checkerspot.com/; https://deepbranch.com/technology/; https://emrod.energy/; https://en.vytal.org/; https://enervenue.com/; https://fullcyclebioplastics.com/; https://lignaenergy.se/; https://polyspectra.com/; https://poseidon.eco/; https://sadako.es/; https://seabenergy.com/; https://solarfoods.fi/ https://tech2impact.com/start-ups/utilis/; https://verv.energy/; https://www.aphea.bio/; https://www.bamomas.com/; https://www.bloomenergy.com/; https://www.bluenalu.com/about; https://www.business.att.com/products/multi-access-edge-computing.html; https://www.caire-solutions.com/; https://www.carbfix.com/; https://www.carbonclean.com/; https://www.carboncure.com/; https://www.climatedatahub.io/; https://www.clingsystems.com/; https://www.crowd4climate.org/; https://www.divergent3d.com/; https://www.freightfarms.com/greenery-s; https://www.ge.com/gas-power/future-of-energy/carbon-capture-storage; https://www.h2arvester.nl/?lang=en; https://www.harvest.london/; https://www.ibm.com/case-studies/energy-block-chain-labs-inc;https://www.meatable.com/; https://www.omniplytech.com/; https://www.ossia.com/; https://www.polybion.mx/materials/; https://www.raisegreen.com/; https://www.riversimple.com/; https://www.samsungsdi.com/business.html; https://www.space4good.com/; https://www.woodlandbiofuels.com/

Table 15.1 Companies and use cases in SDG-13

No	Company info	Value proposal (what?)	Value creation (how?)	Value capture
1	Aker Carbon Capture Norway 13 Carbon capture & storage	It is a company that provides modular carbon capture technology services to its customers.	They provide services to reduce the amount of CO_2 emissions caused by production facilities during production that spread to the atmosphere. This system, which has a modular structure, is delivered to the field to capture CO_2 in the regions where this service is needed.	An increase in the carbon tax will increase the production costs of many companies. By using this service, the amount of CO_2 emitted into the atmosphere during cement, bio/waste-to-energy, steel gas-to-power production can be reduced. This way, the amount of tax that companies have to pay due to CO_2 emissions is reduced. Thanks to this service, the negative environmental effects caused by production are also reduced.
2	AlgaePro Norway 13 Biotech & biomanufacturing, recycling	They use biowaste from municipal waste management, as well as carbon dioxide and waste heat from industrial sources, to develop microalgae in a circular bioeconomy.	They recycle biowaste from municipal waste management to grow microalgae in large ponds or a device named photobioreactor. The grown microalgae can be used as a sustainable feed in fish farms as they contain marine proteins and omega-3 fatty acids.	Savings are captured through low-cost microalgae use in fish farms and help in climate action through microalgae use as a sustainable feed in the fish farming sector.
3	Aphea.Bio Belgium 2, 13, 14, 15 Biotech & biomanufacturing	The company develops microorganisms that are beneficial for plant growth for reducing synthetic fertiliser application and controlling fungal diseases.	They look for microorganisms that are naturally existing and have desirable characteristics for promoting plant development and health. The component bacteria are then isolated and grown into pure colonies from the ambient samples.	Reduction of fertiliser applications and fungal diseases' control are obtained by developing agricultural biologicals. Yield and health of crops can be enhanced by using agricultural biologicals.

(continued)

Table 15.1 (continued)

No	Company info	Value proposal (what?)	Value creation (how?)	Value capture
4	AT&T	It is a telecommunications company providing connectivity solutions and numerous other services.	AT&T Multi-Access Edge Computing (MEC) is a sophisticated traffic controller and data processor that is installed on-site. It connects certain devices to the cellular network and directs data flow according to business specifications. AT&T MEC may be designed to operate as a data bridge across cellular and landline networks since it is data smart.	The company moves computer capacity closer to the edge of the on-premises wireless network, allowing for near-real-time data processing. AT&T MEC, which is powered by AT&T 5G cellular connection, is a cost-effective approach to improve the capabilities of an existing private network via intelligent data routing and traffic prioritisation. Additionally, edge computing improvements and enhanced capacity to undertake data processing tasks, such as artificial intelligence (AI) and data analytics, cause devices to function more efficiently, resulting in lower GHG emissions.
	USA			
	7, 9, 12, 13			
	5G, edge computing			
5	Bamomas	Modern internet of things (IoT) connection and superior battery modelling enable a cloud-based intelligent battery management solution.	The system monitors the batteries' consumption, performance and condition in real time. Using IoT, real-time data may be accessible through the internet or a mobile dashboard. Regular reports, warnings and notifications are supplied, as well as system optimisation and big data. The solutions combine a contemporary cloud infrastructure with AI-powered battery modelling.	The company increases revenue by increasing battery longevity and guaranteeing operational continuity by adhering to maintenance recommendations.
	Finland			
	7, 13			
	AI, big data, energy storage, cloud computing, internet of things			
6	Biome Makers	It is an AI-based agricultural technology company that models soil functionality to improve arable soil productivity.	The company measures the crop quality and delivers agronomic insights such as disease risks, yield improvements and terroir characterisation to optimise farm operations.	Value is created by improving efficiency and sustainability and allowing the traceability of agricultural products by using AI.
	USA			
	2, 13, 15			
	AI			

7	Bloom Energy USA 7, 13 Hydrogen	It is a company that develops products that convert natural gas, biogas and/or hydrogen into electricity without a combustion process.	The fuel (hydrogen, natural gas or biogas) is converted into electricity by electrochemical methods without the combustion process by using the developed solid oxide fuel cell. The developed solid oxide fuel cell has a modular structure and can be scaled according to the customer's request.	Hydrogen technology provides more widespread use of renewable energy technologies. Thanks to this technology, increasing renewable energy systems provide an environmental value by reducing the amount of CO_2 emissions resulting from energy production.
8	BlueNalu USA 2, 13, 14 Cellular agriculture	It is a cell-cultured seafood company, satisfying the global appetite for seafood in a fresh, sustainable and humane way.	The company manufactures seafood using cellular agriculture technology and fish cells. Living cells are isolated from fish tissue, placed in culture media for proliferation, and are thus turned into seafood.	It can prevent overfishing, and thus, marine life can be restored and the amount of CO_2 absorption provided by the plants in the sea can be increased.
9	C-Combinator USA 13, 14, 15 Bioplastics, biotech & biomanufacturing	The company biomanufactures bioplastics, fertilisers and cosmetics from seaweed.	They've created a model for a cascading Sargassum seaweed biorefinery that can produce a variety of goods for agricultural, textile, personal care and cosmetics and bioplastics.	Environmental value is captured by reducing and balancing GHG emissions caused by fossil fuels or conventional plastics.
10	C:aire Austria 3, 11, 13 Biotech & biomanufacturing	It is a filter producer that eliminates unhealthy particles in the air with the help of bacteria.	Plants are grown on a statue-like filter. The unclean air that goes through the hole in the filter becomes purified and odourless, thanks to the degradation of unhealthy particles by bacteria that live in the soil under the plants.	The bacterium captures environmental value by degrading up to 98% of the target contaminants over time.

(continued)

Table 15.1 (continued)

No	Company info	Value proposal (what?)	Value creation (how?)	Value capture
11	Carbfix Iceland 13 Carbon capture & storage	It is a company that converts CO_2 into stone underground to give a natural and permanent storage option.	Carbon is naturally stored in large amounts in rocks. The company mimics and accelerates natural processes in which carbon dioxide is dissolved in water and reacts with reactive rock formations like basalts to generate stable minerals that serve as a long-term carbon sink. CO_2 is captured and permanently removed using the company's technique. The method offers a comprehensive carbon capture and injection solution, in which CO_2 dissolved in water – Sort of like sparkling water – Is pumped into the subsurface, where it combines with suitable rock formations to produce solid carbonate minerals via natural processes in approximately 2 years.	It provides safe underground storage of CO_2 with low investment costs. In this way, the amount of CO_2 released to nature is reduced.
12	Carbon Clean UK 9, 13 Carbon capture & storage	It is a company that provides solutions for carbon capture in a modular structure.	Classical carbon capture technology has been developed and modularised. Thanks to this modular structure, the application area of this technology has been expanded and it has been turned into a system suitable for small-medium or large scales, suitable for fast installation.	This technology provides value creation for carbon capture technology with its fast installation and scalability features. Thus, this technology becomes more accessible for many industries and users. By absorbing CO_2 from the atmosphere, it also reduces existing emission values, thus creating an environmental value.
13	Carbon Cure USA 13 Carbon capture & storage	It is a construction material firm that uses CO_2 in concrete production by using absorbed CO_2 from the atmosphere.	They source CO_2 from industrial production. The used CO_2 gas is collected, purified and distributed by established suppliers. The CO_2 delivered to the concrete plant is used and injected into the concrete mix. Injected CO_2 reacts with calcium ions, and as a result of this reaction, the concrete produced becomes stronger.	Concrete producers using this technology can produce high-quality, strong and more sustainable concrete. Concrete can be produced for green building applications, and value production is realised by developing the customer portfolio. Environmental value is captured by using CO_2 captured from the atmosphere.

#	Company	Description	Environmental value
14	Charm Industrial USA 13 Carbon capture & storage	The company uses plants to capture CO_2 from the atmosphere.	Their pyrolyzers break down biomass into bio-oil, a liquid rich in carbon but low in energy content. This bio-oil is then injected into EPA-regulated injection wells, where the bio-oil sinks and solidifies in place for permanent storage.	They convert biomass into a stable, carbon-rich liquid and then pump it deep underground. Environmental value is captured by removing CO_2 permanently from the atmosphere.
15	Checkerspot USA 13 Advanced materials, biotech & biomanufacturing	It is a biomanufactured advanced materials company for lightweight and high-performance applications.	Sustainable, superior performance advanced materials are obtained by designing and manufacturing convenient raw materials.	Environmental benefits are captured via low raw materials input, high durable, biodegradable and non-petroleum-based advanced materials.
16	Climate Data Hub USA 7, 13 AI, Datahubs	The platform provides rapid access to information about climate through data hubs and AI.	They're working on a dataset exploration and discovery tool that uses artificial intelligence, semantic search and intuitive visualisations to help technologists and data scientists quickly find relevant datasets, explore whitespace and see datasets in their geospatial, bioregional, social and economic contexts.	Environmental revenue is captured through the acceleration of the production of solutions for entrepreneurs and innovators via datahubs against the climate crisis.
17	Cling Systems Sweden 7, 13 Recycling	It is a global B2B market platform that enables the circularity of used rechargeable lithium batteries and their materials by connecting users, manufacturers and recyclers.	Their B2B used battery marketing platform connects workshops/battery dismantlers, recyclers, remanufacturers and OEMs. Used batteries are listed in the platform by used battery owners with all relevant data, verified buyers make bids to batteries or battery materials, the buyer, who proposed the highest price, places the order and cling ensures safe transport of sold goods.	The recyclability of lithium batteries is streamlined by establishing trusted and safe connections between users, manufacturers and recyclers. This way, enormous amounts of toxic lithium batteries can be safely transported and reused.

(continued)

Table 15.1 (continued)

No	Company info	Value proposal (what?)	Value creation (how?)	Value capture
18	Crowd4Climate Austria 7, 13 Crowdfunding	The platform provides a way to make an investment directly in solar energy and climate action projects.	People can select and invest in climate actions and sustainable energy projects that are selected based on ecological and social criteria by using the crowdfunding platform.	Environmental revenue through expansion of utilising renewable energy resources and decreasing greenhouse gases with a crowdfunding platform.
19	Deep Branch UK 13 Advanced materials	It is a carbon dioxide recycling company that uses microorganisms to convert clean CO_2 into high-quality products to enable global sustainable animal nutrition.	CO_2 is captured and converted into proton, a single-cell protein that has been optimised for animal feed. They collaborated with market-leading feed suppliers to create proton-based feeds.	Producing sustainable animal products with up to 60% fewer greenhouse gas emissions adds economic and environmental benefits. Proton-based feeds may be made with 90% reduced carbon intensity using locally obtained components.
20	Divergent Us 13 3D printing	It is a car manufacturing company that aims to develop fully 3D printed automobile production methods	Divergent adaptive production system (DAPS™) aims to transform auto manufacturing economic and environmental impact using a data-driven approach for designing and building vehicle structures	Part productions are optimised with 3D printing, reducing material waste, process length and complexity, so manufacturing operations consume less energy, leaving less carbon footprint. Automobile parts are also optimised to have the least material, reducing the weight of the vehicle and therefore carbon emission of the car
21	Emrod New Zealand 7, 11, 13 Advanced materials, wireless power transfer	They provide long-range wireless power transmission for electricity distribution companies.	Emrod's patented beam shaping, metamaterials and rectenna technologies are used to transport energy across vast distances utilising electromagnetic waves. Tele-energy is a technology that allows for safe, dependable and cost-effective long-range wireless energy transfer.	Monetary and business value is captured by reducing outage risks caused by wire damages and providing lower infrastructure deployment and maintenance time and costs. Remote locations in Africa, the Pacific Islands and other distant communities are supplied with affordable and sustainable energy to power schools and hospitals.
22	EnerVenue USA 7, 13 Energy storage, hydrogen	The company develops and provides metal-hydrogen batteries for heavy-duty energy storage problems with more affordable manufacturing solutions.	They propose affordable and durable energy storage solutions with metal-hydrogen batteries. Since the usage is limited with aerospace and military applications, the company aims to supply demilitarised, low-cost and safe product options.	Metal-hydrogen batteries' high cycle rate, durability, zero-maintenance cost, non-toxic and fire-free characteristics are both user and environmental friendly. Also, products are highly sustainable with a 99% recycling rate.

#	Company	Description	Outcome
23	Flash Forest Canada 13, 15 Drones	It is a drone reforestation company that reforests the earth through UAV hardware, aerial mapping software, automation and biological seed-pod technology.	Drones, rather than people, reforest post-harvest or post-wildfire regions, which speeds up the process by ten times. They work with botanists to find optimum planting locations and give vital follow-up data on ecosystem health using multispectral mapping UAV technology.	Automation of the reforestation process by using drones in planting seeds decreases the time required for the regrowth of trees.
24	Freight Farms USA 7, 13 AI, soilless farming	It is a vertical hydroponic farm company designed and built entirely inside a shipping container.	The greenery S includes a cutting-edge controlled environment system that allows you to have complete control over the elements. Farmhand software employs IoT-connected sensors and auto-updating video feeds to provide full insight into the farm operation, allowing you to grow food 365 days a year.	Revenue is generated by employing artificial intelligence to build containers with improved plant quality, fewer resources, higher yields and 365 days of production.
25	Full Cycle Bioplastics USA 9, 13 Advanced materials, biotech & biomanufacturing, recycling	It is a biotechnology company that tackles plastic pollution and climate change by upcycling organic matter into a compostable alternative to oil-based plastics.	They generate polyhydroxyalkanoate (PHA) biopolymers from organic waste using bacteria-powered alchemy. PHA is a naturally occurring polymer with material properties similar to those of petroleum-derived polymers, making it a versatile, non-toxic and biodegradable alternative to oil-based plastics.	Revenue is captured through producing polyhydroxyalkanoate and then the production of highly adaptable, non-toxic and compostable products.
26	General Electric USA 13 3D printing, carbon capture and storage	It is a long-established company that provides solutions in energy technologies and carbon capture and storage (CCS) methods.	The company develops an advanced carbon capturing and storage facility with the collaboration of UC Berkeley. CCS knowledge of GE and 3D printing studies of UC Berkeley labs will be merged to actualise a fully 3D printed carbon capture facility.	The high raw material efficiency of 3D printed part manufacturing will lower the carbon footprint and eliminate the costs of conventional parts manufacturing methods in CCS facility installations.

(continued)

Table 15.1 (continued)

No	Company info	Value proposal (what?)	Value creation (how?)	Value capture
27	GeoPard Agriculture Germany 2, 13, 15 AI	It is a company that offers smart farming solutions based on AI for the discovery of the potential of agricultural areas and the improvement of agricultural decisions.	The multispectral analysis performed by AI-based algorithms displays the degree of absorption or reflection based on identified wavelengths to offer data on plant vitality, allowing users to see which crops are thriving and which are failing. It enables near-real-time field monitoring and condition evaluation.	The efficiency and sustainability of farming in fertilising, seeding and irrigation applications are increased with the assistance of AI-based prescription maps, thus helping the fight against the rise of carbon footprint resulting from conventional farming methods.
28	Geyser Batteries Finland 7, 13 Energy storage	The firm offers water-based electrolyte-based high-power heavy-duty energy storage technologies.	The batteries are made from common chemical components and use water as an electrolyte solvent. Due to the use of aqueous electrolytes, the batteries take less energy to manufacture than li-ion batteries and superconductors. Their batteries are built with a bipolar architecture that ensures consistent current density on electrodes.	Due to the use of an aqueous electrolyte in the manufacturing process, the batteries are carbon-neutral. Furthermore, their batteries can withstand over a million charge-discharge cycles, making them resilient and, as a result, producing less waste over time.
29	h2arvester Netherlands 7, 13 Autonomous vehicles, hydrogen	It is a circular energy model company manufacturing autonomously moving solar panels and hydrogen storage technologies for local and/or agricultural economies.	The company provides a comprehensive energy system of sun harvesting with autonomously moving PV matrices over arable fields and hydrogen generation for energy storage in fully independent rural microgrids.	Farmers can utilise their agricultural fields for sun harvesting, gain independence from the grid and sell excess power to other locals. Also, they can store energy in hydrogen form for sunless and windless days.
30	Harvest London UK 12, 13 Soilless farming	It is a soilless and vertical farming firm with the goal of producing sustainable food ingredients.	They farm in a controlled environment, ensuring that crops receive just the right amount of light, water, nutrients and humidity to grow. This enables year-round production of vegetables at a scale not achievable in fields, without the use of pesticides and with waste reduction at every stage.	They work directly with customers to produce herbs and vegetables that are hard to source locally in the UK. Thanks to the controlled system, the production efficiency increases, thus minimising the climate effects.

31	IBM Energy Blockchain Labs USA 7, 13 Big data, blockchain	It is a climate change-focused tool for monitoring and purchasing carbon credits in the green energy markets.	They established an efficient, transparent platform using Blockchain technology that allows high-emission firms to track their carbon footprints and satisfy regulations by purchasing carbon credits from low emitters.	It enables enterprises to fulfil government-mandated carbon emission reduction targets in a more effective manner, and as the organisation's usage of IBM Blockchain technology evolves, carbon asset creation timeframes and costs will decrease even more.
32	Ligna Energy Sweden 7, 13 Energy storage	It is an organic material-based battery manufacturing company that uses organic electronic polymers and biopolymers, which are used throughout the lifetime of the energy storage and then recycled and burned as biofuel.	The firm has been working on innovative electrical energy storage technologies and products. Forest waste materials and organic polymers are combined with a water-based electrolyte to create the battery technology.	The reuse of raw battery materials and the manufacture of organic products are used to generate revenue.
33	Mango Materials USA 13, 14, 15 Bioplastics, biotech & biomanufacturing	They create a naturally occurring biopolymer from waste biogas (methane) that is cost-effective when compared to traditional oil-based polymers.	They use methane captured at landfills, wastewater treatment plants or agricultural facilities. Methanotroph bacteria consume methane to produce PHA material. Then, they utilise this PHA as a raw material for their biodegradable YOPP BioPlastic material.	When left in the wild, YOPP BioPlastics contribute to the natural carbon cycle. After being decomposed by microorganisms, they do not leave behind microfibers or microplastics. Approximately the maritime environment, YOPP fibres biodegrade in 6 weeks.
34	Meatable Netherlands 2, 12, 13 Cellular agriculture	It is a company that produces animal-free meat with cellular agriculture technology.	To carry out meat production, a sample is taken from a cow or pig before production is made. Natural muscle and fat growth processes are then copied, and these products are produced. Then, these two products are mixed to produce meat.	It decreases the time required to produce meat from 3 years to 1 week. Environmental value is created by significantly reducing the amount of CO_2 produced in the meat production process.
35	Omniply Canada 7, 13 Flexible electronics & wearables	It is a flexible electronic production company for solar cells, smart packaging, automotive and the like.	Thin-film solar cells with great performance may now be readily incorporated into ordinary things, thanks to advances in technology.	Value is created via cost-effective, easy to integrate, lightweight and controllable technologies

(continued)

Table 15.1 (continued)

No	Company info	Value proposal (what?)	Value creation (how?)	Value capture
36	Ossia USA 7, 13 Wireless power transfer	The company manufactures equipment that charge multiple devices at the same time with wireless electricity transfer.	Cota revolutionises wireless power by reliably supplying remote, focused energy far devices. It automatically charges numerous devices without the need for human involvement, allowing for a more efficient and totally wire-free experience.	The ability to charge multiple devices at a distance reduces the amount of cable materials (plastics and copper) used, resulting in declined carbon emissions.
37	Polybion Mexico 6, 9, 12, 13 Advanced materials, biotech & biomanufacturing, recycling	It is a BioTechnology company that aims to develop substitutable alternative leather and organic foam material by recycling agroindustrial food waste with microorganisms.	The company recycles agroindustrial food waste and manufactures biomaterials that could substitute animal and synthetic leather and commonly used petroleum-derived synthetic polystyrene and polyurethane foam materials. The product, Fungicel, is a biomaterial that is capable of replacing synthetic foams in applications such as insulation and packaging.	Value is captured through keeping the CO_2 footprint and impact on the environment low, such as being plastic-free and animal-free. The manufacturing process is carbon-neutral, cost-efficient and circular.
38	polySpectra USA 13 Advanced materials, 3D printing	It is a 3D printing company that develops advanced printer filaments and prints' durable parts for medical and dental devices, aerospace components, automotive parts, robotics effectors and electrical connectors.	COR alpha, the company's superior 3D printer filament material, outperforms conventional 3D materials in terms of durability, application sectors and carbon footprint. The material's resilience might lead to it being used instead of traditional moulded plastics, which require lengthy manufacturing procedures. In addition, traditional production processes use more energy and have a bigger carbon impact.	PolySpectra COR alpha has less carbon footprint than ABS, nylon 6 and epoxy thermoset 3D printing materials. The rugged structure of the material constitutes a considerable substitution for even high-end demanding industries such as aerospace, medical and robotics and dispenses the need for energy- and time-consuming, unclean conventional moulding and CNC methods.
39	Poseidon Singapore 13 AI, blockchain	It uses AI and blockchain to analyse the carbon footprint of any product or service.	This company analyses the carbon footprint of any goods or service using AI and Blockchain and then processes carbon credits in small fractions to rebalance the product or service at point of sale.	Carbon credits have the ability to overcome the emissions gap and reach the 2 °C limit by pricing CO_2 emissions as well as environmental and social costs of products.

40	Raise Green USA 7, 13 Crowdfunding	It is a crowdfunding platform that gets funding for clean energy and off-grid power need projects.	Crowdfunding allows citizens to participate in renewable energy solutions such as solar energy applications.	Revenue is created via decreasing energy use, enhancing access to locally grown food and increasing energy battery and solar panel systems by supporting the projects.
41	Riversimple UK 11, 13 Hydrogen	It is a vehicle manufacturing company that produces hydrogen-powered cars.	The rasa is a water-emitting electric automobile that runs on hydrogen rather than batteries. The hydrogen is injected into a fuel cell, where it reacts with oxygen in the air to generate electricity. This electricity powers small, lightweight electric motors in each of the car's four wheels, giving it four-wheel drive.	Hydrogen-powered cars create value by shortening the fuel filling process compared to battery-powered electric vehicles, which is another environmentally friendly option. In addition, it creates environmental value by preventing the emission of CO_2 caused by internal combustion engine vehicles.
42	Sadako Technologies Spain 12, 13, 14, 15 AI, recycling, robotics	It is an AI-based waste segregation start-up for the waste and recycling industry.	Artificial intelligence (AI) algorithms can "see" things in highly complicated waste streams in real time, allowing robots to sort and manage waste flows and waste treatment plants to recover more and better from urban rubbish. They're pushing the frontiers of what can be recycled cost-effectively with AI.	With the company's technology, its products can distinguish between pet trays and pet bottles and learn to do this even better with feedback and AI. Further value is captured by improving the efficiency of recycling plants, reducing plastic manufacturing.
43	Samsung SDI South Korea 7, 13 Advanced materials, energy storage	The battery systems production company offers efficient battery solutions by developing advanced polymer-based materials and different types of configurations like cylindrical and prismatic.	With improved, high capacity, lightweight, slimness and cell arrangement, the number of utilised cells in the battery system decreases. The inclusion of Bluetooth to the battery pack allows customers to check residual battery, remaining mileage and other data on their smartphones while riding bicycles.	With a single charge of the battery, the available distance is around 7–80 km, and a single recharge costs 100 won. Because this is a non-fossil fuel mode of transportation, it has no carbon impact. It also offers a number of benefits, including being environmentally friendly and cost-effective and improving one's health, as well as allowing users to avoid traffic congestion and get some exercise by pedalling.

(continued)

Table 15.1 (continued)

No	Company info	Value proposal (what?)	Value creation (how?)	Value capture
44	SEAB Energy UK 7, 13 Recycling	It is an on-site containerised energy producer from waste through anaerobic digestion.	Flexibacter technology is designed to generate energy from food waste. The containerised anaerobic digesters produced are fully automatic and can be monitored remotely. The wastes are transferred to the system, and biogas production is realised, thanks to the processes performed here. Biogas is then used to fuel a combined heat and power (CHP) system engine to provide electricity and heat. Thanks to its built-in pasteurisation process, this system can process various feedstocks in a completely safe and odourless environment.	Flexibuster technology ensures that the garbage produced in the field is eliminated and this garbage is converted into a value, namely, energy. In this way, while protecting the environment, renewable energy is produced cost-effectively.
45	Solar Foods Finland 2, 12, 13 Bioech & biomanufacturing, cellular agriculture	It is a food tech company that develops solutions for food production, such as developing a bioprocess for Solein, a natural protein.	The method, also known as a bioprocess, takes a single bacterium and grows it by fermenting it. The microorganism is fed in the same way that a plant is nourished, except instead of water and fertiliser; it is fed just air and electricity. This is a 20-fold increase in efficiency over photosynthesis (and 200 times more than meat). Unlike traditional protein manufacturing, Solein requires only a fraction of the water found in the air to generate 1 kg.	Environmental value is captured by reducing the land and water used for food production. In this way, CO_2 emissions due to food production are also reduced.

#	Company	Description	Impact
46	SP Group Singapore 7, 9, 11, 13 Digital twins	In Singapore and Australia, the company owns and manages electricity and gas transmission and distribution operations, as well as sustainable energy solutions in Singapore and China.	Grid digital twin is a virtual depiction of the actual power grid assets and network that uses real-time and historical data to function. It consists of two main models. The asset twin can remotely monitor and analyse asset health and performance, allowing for early detection of possible grid threats. As a result, informed decisions on renewal and maintenance plans may be made. The network twin uses modelling and simulations to assess the grid's response to new demands (such as electric car charging) and distributed energy resources (such as solar photovoltaics and energy storage devices).	Improved network planning analysis and remote monitoring of asset status are key advantages of the grid digital twin, which saves labour resources by reducing the need for lengthy physical inspections. Because the grid digital twin gives a more holistic view of the grid, it may help with infrastructure design for a variety of purposes, such as installing electric car chargers and connecting solar PV and energy storage systems. GHG emissions would gradually drop as renewable energy sources and electric vehicles become more commonly used.
47	Space4Good Netherlands 9, 13 AI	They develop AI-based detection, estimation and prediction algorithms for events or states such as deforestation and fire by using satellite image data.	The portfolio of the company comprises detecting and predicting illegal logging activity, fire detection and risk assessments and monitoring groundwater storage. Time-series satellite image data is fed to custom AI algorithms of the company. Any visible disaster related to climate change such as floods or droughts can be monitored and predicted as well.	Disaster prediction and analysis suggest precautions for governments and people, so lives and valuables are protected. Detecting and predicting illegal deforestation and wildfires can trigger counter actions.
48	Utilis Israel 6, 9, 13 AI, spatial computing	Utilis transforms satellite-based synthetic aperture radar (SAR) data into large-scale decision-making tools.	The firm finds essential subsurface water near infrastructure such as water and sewer pipelines, motorways and railways, dams and embankments using microwave imaging from a satellite to assist estimate failure risk, mitigate before a break and limit water loss or pollution.	Emissions of carbon dioxide have decreased by 14,500 metric tonnes. By 2021, there will have been 36,000+ leaks discovered and validated, saving 9200 million gallons of potable water and 21,800 MWH of energy each year.

(continued)

Table 15.1 (continued)

No	Company info	Value proposal (what?)	Value creation (how?)	Value capture
49	Verv Energy UK 7, 13 AI, internet of things	Customers may acquire real-time power and energy statistics, as well as product usage data, from this firm, which helps them manage their energy consumption and waste more effectively and move closer to their net zero goals.	The Verv isolator incorporates Verv's high-resolution predictive energy insights and predictive maintenance technology, which identifies irregularities in the AC's performance. All of the data may be supplied to any NetZero or carbon monitoring reporting tool via APIs or directly from the platform. Each proposal is based on data-driven decision-making for maximising the use of air conditioning across a facility. It detects symptoms of component fatigue and locates defects that are occurring or about to occur using real-time monitoring.	EnergyGenius AI shows energy insights of the performance of the unit, how efficiently it is running and where and how energy and carbon reductions can be made. Using real-time monitoring, Verv can identify signs of component fatigue and locate faults that are occurring or are about to occur, in the utmost detail.
50	VYTAL Germany 12, 13 Advanced materials, recycling	It is a company that provides a digital reusable packaging system that offers people an affordable and sustainable alternative to disposable and plastic waste.	Their reusable containers can be purchased and used from participating system partners without payment. (the reusable system is smart and digital).	Containers are made from durable and recyclable materials and they are reused for food and beverage deliveries that reduce the carbon footprint.
51	Wingtra Switzerland 9, 13, 15 Drones	They create high-precision vertical take-off and landing (VTOL) drones that collect survey-grade aerial data and sell them.	Their drone WingtraOne can work in narrow spaces and rough terrain due to its VTOL design. Also, the VTOL design protects the camera from being scratched as the camera does not contact the ground. It can map eight times faster than a conventional multi-copter drone since it can carry a camera with higher-resolution and fly at roughly two times higher altitude than a multi-copter drone.	It helps climate change researchers in places that are hard to be inspected by humans, such as Antarctica, to analyse a wide area in a relatively short time. Also, it is used to investigate wildfires' consequences from the air. Furthermore, it is used by infrastructure and construction planners to map the construction zone to shorten the time needed by conventional mapping methods.

52	Woodland Biofuels	It is an energy company that produces low-cost ethanol using waste biomass.	To produce ethanol, wastes such as forest industry wastes, agricultural waste, etc., which are continuous and abundant non-food wastes with low cost, are collected. Then, the gasification process of these collected organic wastes is carried out. After this process, catalysts and catalysed processes are carried out and distillation technology is used to purify the produced ethanol.	Value is created by decreasing the automobile fuel cost by using non-food waste biomass. Environmental value is captured by producing an alternative to gasoline.
	Canada			
	7, 13			
	Biotech & biomanufacturing, recycling			

References

S. Abdallah, I. Gough, V. Johnson, J. Ryan-Collins, C. Smith, The distribution of total greenhouse gas emissions by households in the UK, and some implications for social policy. Cent. Anal. Soc. Exclusion **59** (2011)

G.B. Bureau, What Is SDG 13 (Climate Action): How to Apply it in your Business. Green Bus. Bur. (2021) URL https://greenbusinessbureau.com/topics/sdg/sdg-13-climate-action/. Accessed 8.7.21

B.M. Campbell, J. Hansen, J. Rioux, C.M. Stirling, S. Twomlow, E. (Lini) Wollenberg, Urgent action to combat climate change and its impacts (SDG 13): Transforming agriculture and food systems. Curr. Opin. Environ. Sustain. Sustain. Sci. **34**, 13–20 (2018). https://doi.org/10.1016/j.cosust.2018.06.005

Die Welt, *Hochwasser aktuell: Zahl der Toten in Rheinland-Pfalz steigt auf 135 – Mindestens 184 Opfer durch Flut in Deutschland* (Welt, 2021)

F. Doni, A. Gasperini, J.T. Soares, F. Doni, A. Gasperini, J.T. Soares, Practical tools and mechanisms for SDG 13 implementation, in *SDG13 – Climate Action: Combating Climate Change and Its Impacts, Concise Guides to the United Nations sustainable Development Goals,* (Emerald Publishing Limited, 2020), pp. 37–72. https://doi.org/10.1108/978-1-78756-915-720201008

EPSRC, 2003. *Building Knowledge for a Changing Climate the Impacts of Climate Change on the Built Environment*

European Commission (ed.), *Energy Roadmap 2050, Energy* (Publications Office of the European Union, Luxembourg, 2012)

European Floods Are Latest Sign of a Global Warming Crisis, 2021. N. Y. Times

Fankhauser, S., 2019. The costs of adaptation – Fankhauser – 2010 – WIREs climate change – Wiley online library [WWW document]. URL https://wires.onlinelibrary.wiley.com/doi/full/10.1002/wcc.14. Accessed 9.9.21

Global Commission on Adaptation, *Adapt Now: A Global Call for Leadership on Climate Resilience* (World Resources Institute, Washington, DC, 2019). https://doi.org/10.1596/32362

IEA, *SDG7: Data and Projections [WWW Document]* (IEA, 2020) URL https://www.iea.org/reports/sdg7-data-and-projections. Accessed 7.31.21

Imperial College, L., *How Will Acting on Climate Change Affect the Economy? [WWW Document]* (Imp. Coll. Lond, 2021) URL http://www.imperial.ac.uk/grantham/publications/climate-change-faqs/how-will-acting-on-climate-change-affect-the-economy/. Accessed 9.9.21

R.J.T. Klein, Climate change 2014: Impacts. Adapt. Vulner. **34** (2014)

D. Magnay, *Siberia Battles Wildfires After Hottest and Driest June for 133 Years – Releasing High Amounts of Carbon into the Atmosphere* (Sky News, 2021)

D. Malerba, K.S. Wiebe, Analysing the effect of climate policies on poverty through employment channels. Environ. Res. Lett. **16**, 035013 (2021). https://doi.org/10.1088/1748-9326/abd3d3

E.A. Moallemi, S. Malekpour, M. Hadjikakou, R. Raven, K. Szetey, D. Ningrum, A. Dhiaulhaq, B.A. Bryan, Achieving the sustainable development goals requires transdisciplinary innovation at the local scale. One Earth **3**, 300–313 (2020). https://doi.org/10.1016/j.oneear.2020.08.006

G. Montt, K.S. Wiebe, M. Harsdorff, M. Simas, A. Bonnet, R. Wood, Does climate action destroy jobs? An assessment of the employment implications of the 2-degree goal. Int. Labour Rev. **157**, 519–556 (2018). https://doi.org/10.1111/ilr.12118

A. Olhoff, S. Bee, D. Puig, *The Adaptation Finance Gap Update – with Insights from the INDCs (Report), The Adaptation Finance Gap Update - with insights from the INDCs* (United Nations Environment Programme, 2015)

A. Onencan, B. Enserink, B. Van de Walle, J. Chelang'a, Coupling Nile Basin 2050 scenarios with the IPCC 2100 projections for climate-induced risk reduction. Procedia Eng., humanitarian technology: Science, systems and global impact 2016. HumTech2016 **159**, 357–365 (2016). https://doi.org/10.1016/j.proeng.2016.08.212

Organisation for Economic Co-operation and Development (ed.), *Investing in Climate, Investing in Growth* (OECD, Paris, 2017)

SDSN, Goal 13, in *Take Urgent Action to Combat Climate Change and its Impacts – Indicators and a Monitoring Framework [WWW Document],* (Sustain. Dev. Solut. Netw, 2021) URL https://indicators.report/goals/goal-13/. Accessed 8.9.21

S. Solomon, Intergovernmental Panel on Climate Change, Intergovernmental Panel on Climate Change (eds.), *Climate change 2007: The physical science basis: Contribution of working group I to the fourth assessment report of the intergovernmental panel on climate change* (Cambridge University Press, Cambridge/New York, 2007)

UNDP 2021. [WWW Document] https://sdgs.un.org/goals/goal13

UNESCAP, 2019. *SDG13 Goal Profile | ESCAP* [WWW Document]. URL https://www.unescap.org/kp/2019/sdg13-goal-profile-0. Accessed 10.24.21

UNFCCC, 2021. What Is the United Nations Framework Convention on Climate Change? | UNFCCC [WWW Document]. URL https://unfccc.int/process-and-meetings/the-convention/what-is-the-united-nations-framework-convention-on-climate-change. Accessed 10.24.21

United Nations, 2021a. Goal 13 | Department of Economic and Social Affairs [WWW Document]. URL https://sdgs.un.org/goals/goal13. Accessed 10.24.21

United Nations, *Take Urgent Action to Combat Climate Change and its Impacts* (United Nations, 2021b)

United Nations, *The Sustainable Development Goals Report: 2021 [WWW Document]* (U. N, 2021c) URL https://unstats.un.org/sdgs/report/2021/goal-13/. Accessed 7.31.21

R.M. Wise, I. Fazey, M. Stafford Smith, S.E. Park, H. Eakin, E.R.M. Archer Van Garderen, B. Campbell, Reconceptualising adaptation to climate change as part of pathways of change and response. Glob. Environ. Change **28**, 325–336 (2014). https://doi.org/10.1016/j.gloenvcha.2013.12.002

World Bank, *Economics of Adaptation to Climate Change: Synthesis Report* (World Bank, Washington, DC, 2010) https://doi.org/10/01/16436675/economics-adaptation-climate-change-synthesis-report

SDG-14: Life Below Water

16

Abstract

Global systems and processes that assure the supply of rainwater, drinking water and oxygen are regulated by oceanic temperature chemistry, currents and life. Pollution, diminished fisheries and the loss of coastal habitats all have negative impacts on the ocean's sustainability. Such activities have severely impacted around 40% of the world's oceans. SDG-14, Life Below Water, aims to conserve marine ecosystems by establishing regulations for removing pollutants from the sea, decreasing sea acidification and regulating the fishing sector to ensure sustainable fishing. As a result, the major incentive for this goal is to protect and utilise marine ecosystem services sustainably. This chapter presents the business models of 36 companies and use cases that employ emerging technologies and create value in SDG-14. We should highlight that one use case can be related to more than one SDG and it can make use of multiple emerging technologies.

Keywords

Sustainable development goals · Business models · Life below water · Sustainability.

Oceans covering nearly 3/4 of the planet's surface contain 97% of the planet's water and account for 99% of living space by surface area. Nonetheless, nearly 95% of the ocean remains unknown, 91% of oceanic species are still unclassified, and a large number of fish species (1851 as of 2010) are threatened with extinction. Even though human beings live mostly on continents, they are heavily reliant on the oceans. Oceanic processes and biodiversity generate various ecological functions, enabling many species to live on the Earth. Moreover, coastal and marine resources contribute a total of US$28 trillion to the global economy on an annual basis. Noticeably, oceans absorb approximately 40% of the carbon dioxide emitted by humans, thereby mitigating the effects of global warming. Moreover, the oceans are the world's primary protein resources, with nearly 3 billion people relying on them. The estimated market value of such marine and coastal resources and industries is approximately US$3 trillion per year. On the other hand, the quality and biodiversity of marine ecosystems are rapidly declining. Because of primarily human-caused activities, it may be too late to save the oceans if action is not taken quickly. As a result, countries must take precautionary measures to protect marine ecosystems and improve the quality of biodiversity beneath the sea.

S. Küfeoğlu, *Emerging Technologies*, Sustainable Development Goals Series,
https://doi.org/10.1007/978-3-031-07127-0_16

Global systems and processes that assure the supply of rainwater, drinking water and oxygen are regulated by oceanic temperature chemistry, currents and life. Pollution, diminished fisheries and the loss of coastal habitats all have negative impacts on the ocean's sustainability. Such activities have severely impacted around 40% of the world's oceans. There is also the economic impact. Ocean fisheries are now generating $58 billion less per year than they could be due to unregulated fishing (Pandey et al. 2021). Many of today's and future concerns, such as food security and climate change, as well as the availability of energy and natural resources, are recognised as dependent on the oceans (Franco et al. 2020). By increasing fish catches, income and improved health are two ways that Marine Protected Areas help ease poverty. They also contribute to gender equality because women own many small-scale fisheries. The maritime environment is also home to a wide variety of magnificent species, ranging from single-celled organisms to the world's largest animal – the blue whale. Moreover, coral reefs, which are among the world's most diversified ecosystems, also lie within the oceans (Nicklin and Cornwell 2020). The utilisation of the sea and its resources for long-term economic development (blue economy), which contributes to today's and tomorrow's prosperity, is growing rapidly; however, the oceans are under stress. They are already overexploited, contaminated and threatened by global warming (Franco et al. 2020). Debris levels in the world's oceans are rising, causing a severe environmental and economic threat. Entanglement or swallowing of trash by organisms negatively influences biodiversity, as it can kill or prevent species from breeding (Nicklin and Cornwell 2020). The ocean has absorbed a considerable amount of carbon dioxide as carbon emissions have gone up significantly, causing acidification. Rising sea levels and temperatures are causing biodiversity and habitat loss and changes in the composition of fish stocks. Furthermore, approximately 20% of the world's coral reefs have been seriously damaged, with no signs of recovery. Due to human pressures, approximately 24% of the remaining reefs are in impending danger of col-

lapsing, with another 26% facing a longer-term risk of collapsing (Nicklin and Cornwell 2020). In addition to the coral reef problem, overfishing and decreased fish stocks are threatening future ocean development in many parts of the world (Franco et al. 2020). The value of lost economic gains from the fishing industry is estimated to be over $50 billion per year. Poor ocean management practices are estimated to cost the global economy at least US$200 billion each year, according to the United Nations Environment Programme. Climate change will increase the cost of ocean damage by an additional US$322 billion per year by 2050 if no mitigating measures are taken (Nicklin and Cornwell 2020).

SDG-14 aims to conserve marine ecosystems by establishing regulations for removing pollutants from the sea, decreasing sea acidification and regulating the fishing sector to ensure sustainable fishing. As a result, the major incentive for this goal is to protect and utilise marine ecosystem services sustainably. SDG-14 also intends to restrict fishery subsidies that lead to overfishing in specific areas. Several fish species are being rapidly depleted as a result of uncontrolled and subsidised fishing. There is greater competition in markets with limited resources to catch as many fish as possible. Therefore, member states must develop and implement legislation to restrict fishing operations to ensure the fish stock's long-term viability. States must coordinate to control fishing operations in regions where the coast is shared by more than one state. Pollution from land-based activities poses a danger to coastal life. If pollution from the land is poured into the sea without being treated, it will produce eutrophication, characterised by excessive algae development. While eutrophication may appear to be a natural process, it deprives the water of oxygen, which breaks the fish bio-chain. As a result, eutrophication may result in the extinction of living species in the coastal area. Another major issue is ocean acidification. The origins and consequences of ocean acidification are still a source of scientific dispute. However, it is impossible to predict exactly how the ocean food chain would be affected in terms of life. What we do know is that some micro-species are more susceptible than others. As a result, the future of these micro-species may be jeopardised. We need to develop immediate ways to mitigate abnormal levels of acidity as the rate of ocean acidification rises. According to the United Nations, at least 10% of marine and coastal habitats should be legally protected (Gulseven 2020). The long-term benefits largely compensate for the short-term costs of acting. However, while progress is being made, substantial obstacles remain. According to the Convention on Biological Diversity, scaled-up efforts to sustain the global ocean need a one-time public expenditure of US$ 32 billion and ongoing expenses of US$ 21 billion each year. Apart from the need for considerable multi-year fundraising to reach the level of ambition, the ongoing negative aspect of climate change; inadequate industrial, agricultural and household waste management; chemical and plastic pollution; corruption; and a lack of effective governance activities, the alarming rate of biodiversity loss in ecosystems and wilful ignorance of scientific evidence must all be resolved (Nicklin and Cornwell 2020). Imagine how powerful it would be if we collectively harnessed "the ocean in us" as a driving force to increase ocean ambition and enhance ocean action as our planet's "Blue Lung" as we need to see the nexus between the ocean and sustainable human, social and economic development (Nicklin and Cornwell 2020). Figure 16.1 summarises the targets and sub-targets of SDG-14 for 2030, which the United Nations present.

The oceans encompass more than 70% of the Earth's crust. Oceans create more than 50% of the planet's oxygen. They help regulate the climate and offer vital habitats for a wide range of marine and coastal organisms. Oceans also contribute to the global economy and regional life by serving as a means of transport and trading (Kan et al. 2020). Marine fisheries employ 57 million people worldwide and are the major protein source for more than half of the population in LDCs, with over 3 billion people relying on oceans and terrestrial biodiversity for a living. That's why the health of the oceans, the world's water resources and the life below water is important as a matter of being an economic resource

and vital to many of the world's population. The yearly market value of coastal and marine sources and businesses is estimated to be $3 hundred billion or approximately equal to 5% of total global GDP. Nonetheless, human activities, such as pollution, reduced fisheries and coastal habitat loss, are harming up to 40% of the seas. The oceans are the planet's largest source of protein, with over 3 billion human beings dependent on them as their main resource. Nonetheless, the proportion of stocks fished at unsustainable levels was 28.8% in 2011: a slight reduction from the previous high of 32.5% in 2008 but still cause for concern. Fisheries, food, aquaculture and the tourism sector are particularly dependent on clean oceans and coastal areas. They play an important role in dealing with problems to the well-being of our oceans and coastal regions. Notwithstanding, if natural coastal flood protection is destroyed or food security is jeopardised, all sectors may suffer, and all may contribute to reducing marine pollution or the maintenance of sustainable fisheries (PwC 2021).

The ocean is a massive economic resource. Ninety percent of the planet's commodities are traded throughout the seas. Millions of people operate in fishery and mariculture, shipping and docks, tourist industry, offshore energy, medicines and cosmetics, which all depend on marine-related sources (Stuchtey et al. 2021). The ocean food industry itself supports up to 237 million employment, encompassing fishery, mariculture and processing. Thousands of people operate in other ocean industries, such as shipping, docks, energy and the tourist industry, and many others are indirectly related to the marine sector and economy. Coastal ecosystems protect millions of people, foster wildlife, detoxify pollutants that run off the land and serve as nursery grounds for fisheries, boosting food supply and giving jobs. They additionally serve as a source of money. Coral reefs on their own generate $11.5 billion worldwide tourism each year, supporting more than 100 nations and giving food and employment to the locals (Stuchtey et al. 2021).

Investing in a healthy ocean economy benefits more than simply the ocean. They are a fantastic business prospect. Putting money $2.8 hundred

Target 14.1
Reduce Marine Pollution

By 2025, prevent and significantly reduce marine pollution of all kinds, in particular from land-based activities, including marine debris and nutrient pollution.

Target 14.2
Manage and Protect Marine and Coastal Ecosystems

By 2020, sustainably manage and protect marine and coastal ecosystems to avoid significant adverse impacts, including by strengthening their resilience, and take action for their restoration in order to achieve healthy and productive oceans.

Target 14.3
Minimize the Impacts of Ocean Acidification

Minimize and address the impacts of ocean acidification, including through enhanced scientific cooperation at all levels.

Target 14.4
Effectively Regulate Overfishing

By 2020, effectively regulate harvesting and end overfishing, illegal, unreported and unregulated fishing and destructive fishing practices and implement science based management plans, in order to restore fish stocks in the shortest time feasible, at least to levels that can produce maximum sustainable yield as determined by their biological characteristics.

Target 14.5
Conserve Coastal and Marine Areas

By 2020, conserve at least 10 percent of coastal and marine areas, consistent with national and international law and based on the best available scientific information.

Target 14.6
Prohibit Certain Forms of Fisheries

By 2020, prohibit certain forms of fisheries subsidies which contribute to overcapacity and overfishing, eliminate subsidies that contribute to illegal, unreported and unregulated fishing and refrain from introducing new such subsidies, recognizing that appropriate and effective special and differential treatment for developing and least developed countries should be an integral part of the World Trade Organization fisheries subsidies negotiation2

Target 14.7
Increase the Economic Benefits to Small Island Developing Benefits

By 2030, increase the economic benefits to small island developing States and least developed countries from the sustainable use of marine resources, including through sustainable management of fisheries, aquaculture and tourism.

Target 14.a
Develop Marine Technology

Increase scientific knowledge, develop research capacity and transfer marine technology, taking into account the Intergovernmental Oceanographic Commission Criteria and Guidelines on the Transfer of Marine Technology, in order to improve ocean health and to enhance the contribution of marine biodiversity to the development of developing countries, in particular small island developing States and least developed countries

Target 14.b
Provide Access for Small-Scale Fishers

Provide access for small-scale artisanal fishers to marine resources and markets

Target 14.c
Enhance the Sustainable Use of Oceans and Their Resources

Enhance the conservation and sustainable use of oceans and their resources by implementing international law as reflected in the United Nations Convention on the Law of Taking into account ongoing World Trade Organization negotiations, the Doha Development Agenda and the Hong Kong ministerial mandate. the Sea, which provides the legal framework for the conservation and sustainable use of oceans and their resources, as recalled in paragraph 158 of "The future we want".

Fig. 16.1   Targets of SDG-14. (UNDP 2021)

billion presently within only four ocean-based solutions – offshore wind production, sustainable ocean-based production of food, international logistics, decarbonisation and mangrove restoration and production – would yield a real earning of $15.5 hundred billion by 2050, a benefit-cost ratio of more than 5 (Stuchtey et al. 2021). One single source of stress, like overfishing event pollution, can cause significant harm. Moreover, single stressors regionally reinforce each other, with devastating effects on the ecosystem. If nothing is done, these issues might cost the world economy more than $400 billion per year by 2050. The yearly cost might reach $2 trillion by 2100 (Stuchtey et al. 2021).

The "Blue Economy (BE)" or "Oceans/Marine Economy" has been extensively supported by a variety of relevant stakeholders in recent years as a paradigm or strategy for protecting the oceans and water sources. The notion of BE arose from the 2012 United Nations Conference on Sustainable Development in Rio de Janeiro. The concept "Blue Economy" is being used in a variety of contexts, and related topics such as "ocean economy" or "marine economy" are used without clarity (Lee et al. 2020).

Approximately 820 million people rely on fisheries for income, both directly and indirectly, to ensure food security (Steinbach et al. 2017). Furthermore, fisheries offer 20% of the protein consumed by more than three billion people. Fish accounts for 50–60% of total dietary protein in several regions of our planet, including South Asia, Southeast Asia, West Africa and SIDS (small island developing states). Over the last five decades, the worldwide fishing sector has experienced tremendous growth. The yearly fish caught globally increased from approximately 20 metric tonnes in 1950 to more than 90 metric tonnes in 2014. Annual per capita fish intake increased from roughly 10 kilogrammes in the 1960s to nearly 20 kilogrammes in 2013. Production and consumption of fisheries have increased at a cost. According to the FAO (Food and Agriculture Organization), solely 11% of world fish stocks were under-fished in 2013. On the other hand, 58.1% were completely fished,

and 31.4% were fished at biologically unsustainable levels. With a projected worldwide population of 9 billion people by 2050, overfishing has major consequences for the overall health of marine ecosystems, poverty reduction and food security. Millions of livelihoods might be lost if threats to oceans and the services they provide are not addressed, and many people could lose access to a food staple that they rely on to survive (Steinbach et al. 2017). Effects of the climate changes on the oceans, such as sea-level rise, storms and consequences on fisheries, are expected to cost between US 600 million and US 2 trillion dollars by 2100. SIDS and coastal communities in developing countries are threatened by climate change, which poses a threat to their well-being and survival. For example, the climate crisis is expected to boost the intensity and frequency of disasters such as floods and hurricanes, which has already started to cost several small-island developing countries more than 20% of their GDP (gross domestic product). Almost a quarter of the world population is particularly vulnerable since at least 20% are still categorised as least developed countries (LDCs) (Recuero Virto 2018).

SDG-14 advocates for the sustainable use and conservation of marine resources, oceans and seas and achieving sustainable development (Steinbach et al. 2017). SDG-14 indicators rely on present short-term relationships. They will undoubtedly benefit society in the long run, while building marine protected areas, reducing harmful fishing subsidies and ending overfishing may incur short-term costs for individuals. Through suitable mechanisms, policies can be made to reduce these costs. As a result, these trade-offs may be spurious, and achieving decent work and economic growth does not always require giving up aquatic life (Gulseven 2020).

For the future, new data science and engineering approaches offer optimism that data will be acquired for a certain purpose in the first place but subsequently used for various assessments other than the original one, following the principle of "collect once, use many times". By combining data from various old and new sources, researchers will be able to use artificial intelligence techniques, such as machine learning (ML) to acquire fresh perspectives on ocean dynamics. New algorithms have become an effective and efficient tool for accurately analysing oceanographic and environmental datasets. Prediction of ocean weather and climate, habitat modelling, distribution, species detection, coastal water observation, marine resource management, identification of oil spills and pollution and wave modelling are the key applications of machine learning in oceanography. Nonetheless, future advances are projected to expand the number of users and lead to their integration into daily data administration (UNESCO 2020).

16.1   Companies and Use Cases

Table 16.1 presents the business models of 36 companies and use cases that employ emerging technologies and create value in SDG-14. We should highlight that one use case can be related to more than one SDG and it can make use of multiple emerging technologies. In the left column, we present the company name, the origin country, related SDGs and emerging technologies that are included. The companies and use cases are listed alphabetically.[1]

---

[1] For reference, you may click on the hyperlinks on the company names or follow the websites here (Accessed Online – 2.1.2022):

http://apium.com/; http://usvcompany.com/; http://www.aquabyte.ai/; https://bureo.co/; https://calwave.energy/; https://digitaltwinmarine.com/; https://dronesolutionservices.com/wasteshark; https://econcretetech.com/; https://marinelabs.io/; https://monitorfish.com/; https://oceanium.world/; https://seakura.co.il/en/; https://shellcatch.com/; https://sinay.ai/en/; https://symbytech.com/; https://theoceancleanup.com/; https://umitron.com/en/index.html; https://upstream.tech/hydroforecast; https://www.akualogix.com/; https://www.aquaponicsiberia.com/?lang=en; https://www.bioceanor.com/en; https://www.blueoceangear.com/; https://www.blueplanetecosystems.com; https://www.cageeye.com/nb; https://www.ellipsis.earth/; https://www.icoteq.com/; https://www.oceandiagnostics.com/; https://www.projectbb.org/; https://www.ranmarine.io/products/datashark/; https://www.reef.support/; https://www.saildrone.com/; https://www.sofarocean.com/; https://www.vortex-io.fr/; https://www.we4sea.com/; https://www.whaleseeker.com/; https://xpertsea.com/

Table 16.1 Companies and use cases in SDG-14

No	Company info	Value proposal (what?)	Value creation (how?)	Value capture
1	Apium USA 14 Autonomous vehicles, internet of things	It is a software developer that provides swarm software for any autonomous device.	The swarm software they provide enables more than one autonomous device to recognise each other's movements and work as a whole system.	A high number of autonomous devices works like one device, which in turn saves time and money during the processes.
2	Aquabyte Norway 14 AI	It is a company that applies machine learning and computer vision to make aquaculture more efficient with less environmental pollution.	The company's solution consists of a camera module and software platform. The camera is used for detecting sea lice, biomass estimation and appetite detection. The feed optimisation process is also conducted with data that is processed by machine learning. All data can be monitored on the software platform.	Revenue creation is achieved by making the process easier and more sustainable with higher profits. Machine learning provides greater accuracy of the information on which to make decisions.
3	Aquaponis Iberia Portugal 14 Biotech & biomanufacturing, recycling	With the help of biomanufacturing and recycling, they created productive ecosystems that utilise sustainable technologies. They aim to become pioneers for spreading aquaculture awareness and technologies.	They created aquaponics, a product that combines aquaculture and hydroponics, or plant production on water. It's made from fish faeces and gives ocean plants nourishment. As a result, the water is cleaned, and the fish have a healthier habitat to thrive sustainably.	It creates an environmental impact by continuously recycling, preserving marine biodiversity and installing their specialised food production technique that compounds aquaculture and hydroponics.
4	Bioceanor France 14 Big data, cloud computing, internet of things	It is a company that produces underwater weather stations for real-time and predictive water quality monitoring in pounds, open-sea cages, rivers, lakes and beaches.	The company offers two different stations for customers' needs. The first one is a fixed module and the other one is floating. Both modules are IoT-based, can detect up to 14 different parameters, and are autonomous in energy, thanks to solar panels. As a communication protocol, devices use LoRaWan technology to transfer gathered data from modules to servers and clouds.	The utilisation of modules helps the fishery become more efficient by generating revenue via savings. Modules may also be used to monitor coral reefs, which adds to the biosphere's environmental value.

(continued)

Table 16.1 (continued)

No	Company info	Value proposal (what?)	Value creation (how?)	Value capture
5	Blue Ocean Gear USA 14 Internet of things	It is a company that provides real-time tracking and monitoring for deployed fishing gear by using IoT. They have created internet-connected buoys that can track gear in the marine environment, including detecting and locating lost or entangled fishing gear.	From IoT devices, they capture data such as temperature, depth and acceleration that can be later assessed and analysed via their cloud-based web interface or API. The customisable alert system helps clients easily detect the problem on the gears.	Their IoT devices enable faster and more efficient fishing to help fisherpeople save money while protecting the ocean environment.
6	Blue Planet Ecosystems Austria 14 AI, big data, internet of things	The company engages in decoupling fish protein from traditional agricultural supply chains to allow local and environmentally friendly fish production through land-based automated recirculating agriculture.	Through the creation of a natural ecosystem in three-layered fish tanks which are capable of turning sunlight into a food chain for fish, the production is decoupled from marine sources. Water quality sensors, computer vision and machine learning help maintain the optimal physical, environmental and biological conditions of organisms in the ecosystem, thereby constantly informing the process of production.	By decoupling the production of animal protein from the ocean and the agriculture itself, the dependence on fishmeal, soybean and rain is removed. This enables sea-food production to become more sustainable as it is less resource-intensive and independent of the natural ecosystems and potentially limits the impact caused by fishing activities on marine ecosystems.
7	Bureo USA 14 Recycling	It is a company that collects discarded fishing nets from oceans and recycles these nets and produces sunglasses, skateboards and hats using these recycled plastics.	Discarded fishing nets are collected. After cleaning and separating by the material process, these nets are shredded and melted into recycled pellets. Then these recycled pellets are formed into products such as sunglasses, hats, skateboards, etc..	More than 3.4 million lbs. of discarded fishing nets are removed from oceans. Collected nets are recycled and our oceans are protected.
8	CageEye Norway 2, 14 AI	The company specialises in measuring and understanding fish behaviour to optimise feeding in sustainable aquatic food production.	The company uses hydroacoustic technology to observe fish behaviour and to objectively measure fish appetite. The technology then decides on the feeding amount via optimising algorithms calculated with deep learning.	It allows you to follow the fish's natural behavioural patterns and routines and carry out feeding in line with the fish's appetite so that the feeding is as efficient as possible.

(continued)

Table 16.1 (continued)

No	Company info	Value proposal (what?)	Value creation (how?)	Value capture
9	CalWave USA 7, 11, 14 Digital twins, internet of things	They enable the ocean internet of things (IoT) by providing power storing, shipping, deploying and accessing data offshore.	Ocean waves have a 20–60 times higher energy density and are more predictable. The platform uses edge computing to construct a digital duplicate of wave farms, estimates the energy of ocean waves and converts that energy to electricity, providing clean, dependable and local power to coastal communities.	Wave energy has the potential to satisfy 20–30% of global energy demand, and forecasts show that ocean energy can reduce/sequester the equivalent of 1.38 gigatons of CO_2. Assuming wave energy development follows a trajectory similar to offshore wind, production can increase to meet our target of displacing 500 million tonnes of GHG equivalent annually by 2050.
11	Digital Twin Marine USA 14 Digital twins	It is a system provider that creates a digital twin of any vessel and provides necessary training to the user	Using sensors and proper devices, the team creates a digital twin of the system and provides customers with training to use and understand the digital twins program.	Digital twins provide better management of any project, and a well-maintained vessel has less chance of crashing in the sea, which is catastrophic for the ocean wildlife.
12	Drone Solution Services Singapore 14 Drones	Their product "waste shark" is a drone that removes plastic waste from beneath the water's surface. It also gathers real-time data and produces no greenhouse emissions.	Autonomous real-time data harvesting, GPS tagging for accurate measurement, online live-data portal access and data sensors enable the drones to be easily monitored.	It creates value by conserving our waters by removing plastic waste and biomass. Moreover, it produces no emissions.
13	ECOncrete Israel 14 Advanced materials	They created a concrete production technology that meets industry standards while also providing biological, ecological and economic advantages.	They created a product to lessen the impact on the ocean environment by providing a solution for urban, coastal and marine infrastructure. Its products are bio-enhanced concrete additives that assist the ecosystem in thriving and prospering by replacing the ordinarily drab waterfront and coastline with blue-green infrastructure.	With their concrete production technology, they offer shoreline protection for erosion control, flood protection and coastal armouring; waterfront infrastructure for ports, marinas and urban waterfronts; and offshore applications for cable armouring, foundations and scour protection.

(continued)

Table 16.1   (continued)

No	Company info	Value proposal (what?)	Value creation (how?)	Value capture
14	Ellipsis Earth UK 14, 15 AI, drones	The company uses drone imagery and machine learning to identify and track plastic pollution.	Drones take several thousand images, and artificial intelligence (AI) software then blends multiple photos into a master image. It then provides a global heatmap of plastic waste that can be filtered and overlaid against existing data sets and compared geographically and over time. This global-scale resource is continually updated with new imagery from the worldwide data libraryl	Environmental value is captured through the reduction of plastic waste where specific software can identify plastic with 93% accuracy and 95% certaintyl
15	Icoteq UK 6, 14 Cloud computing, internet of things	By providing tracking systems using IoT sensors, they monitor aquatic animals and the path of discarded plastic. They also develop cloud repositories and visualise the captured data using Amazon web Services (AWS) or Google cloud tools.	They use an array of internet-connected environmental sensors such as temperature, pressure and humidity sensors. Data gathered from sensors are then transferred to cloud platforms, and then these data are processed to visualisations and cloud repositories.	By designing a connected tag to track aquatic animals such as green sea turtles at a meagre cost, they monitor endangered species. They can also visualise these data in their cloud platform.
16	MarineLabs Canada 14 Internet of things	It is a company that aims to optimise how society interacts and adapts to the changing ocean. Their IoT devices enable real-time coastal intelligence which quantifies marine hazards and optimises clients' decisions.	They provide data-as-a-service (DaaS) from fleets of their cloud-connected buoys for groups who need to know marine conditions. With MarineLabs, marine operators complement existing marine domain awareness systems with real-time data from multiple locations.	Coastal engineers quantify the marine domain around projects, gain certainty in design loads assessments and minimise the risk of design failures while meeting project budgets. Researchers measure their parts at higher resolution, reduce marine operations costs and achieve research goals.
17	MonitorFish Germany 2, 14 AI, cloud computing, healthcare analytics	It is an AI-based fish health diagnosis company.	It is a cloud-based fish welfare monitoring system. It offers real-time analysis of critical fish growth parameters and water parameters to detect any abnormal development among the fishes, and AI-based decisions suggest recommended actions to the fish farmer.	It automates fish health analytics so that it ensures optimal living conditions. This results in the reduction of fish mortality in ponds so that the farmers' investments are secure.

(continued)

Table 16.1 (continued)

No	Company info	Value proposal (what?)	Value creation (how?)	Value capture
18	Ocean Diagnostics Canada 3, 13, 14 Big data, internet of things	The company collects microplastic data by IoT sensors and creates a database for microplastics in oceans.	In situ sensors can be deployed in the desired area, and that sensor displays real-time microplastic data. Gathered data is stored in databases and is available for any customer.	Tracking microplastic data is critical for any blue action, and making it automated by deployed sensors and storing the data saves effort for anyone interested in taking blue steps.
19	Oceanium UK 12, 13, 14 Biotech & biomanufacturing	It is a seaweed farm company that grows seaweed by using biotech to make it efficient and supply its product to various industries.	They use biotech and new fertilising techniques to improve the quality of their seaweed. The crop they produce becomes raw material for various sectors such as pharmacy and the food industry.	Sustainable seaweed farming and seaweed-based products offer a systemic answer for dealing with climate change, ensuring food security and providing "blue" jobs and alternative livelihoods for fishing communities.
20	Project.BB Netherlands 14 AI, autonomous vehicles, drones	It is a company that produces autonomous robots programmed with AI to efficiently map and collect small pieces of litter on beaches and parks.	The robot uses a convolutional neural network (CNN), a self-learning algorithm that can make connections in terms of AI. A good and large amount of training data is provided to improve the algorithm. It also uses a probabilistic algorithm to indicate litter precisely.	It creates environmental value by cleaning beaches and parks from litter, especially cigarette butts which can pollute the water extensively.
10	Ranmarine Netherlands 14 Autonomous vehicles	The company produces aqua drones that fight plastic pollution and gather environmental data.	Autonomous surface vessels (ASVs) named DataShark are designed to operate in a path that requires waste or data collection. In addition, the drones use GPS routes to navigate to the desired areas and to return home.	Environmental value is captured by efficiently removing 500 kilograms of waste per day from all types of waterways.
21	Reef Support Netherlands 14 AI, big data	Reef.Io, the company's service, is a digital monitoring and predictive maintenance platform that uses AI, satellite imagery and user-driven data to support coral reef health, sustainable and profitable aquaculture and seafood management.	Users upload images of the coral reefs to the web application of the company. The web application uses benthic data collection through machine learning algorithms and delivers the analysed results, categorised by area, time and biodiversity indicators.	It creates environmental value by helping protect coral reefs.

(continued)

Table 16.1 (continued)

No	Company info	Value proposal (what?)	Value creation (how?)	Value capture
22	Saildrone USA 14 AI, autonomous vehicles, drones	The company provides systems like marine security, carbon cycling and weather forecasting which utilises AI and autonomous vehicles.	Saildrone vehicles gather data and inform the users about maritime domain awareness, ocean mapping and ocean data. Advanced maritime intelligence is used in maritime domain awareness while autonomous single- and multibeam data collection is used in ocean mapping.	It offers not only cost-effective solutions to its customers but also creates value for marine environment protection by operating in low or no carbon footprint approaches.
23	Seakura Israel 14 Biotech & biomanufacturing	It is a company that created a way of farming land-based seaweed in order to develop a sustainable system that provides nutritious options without damaging the ocean.	Seawater is piped into enormous pools, where seaweed grows before being released to the sea. Since the pools are placed along the coastline, they help protect the marine ecosystem. The seaweed is planted in organic pools of nutrient-dense seawater, retaining its nutritional value.	This method allows the company to farm at a high density all year round and support the local economy. Locals can consume fresh seaweed. This business model also has a positive environmental impact since it decreases the volume of transportation.
24	Shellcatch Chile 14 AI, internet of things	It is a company whose purpose is to make millions of fishermen's lives better by introducing them to a sustainable and efficient chain of certified ocean tracking systems.	Their three products are eMonitoring, eReporting and eCommerce. In these camera-integrated products, users can track the devices through IoT-based applications. With this, certain occurrences related to the fishery issues can be automatically detected with AI.	Offering an efficient and effective method of fisheries management generates revenue. Fishermen can keep track of species data and configurations. Environmental value is captured by showing the responsible fishing practices to end consumers.
25	Sinay.ai France 14 AI, Datahubs	It is a digital HUB that gathers, analyses and monitors complex maritime data on one platform.	It helps the shipping industry use their data combined with AI algorithms to tackle their key challenges, such as vessel route optimisation or waiting times reduction.	Value is captured by enabling the shipping industry to optimise ship arrivals, better allocate resources and drive operational efficiency
26	Sofar Ocean Technology USA 14 Big data, digital twins	It is a company that uses ocean technologies to provide insight into ocean ecosystems and enables the creation of a comprehensive ocean data network.	With the severity of the world's largest collection of open ocean weather sensors and real-time data, weather forecasting is made along the routes. Also, a digital twin is constructed for the fuel and safety profiles for each vessel.	Products that estimate data can promote safer maritime traffic and more accurate storm forecasting; sensors can monitor mariculture and collect metocean data.

(continued)

Table 16.1 (continued)

No	Company info	Value proposal (what?)	Value creation (how?)	Value capture
27	SymbyTech South Africa 14 Digital twins, drones	It is a company that uses drones for all maritime maintenance and also creates a digital twin of the vessel.	All maintenance processes such as troubleshooting and cleaning are done by drones and digital twins created to provide a way to visit an asset without the need to go to it.	Enabling maritime maintenance efficiently decreases the chance of any marine accidents that would cause enormous harmful effects on sea life.
28	The Ocean Cleanup Netherlands 14 Recycling	It is a non-profit organisation developing and scaling technologies to rid the oceans of plastic.	They create AI models that give information about the massive plastic garbage locations in the ocean. They capture the waste, store them in the ship, recycle this plastic garbage and create valuable materials which people would want to buy.	They clean up the oceans to protect underwater life. To increase societal awareness, they recycle plastic waste and create design products.
29	Umitron Singapore 14 AI, internet of things	They deliver a long-term solution to the well-known challenges of minimising environmental risk, increasing corporate profits and improving fishing efficiency by adopting IoT, an AI-driven platform and remote satellite sensing. They aim to transform agriculture to increase food security and safety by developing humane and environmentally friendly fish farming alternatives.	They integrate IoT, satellite remote sensing and AI to create user-friendly data platforms for aquaculture. Their technology aids farmers in increasing farm efficiency, reducing environmental risks and increasing profits. Their ultimate goal is to use computer models in conjunction with aquaculture to assist the globe in producing protein sustainably and efficiently that is both human and environmentally friendly.	To install sustainable aquaculture on earth, they utilise IoT and satellite remote sensing. Further value is captured by using Machine learning–integrated smart feeders, intelligent cameras to increase seafood production.
30	Upstream Tech USA 14 AI, Datahubs	By enabling informed water market transactions, the intelligent solutions they built promote freshwater conservation. They empower decision-making to become more efficient as more data is provided, resulting in lower costs and more impact.	An AI-powered streamflow forecast system enables optimising source planning while creating a powerful workflow for monitoring with aggregated satellite, aerial and environmental data.	They create value in time and cost savings by offering hydro forecast and AI-based landscape monitoring systems.
31	USV Company Sweden 13, 14 Autonomous vehicles, drones	It is a company that uses wind-powered autonomous sail drones to transport goods with zero emission.	Wind-powered, totally autonomous sail drones are used in logistics. It allows remote monitoring that operators can oversee and control the USV (unmanned surface vehicles) company fleet. It is also possible to equip USVs with solar power propulsion and sensors to allow a larger spectrum of operations.	Revenue is created through more efficient logistics. It also creates environmental value by reducing its carbon footprint. Since sail drones are wind-powered, no waste is dumped into water resources, making seas and oceans cleaner.

(continued)

Table 16.1 (continued)

No	Company info	Value proposal (what?)	Value creation (how?)	Value capture
32	Vertical Oceans Singapore 14 AI, internet of things	It is a company that provides sustainable protein using AI and IoT-based aqua towers to the people living in the cities without using any chemicals or antibiotics.	Ocean water recirculates in a balanced ecosystem where shrimp, fish and algae coexist. Data from sensors and IoT devices flows into AI algorithms to optimise growth, water quality, microbiology and energy usage. The company also reformulated the feed to minimise nutrient loss, using only sustainable ingredients for the feeding process.	Revenue creation is achieved via savings. By the use of IoT and AI algorithms, aqua towers consume less energy. The company also creates environmental value by reducing carbon footprint since food miles are decreased, in comparison with aqua towers based in cities.
33	vorteX.io France 6, 11, 14 Drones, internet of things	It is a platform that measures the level of hydrological systems and alerts communities using remote sensing instruments.	It delivers water level measurements of hydrological systems in real time, with high frequency and precision close to the centimetre, using non-invasive remote sensing sensors based on the small space-based altimeter.	A comprehensive set of small rivers contributing to flooding events are monitored on a budget with the help of drones – a new effective way of addressing climate change and flood risk within the big water cycle.
34	We4Sea Netherlands 14 Big data, digital twins	It is a company that offers solutions to remotely measure, manage and report fuel consumption and related emissions of chartered vessels – In real-time by using digital twin technology, without investing in onboard monitoring equipment.	The company designs a scientifically supported, algorithm-based digital twin for every monitored ship. The algorithm uses over 80 unique characteristics of the vessel to create this digital twin model, which is fed with real-time position and weather data, with up to 480 updates per day and a > 95% global position coverage. It brings full transparency and the possibility to benchmark consumption.	Revenue is created through savings. The company claims that, powered by the digital twin, the speed advisor informs you on the optimal speed to reach your destination, balancing route, ETA, charter rate and fuel costs. This can save 5 to 10% in voyage costs. It also creates environmental value by helping vessels lower fuel consumption and related emissions.
35	Whale Seeker Canada 14 AI, drones	Since the company is formed to evaluate the health of the ocean and analyse its changes, detection and monitoring techniques are essential. They enable decision-makers to implement policies and measures to protect marine mammals by providing access to quick, accurate data.	They track the whales with their aerial tools, satellite systems and infrared imaging technologies. They collect the essential data via image processing through AI and provide them in real time with the decision-makers.	To provide sea mammals a healthy life, they give valuable and fast-accessible information to environmental organisations. They create ecological, social and economic value to the marine ecosystem.

(continued)

Table 16.1 (continued)

No	Company info	Value proposal (what?)	Value creation (how?)	Value capture
36	XpertSea Canada 14 AI	They developed an AI-driven management system that focuses on making aquaculture more efficient and sustainable by providing real-time data. The start-up's mission is to solve the industry's most challenging problems, such as when to harvest and how to stop a disease from spreading.	They use artificial intelligence to assist farmers in modernising their operations and increasing earnings by delivering quick payments, useful production information and verified networks of suppliers and buyers.	Revenue is created through an AI-powered app that enables farmers to optimise aquacultural operations. The app also uses data to point out problems early while supporting SDG-14.

References

I. B. Franco et al. (eds.), Actioning the global goals for local impact: Towards sustainability science, policy. Education and Practice. Singapore: Springer Singapore (Science for Sustainable Societies). (2020). https://doi.org/10.1007/978-981-32-9927-6

O. Gulseven, Measuring achievements towards SDG 14, life below water, in the United Arab Emirates. Mar. Policy **117**, 103972 (2020). https://doi.org/10.1016/j.marpol.2020.103972

D. Kan, E. Meijer, PRé Sustainability, *Linking LCA and SDG 14* (UNEP Life Cycle Initiative, 2020), p. 14

K.-H. Lee, J. Noh, J.S. Khim, The blue economy and the United Nations sustainable development goals: Challenges and opportunities. Environ Int **137**, 6 (2020). https://doi.org/10.1016/j.envint.2020.105528

S. Nicklin, B. Cornwell, *SDG14 - a holistic approach to sustainable development* (2020), p. 110

U.C. Pandey et al., Introduction, in *SDG14 – Life Below Water: Towards Sustainable Management of Our Oceans*, (Emerald Publishing Limited, 2021), pp. 1–15. https://doi.org/10.1108/978-1-80071-709-120211004

PwC, *SDG 14: Life Below Water Conserve and Sustainably Use the Oceans, Seas and Marine Resources for Sustainable Development* (PwC, 2021), p. 12. Available at: www.pwc.com/globalgoals

L. Recuero Virto, A preliminary assessment of the indicators for sustainable development goal (SDG) 14 "Conserve and sustainably use the oceans, seas and marine resources for sustainable development". Mar. Policy **98**, 47–57 (2018). https://doi.org/10.1016/j.marpol.2018.08.036

D. Steinbach, E.Y. Mohammed, P. Steele, A sustainable future for fisheries: How fiscal policy can be used to achieve SDG 14. (2017). https://doi.org/10.13140/RG.2.2.21208.49929

M.R. Stuchtey, et al., Ocean Solutions That Benefit People, Nature and the Economy (2021). oceanpanel.org, p. 32

UNESCO, *Global Ocean Science Report 2020: Charting Capacity for Ocean Sustainability* (UNESCO, 2020). https://doi.org/10.18356/9789216040048

UNDP 2021. [WWW Document] https://sdgs.un.org/goals/goal14

SDG-15: Life on Land

17

Abstract

Population increases, industry, urbanisation, infrastructure development and agricultural expansion influence landscapes, lowering total habitat size and quality and resulting in ecological degradation. SDG-15, Life on Land, aims to maintain, restore and enhance the utilisation of the terrestrial environment and forest management sustainably, struggle with desertification and stop and reverse land degradation, as well as the loss of biodiversity. This chapter presents the business models of 45 companies and use cases that employ emerging technologies and create value in SDG-15. We should highlight that one use case can be related to more than one SDG and it can make use of multiple emerging technologies.

Keywords

Sustainable development goals · Business models · Life on land · Sustainability

Population increases, industry, urbanisation, infrastructure development and agricultural expansion influence landscapes, lowering total habitat size and quality and resulting in ecological degradation. Overall, the pace of extinction caused by human activities is so high that even conservative estimates suggest humankind has entered the sixth major extinction event (Bradshaw et al. 2021). Earth had had a non-aquatic environment throughout its geological history, except when its entire surface was covered with water. Non-aquatic generally means terrestrial environments. Nevertheless, even an entirely aquatic environment like lakes encompasses a broad range of mixed habitats where aquatic and terrestrial areas evolve and interact throughout time (Beraldi-Campesi 2013). Terrestrial ecosystems provide goods and various ecosystem services such as carbon capturing, preserving soil quality, protecting biodiversity, decreasing the risk of natural disasters by regulating water flow, controlling erosion and preserving agricultural systems. Therefore, protecting terrestrial ecosystems significantly contributes to tackling the climate crisis and adaptation efforts. In 2015 United Nations established SDG-15, which is about "Life on Land" to maintain, restore and enhance the utilisation of the terrestrial environment and forest management sustainably, struggle with desertification and stop and reverse land degradation, as well as the loss of biodiversity (Ishtiaque et al. 2020). These initiatives aim to ensure that the advantages of land-based ecosystems, such as sustainable livelihoods, are maintained for future generations (UNEP 2021). The notion that the management of terrestrial ecosystems, especially forests and varied

S. Küfeoğlu, *Emerging Technologies*, Sustainable Development Goals Series,
https://doi.org/10.1007/978-3-031-07127-0_17

biodiversity, is critical for long-term development has gained widespread acceptance. However, the demands of population expansion, development of the economy and greater consumption will only exacerbate the difficulties of maintaining life on land (Sayer et al. 2019). SDG-15 has a vital role in dealing with these problems. As shown in Fig. 17.1, there are 12 targets within the context of SDG-15, and they are measured with 14 indicators.

As explained earlier, SDG-15 targets preserving, restoring and encouraging the use of terrestrial ecosystems in a sustainable manner, sustainably managing forests, fighting desertification and preventing the loss of biodiversity and soil deterioration. According to the Millennium Ecosystem Assessment, the ecological impact of agriculture has grown substantially. As a result, food production is a major contributor to these problems: Expanding agriculture has resulted in habitat loss for 80% of endangered animals (Måren 2019). More than 80% of the human diet is made up of plants, and up to 80% of people in developing nations' rural regions depend on traditional plant-based medicines for basic healthcare. Approximately 2.6 billion people rely directly on agriculture for their livelihoods, and about 1.6 billion people rely on forests for their livelihoods; forests occupy around 30% of the Earth's surface, and they are home to nearly 80% of all terrestrial animals, plants and insect species (UNEP 2021).

Forests assist in preventing climate change by removing CO_2 from the atmosphere; supporting the balance of oxygen, carbon dioxide and humidity in the atmosphere; and conserving watersheds, which provide 75% of the world's freshwater. Natural catastrophes, such as floods, droughts, landslides and other severe occurrences, are also reduced (Kleymann and Mitlacher 2018). The loss of forested areas has a detrimental impact on rural populations' lives since it leads to increased carbon emissions, land degradation (which affects 74% of the world's poor) and biodiversity loss ("SDG 15" 2021). Forests are one of the most biodiverse ecosystems on the planet, supporting more than 80% of all terrestrial animal, plant and insect species. Forests

have a crucial role in people's livelihoods and well-being, particularly among the rural poor, young and women. Forests support roughly 1.6 billion people, including over 2000 indigenous cultures, in addition to providing shelter, income and security for forest-dependent communities (Kleymann and Mitlacher 2018).

Member States of the United Nations stated at the Rio+20 Conference in 2012 that they recognise the social and economic importance of good land use planning, including soil, especially its help to economic growth, biodiversity, sustainable agricultural production, poverty eradication, women's empowerment, climate change mitigation and enhanced water accessibility. They highlight that desertification, land degradation and drought are worldwide concerns that continue to pose major threats to all nations' sustainable development, particularly developing ones. They also underlined the need of taking immediate action to reverse land degradation. In light of this, we shall work to build a land degradation-free world in the context of long-term development. This should operate as a catalyst for mobilising financial resources from both public and private sources (The future we want, 2012).

Forests and biodiversity will almost certainly face challenges in the future years. As a result of this issue, additional protective and long-term strategies should be developed – all efforts to achieve long-term sustainability influence land-based life. The solution-oriented issues discussed will be a big step towards these challenges. There will be a decline in the use of wood for livelihoods, especially if the economy continues to expand, and therefore the problem of poverty may be reduced to a minimum. Health and education level will be positively affected by this situation. As long as forests and wetlands are protected, the services and opportunities offered will also result from these sustainable development goals. If plans progress as intended, the following will occur (Sayer et al. 2019):

- There will be a greater displacement from rural to urban regions.
- By emphasising agriculture, more mechanised farms will be developed. Industrial agriculture

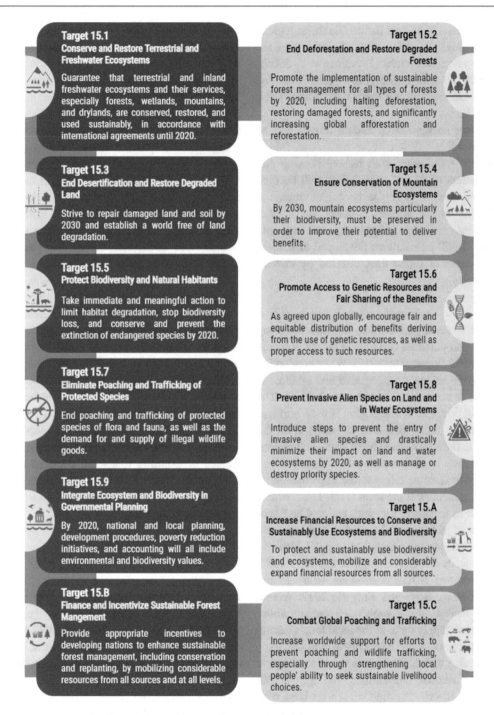

Fig. 17.1 SDG-15 targets and indicators. (UNEP 2021, p. 15)

will increase trees and forests and improve efficiency (Laurance et al. 2013).

- As the purchasing power increases, the demand for agricultural products will

decrease, and the consumption of meat and dairy products will increase.

- To access mineral resources, infrastructure will be expanded into forest regions.

- The desire to be near nature and forest regions will raise the value of biodiversity and the ecosystem it supports (Tyrväinen et al. 2005).
- Countries will transition to a green economy, and expanding forest lands will be added as a result.
- Timber harvests from natural forests will be reduced.
- Forest monitoring with remote sensing and "Internet of things" applications will become more common as new technologies emerge.
- Initiatives to address climate issues will increase as forests are protected and restored.
- There will be solutions developed to address the issue of forest fires induced by climate change.

By 2030, significant progress will be made, even if these targets are not finished. Ongoing studies and practices will ensure that life on land is protected (Sayer et al. 2019). When conducting these applications, an integrated and systematic strategy is necessary (Tremblay et al. 2020). The working principle of integration is used horizontally across policy domains, vertically from the global to the national and local levels and regionally across local governments (Kanuri et al. 2016). Localisation refers to the implementation of SDG practices at the local level. The methods adopted to accomplish global, national and subnational objectives and the process of monitoring these strategies are referred to as SDG localisation (Losada 2014). Although scientific research on cities and SDGs is growing (Barnett and Parnell 2016; Bibri and Krogstie 2017; Graute 2016), there is still a knowledge vacuum about how to best apply them at the local level (Fenton and Gustafsson 2017; Krellenberg et al. 2019).

Together with SDG-15, it is aimed to integrate the international conventions and agreements made for the continuation of life on land with other targets (Sayer et al. 2019). Ecosystems and forests are important from an economic point of view. SDG-15.1 aims to protect ecosystems and their economic values (Dempsey 2016). Payments for ecosystem services (PES), an economic and environmental approach, aim to protect biodiversity by internalising the real value of biodiversity (Pirard 2011). With this approach, the protected ecosystem will provide more services and be good for both the economy and biodiversity in the long run (Pirard 2011). To sustain the business world fed by natural resources, it is necessary to protect the ecosystem. Since institutions take over ecosystems, the focus is on generating income from biodiversity rather than protecting it. The forestry sector is a good example of this situation. The environmental standards determined to protect the ecosystem in public institutions and companies have been stretched under the name of economic incentives (Lovera 2017). Current business models and economic approaches are not sustainable. Evaluating, optimising and minimising the effect and dependency of businesses on the land and ecosystems is a direct way for businesses to support life on land ("SDG 15" 2021). Integrating socio-economic development activities into protection plans as a method of ensuring long-term resource utilisation necessitates a thorough knowledge of the interconnections between humans and natural processes (Bridgewater et al. 2015).

17.1   Companies and Use Cases

Table 17.1 presents the business models of 45 companies and use cases that employ emerging technologies and create value in SDG-15. We should highlight that one use case can be related to more than one SDG and it can make use of multiple emerging technologies. In the left column, we present the company name, the origin country, related SDGs and emerging technolo-

Table 17.1 Companies and use cases in SDG-15

No	Company info	Value proposal (what?)	Value creation (how?)	Value capture
1	Aerium Analytics Canada 9, 15 AI, drones	The company is a provider of bird-shaped remotely piloted aircraft system (RPAS) solutions, RoBird, for wildlife management with drones. It creates real-time and deep data analytics solutions using artificial intelligence (AI) to use geospatial data to make more informed business decisions.	Drones and AI-based RoBird detects animal activity, locates nesting locations, avoids human–wildlife confrontations and keeps industrial zones safe.	Using a mix of AI and drones, environmental value is generated by enabling human tasks to be carried out without hurting animals.
2	Apic.ai Germany 15 AI, Internet of Things	The company, which is the developer of the insect measurement system, aims to preserve biodiversity by using honey bees as biosensors to collect data that causes the extinction of insects using AI and Internet of Things (IoT).	Ecotoxicologists and other scientists can get data on the activity and collection behaviour of bees/bumblebees using the monitoring devices, which can then be analysed. The device captures video at the hive's entrance to identify individuals, their movement patterns and the amount of pollen delivered into the hive. In comparison to prior methodologies, AI-based analysis of recorded picture data from semi-field and field trials provides advantages.	Collecting environmental data using the honey bee generates income from the impact of decisions in urban development and agriculture on the ecosystem.
3	Beewise Israel 15 AI, robotics	The company has developed a gadget that uses robotics and AI to remotely control beekeepers' hives while also providing the essential medication and care for bees.	The system tracks the behaviour and motions of bees using AI and robotics. It automatically intervenes in the event of a problem, but if the issue is not rectified, the device issues a warning. It also saves and recognises frames that are ready to be harvested inside the device. The harvested honey must be dumped after the capacity is filled.	By using AI and robots to protect generations of bees and reduce beekeeper labour, environmental value is obtained.

(continued)

Table 17.1 (continued)

No	Company info	Value proposal (what?)	Value creation (how?)	Value capture
4	Breeze Technologies Germany 15 AI, big data, Internet of Things	The company is a provider of small-scale IoT-based air sensors that can be installed indoors and outdoors measuring common pollutants, air quality data and air quality analytics software using AI and big data.	With an environmental analytics cloud platform, sensors gather real-time data from air quality sensors as well as external data sources. Adaptive Cloud Calibration Engine increases data reliability and accuracy. Environmental Intelligence Cloud uses advanced algorithms for sensor calibration, predictive maintenance and data analytics accessing real-time and historic air quality data from your sensors and additional insights.	Value is created by generating insights about buildings, cities and communities, achieving an arbitrarily high data resolution, providing data to environmental scientists and governments to help cities and businesses create better clean air action plans and predicting potential problems.
5	Cloud Agronomics USA 2, 13, 15 AI, Internet of Things	The platform provides real-time analytics and predictive insights on crop performance using AI and IoT to optimise agriculture.	Hyperspectral imaging aircraft acquire information on soil, disease activity and crop performance. A global geospatial dataset is created using local and satellite data. AI provides real-time carbon monitoring and extracts insights. The grower uses knowledge to produce food in a more sustainable manner. The index can be used by carbon credit exchanges to incentivise carbon farming.	Revenue is captured by enabling farmers to surgically remove or treat the disease before it spreads by detecting them with AI.
6	CoverCress USA 7, 15 Biotech & biomanufacturing	The BioTechnology company manufactures oilseed crops for food, bioenergy and cattle feed.	They develop a new oilseed crop named CoverCress which grows over winter between normal full-season corn and soybeans. It acts as a cover crop while also producing oil and high protein feed that can fit markets similar to canola.	Environmental revenue is captured by preventing erosion and improving soil health with roots, diversifying products for food and biofuels.

#	Company	Description	Environmental value
7	Daumet France 6, 9, 12, 13, 14, 15 Advanced materials	It is a material science company that creates an alloy of white gold as an alternative to gold.	Gold is mixed with tungsten, which is the hardest and has the same density as gold. With this method, an alloy with better properties is obtained without coating with rare elements such as rhodium and palladium.	The advanced material provides less need for gold as raw material and reduces the use of rare elements used in the gold industry. It creates a positive environmental and social impact by ensuring the conscious consumption of raw materials and reducing ecosystem destruction, carbon emissions and water consumption caused by mining activities.
8	Dendra UK 13, 15 AI, drones	The company provides integrated analytics and planting solutions for large-scale ecosystem restoration.	It uses custom-built drones and AI for reforestation. Drones enable the reforestation of regions across the world that have been destroyed by mining, intensive agriculture, fires or other disruptive factors using a combination of ecological expertise, AI and technical gear.	Environmental value is captured by helping the natural resources companies in the world to clean up and replant the degraded land.
9	Descartes Labs USA 15 AI, big data	The AI and Big Data-based geospatial intelligence platform uses scientific analysis of physical world events for the benefit of organisations.	The software as a service platform analyses the collected mineral data and performs mineral screening by using Big Data. Ground stability risks in mining are reduced with models and methodologies that apply deformation studies on a large scale.	The technology developed produces nitrogen fertiliser on the farm, by fixing nitrogen from the air and reacting with ammonia in manure. The reaction stops the ammonia losses, increases the nitrogen content and provides the farmer with essential nutrients for growing crops.
10	Desert Control Norway 2, 13, 15 Biotech & biomanufacturing	The company is a climate technology company that specialises in reclaiming degraded soil and turning desert sand into fertile soil by using Liquid NanoClay (LNC).	LNC increases crop yields, biomass production and carbon uptake while reducing water and fertiliser consumption. The mix, whose name is Liquid NanoClay, sinks into the soil, creating a 40–60 cm deep layer, which retains the water like a sponge. This layer stops water from evaporating and ensures optimal growing conditions for anything planted in it.	Environmental value is captured by using LNC which increases crop yields, biomass production and carbon uptake while reducing water and fertiliser consumption.

(continued)

Table 17.1 (continued)

No	Company info	Value proposal (what?)	Value creation (how?)	Value capture
11	DroneSeed USA 15 Drones	The company offers forestry management services using drones instead of sending out manual work crew.	The company uses drones for fast, efficient, cost-competitive reforestation operations. They offer vertically integrated reforestation services across seed collection and procurement, seedling growth, aerial enhanced seeding with drones, traditional replanting on the ground and funding for reforestation projects from carbon credits.	Environmental value is captured through better forestry management and fast recovery after forest fires.
12	Drylet USA 15 Biotech & biomanufacturing	It is a bioremediation technology company producing dry-to-the-touch nano bioreactor products whose proprietary biocatalysts optimise biological processes in wastewater, lagoon systems and waste-to-energy facilities via biotechnology.	It works by combining selected microbes with a superporous particle, accelerating microbial activity and converting mass to gas in biological processes. Solids, foaming, odour and other irritating problems in pits and lagoons are avoided so that the fertiliser is consistent and suitable for field application.	Generating solutions to biosolid accumulation improves operations, reduces environmental impacts and positively impacts the bottom line, using no genetically modified organisms or chemical ingredients that do not pose any danger to humans, animals and aquatic life, reducing odour and hydrogen sulphide emissions.
13	Ecording Turkey 10, 12, 13, 15 Drones	It is a social enterprise that offers drone solutions for the afforestation of hard-to-reach areas that need to be afforested.	Seed balls increase seed germination in natural circumstances and reduce the negative impacts that seeds may encounter throughout their transformation into trees. Drones are used to drop seed balls in hard-to-reach locations that need to be reforested.	The firm contributes to carbon sequestration, biodiversity preservation and the avoidance of global warming, by deploying drones, and captures social value by hiring women in rural regions to reduce gender inequities.

| 14 | Green City Watch
Netherlands
11, 15
AI, cloud computing, drones, Internet of Things | An open-source collective, geospatial AI (geoAI) builds geospatial software to map, monitor and manage urban trees. | TreeTect is an open-source programme that gathers data on the number and health of urban trees and tree canopy using satellite images and machine learning (ML) and allows users to generate an AI-enabled digital tree inventory. TreeTect can scan extensive regions of metropolitan topography using very high-resolution (VHR) satellite images to monitor both the number and quality of green space in cities. The tree-detection algorithm can figure out not only where the city's parks are but also if they're looking at a green tree, shrubs, bushes or a green garden shed, thanks to this technology, which can measure down to each square metre of land. This means it can detect the position, size, form and even health of individual trees in real time. | Value is captured by tracking health and numbers of urban trees by using AI and storing their data with cloud computing for further applications. |
| 15 | Green Praxis
France
15
AI | The company provides a solution that enables reliable identification of options available to a land or forest owner. Furthermore, it restores viable ecosystem balances in degraded natural environments. | A diagnosis of a bIoTope is made based on the collection of public and private information verified in all dimensions: geography, climate, plants, animals, demographics and regulations. A viable economic and ecological project is structured based on proven and compatible scenarios. The options determined using AI meet both the requirements of the field and the project objectives. | The AI-based solution creates environmental and economic value by providing viable options for degraded natural environments and optimal project planning for land and forest owners. |

(continued)

Table 17.1 (continued)

No	Company info	Value proposal (what?)	Value creation (how?)	Value capture
16	Hortau USA 2, 15 Internet of Things	The company develops a wireless and web-based irrigation management system that uses IoT.	The company places multiple sensors in the root zone at different depths to monitor water movement and plant uptake in real time. The application lets the customer view where necessary action is required and where their crop is in good health. They get quick access to the data to view details about their crops' water needs and, during irrigation, monitor its uptake.	The IoT-based irrigation management solutions system detects plant stress in real time which improves the yield and reduces water/energy use.
17	Internet of Trees UK 15 Internet of Things	The company develops an early warning system that protects assets, monitors forests and prevents the spreading of fires via IoT-based environmental sensors.	The system is made of Internet-connected modules installed in the forest to detect temperature and air humidity variations, which are analysed by an algorithm that then compares the current readings with the ones taken before, also comparing them with data from external weather sources to validate if the area has caught fire or not.	Preventing and slowing deforestation by combining IoT and environmental data and building a team to help in the fight to protect trees, people's lives and their assets while remaining sustainable creates value.
18	Jupiter Intelligence USA 13, 15 AI, Internet of Things	It is a global climate modelling solution that predicts physical climate risk from catastrophic weather events using IoT and AI.	The company uses various data sources to compute probabilistic modelling for predictions via ML using their data and analytics services.	Value is generated by providing a range of analytics critically impacted by climate change which are used for capital planning, enterprise and portfolio risk management, infrastructure resilience engineering, pricing and underwriting, safety and operations and shareholder and regulatory response.
19	Land Life Company Netherlands 15 AI, drones	The drone- and AI-driven reforestation company offers corporations and organisations a sustainable and transparent way to take climate action and compensate carbon emissions through nature restoration.	Using drones and AI, the company attempts to minimise carbon dioxide emissions by detecting and reforesting regions that have been damaged due to over-farming, urbanisation and natural disasters..	Environmental revenue is captured through reforestation projects with drones and AI by regreening the lands that are damaged

#	Company	Description	Value
20	N2 Applied Norway 2, 6, 12, 13, 15 Biotech & biomanufacturing	The company creates a closed-cycle and on-farm system to produce a nitrogen fertiliser using biotechnology.	The technology is developed to produce nitrogen fertiliser on the farm, by fixing nitrogen from the air and reacting with ammonia in manure. The reaction stops the ammonia losses, increases the nitrogen content and provides the farmer with essential nutrients for growing crops.	BioTechnology enables farmers to recycle and obtain fertiliser at a lower cost while generating higher yields using livestock manure, air and renewable energy.
21	NanoFex USA 15 Biotech & biomanufacturing	It is a biotech company that intends to develop a proprietary and environmentally sustainable microparticle formulation that sequesters and breaks down groundwater contaminants more efficiently and cost-effectively than current remediation techniques.	The company's services produce cellulose nanospheres using a customised spray dryer process to produce a nano-particle powder that can be injected into the ground, enabling users to get cleaner groundwater.	Value is created through breaking down groundwater contaminants more efficiently and cost-effectively than current remediation techniques.
22	Nofence Norway 15 Internet of Things	The company produces a virtual fencing system with IoT for cattle, sheep and goats.	Animals can be tracked via GPS using the app, making them easy to locate and monitor. It uses audio warnings and a potential electrical impulse to keep livestock inside the Nofence pasture.	Environmental value is captured by increasing the productivity of pastures and livestock by working with the plant's natural growth pattern.
23	OptiRTC USA 6, 15 Cloud computing, Internet of Things	It is a platform that combines cloud computing and IoT which enables Continuous Monitoring and Adaptive Control (CMAC) of stormwater storage assets.	With their system, rain is collected from roofs and stored in basement cisterns. When the weather forecast calls for rain, by intersensory communication, water is automatically discharged from these cisterns to accommodate the oncoming run-off. The water that's been collected can be reused for agriculture during dry periods.	This system enables stormwater management by data analysis and increases retention times and infiltration by using sensors, reducing downstream erosion and improving water quality. Also, they reduce operating and capital costs.

(continued)

Table 17.1 (continued)

No	Company info	Value proposal (what?)	Value creation (how?)	Value capture
24	OroraTech Germany 3, 9, 11, 12, 13, 15 AI, Internet of Things	The company provides real-time information services for early detection of wildfires with AI and monitoring them by using satellites and sensor technologies.	They have developed a nanosatellite constellation equipped with multispectral imagery in the thermal infrared and visible light range. By combining CubeSat framework with a customised optical system and on-orbit pre-processing capabilities, they will deliver high-quality satellite images and create real-time information for wildfire detection and mapping, severe weather forecasting and many other applications.	Detecting wildfires earlier by using satellite imaging and AI prevents the spreading of these fires in huge areas and provides a chance in controlling wildfires safely.
25	Project Canopy Switzerland 15 Big data	The data-driven firm interacts with Congo Basin conservationists to provide them with the insights they need to make better-informed decisions.	The project gathers, transforms and communicates the data organisations need to end defaunation, deforestation and associated carbon emissions in the Congo Basin rainforest.	The company's big data platform, analytics and newswire provide their audiences with the building blocks they need to make the most impactful programmatic and policy decisions.
26	Robin Radar Systems Netherlands 15 AI, drones	The company develops radar systems that are specifically designed to track small objects like birds, bats (aviation safety and wind farms) and drones (security).	They detect, classify and track birds and drones with advanced radar technology, AI and software to prevent serious incidents. Their products, IRIS and ELVIRA, can detect larger fixed-wing targets at a range of up to 5 kilometres, with smaller multi-rotor drones detected at up to 4 kilometres.	Value is captured by increasing safety for both humans and wildlife. They provide actionable management information for wind farm operators, airports and researchers and keep safety via drone detection by distinguishing drones from birds.
27	Robotto Denmark 13, 15 AI, drones	The company uses AI-powered autonomous smart drones to offer incident controllers with real-time data on the magnitude, position, direction and local weather of wildfires.	The autonomous wildfire recognition and analytics drone, AWRA, will provide the user with accurate data gathering and monitoring of the wildfire. The user will utilise a tablet to mark an area for the drone to search in and will receive real-time feedback regarding the location, size, intensity and direction of the wildfire.	Value is created by saving hours spent in operation and on land that otherwise would have been destroyed, making sure emergency services make the right call, giving the firefighters back the initiative and allowing them for faster smarter decision-making, ultimately extinguishing the wildfires faster.

28	Satelligence Netherlands 15 AI	The company uses satellite data, proves insights about the operation of agricultural production and detects supply chain risks, such as deforestation, land degradation, forest fires and carbon stock loss through AI.	With AI-powered predictive modelling and remote sensing to monitor deforestation, the company identifies environmental risks. Furthermore, where emissions from land-use change are occurring, it reduces and offsets carbon emissions and monitors carbon stock changes in the supply chain with data for historical and current grievances to react to deforestation claims, in real time.	Revenue is captured by using AI to drive key sustainability insights to minimise global environmental footprint, monitor deforestation real-time and protect biodiversity, identify the highest-risk farms and sourcing areas and help make sourcing and investment decisions.
29	SeeTree Israel 15 AI, drones	The company develops an intelligence network that leverages drones, AI and sensor technologies to give farmers actionable insights to assist users in monitoring the crops and assessing the health of the trees.	Each tree in the land is assessed for its health and growth rate, and planting plans are created for each tree variety. This is performed via the analysis of high-resolution photos and data captured by drones and AI.	Environmental revenue is generated through the tree farming revolution with the intelligence-per-tree by providing farmers with an end-to-end service to manage and optimise the health with AI and productivity of their trees.
30	Skymining Sweden 8, 9, 12, 15 Carbon capture & storage	The company is a producer of grass that regenerates degraded soil and facilitates environmental recovery by releasing atmospheric carbon into the soil.	A forest is established from C4 grass on degraded or marginal lands. The grass grows up to 4 metres in 120 days and captures and stores atmospheric carbon.	By capturing carbon, the biosphere is restored, filling the soil with carbon and nutrients. Reclaiming degraded land creates new, arable land and relieves pressure on deforestation.
31	Spinnova Finland 15 Advanced material	It is a sustainable material company that utilises wood, textile waste and agricultural waste such as wheat or barley straw in fibre production.	The business has developed a method that uses a mechanical process to convert cellulosic fibre into textile fibre.	The firm has aided in the resolution of many important environmental issues. For example, stubble is often burnt on-site in agriculture, resulting in pollutants and a health concern.
32	Spoor Norway 15 AI	The company provides software as a service solution that enables continuous monitoring of wildlife for offshore wind farms, pre- and post-construction.	The company is leveraging existing infrastructure and computer vision coupled with AI to provide valuable environmental insights with higher accuracy, resulting in informed decision-making, planning and operations.	Environmental benefits are captured by detecting and preventing dangers that are threatening both biodiversity and being environmentally sustainable.

(continued)

Table 17.1 (continued)

No	Company info	Value proposal (what?)	Value creation (how?)	Value capture
33	Sylvera UK 13, 15 AI, carbon capture & storage	The carbon offsets monitoring firm leverages AI to scale carbon markets.	Raw carbon performance, attachability, permanence, co-benefits and risk are all offered for nature-based offset initiatives. Their own ML-based scoring system is used to grade projects. Projects are reviewed, markets are scaled, and their performance is enhanced over time by tracking their progress.	The ML-based platform creates value by providing transparent, actionable data on nature-based projects to make decisions quickly and confidently while reducing carbon emissions.
34	Terramera Canada 2, 15 AI, biotech & biomanufacturing	The company is an AI-based targeted performance technology provider that makes organic inputs more efficient. This technology increases the effects of plant-based organic activities and provides more efficiency than synthetic chemical pesticides and fertilisers.	To produce natural, plant-based products, many active components are explored. By delivering active chemicals directly into target cells, tailored performance technology improves their efficacy. Laboratory activities, quick phenotyping and chemical discovery are made easier by software, machine learning models and robotics.	The targeted performance technology creates environmental value by preventing water and soil pollution caused by fertilisers by ensuring that most of the fertilisers produced from bioactive components are absorbed by the cells.
35	TerViva US 2, 15 Biotech & biomanufacturing	The food and agriculture company uses biotechnology and biomanufacturing to produce delicious and healthy plant-based food ingredients from the Pongamia tree.	Pongamia genetics, which has been patented, uses sophisticated breeding methods to generate more sustainable food and agriculture with biotechnology and biomanufacturing.	Revenue is generated by restoring farmland to productive use with Pongamia trees, helping farmers produce sustainable food.
36	Tesera Systems Canada 13, 15 AI, big data	The company is a provider of data-driven cloud applications for forest management that are powered by AI and big data.	Data aggregation and analytics supply the clients with the ability to discover hidden and sustainable opportunities and realise new value from their forestry assets.	AI and Big Data help forest managers improve forest management, landscape design, harvest timing and ecosystem services.
37	Tesselo Portugal 15 AI	It is an AI-driven satellite mapping company that creates near real-time satellite imagery to identify environmental challenges, patterns, exposure and risk factors by tracking incremental image changes over time.	Various raw data are absorbed from satellites, drones, radars and multispectral imagery. Clear composite images are created from many photos, eliminating clouds and shadows and adjusting each pixel to atmospheric conditions. Satellite images are created with AI that detects specific problems. Incremental changes over time are tracked: patterns, exposure levels and risk factors.	AI and time-sensitive classification models facilitate value creation by detecting tree species, habitat and infrastructure in a dynamic environment.

38	The Bee Corp USA 15 AI	Verifii is a firm that develops an impartial and objective hive grading system. They utilise infrared image analysis to help growers and beekeepers price pollination by using AI and count the number of bees in a hive and monitor hive health.	The business is working on Verifli, a hive grading system that uses Artificial Intelligence to analyse thermal data included in an infrared (IR) image and translate it into hive strength data while accounting for meteorological variables at the time and place of each image.	By supporting beekeepers in improving pollination quality, maintaining food security and assisting farmers in lowering costs, value is produced.
39	Timbeter Estonia 11, 15 AI, cloud computing, Internet of Things	The forest tech company specialises in timber measuring by using IoT, AI and the management of data that are stored in clouds.	Customers may use their smartphone or tablet to measure log diameters, log count and log pile density. Measurements are saved in the cloud, allowing for a real-time overview of data. Customers can use the storage to evaluate and share their measurements. They have access to inventories, active storage statuses and the ability to generate real-time reports.	A combination of AI, IoT and cloud computing, the platform creates value for the forestry industry by eliminating illegal logging and improving timber supply for tree buyers and sellers.
40	Treevia Brazil 13, 15 AI, big data, Internet of Things	The company is a digital solution provider that uses AI, IoT and big data to monitor and measure forestry.	They have patented an IoT device that allows precision forest management by remotely monitoring growth, quality and forest health, with a series of environmental variables.	Revenue is generated by increasing the scalability and reliability of nature-based solutions, reducing the risk of work-related accidents by working remotely and assisting forest managers by collecting environmental variables such as humidity, precipitation and temperature using ML and AI
41	Virotec Australia 15 Recycling	The company has developed technically engineered solutions called Basecon and Bauxol to environmental and industrial waste, derived from utilising and augmenting the specific physical and chemical properties of alumina refinery residues by recycling.	The technology re-engineers the residue from alumina, refining process into their product enabling it to neutralise acid and reduce the concentration of environmentally hazardous heavy metals. It works by re-engineering the minerals present and recrystallising them to form new minerals which are safe and stable.	Value is captured by creating a positive impact on ecosystem health, biodiversity, heritage, water quality, community, economy and governance. Further value is captured by commercialising environmental remediation and waste treatment technologies derived from the conversion and safe reuse of ARR.

(continued)

Table 17.1 (continued)

No	Company info	Value proposal (what?)	Value creation (how?)	Value capture
42	WasteHero Denmark 9, 11, 12, 15 AI, 5G, Internet of Things, recycling	The company is a provider of end-to-end IoT solutions that optimises waste collection processes using Machine Learning.	The company tracks individual container locations, fill levels and weight information while documenting all past activities associated with assets. Then, they import historical weight data to optimise route schedules. Automated collection routes are created by using AI while optimising based on economic, environmental or service parameters to meet customers' organisational key performance indicators (KPIs) with waste management software.	Value is captured by reducing CO_2 emissions, more efficient waste bin collections and increasing financial savings on collections, thanks to the automated system based on AI and IoT.
43	Watergenics Germany 14, 15 AI, Internet of Things	The firm makes water quality sensors for aqua farming, which use IoT and AI to detect chemical nutrients in surface water and groundwater in plants, fisheries and agricultural drainage systems.	An affordable, accurate and autonomous sensor network is offered to measure water quality. These sensors, using IoT and AI, provide agricultural and aquaculture systems with water quality data, actionable insights and predictive analytics.	Revenue is captured through environmental compliance monitoring using IoT and AI for product optimisation and ecosystem conservation.
44	Wipsea France 15 AI	The company develops AI-powered software that analyses massive amounts of animal imagery captured through camera traps or aerial surveys (planes or drones).	The platform develops image analysis software in a variety of computer programs, and specialist biologists use AI to detect the features of living objects.	Revenue is captured from accessing ultra-high-resolution images to generate distribution maps and extract relative population abundance using AI to automate detection and classification.
45	Xampla UK 2, 14, 15 Advanced materials, bioplastics, biotech & biomanufacturing	The company offers non-synthetic plant protein materials to replace microplastics and single-use plastics for commercial use.	The company invents vegan spider silk, which is very similar to vegetable protein and contains non-covalent hydrogen bonds between its proteins, to replace single-use plastics. A naturally strong, flexible and transparent polymer material is made from 100% plant protein. This material is produced by exposing a plant protein to concentrated vinegar, heat and energy without chemical crosslinking.	By substituting biodegradable polymers for conventional plastics, environmental value is captured. In addition, the pace at which poisons accumulate in the food chain is slowed.

gies that are included. The companies and use cases are listed alphabetically[1].

References

C. Barnett, S. Parnell, Ideas, implementation and indicators: Epistemologies of the post-2015 urban agenda. Environ. Urban. **28**, 87–98 (2016). https://doi.org/10.1177/0956247815621473

H. Beraldi-Campesi, Early life on land and the first terrestrial ecosystems. Ecol. Process. **1** (2013). https://doi.org/10.1186/2192-1709-2-1

S. Bibri, J. Krogstie, Smart sustainable cities of the future: An extensive interdisciplinary literature review. Sustain. Cities Soc. **31** (2017). https://doi.org/10.1016/j.scs.2017.02.016

C.J.A. Bradshaw, P.R. Ehrlich, A. Beattie, G. Ceballos, E. Crist, J. Diamond, R. Dirzo, A.H. Ehrlich, J. Harte, M.E. Harte, G. Pyke, P.H. Raven, W.J. Ripple, F. Saltré, C. Turnbull, M. Wackernagel, D.T. Blumstein, Underestimating the challenges of avoiding a ghastly future. Front. Conserv. Sci. **1**, 9 (2021). https://doi.org/10.3389/fcosc.2020.615419

P. Bridgewater, M. Régnier, R.C. García, Implementing SDG 15: Can large-scale public programs help deliver biodiversity conservation, restoration and management, while assisting human development? Nat. Resour. Forum **39**, 214–223 (2015). https://doi.org/10.1111/1477-8947.12084

A.F. de Losada, *Localizing the Post-2015 Development Agenda- Dialogues on Implementation* (United Nations Development Group, 2014)

J. Dempsey, *Enterprising Nature: Economics, Markets, and Finance in Global Biodiversity Politics* (Wiley, 2016)

P. Fenton, S. Gustafsson, Moving from high-level words to local action—governance for urban sustainability in municipalities. Curr. Opin. Environ. Sustain., Open issue, part II **26–27**, 129–133 (2017). https://doi.org/10.1016/j.cosust.2017.07.009

U. Graute, Local authorities acting globally for sustainable development. Reg. Stud. **50**, 1931–1942 (2016). https://doi.org/10.1080/00343404.2016.1161740

A. Ishtiaque, A. Masrur, Y.W. Rabby, T. Jerin, A. Dewan, Remote sensing-based research for monitoring progress towards SDG 15 in Bangladesh: A review. Remote Sens. (Basel) **12**, 691 (2020). https://doi.org/10.3390/rs12040691

C. Kanuri, A. Revi, J. Espey, H. Kuhle, Getting Started with the SDGs in Cities (2016).

H. Kleymann, G. Mitlacher, *The Role of SDG15 in Underpinning the Achievement of The 2030 Agenda, Global Policy and Advocacy* (World Wide Fund for Nature, 2018)

K. Krellenberg, H. Bergsträßer, D. Bykova, N. Kress, K. Tyndall, Urban sustainability strategies guided by the SDGs—A tale of four cities. Sustainability **11**, 1116 (2019). https://doi.org/10.3390/su11041116

W. Laurance, J. Sayer, K. Cassman, Agricultural expansion and its impacts on tropical nature. Trends Ecol. Evol. **29** (2013). https://doi.org/10.1016/j.tree.2013.12.001

S. Lovera, *Trends in the privatization and corporate capture of biodiversity, Spotlight on Sustainable Development*, vol 15 (University of Amsterdam, Amsterdam, 2017)

I.E. Måren, Food systems for sustainable terrestrial ecosystems (SDG 15). Food Ethics **2**, 155–159 (2019). https://doi.org/10.1007/s41055-018-00032-2

R. Pirard, Payments for environmental services (pes) in the public policy landscape: 'Mandatory' spices in the indonesian recipe. Forest policy and economics. Spec. Issue Glob. Gov **18**, 23–29 (2011)

J. Sayer, D. Sheil, G. Galloway, R.A. Riggs, G. Mewett, K.G. MacDicken, B.J.M. Arts, A.K. Boedhihartono, J. Langston, D.P. Edwards, SDG 15 Life on Land – The Central Role of Forests in Sustainable Development, in *Sustain. Dev. Goals Their Impacts For. People*, (Cambridge University Press, Cambridge, 2019), pp. 482–509. https://doi.org/10.1017/9781108765015.017

SDG 15: Life on Land (2021). https://in.one.un.org/page/sustainable-development-goals/sdg-15/. Accessed 25 Oct 2021

SDG 15: Protect, restore and promote sustainable use of terrestrial ecosystems, sustainably manage forests, combat desertification, halt and reverse land degradation and halt biodiversity loss – SDG Compass [WWW Document] (2021). https://sdgcompass.org/sdgs/sdg-15/. Accessed 22 Aug 2021

---

[1] For reference, you may click on the hyperlinks on the company names or follow the websites here (Accessed Online – 2.1.2022):

http://internetoftrees.tech/; http://www.wipsea.com/; https://apic.ai/; https://dendra.io/; https://descarteslabs.com/; https://droneseed.com/; https://ecording.org/; https://en.greenpraxis.com/; https://hortau.com/; https://jupiterintel.com/; https://landlifecompany.com/; https://n2applied.com/; https://optirtc.com/; https://ororatech.com/; https://satelligence.com/; https://skymining.com/en/skymining-en/; https://spinnova.com/; https://tesselo.com/; https://timbeter.com/; https://treevia.com.br/; https://virotec.com/; https://wastehero.io/; https://www.aeriumanalytics.com/; https://www.beewise.ag/; https://www.breeze-technologies.de/; https://www.cloudagronomics.com/; https://www.covercress.com/; https://www.daumet.com/en/; https://www.desertcontrol.com/; https://www.drylet.com/; https://www.greencitywatch.org/; https://www.nanofexllc.com/; https://www.nofence.no/en/; https://www.projectcanopy.org/; https://www.robin-radar.com/; https://www.robotto.ai/; https://www.seetree.ai/; https://www.spoor.ai/; https://www.sylvera.com/; https://www.terramera.com/; https://www.terviva.com/; https://www.tesera.com/; https://www.thebeecorp.com/; https://www.watergenics.tech/; https://www.xampla.com/

D. Tremblay, F. Fortier, J.-F. Boucher, O. Riffon, C. Villeneuve, Sustainable development goal interactions: An analysis based on the five pillars of the 2030 agenda. Sustain. Dev. **28**, 1584–1596 (2020). https://doi.org/10.1002/sd.2107

L. Tyrväinen, S. Pauleit, K. Seeland, S. de Vries, Benefits and Uses of Urban Forests and Trees, in *Urban Forests and Trees: A Reference Book*, ed. by C. Konijnendijk, K. Nilsson, T. Randrup, J. Schipperijn, (Springer, Berlin, Heidelberg, 2005), pp. 81–114. https://doi.org/10.1007/3-540-27684-X_5

UNEP, 2021. GOAL 15: Life on land [WWW Document]. UNEP – UN Environ. Programme. http://www.unep.org/explore-topics/sustainable-development-goals/why-do-sustainable-development-goals-matter/goal-15. Accessed 18 Aug 2021

SDG-16: Peace, Justice and Strong Institutions

18

Abstract

Institutions and organisations must give due importance to the rule of law, the sanctity of human rights and the effect of stability to ensure sustainable development. SDG-16, Peace, Justice and Strong Institutions, aims to strengthen justice and strong corporate culture to achieve sustainable development and social peace. Greatly reducing crime and conflict through justice and strong institutions, upholding the rule of law and strengthening the presence of developing countries in global governance institutions are essential topics for SDG-16. This chapter presents the business models of eight companies and use cases that employ emerging technologies and create value in SDG-16. We should highlight that one use case can be related to more than one SDG and it can make use of multiple emerging technologies.

Keywords

Sustainable development goals · Business models · Peace, justice and strong institutions · Sustainability

The author would like to acknowledge the help and contributions of Ulaş Özen, Eren Fidan, Uğur Dursun, Büşra Öztürk, Asya Nur Sunmaz and Muhammed Emir Gücer in completing of this chapter.

Institutions and organisations must give due importance to the rule of law, the sanctity of human rights and the effect of stability to ensure sustainable development. Societies that have grown and prospered in the last 300 years are those that adhere to the requirements of democracy, respect human rights and have adopted inclusive economic institutions (Acemoglu et al. 2012). In governments where peace and social reconciliation cannot be achieved, justice is jeopardised, and eventually, conflict and fear dominate. In places with this order, where institutions and justice are not strong, violence and crime rates are high, abuse and exploitation are common, and corruption and bribery are common. This is a problem that exists in many places in the world and must be solved. In an increasingly globalised world, conflict and instability in one region can also affect many parts of the world. SDG-16 aims to strengthen justice and strong corporate culture to achieve sustainable development and social peace. Greatly reducing crime and conflict through justice and strong institutions, upholding the rule of law and strengthening the presence of developing countries in global governance institutions are important topics for SDG-16 (United Nations 2021).

The UN's Department of Economic and Social Affairs defines four different indicators that are observed as a crucial decrease in violence and related death rates worldwide, in Target 16.1. The first indicator is the total number of intentional homicide victims divided by the entire population, expressed per 100,000 people. The total number of conflict-related deaths divided by the entire population stated per 100,000 is stated as the second indicator. Third, the total number of people who have been victims of physical, psychological or sexual violence in the past 12 months is a percentage of the overall population. The fourth indicator measures the percentage of adults who feel comfortable travelling alone in their community (The World Bank 2021a). These indicators are followed by the United Nations Office on Drugs and Crime and the Office of the United Nations High Commissioner for Human Rights.

Three separate indicators are observed under Goal 16.2, which is to end child abuse, exploitation, trafficking and all kinds of violence against and torture of children. These are defined in the United Nations Department of Economic and Social Affairs metadata. The first measure, the percentage of children aged 1–17 years who experienced physical punishment and/or psychological aggression by caregivers in the previous month, is now being defined as the percentage of children between the ages 1 and 14 years who experienced physical punishment and/or psychological aggression by caregivers in the previous month. The second indicator is defined as the ratio of total victims of human trafficking found or residing in a nation to the population resident in the country, expressed per 100,000 people. Thirdly, the percentage of young women and men aged 18–29 who had experienced sexual assault by the age of 18. The United Nations Children's Fund (UNICEF) and the United Nations Office on Drugs and Crime (UNODC) both monitor these metrics (The World Bank 2021a).

Three different indicators are observed under Target 16.3, promoting the rule of law at the national and international levels and ensuring equal access to justice for all. The UN's Department of Economic and Social Affairs defines these in their metadata. The first indicator, the number of victims of violent crime in the previous 12 months who reported their victimisation to competent authorities or other officially recognised conflict resolution mechanisms, is a percentage of all victims of violent crime in the previous 12 months. Second, on a specified date, the total number of persons held in detention who have not yet been sentenced is a percentage of the total number of persons held in detention. Third, by type of mechanism, the number of persons who experienced a dispute during the past 2 years who accessed a formal or informal dispute resolution mechanism is a percentage of all those who experienced a dispute in the past 2 years. These indicators are followed by the United Nations Office on Drugs and Crime (UNODC), United Nations Development Programme (UNDP) and Organisation for Economic Cooperation and Development (OECD) (SDG Tracker 2021).

Under the SDG-16.4 target, organised crime and terrorist organisations continue their existence by creating fear and insecurity in society. While organised crime organisations are for economic profit, the target of terrorism is ideological and political (Bovenkerk and Chakra 2004). Such organisations illegally finance the revenue sources of their actions. For a stronger, more peaceful society, it is essential to have justice and strong security institutions. To achieve this, the security forces' fight against all kinds of crimes is one of the top priorities for social peace. This struggle has reached even more advanced levels with the development of technology, for example, Cybersecurity.

Under SDG-16.5, corruption negatively affects economic growth and society's trust in institutions (Brouthers et al. 2008). Corruption and bribery disrupt the functioning of an institution by doing what is asked instead of what needs to be done. Institutions that do not comply with such laws and regulations have an order dominated by the powerful. They cause an increase in inequalities and a loss of a sense of justice in society. This corruption in authorities and institutions causes a public reaction and damages the culture of democracy.

Under SDG-16.6, making participatory, inclusive decisions with the participants at all levels is one of the ideals of democratic culture. Participatory democracies can increase their understanding of politics and their dialogue with each other, no matter how difficult it is to cope with today's challenges (Collins 2019). In a world where inequality is significantly reduced and women and minorities are more participatory, it is obvious that the decisions taken will be more permanent and more just and will serve more peace. For this to happen, a social consensus and social peace affect each other positively in a two-way manner.

Under SDG-16.7, birth registration implantation ensures that children can access justice and social services and protect children. However, data from 2010 to 2019 shows that one in four children in all populations who are under the age of 5 were never officially recorded by states (United Nations 2021). In 2020, the registration rates of children under the age of 5 in sub-Saharan Africa (46%) and underdeveloped countries (44%) were well below the world average (74%) (UNICEF DATA 2021).

Under SDG-16.8, there is an aim to increase the voting power of developing countries. According to the World Bank, the USA, Germany, UK, France and Japan have 35.21% of all voting power in global economic institutions. However, developing countries don't have enough power in economic institutions (The World Bank 2021b).

Under SDG-16.9, human rights institutions ensure that justice systems are processed fairly in countries. In 2019, 40% of countries in the world had human rights institutions that audit the official institutions of governments. Human rights institutions comply with Paris Principles (European Union Agency for Fundamental Rights 2012). Seventy-eight countries in Eastern and South-Eastern Asia, Latin America and the Caribbean, Oceania and sub-Saharan Africa still have difficulty accessing human rights institutions (United Nations 2020).

Under SDG-16.10, ensuring public access to information and protecting fundamental freedoms under national legislation and international agreements are focused on specifically. SDG-16.10 tries to increase the extent of the state's respect and protection besides citizens' access to information rights (Bolaji-Adio 2015). To achieve the target, adopting and implementing constitutional, regulatory and political measures to guarantee public access to information is essential (Cling et al. 2018). Another indication of SDG-16.10 might be to evaluate if public officers are completely and effectively using the anti-corruption instruments and structures to combat corruption (Bolaji-Adio 2015). SDG-16.10 thus plays a key role in ensuring accountability in the context of the SDGs so that they may be effective.

Target 16.a is to strengthen relevant national institutions including through international cooperation, for preventing violence and combating terrorism and crime. In compliance with the Paris Principles, independent national human rights institutions could be stated as an instance for the target. Appraising the effectiveness of the national institutions in terms of the resources (human, financial and logistics) that have been involved in intra- and inter-state conflict resolution supports the increase of such actions for the target (Bolaji-Adio 2015).

Target 16.b deals with promoting and enforcing non-discriminatory laws and policies. Undoubtedly, non-discrimination must be worked on for the welfare of the world and fair, equitable and timely access to justice. Promoting and protecting the rights of permanently disadvantaged or vulnerable groups, including but not limited to internally displaced persons, refugees and persons with disabilities, is the primary target of 16.b (Bolaji-Adio 2015). When identifying vulnerable groups, transparent, participatory and accountable processes leading would help achieve this target. Assessing the effectiveness of the measures and sharing details of any violation and reports available are important for the accurate determination of the next policies. Figure 18.1 illustrates the targets and sub-targets of SDG-16.

Achieving sustainable development goals requires peaceful, fair and inclusive communities (SDGs). Regardless of their race, religion or sexual orientation, people everywhere deserve to be free from fear of violence and feel secure going about their daily lives. Many studies have shown that peace and development go together. As an example, research by the World Bank and the

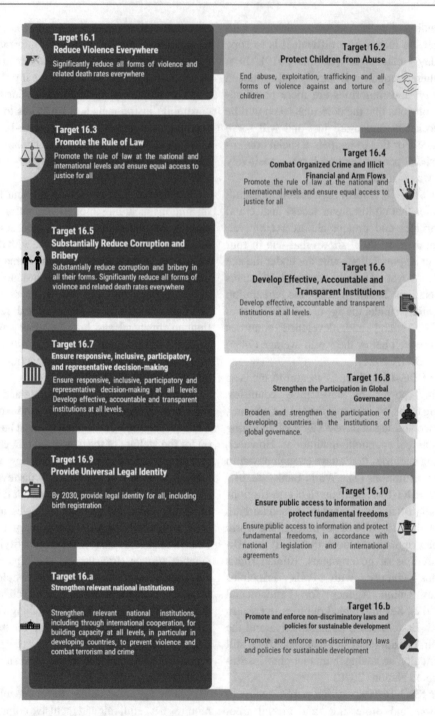

Fig. 18.1 Targets of SDG-16. (United Nations 2021)

United Nations shows that instability and war are key development problems that can stall progress. According to the IEP, increased levels of violence have a detrimental influence on eco- nomic growth by reducing international invest- ment and the financial environment. According to the study's authors, this has an impact on poverty and economic development, expected lifespan and educational achievements and characteristics critical for long-term development, such as new-born mortality and availability of services.

The 2030 Agenda says and maintains that "there can be no sustainable development without peace". Along with people, wealth, the environment and cooperation, peace is regarded as one of five essential areas for humankind. The 2030 Agenda emphasises the importance of building peaceful, just and inclusive communities centred around human rights (along with the right to improvement), an efficient judicial system and effective governance throughout all degrees while also transparent, efficient and responsible institutions. Political goals, such as guaranteeing inclusiveness, strengthening effective governance and ending violence, were seen as equally important as economic, social and environmental goals. It was then that the 2030 Agenda's SDG-16 arose as an "enabler". As a result, SDG-16 is a critical component of the transformational 2030 Agenda. This holds true for all objectives, including those linked to climate change, health, education, economic growth and so forth. Development achievements will be undone in the absence of long-term peace, which includes respect for human rights and the judicial system as well as the absence of violence. Inequalities in poverty reduction and socio-economic development will rise without having access to justice for all and inclusion, and governments' promises to leave no one behind will not be realised.

We may use one of the objectives as an example to show how important SDG-16 is in accomplishing the SDGs and attaining comprehensive sustainable development. Target 16.5 aims, for example, to "significantly eliminate bribery and corruption of any kind". An atmosphere of excellent governance, security and peace is ideal for sustainable development. On the other hand, corruption has a negative influence on long-term growth and frequently results in civil unrest and insecurity. Overall, the empirical evidence and sustainable development measures demonstrate that countries with high rates of corruption have low rates of growth, average life expectancy, mean years of schooling and public policy effectiveness while having high rates of poverty as well as a high number of maternal deaths and high average child mortality rates per 1000 births (UNICEF DATA 2019). As a result, SDG-16 serves as a foundation for the other 16 SDGs, which all depend upon inclusive institutions which can have a responsibility towards public demands in an open and accountable manner. The SDG-16 targets reflect human rights commitment, accountability, transparency and justice, which would be essential for an environment where people can have liberty in life and be safe and prosperous. SDG-16 impacts many elements of society and the 2030 Agenda, from anti-corruption and the judicial system to participatory policy planning, violence minimisation and peace encouragement.

By 2030, the blue planet we live on will not keep up with the increasing population and crisis. As stated by Wahba, 80% of the world' inhabitants will be living in difficult and dangerous conditions (Wahba 2019). The need for SDG-16 will increase further in the coming years. According to Sugg, investment in SDG 16+ should be viewed and emphasised as an investment in the 2030 Agenda as a whole. Civil liberties are dwindling worldwide, with 181 limitations put on non-governmental groups in 82 countries since 2013. Failure to address these issues and invest in SDG16+ leads to further violence, injustice and exclusion. This will result in a reversal of development progress in all SDGs, such as education, health and climate action.

SDG-16, among other things, entails a variety of measures aimed at improving the lives of people with disabilities. Metropolitan areas are predicted to house around 6.25 billion people by 2050, with 15% of them being disabled people (DIAUD 2018). Governments began to create conditions to eliminate discrimination between people and create an inclusive society. According to Radović (2019):

> It is informative to mention positive examples from practice related to the New Zealand Disability Strategy 2016–2026, developed by the Office for Disability Issues in consultation with other government agencies and the disability sector (supported by an Outcomes Framework and a Disability Action Plan), and the Australian National Disability Strategy which includes trend data indicators against each of the focus areas and reports every two years.

It is on the agenda to apply machine learning (ML) while making SDG-16 reach its future goals. According to Dasandi and Mikhaylov, given the limited resources available to help poor and developing countries achieve the SDGs, understanding the interrelationships between the many parts of SDG-16 and other SDG objectives is particularly crucial (Dasandi et al. 2019). If these links are appropriately identified, further country-specific assistance to SDG success can be targeted. Supporting success on some aspects of SDG-16 in one nation, for example, might result in advances on other SDGs in that country. Machine learning may be used to assist and better grasp such linkages in a variety of ways. This will enable the identification of SDG-16 indicators that influence the changes in other SDG indicators such as those related to health, education and poverty. Second, multilayer network models may be utilised to uncover causal linkages between indicators for SDG-16 and other SDG objectives. In other words, thanks to machine learning, it can better understand how governments and institutions affect issues such as health and education (Dasandi et al. 2019).

Like the SDG's aims, their finance must be long term if their goals are to be realised, particularly in developing nations. According to Kempe Ronald Hope Sr., funding is critical for the implementation and success of the 2030 Agenda, as it is for other development programmes (Kempe Ronald Hope Sr. 2019). To begin with, data on SDG-16 objectives shows that violence and war (including persecution and human rights abuses) resulted in the forced relocation of roughly 70.8 million people globally by 2018, with the global economic cost of violence projected to be over $1 trillion. In purchasing power parity terms, it will be $14.1 trillion, or 11.2% of world GDP. Second, according to the Economic Impacts of Child Marriage project's most recent research, child marriages would cost more than $560 billion in welfare losses by 2030. Finally, the International Monetary Fund estimates that bribery costs US$1.5–2 trillion, or around 2% of GDP, with far higher economic and social costs when other kinds of corruption are considered (The World Bank 2017). Overall,

both developed and developing countries will face considerable resource constraints because of the SDGs. Estimates range from US$3.3 trillion to 4.5 trillion for basic infrastructure (roads, railroads, and ports; power plants; water and sanitation), food security (agricultural and rural development), climate change mitigation and adaptation, health and education. Only in undeveloped nations does it fluctuate between US Dollars (Kempe Ronald Hope Sr. 2019).

According to the United Nations Conference on Trade and Development (UNCTAD) and others, there is a total yearly finance gap of roughly US$2.5 trillion in impoverished countries with present levels of public and private investment in SDG-related industries. Considering the global GDP of over US$115 trillion and what attaining the SDGs entails for unlocking human and economic potential and ensuring planetary security, closing such a gap is a massive task (UNCTAD 2014). In the United States, funding for public goods and key services is unquestionably critical. By pursuing fiscal reforms, African nations, for example, might increase their fiscal space by 12–20% of GDP. Fiscal policy in the form of taxes has been a critical component for long-term growth and equity in the Asia-Pacific region. However, both wealthy and developing nations' public financing sources are insufficient to meet the SDGs. As a result, private finance is a critical component of the 2030 Agenda's funding. To address the SDG-16 funding gap, governments must reallocate resources from crisis response to violence prevention while boosting investment in justice and inclusion and lowering resources wasted due to corruption and illicit flows (UNCTAD 2014). Improved governance will boost local resource mobilisation while also improving the efficiency of using resources on the SDGs.

18.1   Companies and Use Cases

Table 18.1 presents the business models of eight companies and use cases that employ emerging technologies and create value in SDG-16. We should highlight that one use case can be related

Table 18.1 Companies and use cases in SDG-16

No	Company info	Value proposal (what?)	Value creation (how?)	Value capture
1	BitGive Foundation USA 1, 2, 10, 16 Blockchain, crowdfunding	The institution provides people with reliable and transparent service in the use of funds, with the ability to track and audit their donations using Bitcoin and blockchain technology.	With the GiveTrack project, donations made by donors are recorded in blocks for charity work. In addition to the donation information transferred to the blockchain, the movements of the donation and the results of the project are added to ensure the traceability of the donation. Thanks to the distributed application, it cannot be manipulated or changed by the authorities.	The institution enables donations made with a transparent system that does not require trust against traditional donation systems with blockchain technology to have a verifiable impact. By eliminating intermediary institutions, it provides a faster, less costly and more reliable process.
2	Blackbird.AI USA 16 AI	It is a threat and perception intelligence platform that enables organisations to proactively protect against misinformation, enhance content safety compliance across digital platforms and uncover the forces behind major events.	It is utilising a variety of artificial intelligence (AI) technologies to address the problem of filtering and understanding emergent narratives from across the Internet in order to identify misinformation threats aimed at its consumers.	Value is captured by identifying which voices are important and which are not, systematically addressing threats, getting down to their roots and networks to prevent their repetition, detecting signals of misinformation early before they have a chance to make an.
3	E-Residency Estonia Estonia 8, 16 Blockchain	The Estonian government's blockchain-based system allows companies and entrepreneurs, regardless of their location, to run their businesses on digital platforms, and provide consultancy or SaaS services, to establish a business in the European Union market by granting digital Estonian citizenship.	Using blockchain's distributed ledger system, the program keeps records of transactions made in its blocks in an efficient and verifiable way. It ensures that all transactions are made from a single platform by keeping the individual's data, contracts such as insurance and rent in these blocks and accepting them as a signature used in the digital environment.	It enables companies and enterprises to integrate into the European Union market more quickly and with less cost, by saving them from European Union customs duties and giving the rights of EU companies.
4	ID2020 USA 1, 9, 10, 16 Blockchain	The company provides digital identity using cryptography and blockchain technology, especially for anonymous people and refugees who are not recognised by official authorities in the world.	The digital identity is turned into a digital fingerprint with the letter and number series algorithm of blockchain technology called a hash. The individual's data is recorded in these hashes, and the accuracy of the information is checked by many nodes.	It gives millions of refugees and unidentified individuals the right to be recognised by official authorities and access to their basic needs. It enables them to use their many rights such as citizenship, voting and access to social services and integrate them into the modern world.

(continued)

Table 18.1

No	Company info	Value proposition (what?)	Value creation (how?)	Value capture
5	Jigsaw USA 9, 10, 16 AI, cloud computing, cybersecurity	The company conducts research and develops products using AI, machine learning and cybersecurity technologies to ensure freedom of expression, cybersecurity and access to accurate information.	With the Perspective project, the company uses AI to identify bad content on the Internet and scores it according to the harm rate. Accordingly, it allows the user to filter the contents according to the damage score. Another project, Project Shield, caches user configuration settings and proxy traffic of institutions and NGOs that play an important role in accessing accurate information with cloud computing and absorbing malicious traffic, thus providing Cybersecurity against DDoS attacks.	It increases Internet literacy by fighting against censorship, disinformation and violence against open societies, by increasing free societies and access to accurate information.
6	Justice Chatbot Uganda 1, 16 AI	It is a software-as-a-service (SaaS) platform providing AI-powered chatbots for organisations and individuals to solve their simple legal issues.	They provide a conversation with an AI-based bot in a social media channel's messaging app. Several legal issues can be solved by talking with the bot via this messaging app, either individually or from a corporate perspective, and if required, a meeting with the most appropriate attorneys can be organised at the client's location.	The chatbot, which was constructed using AI, captures value by providing quick and easy legal consultation services to a larger number of individuals. They also add societal value by providing free legal services to a large number of individuals.
7	Kleros France 9, 16 Blockchain	It is an online settlement and commercial dispute resolution platform for consumers that resolve authorities in different jurisdictions using crowdsourcing and blockchain technology.	The distributed data structure of blockchain technology ensures that the evidence is protected in a transparent and impartial manner and that the juries to be selected for the resolution of the case are selected independently from any authority with smart contracts.	In addition to many transactions and disputes that the judiciary cannot reach, it contributes to the need for quick justice of individuals and companies by filling the justice gap with a fast, independent and efficient solution method against traditional dispute resolution.
8	Provenance UK 3, 12, 16 Blockchain	The company transparently presents the production stages of the products offered for sale and the impact it creates on the consumers using the distributed data ledger of blockchain technology.	It records the production and supply chain mapping data of the product on blocks with the immutable data ledger feature of the blockchain, allowing the data to reach the consumer in a transparent way without allowing the company to be manipulated.	By making the social and environmental impact of the product more accessible and reliable, it creates an awareness in the consumer and the producer whether the product produced harms the environment and society. It reduces the consumer's indifference towards inaccurate and opaque supply chain methods, especially in the food and fashion industry.

to more than one SDG and it can make use of multiple emerging technologies. In the left column, we present the company name, the origin country, related SDGs and emerging technologies that are included. The companies and use cases are listed alphabetically.[1]

References

D. Acemoglu et al., The environment and directed technical change. Am. Econ. Rev. **102**(1), 131–166 (2012). https://doi.org/10.1257/AER.102.1.131

A. Bolaji-Adio, in *European Centre for Development Policy Management Discussion Paper*. The Challenge of Measuring SDG 16: What Role for African Regional Frameworks? (2015). Available at: www.ecdpm.org/dp175. Accessed 2 Nov 2021

F. Bovenkerk, B.A. Chakra, *Terrorism and Organized Crime* (Research Gate, 2004). Available at: https://www.researchgate.net/publication/27694841_Terrorism_and_organized_crime. Accessed 2 Nov 2021

K.D. Brouthers, L.E. Brouthers, S. Werner, Real options, international entry mode choice and performance. J. Manag. Stud. **45**(5), 936–960 (2008). https://doi.org/10.1111/J.1467-6486.2007.00753.X

J.-P. Cling, M. Razafindrakoto, F. Roubaud, *SDG 16 on Governance and its Measurement: Africa in the Lead, Working Papers* (2018). Available at: https://ideas.repec.org/p/dia/wpaper/dt201802.html. Accessed 2 Nov 2021

P.H. Collins, The Difference that Power Makes: Intersectionality and Participatory Democracy, in *The Palgrave Handbook of Intersectionality in Public Policy*, (Palgrave Macmillan, Cham, 2019), pp. 167–192. https://doi.org/10.1007/978-3-319-98473-5_7

N. Dasandi, Slava, †, J. Mikhaylov, *AI for SDG 16 on Peace, Justice, and Strong Institutions: Tracking Progress and Assessing Impact* * (Sjankin, 2019). Available at: https://sjankin.com/files/ijcai19-sdg16.pdf. Accessed 2 Nov 2021

DIAUD, *The Inclusion Imperative: Towards Disability-Inclusive and Accessible Urban Development Key Recommendations for an Inclusive Urban Agenda* (Global Network on Disability Inclusive and Accessible Urban Development, 2018). Available at: http://disabilityrightsfund.org/wp-content/uploads/2016/11/The_Inclusion_Imperative__Towards_Disability-Inclusive_Development_and_Accessible_Urban_Development.pdf. Accessed 2 Nov 2021

European Union Agency for Fundamental Rights, F., Handbook on the establishment and accreditation of National Human Rights Institutions in the European Union (2012). https://doi.org/10.2811/14554

Kempe Ronald Hope Sr, Peace, justice and inclusive institutions: Overcoming challenges to the implementation of Sustainable Development Goal 16. Sci-Hub: Glob. Chang. Peace Secur., 1–21 (2019). Available at: https://sci-hub.se/https://www.tandfonline.com/doi/abs/10.1080/14781158.2019.1667320. Accessed 2 Nov 2021

V. Radović, *SDG16 – Peace and Justice, Concise Guides to the United Nations Sustainable Development Goals* (Emerald Publishing Limited, Bingley, 2019). https://doi.org/10.1108/9781789734775

SDG Tracker, Goal 16 (2021). Available at: https://www.sdg.gov.bd/page/indicator-wise/5/458/3/0#1. Accessed 2 Nov 2021

The World Bank, *Child Marriage Will Cost Developing Countries Trillions of Dollars by 2030, Says World Bank/ICRW Report* (2017)

The World Bank, DataBank: Metadata glossary (2021a). Available at: https://databank.worldbank.org/metadataglossary/world-development-indicators/series/VC.IHR.PSRC.MA.P5. Accessed 2 Nov 2021

The World Bank, *International Bank for Reconstruction and Development Subscriptions and Voting Power Of Member Countries Total Subscriptions Voting Power Member Amount (*) Percent Of Total No. Of Votes Percent of Total* (2021b)

UNCTAD, *World Investment Report 2014 New York and Geneva, 2014 Investing in the SDGs: An Action Plan*, in (2014)

UNICEF DATA, Maternal mortality rates and statistics (2019). Available at: https://data.unicef.org/topic/maternal-health/maternal-mortality/. Accessed 2 Nov 2021

UNICEF DATA, Birth registration (2021). Available at: https://data.unicef.org/topic/child-protection/birth-registration/. Accessed 2 Nov 2021

United Nations, Economic and Social Council: Progress towards the Sustainable Development Goals (2020). Available at: https://undocs.org/en/E/2020/57. Accessed 2 Nov 2021

United Nations, Peace, justice and strong institutions (2021). Available at: https://www.un.org/sustainabledevelopment/peace-justice/. Accessed 2 Nov 2021

M. Wahba, SDG 16+ and the future we want (United Nations Development Programme, 2019). Available at: https://www.undp.org/speeches/sdg-16-and-future-we-want. Accessed 2 Nov 2021

---

[1]For reference, you may click on the hyperlinks on the company names or follow the websites here (Accessed Online – 2.1.2022):

http://blackbird.ai/; https://e-resident.gov.ee/; https://id2020.org/; https://jigsaw.google.com; https://justice-chatbot.org/; https://kleros.io/; https://www.bitgivefoundation.org/; https://www.provenance.org/

SDG-17: Partnerships for the Goals

19

Abstract

Global partnerships have been rapidly increased due to the transition to digitalisation, and an event at one end of the world causes different circumstances in many other regions. SDG-17, Partnerships for the Goals, fundamentally calls for strengthening the global cooperation on sustainable development goals in the agenda 2030. SDG-17 has a crucial role in advancing the global partnership and implementation tools in reaching the solutions to social and ecological problems. This chapter presents the business model of one company and use case that employ emerging technologies and create value in SDG-17.

Keywords

Sustainable development goals · Business models · Partnerships for the goals · Sustainability

Global partnerships have been rapidly increased due to the transition to digitalisation, and an event at one end of the world causes different circumstances in many other regions. The cause of con-

sequences such as wars, natural disasters, climate disasters and humanitarian crises does not come from a single location, but it is a global cause. In this respect, world leaders are now meticulous about whether a problem is local or global. It was accepted that global partnership and cooperation in line with such sustainable development goals can only be achieved through close solidarity (MacDonald et al. 2018). At this point, developed countries have committed to helping other countries where they are strong. They even aim to increase development aid to enhance growth and welfare in many countries. Based on these, the development of international trade and financial restructuring of less developed countries can be given as examples to implement the global partnership for the goals.

SDG-17 fundamentally calls for strengthening the global cooperation on sustainable development goals in the agenda 2030. SDG-17 has a crucial role in advancing the global partnership and implementation tools in reaching the solutions to social and ecological problems. Regarding the goal, partnerships between governments, the private sector and civil society are planned to be deepened and coordinated. Additionally, the need to ensure the consistency of the policies of the sustainable development goals at the domestic and international levels is met within the SDG-17. SDG-17 could be considered a bridge for achieving all SDGs. In other words, it is extremely necessary to fulfil the

The author would like to acknowledge the help and contributions of Ulaş Özen, Eren Fidan, Uğur Dursun, Büşra Öztürk, Asya Nur Sunmaz and Muhammed Emir Gücer in completing of this chapter.

objectives and goals of SDG-17 for successfully advancing and executing the SDGs at all levels (Franco and Abe 2020). SDG 17 has 19 selected targets to be achieved upon the 2030 agenda. While the 19 targets covered a vast range of affairs, these are mainly associated with the targets from SDG-16 and SDG-9 (Maltais et al. 2018). This relation is in terms of enhancing the quality of government and public administration and access to technology, respectively. To increase the possibility of implementation of the targets, the objectives of SDG-17 are classified into more detailed key themes in the studies. Revitalising global partnerships in five broad categories such as finance, technology, capacity building, trade, policy and institutional coherence is the most common classification for targets of SDG-17 (United Nations 2017). Finance, encompassing targets 17.1–17.5, focuses on developing countries partnering with and assisting developing countries in revenue collection, mobilising aid, long-term debt sustainability and promoting investment.

The targets of SDG-17 between 17.6 and 17.8 are about technology which focuses on the technological distinction between North and South. The divide includes enhancing global partnership, improving coordination to accessing technology and innovation and improving sound technologies to enhance information and communication technology usage. According to World Bank data, the average world fixed broadband subscription per 100 people is 15.87. However, the average of the least developed countries is 1.39 per 100 people (International Telecommunication Union Database 2020). So, it is understood that people of the least developed countries cannot regularly reach an Internet infrastructure. But fixed broadband subscriptions are increasing year by year. During the COVID-19 pandemic, the Internet connection has been crucially important for people. A lack of Internet infrastructure has a high cost for the least developed and developing countries, especially in health, economic and social life.

Under the SDG-17.9 target, it aims to build capacity in developing countries and enhance international support for adapting their national plan to sustainable development goals. According to OECD, official development assistance (ODA) has reached 161 billion dollars in 2019 (OECD Statistics 2021).

Trade targets 17.10–17.12, draw attention to the importance of rules-based and equitable trading and seek to increase the share in international trade by developing countries through the use of multinational organisations and frameworks. Targets 17.13–17.15 focus on policy and institutional coherence to enhance macroeconomic stability and sustainable development, intertwined with an approach that respects country-specific modalities. Targets 17.16 and 17.17 address the necessity of multi-stakeholder partnerships for coordinating and sharing resources, knowledge, expertise and technology in support of the SDGs, in developing countries. Targets 17.18 and 17.19 cover data, monitoring and accountability, supporting and enhancing countries' capacity to increase the availability of high-quality data and developing countries' statistical capacity (United Nations Sustainable Development 2021). Figure 19.1 presents the targets and sub-targets of SDG-17.

While introducing all the sustainable development goals, achieving the goal in collaboration stands out as the real challenge. The core of the UNDP SDGs is reducing a variety of inequalities within different nations and areas while trying to keep the world healthy and able for the next generations who will be suffering from past and current mistakes of humanity about the world and the environment. Sustainable development goals from 1 to 16 all try to achieve this core philosophy while focusing on different needs and grounds. Goal 17, on the other hand, acts as the backbone for all others. To fully integrate sustainable development goals into real-life functioning improvements, establishing partnerships among governments, both the public and private sector and the society itself, is crucial. At all degrees that SDG applications take place, strong and mutually generated partnerships are required that evolve around prioritising people and the environment (Earth Changers 2020).

This sustainable development goal aims to secure the collaborative act of these partnerships

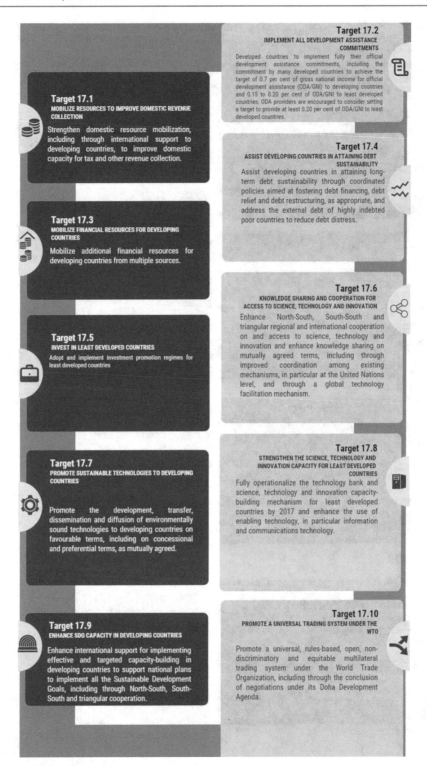

Fig. 19.1   Targets of SDG-17. (United Nations 2021)

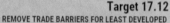

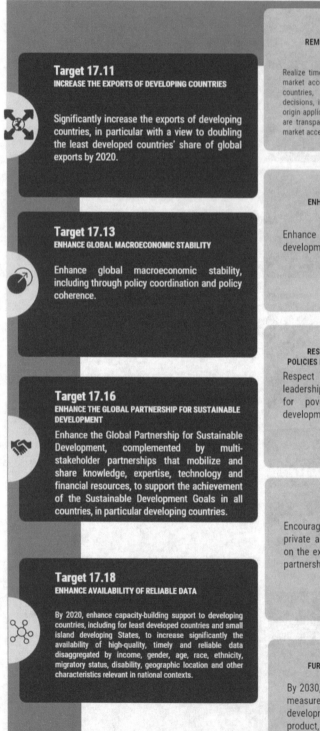

Target 17.12
REMOVE TRADE BARRIERS FOR LEAST DEVELOPED COUNTRIES

Realize timely implementation of duty-free and quota-free market access on a lasting basis for all least developed countries, consistent with World Trade Organization decisions, including by ensuring that preferential rules of origin applicable to imports from least developed countries are transparent and simple, and contribute to facilitating market access.

Target 17.11
INCREASE THE EXPORTS OF DEVELOPING COUNTRIES

Significantly increase the exports of developing countries, in particular with a view to doubling the least developed countries' share of global exports by 2020.

Target 17.14
ENHANCE POLICY COHERENCE FOR SUSTAINABLE DEVELOPMENT

Enhance policy coherence for sustainable development.

Target 17.13
ENHANCE GLOBAL MACROECONOMIC STABILITY

Enhance global macroeconomic stability, including through policy coordination and policy coherence.

Target 17.15
RESPECT NATIONAL LEADERSHIP TO IMPLEMENT POLICIES FOR THE SUSTAINABLE DEVELOPMENT GOALS

Respect each country's policy space and leadership to establish and implement policies for poverty eradication and sustainable development.

Target 17.16
ENHANCE THE GLOBAL PARTNERSHIP FOR SUSTAINABLE DEVELOPMENT

Enhance the Global Partnership for Sustainable Development, complemented by multi-stakeholder partnerships that mobilize and share knowledge, expertise, technology and financial resources, to support the achievement of the Sustainable Development Goals in all countries, in particular developing countries.

Target 17.17
ENCOURAGE EFFECTIVE PARTNERSHIPS

Encourage and promote effective public, public-private and civil society partnerships, building on the experience and resourcing strategies of partnerships.

Target 17.18
ENHANCE AVAILABILITY OF RELIABLE DATA

By 2020, enhance capacity-building support to developing countries, including for least developed countries and small island developing States, to increase significantly the availability of high-quality, timely and reliable data disaggregated by income, gender, age, race, ethnicity, migratory status, disability, geographic location and other characteristics relevant in national contexts.

Target 17.19
FURTHER DEVELOP MEASUREMENTS OF PROGRESS

By 2030, build on existing initiatives to develop measurements of progress on sustainable development that complement gross domestic product, and support statistical capacity-building in developing countries.

Fig. 19.1   (continued)

mentioned above where they do not have to perform or handle the possible crisis about poverty and environmental degradation by themselves. As well as promoting collaboration at all degrees (global, regional, national and local), SDG-17 also promotes the need to obtain new financial resources to accomplish the rest of the SDGs. It would be very hard to execute the other 16 SDGs without significant progress on SDG-17.

Governmental collaborations with a range of public, private and civil society associates might help raise funds for various developments and encourage greater inclusivity all through their execution (ICLEI 2015). The nature of the goal might be explained as: It encourages wealthier nations to take on more responsibilities, such as influencing cohesive decision-making (SDG-17.14), endorsing infrastructure construction in developing states (SDG-17.9) or enhancing developing states' access to sustainable and greener technology (SDG-17.9) (ICLEI 2015).

In a world where working together has become more important in recent years, it was inevitable for the SDGs to work together. The globe is more interconnected than ever before because of the Internet, travel and global organisations. The need to act together to combat climate change is becoming increasingly obvious. The sustainable development goals are also a big deal. 193 nations agreed upon these goals. The ultimate objective establishes a framework for nations to collaborate to obtain all other goals (United Nations 2021). Transformative change is needed around the globe and by multi-stakeholder partnerships. Many partnerships have emerged expressly to solve international problems in recent years, resulting in literally hundreds of multi-stakeholder collaborations worldwide. Finding and developing transformational collaborations provides the best chance to tackle these issues and drive significant shifts successfully.

Consequently, the United Nations has called 2018 the "Decade of Action" devoted to delivering on the 2030 agenda and SDGs (Li et al. 2020). We need fresh dedication and investment to achieve the SDGs and take meaningful action on climate change. Nevertheless, multi-stakeholder partnerships can only be effective if they are willing to take on the task of beginning a transformative journey. The need for global collaboration has never been more conspicuous than in the year 2020. The spread of the COVID-19 virus was global, and its effects were felt globally, but this virus shed light on the interconnectedness of the world. Goal 17 of the sustainable development goals, the necessity of partnerships for the goals, was further highlighted by the crisis. Partnering with governments, the business sector and civil society is a requirement for the first 16 objectives, according to SDG-17 (Pierce 2018; European Commission 2021). The COVID-19 pandemic is a stark warning that achieving the 2030 agenda would need a concerted and cooperative effort. Today, multilateralism and global cooperation are becoming more and more vital.

The SDGs can only be accomplished via strong global partnerships and collaborations; thus, the SDGs must collaborate. "A successful development agenda requires inclusive partnerships—at the global, regional, national, and local levels—built upon principles and values, and upon a shared vision and shared goals placing people and the planet at the center" (United Nations Sustainable Development 2021). Many nations require official development aid to boost growth and trade. Nonetheless, assistance levels are declining, and donor nations are failing to achieve their pledges to boost development money. For nations to recover from the pandemic, rebuild properly, and accomplish sustainable development goals, strong international collaboration is needed today more than ever before.

The European Union (EU) is the world's largest source of official development aid (ODA), providing more than half of all ODA to developing nations. In 2018, the EU's total ODA was $74.4 billion. The EU and its member states raised their assistance for local revenue mobilisation in developing countries considerably in 2016, boosting pledges from €112.7 million in 2015 to €197.9 million in 2016 (European Commission 2021). The EU and its member states promote sharing information for tax reasons, anti-corruption, tax evasion and illicit flows. With its contribution to the World Bank's Debt Reduction Trust Fund, which funds the

High Debt Poor Countries program, the EU stands at the forefront of debt relief. Until the end of August 2016, the EU's collective funding accounted for 41% of the overall contribution throughout this time. In absolute terms, remittances constitute a far greater source of development funding than ODA. There is a concerted effort by the European Union and its member states to boost the impact of remittance. As a result, remittance fees are reduced. The European Union also encourages research and development. A Research and Innovation Partnership in the Mediterranean Region has been formed to strengthen cooperation with emerging nations on research, technology and innovation. It supports sustainable food systems and integrated water management through creative solutions. In 2017, the EU established the EU-CELAC policy advisory mechanism in research and innovation to assist Latin American and Caribbean nations in meeting the Sustainable Development Goals (European Commission 2019). Facility for policy discourse E-READI supports EU-ASEAN scientific and technology collaboration. Capacity building is an essential component of nearly all development cooperation. The EU helps developing countries build their capacity to create and execute inclusive sustainable development policies at the national level and improve account-

ability and sensitivity to their populations. The EU has created dialogues with partner nations to debate and evaluate progress on the SDGs. The success of the SDGs depends on the development of multi-stakeholder partnerships that exchange information, experience, technology and financial assistance. Therefore, SDG-17 is essential to accomplishing all 16 SDGs and the aims of Addis Ababa's2030 Agenda for Sustainable Development (European Commission 2021).

19.1　Companies and Use Cases

Table 19.1 presents the business model of one company and use case that employ emerging technologies and create value in SDG-17. We should highlight that one use case can be related to more than one SDG and it can make use of multiple emerging technologies. In the left column, we present the company name, the origin country, related SDGs and emerging technologies that are included. The companies and use cases are listed alphabetically[1].

---

[1]For reference, you may click on the hyperlink on the company names or follow the website here (Accessed Online – 2.1.2022):

https://www.stambol.com/

Table 19.1 Companies and use cases in SDG-17

No	Company info	Value proposal (what?)	Value creation (how?)	Value capture
1	Stambol Canada 17 Spatial computing	It's a creative technology studio that specialises in extremely immersive content and interactive apps to help businesses innovate by combining creativity with the disruptive force of technology	Through Virtual reality (VR) for architecture and real estate services, the company enables remote collaboration via VR and Augmented Reality (AR) at home	Value is created by easier, cheaper and faster remote meetings and partnerships via enhanced collaborations with VR and AR. Furthermore, environmental value is captured by reducing travel and hence resulting in a reduced carbon footprint

References

Earth Changers, SDG 17 Global Partnerships Purpose Tourism For Sustainable Development (2020), Available at: https://www.earth-changers.com/purpose/partnerships. Accessed 1 Nov 2021

European Commission, Progress of the EU and its Member States-Goal by Goal. Brussels (2019), Available at: https://www.ilo.org/wcmsp5/groups/public/dgreports/. Accessed 1 Nov 2021

European Commission, Partnerships for the Goals | International Partnerships (2021), Available at: https://ec.europa.eu/international-partnerships/sdg/partnerships-goals_en. Accessed 1 Nov 2021

I.B. Franco, M. Abe, SDG 17 Partnerships for the Goals, in *Actioning the Global Goals for Local Impact*, (Springer, Singapore, 2020), pp. 275–293. https://doi.org/10.1007/978-981-32-9927-6_18

ICLEI, ICLEI World Congress 2015 Report: 25 years of impact – one world of local action, in *Local Governments for Sustainability*, ed. by E. O'loughlin et al., (2015). Available at: www.iclei.org/ Accessed 1 Nov 2021

International Telecommunication Union Database, Fixed broadband subscriptions | Data (2020), Available at: https://data.worldbank.org/indicator/IT.NET.BBND. Accessed 1 Nov 2021

S. Li, E. Gray, M. Dennis, How Partnerships Can Turbo Charge Progress on the SDGs (2020), Available at: https://www.wri.org/insights/how-partnerships-can-turbo-charge-progress-sdgs. Accessed 1 Nov 2021

A. MacDonald et al., Multi-stakeholder Partnerships (SDG #17) as a Means of Achieving Sustainable Communities and Cities (SDG #11), in *World Sustainability Series*, (Springer, Cham, 2018), pp. 193–209. https://doi.org/10.1007/978-3-319-63007-6_12

A. Maltais, N. Weitz, Å. Persson, *SDG 17: Partnerships for the Goals* (Stockholm Environment Institute, 2018). Available at: https://www.sei.org/wp-content/uploads/2020/01/sdg-17-review-of-research-needs-171219.pdf. Accessed 1 Nov 2021

OECD Statistics, Total Flows by Donor (ODA+OOF+Private) [DAC1] (2021), Available at: https://stats.oecd.org/Index.aspx?DataSetCode=TABLE1. Accessed 1 Nov 2021

A. Pierce, Why SDG 17 is the Most Important UN SDG? (2018), Available at: https://www.sopact.com/perspectives/sdg17-most-important-sdg. Accessed 1 Nov 2021

United Nations, General Assembly: Work of the Statistical Commission pertaining to the 2030 Agenda for Sustainable Development (2017), Available at: https://ggim.un.org/documents/A_RES_71_313.pdf. Accessed 1 Nov 2021

United Nations, The Sustainable Development Goals Report (2021), Available at: https://unstats.un.org/sdgs/report/2021/The-Sustainable-Development-Goals-Report-2021.pdf. Accessed 1 Nov 2021

United Nations Sustainable Development, Partnerships for the Goals (2021), Available at: https://www.un.org/sustainabledevelopment/globalpartnerships/. Accessed 1 Nov 2021

Conclusions

20

Abstract

This chapter brings a concise conclusion for the book by presenting the list of 34 emerging technologies, 17 United Nations (UN) Sustainable Development Goals (SDGs), countries and number of companies that are included in this book, the distribution of 650 companies in the world, emerging technology use cases per each sustainable development goal, the share of emerging technology use cases and the share of sustainable development goals number of applications. The chapter also includes a brief discussion about the findings of this book.

Keywords

Sustainable development · Emerging technologies · Value creation · Use cases · Business and Model

Tackling climate change, building a sustainable future, protecting economies and creating new jobs are the common desires of academia, industry, business and policymakers. We all know that this is not an easy task. There are numerous challenges and barriers to achieving this radical transformation from a traditional carbon-based economy to a sustainable one. At this point, emerging technologies step forth as remedies for this challenge. On the other hand, entrepreneurs foster innovative solutions and utilise these emerging technologies to create novel products and services to support sustainable development. Some traditional jobs and businesses will be phased out and replaced by these novel ones. At this point, we believe it is imperative to present "a guidebook" to illuminate how all these efforts came true so that others can learn, reflect and even advance in the business development with emerging technologies to help contribute to a sustainable future.

This book first presents an in-depth assessment of the innovation concept, funding and financing, supporting mechanisms for innovation and an impact assessment from various perspectives. The second chapter includes information on 34 different emerging technologies with market diffusion. This means we deliberately excluded the technologies either in the research and development phase or the ones not commercially available in the market. We then continue with the 17 United Nations Sustainable Development Goals (SDGs). With each SDG, we present a brief introduction about the concept and then provide a company and use cases table. The tables include companies worldwide that utilise 34 emerging technologies and create economic, environmental and social value to reinforce sustainable development. The 34 emerging technologies and 17 United Nations (UN) Sustainable

Table 20.1 List of 34 emerging technologies

3D printing	Blockchain	Drones	Natural language processing
5G	Carbon capture & storage	Edge computing	Quantum computing
Advanced materials	Cellular agriculture	Energy storage	Recycling
Artificial intelligence	Cloud computing	Flexible electronics & wearables	Robotic process automation
Autonomous vehicles	Crowdfunding	Healthcare analytics	Robotics
Big data	Cybersecurity	Hydrogen	Soilless farming
Biometrics	Datahubs	Internet of Behaviours	Spatial computing
Bioplastics	Digital twins	Internet of Things	Wireless power transfer
Biotech & biomanufacturing	Distributed computing		

Table 20.2 List of 17 United Nations (UN) sustainable development goals (SDGs)

1	No Poverty
2	Zero Hunger
3	Good Health and Well-being
4	Quality Education
5	Gender Equality
6	Clean Water and Sanitation
7	Affordable and Clean Energy
8	Decent Work and Economic Growth
9	Industry, Innovation and Infrastructure
10	Reduced Inequality
11	Sustainable Cities and Communities
12	Responsible Consumption and Production
13	Climate Action
14	Life Below Water
15	Life On Land
16	Peace, Justice, and Strong Institutions
17	Partnerships for the Goals

Table 20.3 List of countries and number of companies

Country and number of companies					
USA	245	China	8	Costa Rica	1
UK	57	Italy	8	Czechia	1
Netherlands	42	Austria	7	Ghana	1
Germany	33	South Korea	7	Greece	1
France	27	Belgium	5	Guinea	1
Israel	25	Kenya	5	Iceland	1
Canada	24	Japan	4	Ivory Coast	1
India	14	Nigeria	4	Lebanon	1
Norway	13	Portugal	4	Malawi	1
Sweden	13	Estonia	3	Malaysia	1
Finland	12	Slovakia	3	Mexico	1
Switzerland	11	Ukraine	3	Pakistan	1
Denmark	10	Brazil	2	Romania	1
Turkey	10	Ireland	2	Slovenia	1
Spain	10	New Zealand	2	South Africa	1
Singapore	9	Cambodia	1	Taiwan	1
Australia	8	Chile	1	Thailand	1
				Uganda	1

Table 20.4 SDGs and number of companies

SDG-1	SDG-2	SDG-3	SDG-4	SDG-5	SDG-6
39	40	55	49	16	36
SDG-7	SDG-8	SDG-9	SDG-10	SDG-11	
60	37	59	21	50	
SDG-12	SDG-13	SDG-14	SDG-15	SDG-16	SDG-17
46	52	36	45	8	1

Development Goals (SDGs) that we review are listed in Tables 20.1 and 20.2, respectively.

We made a comprehensive review and market scanning and inspected thousands of companies worldwide. Finally, we managed to compile 650 noteworthy and innovative companies from 51 countries. This does not mean that the companies presented here are the only interesting and innovative ones in the world. Indeed, we cannot claim that we reviewed all companies in the world due to time and labour constraints. Moreover, there are many companies, which have similar business models to the ones on our lists. We had to exclude these due to the space of the book. Table 20.3 shows these countries and the corresponding number of companies. Similarly, Table 20.4 gives the SDGs and num-

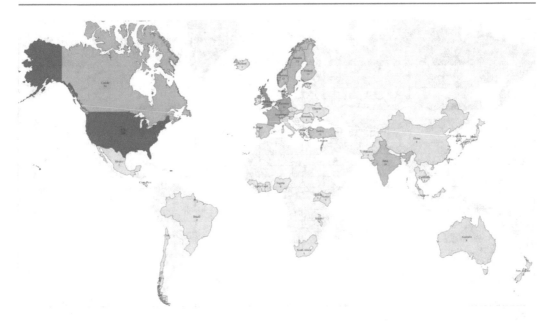

Fig. 20.1 Distribution of 650 companies in the world

ber of companies that are inspected in each field. Figure 20.1 illustrates the distribution of these companies on a world map. Figure 20.2 summarises the emerging technology use case distribution per each sustainable development goal on a matrix. Figures 20.3 and 20.4 are pie charts showing the distribution of use cases per emerging technologies and sustainable development goals

The USA turned out to be the most innovative country on our list by far, with 245 companies out of 650. The UK and the Netherlands follow the USA, and Nordic countries also perform exceptionally well. All these countries are well-known for their vivid innovative and entrepreneurial ecosystems. There is no surprise that these countries take the lead in adopting emerging technologies and converting these into business cases. Perhaps we should mention Germany here as one might expect a "better performance" in business and value generation with technology. However, we might state that there is an ongoing debate about Germany's success with digitalisation in general. Thus, we might claim that Germany could and should do much better, especially with digital technologies. We acknowledge that the distribution of companies worldwide might be "western biased" as most of the use cases are

located either in North America or in Europe, especially in Western and Northern Europe. Language could be one of the barriers. For example, we wanted to inspect one Japanese company to see how they work on innovative solutions by using advanced materials and biotech and bio-manufacturing. Nonetheless, the company's website does not provide sufficient information in English; thus, we did not include any of the products or services from them.

The matrix shown in Fig. 20.2 summarises our findings in this book. We can see each technology's number of use cases per SDG here. Artificial intelligence (AI) tops as the most utilised emerging technology with 473 use cases. Similarly, SDG-9: Industry, Innovation and Infrastructure is the most preferred field with 240 applications. On the other hand, Internet of Behaviours and SDG-17: Partnerships for the Goals have the least number of use cases. This matrix shows which technology is concentrated in what field and which technologies have no application in what fields. To better articulate these findings, we prepared the following figures.

As seen in Fig. 20.3, more than one-third of all use cases are done with AI, IoT (Internet of Things) and big data. One might argue that these technologies are no longer emerging ones;

Emerging Technologies	Sustainable Development Goals and Number of Use Cases																	Total
	SDG-1	SDG-2	SDG-3	SDG-4	SDG-5	SDG-6	SDG-7	SDG-8	SDG-9	SDG-10	SDG-11	SDG-12	SDG-13	SDG-14	SDG-15	SDG-16	SDG-17	
1 Artificial Intelligence	19	37	43	38	16	13	28	36	47	29	34	30	33	19	46	4	1	473
2 Internet of Things	8	14	10	7	2	17	17	13	24	2	34	20	18	13	17		2	218
3 Big Data	17	14	14	9	7	9	22	20	15	9	12	11	16	5	13		1	194
4 BioTech & Biomanufacturing	5	20	12			9	9		6		4	19	20	13	12			129
5 Recycling	1		1			9	12	2	13		13	32	22	8	4	1	2	120
6 Advanced Materials		2	5	1		10	16		10		10	14	18	6	4	1		97
7 Cloud Computing	5	5	7	2	3	3	8	10	18	4	11	5	3	3	5	2	1	95
8 Blockchain	11	3	5	3	2	1	6	11	12	4	4	6	5	1	3	7	1	85
9 Drones	6	7	3	1		2	3	4	4	1	6	3	12	8	16			76
10 Energy Storage	4	1	1			1	27	3	5	1	5	3	12					63
11 Autonomous Vehicles	1	2	5	1		1	4	2	8	1	12	3	4	5				49
12 Spatial Computing	2		12	11	1	2	1	2	7	1	4	2	1	1			2	49
13 Robotics		4	5	5		2	1	1	5	5	2	7	4	2	4			47
14 3D Printing	3	4	8	3	1	2	4	1	4	1	2	3	3	1	1		1	42
15 Flexible Electronics & Wearables	2		8	3	2		4	2	5	4	1	4						35
16 Crowdfunding	6	3	1	5	1	1	3	2	1	2	1		3			2	3	34
17 Digital Twins	2		2			2	3	2	7		4	2	3	6	1			34
18 Cellular Agriculture	3	11	2						2			5	3	4				30
19 5G	1	2	2	2		1	3	2	8		4	2	1		1			29
20 Natural Language Processing	1	1	5	3	5		2	2	5	2	2		1		1	1		29
21 Carbon Capture & Storage							3	1	5		1	5	9		2			26
22 Hydrogen		1					9	1	2		1	2	8					24
23 Cybersecurity	2	1	1		1		2	3	5	2	2		1		1	1		22
24 Healthcare Analytics		1	11			1	1		4	2	1			1				22
25 BioPlastics		2	1			1	1		1			3	3	3				18
26 Datahubs			1		1	2	1	4	3		1		2	3				18
27 Edge Computing				1		1	2	2	6		3	2	1					18
28 Soilless Farming	3	4	1			1	1	2		2		1	2		1			18
29 Wireless Power Transfer				1			4		1		3		3					12
30 Quantum Computing							1	2	3			2	1					9
31 Biometrics			1						2		2	1						6
32 Robotic Process Automotation		1				2			1		1							5
33 Distributed Computing								1	3									4
34 Internet of Behaviours				1	1				1									3
Total	102	139	170	96	42	93	196	131	240	75	179	184	216	102	135	19	14	

Fig. 20.2 The emerging technology use cases per each sustainable development goal

instead, they are mature technologies. However, we should remember the definition of emerging technologies: "development and application areas are still expanding rapidly, and their technical and value potential is still largely unrealised". We are positive that AI, IoT and big data's application areas are still expanding rapidly, and their technical and value potential is still largely unrealised. We will be seeing fascinating and promising developments with these 34 technologies, some of which are still at their initial stages of market diffusion.

Figure 20.4 shows us that some SDGs are popular among technology companies, whereas some have limited applications. These are SDG-7: Affordable and Clean Energy, SDG-9: Industry, Innovation and Infrastructure, SDG-12: Responsible Consumption and Production and SDG-13: Climate Action. Of course, it is unrealistic to expect an even number of applications at each SDG. However, we hope that the technology providers will focus more on SDG-5: Gender Equality, SDG-10: Reduced Inequality and

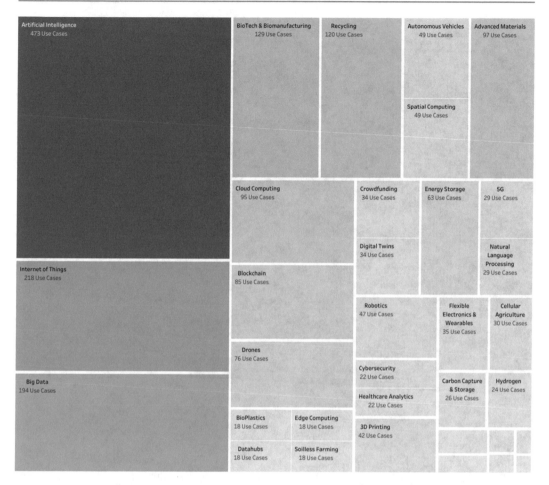

Fig. 20.3 The share of emerging technology use cases

SDG -16: Peace, Justice, and Strong Institutions.

This book attempts to compile as many relevant innovative business models as possible in employing emerging technologies to create value for sustainable development. However, it is impossible to present all valuable companies and use cases here. Besides, our aim is not to advertise technology companies. Instead, we solely focused on the value captured by technologies rather than the companies' revenue models. This way, entrepreneurs and other companies who wish to expand their businesses into one of the sustainable development goal fields can check and inspect how emerging technologies create value for fostering sustainability in general. We hope that this book will be helpful as a guide for those interested in innovation, emerging technologies, new business models, value creation and sustainable development.

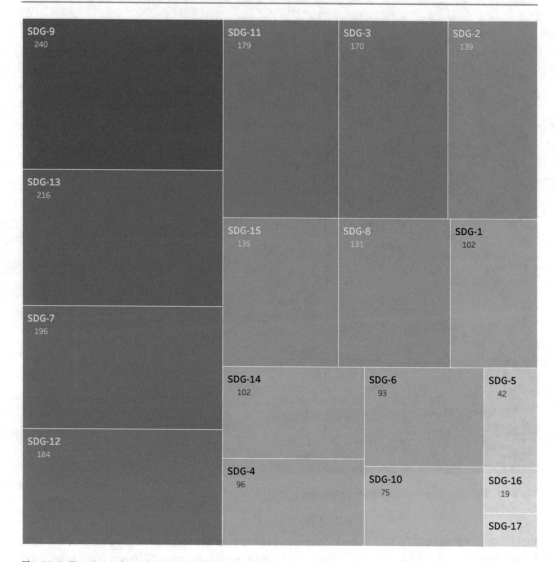

Fig. 20.4 The share of sustainable development goals number of applications

Index

© The Editor(s) (if applicable) and The Author(s) 2022
S. Küfeoğlu, *Emerging Technologies*, Sustainable Development Goals Series,
https://doi.org/10.1007/978-3-031-07127-0

Printed in the United States
by Baker & Taylor Publisher Services